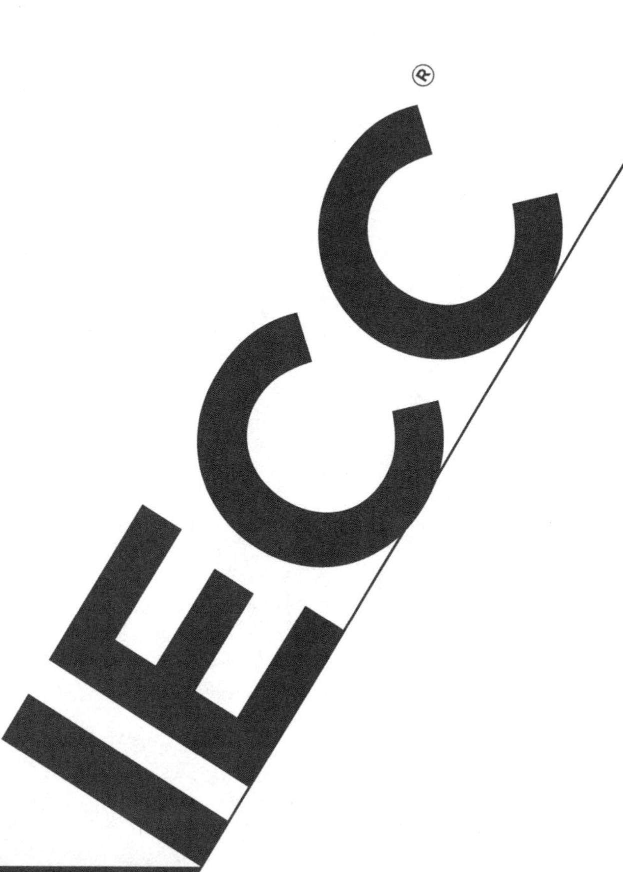

2015 CODE AND COMMENTARY

The complete IECC with commentary after each section

2015 International Energy Conservation Code® and Commentary

First Printing: October 2015

ISBN: 978-1-60983-287-2 (soft-cover edition)

With the cooperation of:
Jack Bailey
Jim Edelson
Steve Ferguson
Mark Heizer
Amanda Hickman
Duane Jonlin
Erick Makala

PRINTED IN THE U.S.A.

PREFACE

The principal purpose of the Commentary is to provide a basic volume of knowledge and facts relating to building construction as it pertains to the regulations set forth in the 2015 *International Energy Conservation Code®*. The person who is serious about effectively designing, constructing and regulating buildings and structures will find the Commentary to be a reliable data source and reference to almost all components of the built environment.

As a follow-up to the *International Energy Conservation Code*, we offer a companion document, the *International Energy Conservation Code Commentary*. The basic appeal of the Commentary is thus: it provides in a small package and at reasonable cost thorough coverage of many issues likely to be dealt with when using the *International Energy Conservation Code*—and then supplements that coverage with historical and technical background. Reference lists, information sources and bibliographies are also included.

Throughout all of this, strenuous effort has been made to keep the vast quantity of material accessible and its method of presentation useful. With a comprehensive yet concise summary of each section, the Commentary provides a convenient reference for regulations applicable to the construction of buildings and structures. In the chapters that follow, discussions focus on the full meaning and implications of the code text. Guidelines suggest the most effective method of application and the consequences of not adhering to the code text. Illustrations are provided to aid understanding; they do not necessarily illustrate the only methods of achieving code compliance.

The format of the Commentary includes the full text of each section, table and figure in the code, followed immediately by the commentary applicable to that text. At the time of printing, the Commentary reflects the most up-to-date text of the 2015 *International Energy Conservation Code*. Each section's narrative includes a statement of its objective and intent and usually includes a discussion about why the requirement commands the conditions set forth. Code text and commentary text are easily distinguished from each other. All code text is shown as it appears in the *International Energy Conservation Code*, and all commentary is indented below the code text and begins with the symbol ❖.

Readers should note that the Commentary is to be used in conjunction with the *International Energy Conservation Code* and not as a substitute for the code. **The Commentary is advisory only;** the code official alone possesses the authority and responsibility for interpreting the code.

Comments and recommendations are encouraged, for through your input, we can improve future editions. Please direct your comments to the Codes and Standards Development Department at the Central Regional Office.

TABLE OF CONTENTS

IECC—COMMERCIAL PROVISIONS

IECC—RESIDENTIAL PROVISIONS

IECC—COMMERCIAL PROVISIONS

TABLE OF CONTENTS

Chapter 1 [CE]: Scope and Administration

General Comments

The 2015 edition of the *International Energy Conservation Code®* (IECC®) is the result of aggressive efforts to increase commercial and residential energy efficiency requirements. Construction enhancements include required energy savings for windows, doors and skylights; thermal envelope efficiency; and increased efficiencies for installed heating, ventilating and air-conditioning (HVAC) equipment for commercial buildings three stories or greater in height. The 2015 edition represents a modest increase in required energy-efficient equipment and design over that of the 2012 edition. The code provides efficiency requirements for many systems not previously in the code. The 2012 edition provided a significant increase in energy efficiency levels over the 2009 edition of the code, which represented a significant increase over 2006 levels. The aggressive code change proposals are reflective of a national focus on reduction in energy consumption that stems not only from concerns about our oil reserves, but also from growing concerns over global warming.

Purpose

Though not stated specifically, the code is applicable to all buildings and structures, and their components and systems that use energy primarily for human comfort. The requirements are specified individually for commercial buildings and residential buildings. This portion of the code addresses commercial buildings. The code does not regulate the energy for industrial equipment for manufacturing or for items such as computers or coffeepots. The code addresses the design of energy-efficient building envelopes, and the selection and installation of energy-efficient mechanical, service water-heating, electrical distribution and illumination systems and equipment in residential and commercial buildings alike.

PART 1—SCOPE AND APPLICATION

SECTION C101 SCOPE AND GENERAL REQUIREMENTS

C101.1 Title. This code shall be known as the *International Energy Conservation Code* of **[NAME OF JURISDICTION]**, and shall be cited as such. It is referred to herein as "this code."

❖ This section directs the adopting jurisdiction to insert the name of the jurisdiction into the code. Because the IECC is a "model" code, it is not an enforceable document until it is adopted by a jurisdiction or agency that has enforcement powers.

C101.2 Scope. This code applies to *commercial buildings* and the buildings' sites and associated systems and equipment.

❖ This portion of the code applies to commercial buildings, commercial building sites and associated systems and equipment. The definitions for "Residential building," "Commercial building," and "Building site" will be important in correctly applying the provisions of the code. See the commentary related to the definitions in Chapter 2 [CE]. Additional discussion can be found in the commentary to Chapter 4 [CE].

C101.3 Intent. This code shall regulate the design and construction of buildings for the use and conservation of energy over the life of each building. This code is intended to provide flexibility to permit the use of innovative approaches and techniques to achieve this objective. This code is not intended to abridge safety, health or environmental requirements contained in other applicable codes or ordinances.

❖ The code is broad in its application, yet specific to regulating the use of energy in buildings where that energy is used primarily for human comfort, or heating and cooling of a building to protect the contents. Thus, energy used for commercial or industrial processing is to be considered exempt from the code because that energy is not used for human comfort or conditioning the space. The code also addresses efficiency of other systems that relate to the use of the space for human "habitation." In general, the requirements of the code address the design of all building systems that affect the comfort of the occupants and their use of the building, including:

- Lighting systems and controls.
- Wall, roof and floor insulation.
- Windows and skylights.

- Cooling equipment (air conditioners, chillers and cooling towers).
- Heating equipment (boilers, furnaces and heat pumps).
- Pumps, piping and liquid circulation systems.
- Supply and return fans.
- Service hot water systems (kitchens and lavatories).
- Kitchen exhaust systems.
- Refrigeration systems and refrigerated spaces.
- Permanent electric motors (e.g., elevators and escalators).

It does not address the energy used by office equipment such as personal computers, copy machines, printers, fax machines and coffee makers. Nor does it address kitchen equipment in restaurants, commercial kitchens and cafeterias, although water heating, lighting and HVAC energy uses in these types of spaces are covered.

The code is intended to define requirements for the portions of a building and building systems that affect energy use in new construction and to promote the effective use of energy. Where code application for a specific situation is in question, the code official should favor the action that will promote effective energy use. The code official may also consider the cost of the required action compared to the energy that will be saved over the life of that action.

This statement supports flexibility in applying code requirements. Although many of the requirements are given in a prescriptive format for ease of use, the code is not intended to stifle innovation—especially techniques that conserve energy. Innovative approaches that lead to energy efficiency should be encouraged, even if the approach is not specifically listed in the code or does not meet the strict letter of the code. This principle should be applied to building construction techniques used to meet the code and the methods used to approve them.

Any design should first be evaluated to see whether it meets the code requirements directly. Where the literal code requirements have not been satisfied but the applicant claims an innovative approach meets the intent, the code official will likely have to exercise professional judgment to determine the accuracy of that claim (see commentary, Section C103).

C101.4 Applicability. Where, in any specific case, different sections of this code specify different materials, methods of construction or other requirements, the most restrictive shall govern. Where there is a conflict between a general requirement and a specific requirement, the specific requirement shall govern.

❖ In cases where the code establishes a specific requirement for a certain condition, that requirement is applicable even if it is less restrictive than a general requirement elsewhere in the code. The most restrictive requirement is to apply where there may be different requirements in the code for a specific issue.

C101.4.1 Mixed occupancy. Where a building includes both *residential* and *commercial* occupancies, each occupancy shall be separately considered and meet the applicable provisions of IECC—Commercial Provisions or IECC—Residential Provisions.

❖ A mixed-occupancy building is one that contains both residential and commercial uses (see definitions, Chapter 2 [CE]). When residential and commercial uses coexist in a building and a portion of the building otherwise meets the definition of "Residential building," each occupancy must be evaluated separately.

Commentary Figure C101.4.1 provides an example of a mixed-occupancy building. The figure shows a three-story building with the first story occupied by a convenience store (a commercial use). The top two stories are shown as apartments that are classified as Group R-2 occupancy by the *International Building Code®* (IBC®). As the total building is not over three stories, its dwelling units meet the definition of "Residential building." The first story must be evaluated under the commercial provisions and the upper two stories under the residential provisions of the code. The treatment of each segment of the building by the appropriate portion of the code applies regardless of how much of the building meets the commercial versus residential definitions.

However, confusion sometimes arises when considering buildings taller than three stories. If, for example, the building in Commentary Figure C101.4.1 were four stories and the Group R-2 apartments occupied the top three floors, would it be a commercial building because it is over three stories high or is it a residential building because it has three stories of dwelling units? In such an example, the definition of "Residential building" dictates that the entire building would be considered as commercial and be subject to the requirements of Chapter 4 [CE]. That approach is based on the fact that the patterns of energy use generally change in buildings four stories or greater in height, and that the code limits residential buildings to a maximum height of three stories above grade. Any structure over three stories is considered a commercial building for purposes of apply-

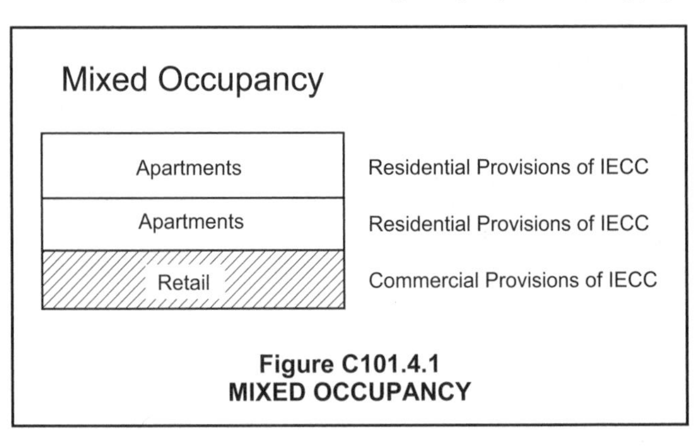

Figure C101.4.1
MIXED OCCUPANCY

ing the code, regardless of the occupancy classification of the structure. The only exception to this distinction would be single-family or duplex detached residences and townhouses four stories or greater in height. See also the definitions and commentary for "Commercial building" and "Residential building" to help clarify the application of the code to mixed-occupancy buildings.

C101.5 Compliance. *Residential buildings* shall meet the provisions of IECC—Residential Provisions. *Commercial buildings* shall meet the provisions of IECC—Commercial Provisions.

❖ For commercial buildings, the technical provisions of Chapter 4 [CE] offer three options for compliance. Two of the options are in Chapter 4 [CE]. The third option is to comply with the requirements of ANSI/ASHRAE/IESNA 90.1. The compliance approach is at the discretion of the building owner and the owner's designer.

C101.5.1 Compliance materials. The *code official* shall be permitted to approve specific computer software, worksheets, compliance manuals and other similar materials that meet the intent of this code.

❖ As mentioned in Section C101.3, the code is intended to allow the use of innovative approaches and techniques, provided that they result in the effective use of energy. This section recognizes that there are many federal, state and local programs, as well as computer software, that deal with energy efficiency. Therefore, the code simply states that the code official has the authority to accept those methods of compliance, provided that they meet the intent of the code. Some of the easiest examples to illustrate this provision are the REScheck™ and COMcheck™ software that are put out by the U.S. Department of Energy (DOE).

SECTION C102 ALTERNATE MATERIALS—METHOD OF CONSTRUCTION, DESIGN OR INSULATING SYSTEMS

C102.1 General. This code is not intended to prevent the use of any material, method of construction, design or insulating system not specifically prescribed herein, provided that such construction, design or insulating system has been *approved* by the *code official* as meeting the intent of this code.

❖ This section reinforces Section C101.3, which states that the code is meant to be flexible, as long as the intent of the proposed alternative is to promote the effective use of energy. The code is not intended to inhibit innovative ideas or technological advances. A comprehensive regulatory document, such as an energy code, cannot envision and then address all future innovations in the industry. As a result, a performance code must be applicable to and provide a basis for the approval of an increasing number of newly developed, innovative materials, systems and methods for which no code text or referenced standards yet exist. The fact that a material, product or method of construction is not addressed in the code is not an indication that the material, product or method is prohibited.

The code official is expected to apply sound technical judgment in accepting materials, systems or methods that, while not anticipated by the drafters of the current code text, can be demonstrated to offer equivalent or better performance. By virtue of its text, the code regulates new and innovative construction practices while addressing the relative safety of building occupants. The code official is responsible for determining whether a requested alternative protects the public health, safety and welfare in a manner consistent with code requirements.

C102.1.1 Above code programs. The *code official* or other authority having jurisdiction shall be permitted to deem a national, state or local energy efficiency program to exceed the energy efficiency required by this code. Buildings *approved* in writing by such an energy efficiency program shall be considered in compliance with this code. The requirements identified as "mandatory" in Chapter 4 shall be met.

❖ The purpose of this section is to specifically state that the code official has the authority to review and accept compliance with another energy program that may exceed code requirements, as long as the minimum "mandatory" requirements of the code are met. This provision is really a continuation of those stated in Sections C101.3 and C103.1, and the fact that the code is intended to allow alternatives, as long as the end result is an energy-efficient building comparable to or better than that required by the code.

This also is a good section to help reinforce the fact that the IECC as a model code is a "minimum" code. Therefore, it establishes the minimum requirement that must be met. Anything that exceeds that level is permitted.

While "above code programs" are acceptable because they do exceed the "minimum" requirements of the code, it would not be proper to require compliance with such "above code" programs. Besides the code being the minimum level of acceptable energy efficiency, it is also the maximum efficiency that the code official can require. A building built to the absolute minimum requirement is also the maximum that the code official can demand. It is perfectly acceptable for a designer or builder to exceed the code requirements, but it is not proper for the code official to demand such higher performance. Since the International Code Council (ICC)® has deemed that the mandatory requirements should apply to all buildings, it is reasonable that "above code programs" not be allowed to bypass these requirements.

PART 2—ADMINISTRATION AND ENFORCEMENT

SECTION C103
CONSTRUCTION DOCUMENTS

C103.1 General. Construction documents and other supporting data shall be submitted in one or more sets with each application for a permit. The construction documents shall be prepared by a registered design professional where required by the statutes of the jurisdiction in which the project is to be constructed. Where special conditions exist, the *code official* is authorized to require necessary construction documents to be prepared by a registered design professional.

> **Exception:** The *code official* is authorized to waive the requirements for construction documents or other supporting data if the *code official* determines they are not necessary to confirm compliance with this code.

❖ In most jurisdictions, the permit application must be accompanied by not less than two sets of construction documents. The code official can waive the requirements for filing construction documents when the scope of the work is minor and compliance can be verified through other means. When the quality of the materials is essential for conformity to the code, specific information must be given to establish that quality.

The code must not be cited, or the term "legal" or its equivalent used as a substitute for specific information. For example, it would be improper for the plans to simply state "windows in accordance with IECC requirements."

A detailed description of the work covered by the application must be submitted. When the work is "minor," either in scope or needed description, the code official may use judgment in determining the need for a detailed description. An example of "minor" work that may not involve a detailed description is the replacement of an existing 60-amp electrical service in a single-family residence with a 100-amp service.

The exception permits the code official to determine that construction documents are not necessary when the code official determines that compliance can be obtained and verified without the documents.

C103.2 Information on construction documents. Construction documents shall be drawn to scale upon suitable material. Electronic media documents are permitted to be submitted where *approved* by the *code official*. Construction documents shall be of sufficient clarity to indicate the location, nature and extent of the work proposed, and show in sufficient detail pertinent data and features of the building, systems and equipment as herein governed. Details shall include, but are not limited to, the following as applicable:

1. Insulation materials and their *R*-values.
2. Fenestration *U*-factors and solar heat gain coefficients (SHGCs).
3. Area-weighted *U*-factor and solar heat gain coefficient (SHGC) calculations.
4. Mechanical system design criteria.
5. Mechanical and service water heating system and equipment types, sizes and efficiencies.
6. Economizer description.
7. Equipment and system controls.
8. Fan motor horsepower (hp) and controls.
9. Duct sealing, duct and pipe insulation and location.
10. Lighting fixture schedule with wattage and control narrative.
11. Location of *daylight* zones on floor plans.
12. Air sealing details.

❖ For a comprehensive plan review that will enable verification of compliance with the code, all code requirements need to be incorporated in the construction documents. Adequate details must be included to allow the code official to verify compliance. A statement on the construction documents, such as, "All insulation levels shall comply with the 2015 edition of the IECC," is not an acceptable substitute for showing the required information. Note also that the code official is authorized to require additional code-related information as necessary.

For example, insulation *R*-values and glazing and door *U*-factors must be clearly marked on the building plans, specifications or forms used to show compliance. Where two or more different insulation levels exist for the same component (two insulation levels are used in ceilings), record each level separately on the plans or specifications and clarify where in the building each level of insulation will be installed.

In addition to the 12 items listed above, Section C103.2.1 specifically requires that the thermal envelope of the building be clearly shown on the construction drawings submitted for permit.

The following discussion is presented for the benefit of both the applicant and the plans examiner.

Permit Applicant's Responsibilities. At permit application, the goal of the applicant is to provide all necessary information to show compliance with the code. If the plans examiner is able to verify compliance in a single review, the permit can be issued and construction may be started without delay.

Depending on whether the prescriptive or performance methods of compliance are used, the amount and detail of the required information may vary. For example, if using the prescriptive method of compliance, the *U*-factor and SHGC may be the only information needed to verify fenestration compliance. If the assembly *U*-factor method (see Section C402.1.4) or the total building performance method (TBP) (see Section C407) is used, then additional information, such as the fenestration sizes and orientation, may be needed to demonstrate compliance.

Plans Examiner's Responsibilities. The plans examiner must review each application for code compliance before a permit is issued. By the owner, designer and contractor knowing what is expected of them early in the process, the building department can increase the likelihood that the approved drawings will comply with the code. This helps the inspector of the construction work.

Often the biggest challenges for the plans examiner are determining where the necessary information is and whether the drawings are complete.

A good plan review is essential to ensure code compliance and a successful project. Construction documents are graphic depictions of a legal contract. The more precise and code compliant the documents are, the less room for dispute or conflict. The design professional can then be confident that the intent of the design is communicated. The owner can understand what is to be built. The contractor can scope the contract appropriate to the project—without unexpected changes or additions. The building department can know what is being approved and what to look for during inspections.

C103.2.1 Building thermal envelope depiction. The *building's thermal envelope* shall be represented on the construction drawings.

❖ The building's thermal envelope is the key design feature in efficient use of energy to condition the indoor environments of a building. Without a clear indication of the location of the thermal envelope, the plans examiner may not be able to quickly review the building's assemblies for compliance with Section C402.

C103.3 Examination of documents. The *code official* shall examine or cause to be examined the accompanying construction documents and shall ascertain whether the construction indicated and described is in accordance with the requirements of this code and other pertinent laws or ordinances. The *code official* is authorized to utilize a registered design professional, or other *approved* entity not affiliated with the building design or construction, in conducting the review of the plans and specifications for compliance with the code.

❖ This section describes the required action of the code official in response to a permit application. The code official can delegate review of the construction documents to subordinates as well as third-party reviewers.

C103.3.1 Approval of construction documents. When the *code official* issues a permit where construction documents are required, the construction documents shall be endorsed in writing and stamped "Reviewed for Code Compliance." Such *approved* construction documents shall not be changed, modified or altered without authorization from the *code official*. Work shall be done in accordance with the *approved* construction documents.

One set of construction documents so reviewed shall be retained by the *code official*. The other set shall be returned to the applicant, kept at the site of work and shall be open to inspection by the *code official* or a duly authorized representative.

❖ The code official must stamp or otherwise endorse as "Reviewed for Code Compliance" the construction documents on which the permit is based. One set of approved construction documents must be kept on the construction site to serve as the basis for all subsequent inspections. To avoid confusion, the construction documents on the site must be the documents that were approved and stamped. This is because inspections are to be performed with regard to the approved documents, not the code itself. Additionally, the contractor cannot determine compliance with the approved construction documents unless they are readily available. If the approved construction documents are not available, the inspection should be postponed and work on the project halted.

C103.3.2 Previous approvals. This code shall not require changes in the construction documents, construction or designated occupancy of a structure for which a lawful permit has been heretofore issued or otherwise lawfully authorized, and the construction of which has been pursued in good faith within 180 days after the effective date of this code and has not been abandoned.

❖ If a permit is issued and construction proceeds at a normal pace and a new edition of the code is adopted by the legislative body, requiring that the building be constructed to conform to the new code is unreasonable. This section provides for the continuity of permits issued under previous codes, as long as such permits are being "actively prosecuted" subsequent to the effective date of the ordinance adopting this edition of the code.

C103.3.3 Phased approval. The *code official* shall have the authority to issue a permit for the construction of part of an energy conservation system before the construction documents for the entire system have been submitted or *approved*, provided adequate information and detailed statements have been filed complying with all pertinent requirements of this code. The holders of such permit shall proceed at their own risk without assurance that the permit for the entire energy conservation system will be granted.

❖ The code official has the authority to issue a partial permit to allow for the practice of "fast-tracking" a job. Any construction under a partial permit is "at the holder's own risk" and "without assurance that a permit for the entire structure will be granted." The code official is under no obligation to accept work or issue a complete permit in violation of the code, ordinances or statutes simply because a partial permit had been issued. Fast-tracking puts unusual administrative and technical burdens on the code official. The purpose is to proceed with construction while the design continues for other aspects of the work. Coordinating and correlating the code aspects into the project in phases requires attention to detail and project tracking so that all code issues are addressed. The coordination of these submittals is the responsibility of the registered design professional in responsible charge.

C103.4 Amended construction documents. Changes made during construction that are not in compliance with the *approved* construction documents shall be resubmitted for approval as an amended set of construction documents.

❖ The code requires that all work be done in accordance with the approved plans and other construction documents. Where the construction will not conform to the approved construction documents, the documents must be revised and resubmitted to the code official for review and approval. Code officials should maintain a policy that all amendments be submitted for review. Otherwise, a significant change that is not approved could result in an activity that is not in compliance with the code, and therefore cause needless delay and extra expense. The code official must retain one set of the amended and approved plans. The other set is to be kept at the construction site, ready for use by the jurisdiction's inspection staff.

C103.5 Retention of construction documents. One set of *approved* construction documents shall be retained by the *code official* for a period of not less than 180 days from date of completion of the permitted work, or as required by state or local laws.

❖ Construction documents must be retained in case a question or dispute arises after completion of the project. Unless modified because of state or local statutes, the retention period for the approved construction documents is a minimum of 180 days following the completion of the work, typically the date the certificate of occupancy is issued. Any further retention of plans by the jurisdiction as an archival record of construction activity in the community is not required by the code.

SECTION C104 INSPECTIONS

C104.1 General. Construction or work for which a permit is required shall be subject to inspection by the *code official* or his or her designated agent, and such construction or work shall remain accessible and exposed for inspection purposes until *approved*. It shall be the duty of the permit applicant to cause the work to remain accessible and exposed for inspection purposes. Neither the *code official* nor the jurisdiction shall be liable for expense entailed in the removal or replacement of any material, product, system or building component required to allow inspection to validate compliance with this code.

❖ Where a permit is required by state or local law, the building is subject to an inspection. The code official must determine whether appropriate energy-efficient features and equipment are installed in accordance with the approved construction documents and applicable code requirements.

The contractor, builder, owner or owner's authorized agent is responsible for arranging and coordinating required inspections to prevent work from being concealed prior to inspection. For example:

- Insulation must be inspected prior to concealment. Where the insulation is concealed prior to inspection and approval, the code official has the authority to require removal of the concealing components.
- Basement wall insulation may be installed on the exterior of a below-grade basement wall. Where the insulation application is not confirmed prior to backfilling, reinspection is necessary.
- Glazing assembly *U*-factor labels are to be left on until after the building has been inspected for compliance. The applicant is responsible for giving the inspector adequate information on site to verify code-related features, such as window *U*-factor and equipment efficiencies.

C104.2 Required inspections. The *code official* or his or her designated agent, upon notification, shall make the inspections set forth in Sections C104.2.1 through C104.2.6.

❖ This section lists six specific inspections: five during the progressive stages of construction, followed by a final inspection. Because the majority of energy-efficient construction occurs in steps or phases, periodic inspections are often necessary before portions of these systems are covered by further construction. Depending on the size of the building project, the listed inspections may actually require multiple inspection visits. The exact number of required inspections cannot always be specified. Reinspection may be necessary if violations are noted and corrections are required (see commentary, Section C104.3). Where time permits, frequent inspections of some job sites, especially where the work is complex, can be beneficial to detect potential problems before they become too difficult to correct.

An inspector's ongoing challenge is responding to change orders during construction. In any construction project there will be field changes. The call may be easy where a more efficient piece of equipment is being substituted for a less efficient one; where more insulation is provided, or where fenestration with a lower *U*-factor is installed. Section C103.4 requires that field changes be reflected on revised and approved construction drawings. Compliance with the energy conservation requirements of the code often require a specific "package" of installations; changing one, even where it appears to improve energy efficiency, may upset the balance of the approved design. The amount of information and the ease of confirming compliance will depend on whether the prescriptive or performance approach was used initially. In these cases, compliance is based on a combination of the fenestration area, *U*-factor, SHGC, the projection factor, and (if a performance-based analysis has been used) even the opaque wall characteristics. Although there may be enough latitude to decrease the efficiency somewhat, it is not possible to make such a determination without reviewing all the elements and how compliance was initially demonstrated. Whenever there are significant changes such as described in this paragraph, the inspector is expected to request that the applicant submit revised plans, so the plans examiner can verify compliance and ensure there is a correct record on file in the building department.

C104.2.1 Footing and foundation inspection. Inspections associated with footings and foundations shall verify compliance with the code as to *R*-value, location, thickness, depth of burial and protection of insulation as required by the code and *approved* plans and specifications.

❖ See the commentary to Section C104.2.

C104.2.2 Framing and rough-in inspection. Inspections at framing and rough-in shall be made before application of interior finish and shall verify compliance with the code as to types of insulation and corresponding *R*-values and their correct location and proper installation; fenestration properties (*U*-factor, SHGC and VT) and proper installation; and air leakage controls as required by the code and approved plans and specifications.

❖ See the commentary to Section C104.2.

C104.2.3 Plumbing rough-in inspection. Inspections at plumbing rough-in shall verify compliance as required by the code and *approved* plans and specifications as to types of insulation and corresponding *R*-values and protection; required controls; and required heat traps.

❖ See the commentary to Section C104.2.

C104.2.4 Mechanical rough-in inspection. Inspections at mechanical rough-in shall verify compliance as required by the code and *approved* plans and specifications as to installed HVAC equipment type and size; required controls, system insulation and corresponding *R*-value; system and damper air leakage; and required energy recovery and economizers.

❖ See the commentary to Section C104.2.

C104.2.5 Electrical rough-in inspection. Inspections at electrical rough-in shall verify compliance as required by the code and *approved* plans and specifications as to installed lighting systems, components and controls; and installation of an electric meter for each dwelling unit.

❖ See the commentary to Section C104.2.

C104.2.6 Final inspection. The building shall have a final inspection and shall not be occupied until *approved*. The final inspection shall include verification of the installation and proper operation of all required building controls, and documentation verifying activities associated with required *building commissioning* have been conducted and findings of noncompliance corrected. Buildings, or portions thereof, shall not be considered for a final inspection until the *code official* has received a letter of transmittal from the building owner acknowledging that the building owner has received the Preliminary Commissioning Report as required in Section C408.2.4.

❖ To establish compliance with all previously issued correction orders and to determine whether subsequent violations exist, a final inspection is required. The final inspection is conducted after all work is completed. Typically, the final inspection includes all items installed after the rough-in inspection and not concealed in the building construction. Subsequent reinspection is necessary if the final inspection generates a notice of violation (see commentary, Section C104.3). All violations observed during the final inspection must be noted and the permit holder must be advised of them.

Final approval is required prior to issuing the certificate of occupancy and, therefore, before the building may be occupied.

C104.3 Reinspection. A building shall be reinspected when determined necessary by the *code official*.

❖ The provisions for reinspection could affect the entire structure or a portion of the structure. As an example, under the circumstance where no approval was given to apply interior finish that conceals ducts in an exterior wall, the code official must require removal of the interior finish to verify the ducts are insulated to code.

Reinspections generally occur when some type of violation or correction notice was issued during one of the previous inspections or where the work was not ready for the inspection. For example, if the inspector went to the project to conduct an insulation inspection and not all of the insulation was installed at that point, the inspector would need to go back to the project and "reinspect" the insulation to verify that it had been completed.

C104.4 Approved inspection agencies. The *code official* is authorized to accept reports of third-party inspection agencies not affiliated with the building design or construction, provided such agencies are *approved* as to qualifications and reliability relevant to the building components and systems they are inspecting.

❖ As an alternative to building department staff conducting the inspections, the code official is permitted to accept inspections of and reports by approved inspection agencies. The code provides general guidance under which the code official can review and approve such agencies. More extensive criteria can be found in IBC Section 1703.

C104.5 Inspection requests. It shall be the duty of the holder of the permit or their duly authorized agent to notify the *code official* when work is ready for inspection. It shall be the duty of the permit holder to provide access to and means for inspections of such work that are required by this code.

❖ It is the responsibility of the permit holder or other authorized person, such as the contractor performing the work, to arrange for the required inspections when completed work is ready and to allow for sufficient time for the code official to schedule a visit to the site to prevent work from being concealed prior to being inspected. Access to the work to be inspected must be provided, including any special means such as a ladder.

C104.6 Reinspection and testing. Where any work or installation does not pass an initial test or inspection, the necessary corrections shall be made to achieve compliance with this code. The work or installation shall then be resubmitted to the *code official* for inspection and testing.

❖ This section provides for necessary actions in the event that a tested or inspected item is not originally in compliance with the code.

C104.7 Approval. After the prescribed tests and inspections indicate that the work complies in all respects with this code, a notice of approval shall be issued by the *code official*.

❖ This section provides a needed administrative tool in the form of a notice of approval that the code official issues to indicate completion of an energy conservation installation. While certificates of occupancy for construction are traditionally under the purview of one of the construction codes, the notice of approval will fill a need with regard to application and enforcement of nonbuilding codes.

C104.7.1 Revocation. The *code official* is authorized to, in writing, suspend or revoke a notice of approval issued under the provisions of this code wherever the certificate is issued in error, or on the basis of incorrect information supplied, or where it is determined that the *building* or structure, premise, or portion thereof is in violation of any ordinance or regulation or any of the provisions of this code.

❖ This section provides an important administrative tool by giving the code official the authority to revoke a certificate of completion for the reasons indicated in the text. The code official also may suspend the certificate until any code violations are corrected.

SECTION C105 VALIDITY

C105.1 General. If a portion of this code is held to be illegal or void, such a decision shall not affect the validity of the remainder of this code.

❖ This section is applicable when a court of law rules that a portion of the code is invalid. Only invalid sections of the code (as established by the court of jurisdiction) can be set aside. This is essential to safeguard the application of the code text to situations in which a provision of the code is declared illegal or unconstitutional. This section preserves the original legislative action that put the legal requirements (the code) in place.

All sections of the code not judged invalid must remain in effect. Although a dispute over a particular issue may have precipitated the litigation causing the requirement to be found invalid, the remainder of the code must still be considered as being applicable. This is sometimes called the "severability clause" and simply means that the invalid section can be removed from the code without affecting the entire document.

SECTION C106 REFERENCED STANDARDS

C106.1 Referenced codes and standards. The codes and standards referenced in this code shall be those listed in Chapter 6, and such codes and standards shall be considered as part of the requirements of this code to the prescribed extent of each such reference and as further regulated in Sections C106.1.1 and C106.1.2.

❖ The code references many standards promulgated and published by other organizations. A complete list of these standards appears in Chapter 6 [CE]. The wording of this section was carefully chosen to establish the edition of the standard that is enforceable under the code.

Although a standard is referenced, its full scope and content are not necessarily applicable. The standard is applicable only to the extent indicated in the text in which the standard is specifically referenced. A referenced standard or the portion cited in the text is an enforceable extension of the code as if the content of the standard was included in the body of the code. The use and applicability of referenced standards are limited to those portions of the standards that are specifically identified.

C106.1.1 Conflicts. Where conflicts occur between provisions of this code and referenced codes and standards, the provisions of this code shall apply.

❖ The use of referenced codes and standards to cover certain aspects of various occupancies and operations rather than write parallel or competing requirements into the code is a long-standing code development principle. In general, the code takes precedence when the requirements of a standard conflict with, or are less stringent than, those of the code. Although the code is intended to be in harmony with referenced standards, the code text generally governs should a conflict occur. This section establishes that, where questions and potential conflicts in the use of referenced codes and standards arise, the provisions of the code prevail regardless of the level of stringency.

C106.1.2 Provisions in referenced codes and standards. Where the extent of the reference to a referenced code or standard includes subject matter that is within the scope of this code, the provisions of this code, as applicable, shall take precedence over the provisions in the referenced code or standard.

❖ This section expands on the provisions of Section C106.1.1 by making it clear that even if a referenced standard contains requirements that parallel the code (or the other referenced sections), the provisions of the code [or the other referenced *International Codes*® (I-Codes®)] will always take precedence. This section is not intended to take the place of carefully scoped and referenced text for written standards for the I-Codes but, rather, provides the policy underpinnings on which sound code change proposals can be based.

C106.2 Application of references. References to chapter or section numbers, or to provisions not specifically identified by number, shall be construed to refer to such chapter, section or provision of this code.

❖ This section outlines the conventions used in making references to other portions of the code. By implication then, references to other codes would be required to specify those codes.

C106.3 Other laws. The provisions of this code shall not be deemed to nullify any provisions of local, state or federal law.

❖ This provision is intended to assist the code official in dealing with situations where other laws enacted by the jurisdiction or the state or federal government may be applicable to a condition also governed by a requirement in the code. In such circumstances, the requirements of the code would be in addition to that other law, although the code official may not be responsible for its enforcement.

SECTION C107 FEES

C107.1 Fees. A permit shall not be issued until the fees prescribed in Section C107.2 have been paid, nor shall an amendment to a permit be released until the additional fee, if any, has been paid.

❖ This section requires that all fees be paid prior to permit issuance or release of an amendment to a permit. Since department operations are usually intended to be supported by fees paid by the user of department activities, it is important that these fees are received before incurring any expense.

C107.2 Schedule of permit fees. A fee for each permit shall be paid as required, in accordance with the schedule as established by the applicable governing authority.

❖ This section authorizes the establishment of a schedule of fees by the jurisdiction. The fees are usually established by law, such as in an ordinance adopting the code, a separate ordinance or legally promulgated regulation, as required by state or local law and are often based on a valuation of the work to be performed.

C107.3 Work commencing before permit issuance. Any person who commences any work before obtaining the necessary permits shall be subject to an additional fee established by the *code official* that shall be in addition to the required permit fees.

❖ The department will incur certain costs (i.e., inspection time and administrative) when investigating and citing a person who has commenced work without obtaining a permit. This section authorizes the code official to recover those costs by establishing a fee, in addition to that collected when the required permit is issued, to be imposed on the responsible party.

C107.4 Related fees. The payment of the fee for the construction, *alteration*, removal or demolition of work done in connection to or concurrently with the work or activity authorized by a permit shall not relieve the applicant or holder of the permit from the payment of other fees that are prescribed by law.

❖ This provision gives the code official a useful administrative tool that makes it clear that all applicable fees of the jurisdiction for regulated work collateral to the work being done under the code's permit, such as sewer connections, water taps, driveways or signs, must be paid.

C107.5 Refunds. The *code official* is authorized to establish a refund policy.

❖ This section authorizes the code official to establish a policy to regulate the refund of fees, which may be full or partial, typically resulting from the revocation, abandonment or discontinuance of a building project for which a permit has been issued and fees have been collected.

SECTION C108 STOP WORK ORDER

C108.1 Authority. Where the *code official* finds any work regulated by this code being performed in a manner either contrary to the provisions of this code or dangerous or unsafe, the *code official* is authorized to issue a stop work order.

❖ This section provides for the suspension of work for which a permit was issued, pending the removal or correction of a severe violation or unsafe condition identified by the code official. Stop work orders are issued when enforcement can be accomplished no other way or when a dangerous condition exists.

C108.2 Issuance. The stop work order shall be in writing and shall be given to the owner of the property involved, the owner's authorized agent, or to the person doing the work. Upon issuance of a stop work order, the cited work shall immediately cease. The stop work order shall state the reason for the order and the conditions under which the cited work will be permitted to resume.

❖ This section makes it clear that, upon receipt of a violation notice from the code official, all construction activities identified in the notice must immediately cease, except as expressly permitted to correct the violation.

C108.3 Emergencies. Where an emergency exists, the *code official* shall not be required to give a written notice prior to stopping the work.

❖ This section gives the code official the authority to stop the work in dispute immediately when, in his or her opinion, there is an unsafe emergency condition that has been created by the work. The need for the written notice is suspended for this situation so that the work can be stopped immediately.

C108.4 Failure to comply. Any person who shall continue any work after having been served with a stop work order, except such work as that person is directed to perform to remove a violation or unsafe condition, shall be liable to a fine as set by the applicable governing authority.

❖ This section establishes consequences for when the stop work order is disregarded and the person responsible continues the work at issue, other than abatement work. The dollar amounts for the minimum and maximum fines are to be specified by the adopting jurisdiction.

SECTION C109 BOARD OF APPEALS

C109.1 General. In order to hear and decide appeals of orders, decisions or determinations made by the *code official* relative to the application and interpretation of this code, there shall be and is hereby created a board of appeals. The *code official* shall be an ex officio member of said board but shall not have a vote on any matter before the board. The board of appeals shall be appointed by the governing body and shall hold office at its pleasure. The board shall adopt rules of procedure for conducting its business, and shall render all decisions and findings in writing to the appellant with a duplicate copy to the *code official*.

❖ This section provides an aggrieved party having a material interest in the decision of the code official with an appeals process. This provides a forum, other than the court of jurisdiction, in which to review the code official's actions. The intent of the appeal process is not to waive or set aside a code requirement; rather, it is intended to provide a means of reviewing a code official's decision on an interpretation or application of the code.

C109.2 Limitations on authority. An application for appeal shall be based on a claim that the true intent of this code or the rules legally adopted thereunder have been incorrectly interpreted, the provisions of this code do not fully apply or an equally good or better form of construction is proposed. The board shall not have authority to waive requirements of this code.

❖ This section establishes the grounds for an appeal that claims that the code official has misinterpreted or misapplied a code provision. The board is not allowed to set aside any of the technical requirements of the code; however, it is allowed to consider alternative methods of compliance with the technical requirements.

C109.3 Qualifications. The board of appeals shall consist of members who are qualified by experience and training and are not employees of the jurisdiction.

❖ This section requires that the members of the appeals board are to have experience in building construction and system matters because the decisions of the appeals board are to be based purely on the technical merits involved in an appeal.

Bibliography

The following resource materials were used in the preparation of the commentary for this chapter of the code.

ASHRAE-97, *Handbook of Fundamentals*. Atlanta, GA: American Society of Heating, Refrigerating and Air-Conditioning Engineers, Inc.,1997.

ANSI/ASHRAE/IESNA 90.1-13, *Energy Code for Commercial and High-rise Residential Buildings*—Atlanta, GA: American Society of Heating, Refrigerating and Air-Conditioning Engineers, Inc., 2013.

Chapter 2 [CE]: Definitions

General Comments

All terms defined in the code are listed alphabetically in Chapter 2 [CE]. The words or terms defined in this chapter are considered to be of prime importance in either specifying the subject matter of code provisions or in giving meaning to certain terms used throughout the code for administrative or enforcement purposes. The code user should be familiar with the terms found in this chapter because the definitions are essential to the correct interpretation of the code and because the user might not be aware of the fact that a particular term found in the text is defined.

Purpose

Codes, by their nature, are technical documents. Every word, term and punctuation mark can alter a sentence's meaning and, if misused, muddy its intent. Further, the code, with its broad scope of applicability, includes terms inherent in a variety of construction disciplines. These terms can often have multiple meanings, depending on the context or discipline in which they are being used.

For these reasons, maintaining a consensus on the specific meaning of terms contained in the code is essential. Chapter 2 [CE] performs this function by stating clearly what specific terms mean for the purpose of the code.

SECTION C201 GENERAL

C201.1 Scope. Unless stated otherwise, the following words and terms in this code shall have the meanings indicated in this chapter.

❖ For the purposes of the code, certain abbreviations, terms, phrases, words and their derivatives have the meanings given in Chapter 2 [CE]. The code, with its broad scope of applicability, includes terms used in a variety of construction and energy-related disciplines. These terms can often have multiple meanings, depending on their context or discipline. Therefore, Chapter 2 [CE] establishes specific meanings for these terms.

C201.2 Interchangeability. Words used in the present tense include the future; words in the masculine gender include the feminine and neuter; the singular number includes the plural and the plural includes the singular.

❖ Although the definitions contained in Chapter 2 [CE] are to be taken literally, gender, number and tense are considered to be interchangeable.

C201.3 Terms defined in other codes. Terms that are not defined in this code but are defined in the *International Building Code*, *International Fire Code*, *International Fuel Gas Code*, *International Mechanical Code*, *International Plumbing Code* or the *International Residential Code* shall have the meanings ascribed to them in those codes.

❖ When a word or term that is not defined in this chapter appears in the code, other references may be used to find its definition, such as other *International Codes®* (I-Codes®). Definitions that are applicable in other I-Codes are applicable everywhere the term is used in the code. As stated in both the "Purpose" section above and in the commentary to Section C201.1, a bit of caution is needed when looking at definitions from other codes. Because the context and discipline can vary, it is important to determine that the term does fit within the code context. As an example, the term "accessible" would have a different meaning in the *International Plumbing Code®* (IPC®) and the *International Mechanical Code®* (IMC®) versus that of the *International Building Code®* (IBC®).

C201.4 Terms not defined. Terms not defined by this chapter shall have ordinarily accepted meanings such as the context implies.

❖ Another option for defining words or terms not defined here or in other codes is their "ordinarily accepted meanings." The intent of this statement is that a dictionary definition may suffice if the definition is in context. Often, construction terms used throughout the code may not be defined in Chapter 2 [CE] or a dictionary. In such a case, the definitions contained in the referenced standards (see Chapter 6 [CE]) and published textbooks on the subject in question are good resources.

SECTION C202 GENERAL DEFINITIONS

ABOVE-GRADE WALL. See "Wall, above-grade."

ACCESSIBLE. Admitting close approach as a result of not being guarded by locked doors, elevation or other effective means (see "Readily *accessible*").

❖ Providing access to mechanical equipment and appliances is necessary to facilitate inspection, observation, maintenance, adjustment, repair or replacement. Access to equipment means the equipment can be physically reached without having to remove a permanent portion of the structure. It is acceptable, for example, to install equipment in an interstitial space that would require removal of lay-in suspended ceiling panels to gain access. Equipment would not be considered accessible if it were necessary to remove or open any portion of a structure other than panels, doors, covers or similar obstructions intended to be removed or opened (see the definition of "Readily accessible"). Access can be described as the capability of being reached or approached for the purpose of inspection, observation, maintenance, adjustment, repair or replacement. Achieving access may first require the removal or opening of a panel, door or similar obstruction, and may require the overcoming of an obstacle such as elevation.

ADDITION. An extension or increase in the *conditioned space* floor area or height of a building or structure.

❖ The code uses this term to reflect new construction that is being added to an existing building. This definition is important when determining the applicability of the code provisions (see Section C502).

AIR BARRIER. Materials assembled and joined together to provide a barrier to air leakage through the building envelope. An air barrier may be a single material or a combination of materials.

❖ Building tightness against air infiltration is an important aspect of energy conservation. The term "air barrier" is defined to support the provisions of Section C402.5 regarding air leakage and building tightness. Note that an air barrier is not a single membrane, but rather the system of sealants, seals, insulation and wall sheathing that prevent air infiltration.

AIR CURTAIN. A device, installed at the building entrance, that generates and discharges a laminar air stream intended to prevent the infiltration of external, unconditioned air into the conditioned spaces, or the loss of interior, conditioned air to the outside.

❖ Air curtains are specifically allowed as an alternative to providing a lobby at a door between conditioned and nonconditioned space or the outside of a building. An air curtain installed on the interior of a building provides a coherent sheet of air created by an air stream and the surrounding entrained air. This sheet of air is able to bend and resist thermal exchange over an opening by way of support from the building's interior pressure and the stability created as the air stream meets a return grill or splits when it meets a surface such as a floor or another air stream.

ALTERATION. Any construction, retrofit or renovation to an existing structure other than repair or addition that requires a permit. Also, a change in a building, electrical, gas, mechanical or plumbing system that involves an extension, addition or change to the arrangement, type or purpose of the original installation that requires a permit.

❖ This definition actually includes two separate definitions: one that applies to building construction and the other to building systems. An alteration is any modification or change made to an existing installation. For example, changing refrigerant types or heat transfer fluids in a system would be considered an alteration. This definition specifically excludes additions or repairs, and also ties the term to situations where a permit is required. See Chapter 1 of the IBC, IMC, IPC and the *International Fuel Gas Code*® (IFGC®) for the information regarding when a permit is required.

APPROVED. Approval by the *code official* as a result of investigation and tests conducted by him or her, or by reason of accepted principles or tests by nationally recognized organizations.

❖ As related to the process of accepting envelope, mechanical, service water heating, lighting and electrical power installations, including materials, equipment and construction systems, this definition identifies where ultimate authority rests. Whenever this term is used, it intends that only the enforcing authority can accept a specific installation or component as complying with the code.

The definition does not force the code official to accept any third-party test. It merely permits the acceptance of the test so that the code official does not have to personally test the item. One example that demonstrates this process is an ICC Evaluation Report. An evaluation report prepared and published by the International Code Council (ICC)® is permitted to be used by a code official to aid in the review and approval of the material or method described in the report. Although the evaluation and report are prepared by the ICC, the code official is not obligated to accept or approve the product based on the evaluation report. The term "approved" is always tied to the code official's approval of the product or project. That the ICC published an evaluation report does not supersede the fact that the approval of the code official is still needed for the material or method described in the report.

APPROVED AGENCY. An established and recognized agency regularly engaged in conducting tests or furnishing inspection services, when such agency has been approved by the *code official*.

❖ In order to identify basic criteria or understand what agencies are being referred to in the code, there is a need to define approved agency. The word "approved" means "acceptable to the code official" (see definition of "Approved"). The basis for the code official's approval of any agency for a particular activity may include, but is not necessarily limited to, the capacity of the agency to perform the work in accordance with the code. This is typically done through a review of resumés and references of the agency and its personnel.

AUTOMATIC. Self-acting, operating by its own mechanism when actuated by some impersonal influence, as, for example, a change in current strength, pressure, temperature or mechanical configuration (see "Manual").

❖ Operation or control devices, or systems operating automatically as opposed to manually, are designed to operate safely with only periodic human intervention or supervision. A thermostat would be an example of something that is automatic. While a person would set the thermostat to the desired temperature, the thermostat would cycle the heating or cooling system on or off on its own once the temperature hits the established setpoints.

BELOW-GRADE WALL. See "Wall, below-grade."

BOILER, MODULATING. A boiler that is capable of more than a single firing rate in response to a varying temperature or heating load.

❖ Modulating boilers are one type of boiler. The minimum efficiency of individual boilers is specified in the various tables in Section C403.2. In addition, the controls of boilers and boiler systems are required to provide efficiencies in the operation of this equipment. Modulating boilers are called out in Section C403.4.2.5 as a method of compliance for the requirement of boiler system turndown.

BOILER SYSTEM. One or more boilers, their piping and controls that work together to supply steam or hot water to heat output devices remote from the boiler.

❖ Boilers are defined in the IMC as "a closed heating appliance intended to supply hot water or steam for space heating, processing or power purposes." The boilers regulated in the code are those used for space heating. The minimum efficiency of individual boilers is specified in the various tables in Section C403.2. Boiler systems are required by Section C403.4.2.5 to provide a turndown function. Additional controls of boilers and boiler systems are specified in Sections C403.2.5, C403.4.2 and C403.4.2.6.

BUBBLE POINT. The refrigerant liquid saturation temperature at a specified pressure.

❖ One of two temperatures used to determine the saturated condensing temperature in a refrigeration system.

BUILDING. Any structure used or intended for supporting or sheltering any use or occupancy, including any mechanical systems, service water heating systems and electric power and lighting systems located on the building site and supporting the building.

❖ This definition indicates that where this term is used in the code, it means a structure intended to provide shelter or support for some activity or occupancy. Though not addressed in the code, it is important to note that the IBC does permit that a fire wall forms a demarcation between two separate, structurally independent buildings. Therefore, the code provisions could be applied to each building separately or to the structure as a whole. This would be a designer's decision to make.

BUILDING COMMISSIONING. A process that verifies and documents that the selected building systems have been designed, installed, and function according to the owner's project requirements and construction documents, and to minimum code requirements.

❖ Building commissioning verifies that building systems are operating as designed. This is important for energy conversation; a system not operating correctly can waste energy.

BUILDING ENTRANCE. Any door, set of doors, doorway, or other form of portal that is used to gain access to the building from the outside by the public.

❖ A building entrance is used to gain access to the building from the outside by the public.

BUILDING SITE. A continguous area of land that is under the ownership or control of one entity.

❖ "Building site" is a key term used throughout the code. It is the area of land that is under the ownership or control of one entity. A building site may be a combination of many parcels of land but is treated as a single unit for the purposes of compliance with this code.

BUILDING THERMAL ENVELOPE. The basement walls, exterior walls, floor, roof and any other building elements that enclose *conditioned space* or provide a boundary between *conditioned space* and exempt or unconditioned space.

❖ "Building thermal envelope" is a key term and resounding theme used throughout the code. It defines what portions of the building form a structurally bound conditioned space, and, thereby, are covered by the insulation and infiltration (air leakage) requirements of the code. The building thermal envelope includes all building components separating conditioned spaces (see commentary, "Conditioned space") from unconditioned spaces or outside ambient conditions and through which heat is transferred. For example, the walls and doors separating an unheated garage (unconditioned space) from an office or living area (conditioned space) are part of the building thermal envelope. The walls and doors separating an unheated garage from the outdoors are not part of the building thermal envelope. Walls, floors and other building components separating two conditioned spaces are not part of the building thermal envelope. For example, interior partition walls, the common or party walls separating dwelling units in multiple-family buildings and the wall between a new conditioned addition and the existing conditioned space are not considered part of the building thermal envelope. Unconditioned spaces (areas having no heating or cooling sources) are placed outside the building envelope. A space is conditioned if it is heated or cooled directly or where a space is indirectly supplied with heating or cooling through uninsulated walls, floors or uninsulated ducts or heating, ventilating and air-conditioning (HVAC) piping. Boundaries that define the building envelope include the following:

- Building assemblies separating a conditioned space from outdoor ambient weather conditions.
- Building assemblies separating a conditioned space from the ground under or around that space, such as the ground around the perimeter of a slab or the soil at the exterior of a conditioned basement wall. Note that the code does not specify requirements for insulating basement floors or underneath slab floors (except at the perimeter edges).
- Building assemblies separating a conditioned space from an unconditioned garage, unconditioned sunroom or similar unheated/cooled area.

The code specifies requirements for ceiling, wall, floor, basement wall, slab-edge and crawl space wall components of the building envelope. In some cases, it may be unclear how to classify a particular part of a building. For example, skylight shafts have properties of a wall assembly but are located in the ceiling assembly. Because many of these items are not addressed specifically in the code, the code official should make the determination as to the appropriate classification and construction. When no distinction exists between roof and wall, such as in an A-frame structure, the code official should determine the appropriate classification. Historically, some codes have designated a wall as having a slope of 60 degrees or greater from the horizontal plane. In such situations, if the wall slope is less than 60 degrees, then classification as a "roof" is appropriate. Because the code is silent on this issue, other options, such as stating that the roof could be considered to begin at a point 8 feet (2439 mm) above the floor surface of the uppermost story, could be used. The phrase "exempted spaces" in the definition of "Building thermal envelope" refers to spaces identified as exempt from this scope of the code or exempted from the thermal envelope requirements of the code (see commentary, Chapter 5 [CE] and Sections C402.1.1 and C402.1.2).

***C*-FACTOR (THERMAL CONDUCTANCE).** The coefficient of heat transmission (surface to surface) through a building component or assembly, equal to the time rate of heat flow per unit area and the unit temperature difference between the warm side and cold side surfaces (Btu/h • ft^2 • °F) [W/(m^2 • K)].

❖ This definition addresses a term needed in the provisions of the code for the *U*-factor method given in Section C402.1.4. The term appears in Table C402.1.4.

CIRCULATING HOT WATER SYSTEM. A specifically designed water distribution system where one or more pumps are operated in the service hot water piping to circulate heated water from the water-heating equipment to the fixture supply and back to the water-heating equipment.

❖ The definition specifies how a circulating hot water system should be designed and operated. These systems are not to be confused with those hot water systems used for space conditioning. The definition is consistent with industry standards.

CLIMATE ZONE. A geographical region based on climatic criteria as specified in this code.

❖ Climate zones are used throughout the code to provide regionally appropriate design standards. The climate zones are specified in Chapter 3 [CE] of the code.

CODE OFFICIAL. The officer or other designated authority charged with the administration and enforcement of this code, or a duly authorized representative.

❖ The statutory power to enforce the code is usually vested in a building department of a state, county or municipality with a designated enforcement officer who is termed the "code official."

COEFFICENT OF PERFORMANCE (COP) – COOLING. The ratio of the rate of heat input, in consistent units, for a complete refrigerating system or some specific portion of that system under designated operating conditions.

❖ This term is from ANSI/ASHRAE/IESNA 90.1 as listed in Chapter 6 [CE]. It is a measure of performance used in the code for refrigeration systems.

COEFFICIENT OF PERFORMANCE (COP) – HEATING. The ratio of the rate of heat delivered to the rate of energy input, in consistent units, for a complete heat pump system, including the compressor and, if applicable, auxiliary heat, under designated operating conditions.

❖ This term is from ANSI/ASHRAE/IESNA 90.1 as listed in Chapter 6 [CE]. It is a measure of performance used in the code for heat pump systems.

COMMERCIAL BUILDING. For this code, all buildings that are not included in the definition of "Residential building."

❖ Commercial buildings include, among others, occupancies for assembly, educational, business, institutional, mercantile, factory/industrial, hazardous storage and utility occupancies (see definition and commentary, "Residential building").

One item that may easily be overlooked is that a Group R-1 occupancy building (a hotel or motel) would be classified as a "commercial building" because it is not included in the definition of "Residential building." Although classified as a residential occupancy by the IBC, hotels and motels tend to be more closely associated with commercial buildings as far as energy usage and operation. In addition, Group

R-2, R-3 and R-4 occupancies that are four or more stories in height (other than one- and two family-dwelling structures and townhouses) are also classified as commercial buildings for the purposes of the code.

COMPUTER ROOM. A room whose primary function is to house equipment for the processing and storage of electronic data and that has a design electronic data equipment power density exceeding 20 watts per square foot of conditioned floor area.

❖ Computer rooms, due to the unique nature of the space, have a significant level of internal heat generation that must be addressed to ensure the equipment therein functions properly. This is not intended to apply to office spaces where a computer may be in use at each employee's desk, but is aimed at rooms that are dedicated to operating computer equipment. Table C403.2.3(9) provides minimum efficiency standards for air conditioners used to cool these rooms.

CONDENSING UNIT. A factory-made assembly of refrigeration components designed to compress and liquefy a specific refrigerant. The unit consists of one or more refrigerant compressors, refrigerant condensers (air-cooled, evaporatively cooled, or water-cooled), condenser fans and motors (where used) and factory-supplied accessories.

❖ Condensing units are one type of equipment installed for air-conditioning and refrigeration purposes. The minimum efficiency of condensing units is established in Section C403.2. Condensers serving refrigeration systems must also comply with Section C403.5.1.

CONDITIONED FLOOR AREA. The horizontal projection of the floors associated with the *conditioned space*.

❖ The conditioned floor area is the total area of all floors in the conditioned space of the building.

CONDITIONED SPACE. An area, room or space that is enclosed within the building thermal envelope and is directly or indirectly heated or cooled. Spaces are indirectly heated or cooled where they communicate through openings with conditioned spaces, where they are separated from conditioned spaces by uninsulated walls, floors or ceilings, or where they contain uninsulated ducts, piping or other sources of heating or cooling.

❖ A conditioned space typically is any space that does not communicate directly to the outside; that is, a space not directly ventilated to the outdoors and meets one of the following criteria:

1. The space has a heating or cooling supply register.
2. The space has heating or cooling equipment designed to heat or cool the space, or both, such as a radiant heater built into the ceiling, a baseboard heater or a wall-mounted gas heater.
3. The space contains uninsulated ducts or uninsulated hydronic heating surfaces.
4. The space is inside the building thermal envelope. For example:
   - A basement with insulated walls but without insulation on the basement ceiling.
   - A space adjacent to and not physically separated from a conditioned space (such as a room adjacent to another room with a heating duct, but without a door that can be closed between the rooms).
   - A room completely surrounded by conditioned spaces.

The builder/designer has some flexibility in defining the bounds of the conditioned space as long as the building envelope requirements are met. Spaces that are not conditioned directly but have uninsulated surfaces separating them from conditioned spaces are included within the insulated envelope of the building. For example, an unventilated crawl space below an uninsulated floor is considered part of the conditioned space, even where no heat is directly supplied to the crawl space area. Where the crawl space is included as a conditioned space, the builder must insulate the exterior crawl space walls instead of the floor above. The task of defining the building thermal envelope is left to the permit applicant and must be depicted on the construction documents in accordance with Section C103.2.1.

Examples of unconditioned spaces include parking garages and penthouses that are neither heated nor cooled if all duct surfaces running through these spaces are insulated. Note that the boundary between the conditioned and unconditioned space is subject to the infiltration control requirements of the code (see commentary, Section C402.5).

Historically, the code tied this definition to "conditioning for human comfort" or by specifying that the conditioning fall within a specific range. Starting with the 2006 edition, the code considers any type of conditioning as creating a "conditioned space." Therefore, providing heating within a storage building to keep the stock from freezing would still be considered as creating a conditioned space.

CONTINUOUS AIR BARRIER. A combination of materials and assemblies that restrict or prevent the passage of air through the building thermal envelope.

❖ This definition is necessary to clarify the options for meeting continuous air barrier requirements.

CONTINUOUS INSULATION (ci). Insulating material that is continuous across all structural members without thermal bridges other than fasteners and service openings. It is installed on the interior or exterior or is integral to any opaque surface of the building envelope.

❖ Continuous insulation is essential to building envelope assemblies in nearly all climate zones and most types of construction. The definition is generic in the sense that the type of insulating material is not limited. Provided the material used as insulation provides the required insulation value, it can meet the criteria of being continuous insulation. The requirement can be met with the insulation in any location of the wall, on the exterior of the wall, on the interior side of the wall adjoining the conditioned space or an element in the wall assembly located between other elements of the wall.

CRAWL SPACE WALL. The opaque portion of a wall that encloses a crawl space and is partially or totally below grade.

❖ Crawl spaces are more common in residential construction than in commercial buildings. The residential portion of the code addresses insulation of crawl space walls. Insulation on crawl space walls must be protected in accordance with Section C303.2.1.

CURTAIN WALL. Fenestration products used to create an external nonload-bearing wall that is designed to separate the exterior and interior environments.

❖ This definition is included to help classify and properly apply the code requirements to these products. The industry uses the term to help establish the requirements and separate the products into different groupings. Without defining the curtain wall as a fenestration product, it would be difficult to determine how to properly apply the building envelope requirements.

DAYLIGHT RESPONSIVE CONTROL. A device or system that provides automatic control of electric light levels based on the amount of daylight in a space.

❖ The definition clarifies that these controls must be designed to control the level of electric illumination as a response to the amount of daylight being received.

DAYLIGHT ZONE. That portion of a building's interior floor area that is illuminated by natural light.

❖ Daylight zones are areas that are often able to receive adequate natural light without any artificial lighting. Daylight zones are specifically described in Sections C405.2.3.2 and C405.2.3.3. Section C405.2.3 provides criteria for lighting control in daylight zones to allow occupants to take advantage of this natural light.

DEMAND CONTROL VENTILATION (DCV). A ventilation system capability that provides for the automatic reduction of outdoor air intake below design rates when the actual occupancy of spaces served by the system is less than design occupancy.

❖ This definition is included to support the requirements of Section C403.2.6.1, which requires automatic ventilation controls based on occupancy loads in spaces larger than 500 square feet (46.5 m^2) and with an average occupant load of 25 people per 1,000 square feet (93 m^2).

DEMAND RECIRCULATION WATER SYSTEM. A water distribution system where pumps prime the service hot water piping with heated water upon demand for hot water.

❖ Demand recirculation water systems are more energy efficient than other recirculation systems. Section C404.7 specifies that the water returning to the heating source return via the cold water piping.

DUCT. A tube or conduit utilized for conveying air. The air passages of self-contained systems are not to be construed as air ducts.

❖ Ducts can be factory manufactured or field constructed of sheet metal, gypsum board, fibrous glass board or other approved materials. Ducts are used in air distribution systems, exhaust systems, smoke control systems and combustion air-supply systems. Air passageways that are integral parts of an air handler, packaged air-conditioning unit or similar piece of self-contained, factory-built equipment are not considered ducts in the context of the code.

DUCT SYSTEM. A continuous passageway for the transmission of air that, in addition to ducts, includes duct fittings, dampers, plenums, fans and accessory air-handling equipment and appliances.

❖ Duct systems are part of an air distribution system and include supply, return, transfer and relief/exhaust air systems.

[B] DWELLING UNIT. A single unit providing complete independent living facilities for one or more persons, including permanent provisions for living, sleeping, eating, cooking and sanitation.

❖ A dwelling unit, as stated, is a residential unit that contains all of the necessary facilities for independent living. This provides a single, independent unit that serves a single family or single group of individuals. Dwelling units can be found in Group R-2 and R-3 occupancies as established in the IBC. Where such occupancies are more than three stories, they are considered commercial buildings under this part of the code. The code requirements are applied consistently to all dwelling units whether owner occupied, rented or leased. A dwelling unit is distinguished from a sleeping unit, which does not have all the features of a dwelling unit.

DYNAMIC GLAZING. Any fenestration product that has the fully reversible ability to change its performance properties, including *U*-factor, solar heat gain coefficient (SHGC), or visible transmittance (VT).

❖ Dynamic glazing allows the user to change the performance of the window, such as SHGC ratings to allow additional light in when it is desired or less light. This permits adjustments from season to season, or even daily (see Section C402.4.3.3).

ECONOMIZER, AIR. A duct and damper arrangement and automatic control system that allows a cooling system to supply outside air to reduce or eliminate the need for mechanical cooling during mild or cold weather.

❖ Also called air-side economizers, air economizers use controllable dampers to increase the amount of outside air drawn into the building when the outside air is cool or cold and the system requires cooling. To meet the code, economizer systems must be able to supply up to 100 percent of the design supply-air quantity as outside air. The economizer (such as outside air dampers) need only be designed for this flow rate (see commentary, Sections C403.3 and C403.3.3).

As the outside air warms up, there will be a point where outside air intake will increase energy usage. At this point, the economizer must be shut off and the system operated at the minimum outside air volume required for ventilation. The controller that causes this to occur is called the "economizer high-limit shutoff control" or "changeover limit switch."

ECONOMIZER, WATER. A system where the supply air of a cooling system is cooled indirectly with water that is itself cooled by heat or mass transfer to the environment without the use of mechanical cooling.

❖ Although air-side economizers use cool air directly to reduce the cooling load, a water economizer (also called a water-side economizer) uses cool outside air indirectly first to cool water that then cools supply air through a cooling coil, thus reducing the mechanical cooling load. The water economizer is essentially an indirect evaporative cooler. Water is circulated through a cooling tower where it is evaporatively cooled before being circulated through cooling coils to cool supply air indirectly.

To meet the code requirements, water economizers must be able to satisfy the system's entire expected cooling load when outside air temperatures are 50°F (10°C) dry bulb and 45°F (7°C) wet bulb, and below. This design criterion is specified because, unlike air economizers that use cold outside air directly for cooling, the performance of water economizers depends greatly on the selection of components such as cooling towers and heat exchangers (see commentary, Section C403.3.4.1).

ENCLOSED SPACE. A volume surrounded by solid surfaces such as walls, floors, roofs, and openable devices such as doors and operable windows.

❖ This definition supports the requirements of the code regarding daylighting.

ENERGY ANALYSIS. A method for estimating the annual energy use of the *proposed design* and *standard reference design* based on estimates of energy use.

❖ Designs founded on total building performance (see Section C407) use energy analysis when the proposed building cannot satisfy the prescriptive criteria in Section C402, C403, C404 or C405 or when the design professional requires more flexibility for a sophisticated or innovative design. Using the total-building-performance design methodology, the proposed design is evaluated based on the cost of various types of energy used rather than the units of energy used (Btu, kWh). That cost must be established using an hour-by-hour, full-year (8,760 hours) simulation tool capable of simulating the performance of both the proposed and standard designs (see commentary, "Energy simulation tool"). The simulation must be capable of converting calculated energy demand and consumption into utility costs using the actual utility rate schedules rather than the average cost of electricity or gas.

ENERGY COST. The total estimated annual cost for purchased energy for the building functions regulated by this code, including applicable demand charges.

❖ The total annual cost for purchased energy includes demand, power and fuel adjustment charges and the impact of special rate programs for large volume customers. A thorough evaluation of existing tariffs and fee schedules may uncover substantial savings opportunities. In some states, for example, manufacturing customers are exempt from sales taxes for energy. In other states, utilities may have multiple tariff options.

[M] ENERGY RECOVERY VENTILATION SYSTEM. Systems that employ air-to-air heat exchangers to recover energy from exhaust air for the purpose of preheating, precooling, humidifying or dehumidifying outdoor ventilation air prior to supplying the air to a space, either directly or as part of an HVAC system.

❖ Energy recovery systems allow for the exchange of heat from air that is already in the building and near the desired temperature to the outside supply air that is coming into the building. Using the existing conditioned air to help condition or partially condition the outside supply air helps reduce the temperature difference in the supply air so that there is not as great of a demand for mechanical conditioning of the air. As an example, a building that keeps the indoor air at 68°F (20°C) during the winter may be bringing in an outdoor air supply where the outdoor temperature is 20°F (-7°C). Under these conditions, the supply air normally would have to be conditioned to bring it up to the indoor conditions. By using an energy recovery system to exchange some of the heat from the return air that is in the building to the incoming supply air, the outside supply air is partially conditioned so that there is less of a demand for mechanically conditioning the air to the desired temperature (see commentary, Section C403.2.7).

The definition is based on and consistent with the definition in AHRI 1060, which is the standard under which energy recovery ventilation systems are rated and certified for performance.

ENERGY SIMULATION TOOL. An *approved* software program or calculation-based methodology that projects the annual energy use of a building.

❖ An energy simulation tool typically is a software package incorporating, among other features, an hour-by-hour, full-year (8,760 hours), multiple-zone program to simulate the performance of both proposed and standard design buildings. It is possible to use other types of simulation tools to approximate the dynamics of hourly energy programs and that can be shown to produce equivalent results for the type of building and HVAC systems under consideration. However, the simulation must be capable of converting calculated energy demand and consumption into utility costs using the actual utility rate schedules (rather than average cost of electricity or gas). Some examples of when an hour-by-hour, full-year type of program is required are:

- When the features intended to reduce energy consumption require time-of-day interactions between weather, loads and operating criteria. Examples include: night ventilation or building thermal storage; chilled water or ice storage; heat recovery; daylighting; and water economizer cooling.
- When utility rates are time-of-day sensitive, and the proposed design uses time-of-day load shifting between different types of mechanical plant components.

ENTRANCE DOOR. Fenestration products used for ingress, egress and access in nonresidential buildings, including, but not limited to, exterior entrances that utilize latching hardware and automatic closers and contain over 50-percent glass specifically designed to withstand heavy use and possibly abuse.

❖ This definition is included to help classify and properly apply the code requirements to these products. The commercial fenestration requirements vary depending on the specific type of product (see Table C402.4). The industry uses the various terms shown in the table (fixed fenestration, operable fenestration, entrance door) to help establish the requirements and separate the products into different groupings.

EQUIPMENT ROOM. A space that contains either electrical equipment, mechanical equipment, machinery, water pumps or hydraulic pumps that are a function of the building's services.

❖ A space where electrical equipment, mechanical equipment, machinery, water pumps or hydraulic pumps are located to support the building's services.

EXTERIOR WALL. Walls including both above-grade walls and basement walls.

❖ Wall insulation requirements defined in the code include almost all opaque exterior construction bounding conditioned space. Depending on the compliance method, the wall type, glazing percentage and whether the wall is on the exterior or just separating conditioned from unconditioned space can affect the wall insulation requirement. Note also that doors (both glazed and opaque) are considered fenestration (see commentary, "Fenestration").

In earlier editions, the code included the limitation that an exterior wall is vertical or sloped at an angle of 60 degrees (1.1 rad) or greater from the horizontal. This limitation may still be helpful to consider if dealing with unusual situations, such as an A-frame building. Where a determination is needed to decide whether the roof or wall provisions are appropriate, this limitation could be helpful. This angle is still included in the definition of vertical fenestration.

FAN BRAKE HORSEPOWER (BHP). The horsepower delivered to the fan's shaft. Brake horsepower does not include the mechanical drive losses (belts, gears, etc.).

❖ This definition supports code requirements for the design of HVAC fan systems that regulate the motor energy use for these systems (see Section C403.2.12.1).

FAN EFFICIENCY GRADE (FEG). A numerical rating identifying the fan's aerodynamic ability to convert shaft power, or impeller power in the case of a direct-driven fan, to air power.

❖ Fan efficiency grade (FEG) is one of two efficiency measures for fans. FEG's are based on fan peak (optimum) energy efficiency that indicates the quality of fan energy usage and the potential for minimizing the fan's energy usage. This second minimum for fan efficiency allows for improved efficiency in the whole HVAC design. It counters temptations to use inexpensive fans over the more efficient.

FAN SYSTEM BHP. The sum of the fan brake horsepower of all fans that are required to operate at fan system design conditions to supply air from the heating or cooling source to the *conditioned spaces* and return it to the source or exhaust it to the outdoors.

❖ This definition supports code requirements for the design of HVAC fan systems that regulate the motor energy use for these systems (see Section C403.2.12.1).

FAN SYSTEM DESIGN CONDITIONS. Operating conditions that can be expected to occur during normal system operation that result in the highest supply fan airflow rate to conditioned spaces served by the system.

❖ See the commentary to "Fan system BHP."

FAN SYSTEM MOTOR NAMEPLATE HP. The sum of the motor nameplate horsepower of all fans that are required to operate at design conditions to supply air from the heating or cooling source to the *conditioned spaces* and return it to the source or exhaust it to the outdoors.

❖ See the commentary to "Fan system BHP."

FENESTRATION. Products classified as either vertical fenestration or skylights.

❖ The term "fenestration" refers both to opaque and glazed doors and the light-transmitting areas of a wall or roof, but primarily windows and skylights. The code sets performance requirements for fenestration by establishing requirements that differ from the wall and roof requirements based on the type of fenestration and, in the case of the prescriptive commercial requirements, by limiting the fenestration area. In some of the compliance options the fenestration type and area allowed depend on the shading coefficient, the size of overhangs, the thermal performance (*U*-factor) and whether daylight responsive controls are installed.

Skylight. Glass or other transparent or translucent glazing material installed at a slope of less than 60 degrees (1.05 rad) from horizontal.

❖ Fenestration falls into two categories: Vertical fenestration and skylights. A skylight is a glazed opening in a roof to admit daylight. Section C402.4 sets limits on the amount of skylights permitted as well as minimum energy efficiency for skylights installed in various climate zones.

Vertical fenestration. Windows (fixed or moveable), opaque doors, glazed doors, glazed block and combination opaque/glazed doors composed of glass or other transparent or translucent glazing materials and installed at a slope of at least 60 degrees (1.05 rad) from horizontal.

❖ Vertical fenestration is that which is typically installed on exterior walls. The required thermal efficiency of fenestration varies by climate zone (see Section C402.4).

FENESTRATION PRODUCT, FIELD-FABRICATED. A fenestration product whose frame is made at the construction site of standard dimensional lumber or other materials that were not previously cut, or otherwise formed with the specific intention of being used to fabricate a fenestration product or exterior door. Field fabricated does not include site-built fenestration.

❖ Field-fabricated fenestration products are not required to be tested to determine their rate of air leakage but must be sealed as required for all assemblies (see Section C402.5.2).

FENESTRATION PRODUCT, SITE-BUILT. A fenestration designed to be made up of field-glazed or field-assembled units using specific factory cut or otherwise factory-formed framing and glazing units. Examples of site-built fenestration include storefront systems, curtain walls, and atrium roof systems.

❖ Fenestration products can be site built, not to be confused with field-fabricated fenestration products. This definition is installed to make that distinction.

***F*-FACTOR.** The perimeter heat loss factor for slab-on-grade floors (Btu/h · ft · °F) [W/(m · K)].

❖ As defined, *F*-factor is needed in the provisions of the code for the assembly *U*-factor method of determining the energy conservation of a building's thermal envelope given in Section C402.1.4. The term appears in Table C402.1.4.

FLOOR AREA, NET. The actual occupied area not including unoccupied accessory areas such as corridors, stairways, toilet rooms, mechanical rooms and closets.

❖ This area is intended to be only the room areas that are used for specific occupancy purposes and does not include circulation areas such as corridors, ramps or stairways, or service and utility spaces such as toilet rooms and mechanical and electrical equipment rooms. Net floor area is typically measured between inside faces of walls in a room.

GENERAL LIGHTING. Lighting that provides a substantially uniform level of illumination throughout an area. General lighting shall not include decorative lighting or lighting that provides a dissimilar level of illumination to serve a specialized application or feature within such area.

❖ The energy requirements for lighting apply in large part to general lighting. Therefore, this definition is needed to clarify code applications.

GENERAL PURPOSE ELECTRIC MOTOR (SUBTYPE I). A motor that is designed in standard ratings with either of the following:

1. Standard operating characteristics and standard mechanical construction for use under usual service conditions, such as those specified in NEMA MG1, paragraph 14.02, "Usual Service Conditions," and without restriction to a particular application or type of application.
2. Standard operating characteristics or standard mechanical construction for use under unusual service conditions, such as those specified in NEMA MG1, paragraph 14.03, "Unusual Service Conditions," or for a particular type of application, and that can be used in most general purpose applications.

General purpose electric motors (Subtype I) are constructed in NEMA T-frame sizes or IEC metric equivalent, starting at 143T.

❖ The code regulates the efficiency of electric motors. The efficiency is established based on three categories of motors as defined in this chapter: General purpose electric motor (Subtype I); General purpose electric motor (Subtype II); and Small electric motor.

GENERAL PURPOSE ELECTRIC MOTOR (SUBTYPE II). A motor incorporating the design elements of a general purpose electric motor (Subtype I) that is configured as one of the following:

1. A U-frame motor.
2. A Design C motor.
3. A close-coupled pump motor.
4. A footless motor.
5. A vertical, solid-shaft, normal-thrust motor (as tested in a horizontal configuration).
6. An 8-pole motor (900 rpm).

7. A polyphase motor with voltage of not more than 600 volts (other than 230 or 460 volts).

❖ See the commentary to "General purpose electric motor (Subtype I)."

GREENHOUSE. A structure or a thermally isolated area of a building that maintains a specialized sunlit environment exclusively used for, and essential to, the cultivation, protection or maintenance of plants.

❖ This term is used to apply to a variety of buildings or parts of buildings, however for the purposes of the code, a definition is provided to clarify application of Section C402.1.1. Greenhouses meeting the definition are exempt from the thermal envelope requirements of the code. Greenhouses used for other purposes such as habitation, a sunroom or for retail business would not meet the definition.

HEAT TRAP. An arrangement of piping and fittings, such as elbows, or a commercially available heat trap that prevents thermosyphoning of hot water during standby periods.

❖ Storage hot water tanks with vertical risers in noncirculating systems are required to have either integral heat traps or external heat traps on both the inlet and outlet piping located as close as practicable to the water heater (see commentary, Section C404.3). A heat trap is a device or an arrangement of piping that keeps the buoyant hot water from circulating through a piping distribution system through natural convection (or thermal-siphoning effects). Heat traps can be made by offsetting the device or piping vertically 24 inches (610 mm) down from the cold water inlet on the tank, then horizontally about 6 inches (152 mm), then back up to the cold water supply to the tank [see Commentary Figure C404.4(2)]. The same design can be applied to the hot water outlet side of the tank. Integral heat traps are designed into the water heater by the manufacturer. Heat traps are not required on tanks without vertical risers; however, systems without heat traps, regardless of whether they have circulation pumps, must meet the insulation requirements for circulating systems (see commentary, Section C404.4). There are effective heat trap fittings that use either a teflon/polypropylene ball and seat arrangement or a flexible polypropylene flapper. When water is drawn, the ball checks are moved from their standby position or the flapper is disengaged for proper water flow to and from the tank.

HEATED SLAB. Slab-on-grade construction in which the heating elements, hydronic tubing, or hot air distribution system is in contact with, or placed within or under, the slab.

❖ The space above a heated slab is always conditioned (heated). The definition clarifies that certain slabs are the heating source and, as covered in the code, may require more insulation than unheated slabs. The installation of a radiant heat source in a space does not, in and of itself, qualify a slab as heated.

HIGH SPEED DOOR. A nonswinging door used primarily to facilitate vehicular access or material transportation, with a minimum opening rate of 32 inches (813 mm) per second, a minimum closing rate of 24 inches (610 mm) per second and that includes an automatic-closing device.

❖ High speed doors are afforded a higher air leakage (infiltration) rate than other doors. The nature of the door design to achieve quick opening and closing makes it difficult to limit air leakage. These doors are used where maintaining a specific indoor environment is essential to the use, yet a doorway through which materials can be moved is also critical.

HISTORIC BUILDING. Any building or structure that is one or more of the following:

1. Listed, or certified as eligible for listing by the State Historic Preservation Officer or the Keeper of the National Register of Historic Places, in the National Register of Historic Places.
2. Designated as historic under an applicable state or local law.
3. Certified as a contributing resource within a National Register-listed, state-designated or locally designated historic district.

❖ This definition serves as the baseline to determine whether a building is considered historic. The code in Section C501.6 waives application of the code requirements where such action would harm the historic nature of the building.

HUMIDISTAT. A regulatory device, actuated by changes in humidity, used for automatic control of relative humidity.

❖ A humidistat is a sensing device that measures the amount of moisture vapor in the air. Humidistats are used to control humidifying and dehumidifying equipment.

INFILTRATION. The uncontrolled inward air leakage into a building caused by the pressure effects of wind or the effect of differences in the indoor and outdoor air density or both.

❖ Air leakage is random movement of air into and out of a building through cracks and holes in the building envelope. In technical terms, air leakage is called "infiltration" (air moving into a building) or "exfiltration" (air moving out of a building). In nontechnical terms, air leaks are often referred to as "drafts." Infiltration may be reduced by either reducing the sources of air leakage (joints, penetrations and holes in the building envelope) or by reducing the pressures driving the airflow.

INTEGRATED PART LOAD VALUE (IPLV). A single-number figure of merit based on part-load EER, COP or kW/ton expressing part-load efficiency for air-conditioning and heat pump equipment on the basis of weighted operation at various load capacities for equipment.

❖ This is an efficiency measure for electrically operated condensing units, chillers and other cooling equipment used throughout the code.

LABELED. Equipment, materials or products to which have been affixed a label, seal, symbol or other identifying mark of a nationally recognized testing laboratory, inspection agency or other organization concerned with product evaluation that maintains periodic inspection of the production of the above-labeled items and whose labeling indicates either that the equipment, material or product meets identified standards or has been tested and found suitable for a specified purpose.

❖ When a product is labeled, the label indicates that the equipment or material has been tested for conformance to an applicable standard and that the component is subject to third-party inspection, which verifies that the minimum level of quality required by the applicable standard is maintained. Labeling is a readily available source of information that is useful for field inspection of installed products. The label identifies the product or material and provides other information that can be further investigated if there is a question concerning the suitability of the product or material for the specific installation. The labeling agency performing the third-party inspection must be approved by the code official and the basis for this approval may include, but is not necessarily limited to, the capacity and capability of the agency to perform the specific testing and inspection. The applicable referenced standard often states the minimum identifying information that must be on a label. The data contained on a label typically includes, but is not necessarily limited to, the name of the manufacturer, the product name or serial number, installation specifications, applicable tests and standards, the testing approved and the labeling agency.

LINER SYSTEM (Ls). A system that includes the following:

1. A continuous vapor barrier liner membrane that is installed below the purlins and that is uninterrupted by framing members.
2. An uncompressed, unfaced insulation resting on top of the liner membrane and located between the purlins.

For multilayer installations, the last rated *R-value* of insulation is for unfaced insulation draped over purlins and then compressed when the metal roof panels are attached.

❖ A liner system is a method of insulation installation used in metal buildings. The method is specified in Table C402.1.3.

LISTED. Equipment, materials, products or services included in a list published by an organization acceptable to the *code official* and concerned with evaluation of products or services that maintains periodic inspection of production of *listed* equipment or materials or periodic evaluation of services and whose listing states either that the equipment, material, product or service meets identified standards or has been tested and found suitable for a specified purpose.

❖ When a product is listed and labeled, it indicates that it has been tested for conformance to an applicable standard and is subject to a third-party inspection quality assurance (QA) program. The QA verifies that the minimum level of quality required by the appropriate standard is maintained. The listing states either that the equipment or material meets nationally recognized standards or it has been found suitable for use in a specified manner. The listing becomes part of the documentation that the code official can use to approve or disapprove the equipment or appliance.

LOW-SLOPED ROOF. A roof having a slope less than 2 units vertical in 12 units horizontal.

❖ Solar reflectance and thermal emittance requirements of Section C402.3 apply only to low-sloped roofs in specified climate zones.

LOW-VOLTAGE DRY-TYPE DISTRIBUTION TRANSFORMER. A transformer that is air-cooled, does not use oil as a coolant, has an input voltage less than or equal to 600 volts and is rated for operation at a frequency of 60 hertz.

❖ The code sets minimum efficiency requirements for transformers in Section C405.7.

LOW-VOLTAGE LIGHTING. Lighting equipment powered through a transformer such as a cable conductor, a rail conductor and track lighting.

❖ The wattage of the transformer supplying power to low-voltage lighting is included in the equation determining the total connected lighting power of a building.

MANUAL. Capable of being operated by personal intervention (see "Automatic").

❖ Devices, systems or equipment having manual controls or overrides are designed to operate safely with only human intervention instead of having an automatic operation or control system (see commentary, "Automatic").

Manual switch controls stay either off or on until a person flips the switch to a different position.

NAMEPLATE HORSEPOWER. The nominal motor horsepower rating stamped on the motor nameplate.

❖ This definition supports code requirements for the design of HVAC fan systems that regulate the motor energy use for these systems (see Section C403.2.12.1).

NONSTANDARD PART LOAD VALUE (NPLV). A single-number part-load efficiency figure of merit calculated and referenced to conditions other than IPLV conditions, for units that are not designed to operate at ARI standard rating conditions.

❖ This is an efficiency measure for electrically operated condensing units, chillers and other cooling equipment used throughout the code.

OCCUPANT SENSOR CONTROL. An automatic control device or system that detects the presence or absence of people within an area and causes lighting, equipment or appliances to be regulated accordingly.

❖ Occupant sensor controls are one type of lighting control required in Sections C405.2 and C405.2.1. The controls allow lights to be automatically turned off when people are not in an area, thus reducing energy use for lighting in such rooms.

ON-SITE RENEWABLE ENERGY. Energy derived from solar radiation, wind, waves, tides, landfill gas, biomass or the internal heat of the earth. The energy system providing on-site renewable energy shall be located on the project site.

❖ The code requires one of six additional energy efficiency packages in Section C406, one of which is on-site renewable energy. This definition attempts to clarify what that is.

OPAQUE DOOR. A door that is not less than 50-percent opaque in surface area.

❖ Opaque doors are included in the definition of fenestration but are regulated as part of the opaque envelope assemblies, not like other fenestration in a building's thermal envelope. Specific insulation and thermal resistances standards are provided in Tables C402.1.3 and C402.1.4.

POWERED ROOF/WALL VENTILATORS. A fan consisting of a centrifugal or axial impeller with an integral driver in a weather-resistant housing and with a base designed to fit, usually by means of a curb, over a wall or roof opening.

❖ Powered roof and wall ventilators are exempt from the minimum fan efficiency requirements specified in Section C403.2.12.3.

PROPOSED DESIGN. A description of the proposed building used to estimate annual energy use for determining compliance based on total building performance.

❖ The proposed design is simply a description of the proposed building, used to estimate annual energy costs for determining compliance based on total building performance. The proposed design is effectively the subject building intended to be built. The performance of the proposed design (the building exactly as it is anticipated to be constructed) is then compared to the standard reference design (a similar building that is assumed to be built to a prescriptive set of minimum code requirements).

RADIANT HEATING SYSTEM. A heating system that transfers heat to objects and surfaces within a conditioned space, primarily by infrared radiation.

❖ Radiant heating systems are most commonly installed in floors, but radiant heating panels can also be installed in walls and ceilings. There are three types used in floors: radiant air floors; electric radiant floors; and hot water (hydronic radiant) floors. The systems are installed between layers of other materials comprising the floor assembly. Radiant panels in walls and ceilings are usually either electric or hydronic. The code specifies minimum insulation standards where these heating systems are installed in a building's thermal envelope (see Section C402.2.6).

READILY ACCESSIBLE. Capable of being reached quickly for operation, renewal or inspection without requiring those to whom ready access is requisite to climb over or remove obstacles or to resort to portable ladders or access equipment (see "*Accessible*").

❖ Readily accessible can be described as able to be quickly reached or approached for operation, inspection, observation or emergency action. Ready access does not require the removal or movement of any door panel or similar obstruction, or overcoming physical obstructions or obstacles, including differential elevation.

REFRIGERANT DEW POINT. The refrigerant vapor saturation temperature at a specified pressure.

❖ One of two temperatures used to determine the saturated condensing temperature in a refrigeration system.

REFRIGERATED WAREHOUSE COOLER. An enclosed storage space capable of being refrigerated to temperatures above 32°F (0°C), that can be walked into and has a total chilled storage area of not less than 3,000 square feet (279 m^2).

❖ There are four related cooled spaces regulated by the code: Refrigerated warehouse cooler; Refrigerated warehouse freezer; Walk-in cooler and Walk-in freezer. The "walk-in spaces" are distinguished by being less than 3,000 square feet (279 m^2). The freezer spaces are distinguished from the coolers by having temperatures maintained at or below 32°F (0°C). The regulations for these four items are fairly consistent.

REFRIGERATED WAREHOUSE FREEZER. An enclosed storage space capable of being refrigerated to temperatures at or below 32°F (0°C), that can be walked into and has a total chilled storage area of not less than 3,000 square feet (279 m^2).

❖ See the commentary to "Refrigerated warehouse cooler."

REFRIGERATION SYSTEM, LOW TEMPERATURE. Systems for maintaining food product in a frozen state in refrigeration applications.

❖ One of two refrigeration systems for which the code sets standards for fan-powered condensers remotely located from the condensing unit or the feature being served.

REFRIGERATION SYSTEM, MEDIUM TEMPERATURE. Systems for maintaining food product above freezing in refrigeration applications.

❖ One of two refrigeration systems for which the code sets standards for fan-powered condensers remotely located from the condensing unit or the feature being served.

REGISTERED DESIGN PROFESSIONAL. An individual who is registered or licensed to practice their respective design profession as defined by the statutory requirements of the professional registration laws of the state or jurisdiction in which the project is to be constructed.

❖ Legal qualifications for engineers and architects are established by the state having jurisdiction. Licensing and registration of engineers and architects are accomplished by written or oral examinations offered by states, or by reciprocity (licensing in other states).

REPAIR. The reconstruction or renewal of any part of an existing building for the purpose of its maintenance or to correct damage.

❖ The repair of an appliance, energy-using subsystem or other piece of equipment typically does not require a permit. This definition makes it clear that a repair is limited to work on the item and does not include complete or substantial replacement or other new work.

This definition is important when applying the provisions of Chapter 5 [CE] and determining how the code will apply to existing buildings and installations.

REROOFING. The process of recovering or replacing an existing roof covering. See "Roof recover" and "Roof replacement."

❖ Reroofing is the general term that includes both subcategories of roof recovering and roof replacement. The subcategories are separately defined. This term refers to the process of covering an existing roof system or replacing an existing roof system with a new roofing system. See definitions of "Roof recover," "Roof replacement" and "Roof repair."

RESIDENTIAL BUILDING. For this code, includes detached one- and two-family dwellings and multiple single-family dwellings (townhouses) as well as Group R-2, R-3 and R-4 buildings three stories or less in height above grade plane.

❖ The definition of a residential building is important not only for what it does include, but also for what it does not. One of the primary limitations of this definition is the fact that this term will only include the Group R-2 and R-4 occupancies when they are three stories or less in height. Therefore, if a Group R-2 or R-4 occupancy is over three stories in height, it would be defined as a "Commercial building" (see definition, "Commercial building") and be required to comply with the requirements of Chapter 4[CE] instead of the residential provisions of Chapter 4 [RE]. Detached one- and two-family dwellings as well as townhouses are considered residential buildings regardless of the number of stories. However, the Group R-3 category as outlined in IBC Section 310.5 includes uses in addition to those containing only one or two dwelling units and townhouses. These other R-3 uses [e.g., boarding houses, congregate living facilities, lodging houses and small care facilities (five or fewer care recipients)] if over three stories would fall into the commercial building category.

Group R-1 buildings are not included in this definition. Therefore, any hotel, motel or similar use classified as a Group R-1 occupancy would need to comply with the commercial building requirements. This would apply even if the hotel or motel was three stories or less in height.

Though not specifically addressed in this definition, any building built under the provisions of the *International Residential Code*® (IRC®) would also be considered a "Residential building."

The Group R-2 classification of residential buildings includes apartments and condominiums where three or more units are physically attached. In addition, it includes certain boarding houses, convents, dormitories, fraternities or other such facilities. The code defines a dwelling unit as a single housekeeping unit of one or more rooms providing complete independent living facilities, including permanent provisions for living, sleeping, cooking and sanitation (see commentary, "Dwelling unit"). Many of the code provisions address dwelling units specifically. All of the elements listed in the definition must be present for a dwelling unit to exist. Guest rooms (sleeping units) in hotels and motels, patient care rooms in nursing homes and larger group care facilities do not meet these criteria. Therefore, these buildings are considered commercial uses, even if they are three stories or less in height (see the IBC for additional information on dwelling units). Although this definition closely follows the Group R-2 occupancy classification of the IBC, it is important to note there are differences. The main distinction is the three-story limitation, which is found only in the code.

The Group R-4 classification is included in order to coordinate the provisions of the code with the IBC. Group R-4 includes facilities where between six and 16 people reside in a supervised residential environment and receive custodial care. This distinction is important because larger facilities are classified by the IBC as Group I occupancies and, therefore, would be considered commercial buildings by the code (see definition, "Commercial building").

As mentioned earlier, it is important to note the application of the three-story limitation, which is found only in the code. This issue of defining a residential building over three stories in height as a commercial building coordinates the code with ANSI/ASHRAE/IESNA 90.1.

ROOF ASSEMBLY. A system designed to provide weather protection and resistance to design loads. The system consists of a roof covering and roof deck or a single component serving as both the roof covering and the roof deck. A roof assembly includes the roof covering, underlayment, roof deck, insulation, vapor retarder and interior finish.

❖ This section basically defines the cover or protection over a building, but it will be important to the energy requirements in several sections. The roof assembly generally serves as the building thermal envelope for the top of the building. By defining the various elements that are included in the definition, the phrase can be applied to situations where the insulation is installed above the roof deck, beneath the roof deck or above a ceiling that occurs below.

This definition will be important not only when applying the prescriptive insulation requirements of the building thermal envelope, but also when applying the performance requirements where the type of roof covering and ventilation can affect the energy efficiency.

Numerous horizontal and sloped surfaces may be associated with the roof or roof/ceiling assembly, including flat and cathedral ceilings, dormer roofs, bay-window roofs, overhead portions of an interior stairway to an attic, or other unconditioned space, attic hatches and skylights. When determining the area of the assembly under the performance options, ceiling assembly areas should be measured on the slope of the finished interior surface.

In earlier editions, the code included the limitation that a roof is generally horizontal or sloped at an angle of less than 60 degrees (1.1 rad) from the horizontal. This limitation still may be helpful to consider if dealing with unusual situations, such as an A-frame building. Where a determination is needed to decide whether the roof or wall provisions are appropriate, this limitation could be helpful (see commentary, "Skylights," "Fenestration" and "Exterior wall").

ROOF RECOVER. The process of installing an additional roof covering over an existing roof covering without removing the existing roof covering.

❖ This term refers to the process of adding a new roof covering on top of the existing roof system without removing the existing roofing system. See definitions of "Roof replacement" and "Roof repair." See Chapter 5 [CE] for requirements for roof recovers.

ROOF REPAIR. Reconstruction or renewal of any part of an existing roof for the purpose of its maintenance.

❖ Roof repair is distinguished from roof recovering and roof replacement. It is limited to actions aimed at maintaining an existing roof and restoring portions that have been damaged. Roof repairs are typically exempt from permit. See definitions of "Roof recover" and "Roof replacement." See Chapter 5 [CE] for requirements for roof repairs.

ROOF REPLACMENT. The process of removing the existing roof covering, repairing any damaged substrate and installing a new roof covering.

❖ This term refers to the process of removing an existing roof covering and replacing it with a new covering. This is distinguished from roof recovering, where the new cover is added on top of the existing roof cover. See definitions of "Roof recover" and "Roof repair." See Chapter 5 [CE] for requirements for roof replacements.

ROOFTOP MONITOR. A raised section of a roof containing vertical fenestration along one or more sides.

❖ Where rooftop monitors are present, they help establish the daylight zone in a space. Where there is sufficient daylight zone area provided under rooftop monitors, skylights are not required (see Section C402.4.2).

***R*-VALUE (THERMAL RESISTANCE).** The inverse of the time rate of heat flow through a body from one of its bounding surfaces to the other surface for a unit temperature difference between the two surfaces, under steady state conditions, per unit area ($h \cdot ft^2 \cdot °F/Btu$) [$(m^2 \cdot K)/W$].

❖ Thermal resistance measures how well a material or series of materials retards heat flow. Insulation thermal resistance is rated using *R*-values. As the *R*-value of an element or assembly increases, the heat loss or gain through that element or assembly decreases. Thus, a higher *R*-value is considered better than a lower *R*-value.

SATURATED CONDENSING TEMPERATURE. The saturation temperature corresponding to the measured refrigerant pressure at the condenser inlet for single component and azeotropic refrigerants, and the arithmetic average of the dew point and *bubble point* temperatures corresponding to the refrigerant pressure at the condenser entrance for zeotropic refrigerants.

❖ Fan-powered condensers in refrigeration systems must comply with various standards set in Section C403.5. One of these standards is a maximum saturated condensing temperature in both low- and medium-temperature refrigeration systems.

SCREW LAMP HOLDERS. A lamp base that requires a screw-in-type lamp, such as a compact-fluorescent, incandescent or tungsten-halogen bulb.

❖ Holders come in various sizes, including candelabra base, intermediate base and medium base. The definition is specific to screw-in-type lamps. The type of bulb used in the holder does not make a difference. It is simply the type of base on the bulb that is important under this definition.

SERVICE WATER HEATING. Supply of hot water for purposes other than comfort heating.

❖ Although the definition makes it clear that the code requirement applies to equipment used to produce and distribute hot water for purposes other than comfort heating, the definition also applies to energy-efficient process water-heating systems and equipment. Equipment providing or distributing hot water for uses such as restrooms, showers, laundries, kitchens, pools and spas, defrosting of sidewalks and driveways, carwashes, beauty salons and other commercial enterprises are included. Space-conditioning boilers and distribution systems are not considered service water-heating components.

[B] SLEEPING UNIT. A room or space in which people sleep, which can also include permanent provisions for living, eating, and either sanitation or kitchen facilities but not both. Such rooms and spaces that are also part of a dwelling unit are not *sleeping units*.

❖ This definition is provided as a distinction between these types of spaces and dwelling units.

In general, sleeping units will be found in commercial buildings such as hotels and motels that are classified by the IBC as a Group R-1 occupancy and Group I-2 patient care facilities (e.g., hospitals and nursing homes). However, they can be found in dormitories, fraternities and other uses that would be classified as Group R-2 occupancies and, therefore, could be considered a "residential building," depending on the building's height.

SMALL ELECTRIC MOTOR. A general purpose, alternating current, single speed induction motor.

❖ See the commentary to "General purpose electric motor (Subtype I)."

SOLAR HEAT GAIN COEFFICIENT (SHGC). The ratio of the solar heat gain entering the space through the fenestration assembly to the incident solar radiation. Solar heat gain includes directly transmitted solar heat and absorbed solar radiation which is then reradiated, conducted or convected into the space.

❖ The SHGC is the fraction of incident solar radiation admitted through a window or skylight. This includes the solar radiation that is directly transmitted, and that which is absorbed and subsequently released inward. Therefore, the SHGC measures how well a window blocks the heat from sunlight. The SHGC is the fraction of the heat from the sun that enters through a window. SHGC is expressed as a number between 0 and 1. The lower a window's SHGC, the less solar heat it transmits.

In the warmer climate zones, where cooling is the dominant requirement, the code generally will impose a limitation on the amount of solar heat gain permitted. In colder climate zones (7 and 8), the code generally will not place any SHGC requirements on the fenestration. Thus, it will depend on the climate zone as to whether a higher or lower SHGC is best. While windows with lower SHGC values reduce summer cooling and overheating, they also reduce free winter solar heat gain.

STANDARD REFERENCE DESIGN. A version of the *proposed design* that meets the minimum requirements of this code and is used to determine the maximum annual energy use requirement for compliance based on total building performance.

❖ The standard reference design is simply the same building design as that intended to be built (see commentary, "Proposed design"), except the energy conservation features required by the code (insulation, windows, infiltration, mechanical, lighting and service water-heating systems) are modified to meet the minimum prescriptive requirements, as applicable. Note that the standard reference design is not truly a separate building, and it is never actually built. It is the baseline against which the proposed design is measured.

The performance of the proposed design (the building exactly as it is anticipated to be constructed) is then compared to the standard reference design (a similar building that is assumed to be built to a prescriptive set of minimum code requirements). Under the performance paths, if the energy efficiency of the proposed design is equal to or better than that of the standard reference design, then the proposed design is acceptable and in compliance with the code requirements.

STOREFRONT. A nonresidential system of doors and windows mulled as a composite fenestration structure that has been designed to resist heavy use. *Storefront* systems include, but are not limited to, exterior fenestration systems that span from the floor level or above to the ceiling of the same story on commercial buildings, with or without mulled windows and doors.

❖ This definition provides the description for this specific type of commercial glazing system. Storefront systems, due to their size and required structural requirements, are treated differently than other commercial fenestration systems.

THERMOSTAT. An automatic control device used to maintain temperature at a fixed or adjustable set point.

❖ Thermostats combine control and sensing functions in a single device. Thermostats may signal other control devices to trigger an action when certain temperatures are reached or surpassed. Because thermostats are so prevalent, the various types and their operating characteristics are described here.

The occupied/unoccupied or dual-temperature room thermostat reduces temperature at night. It may be indexed (changed from occupied to unoccupied operation or vice versa) individually from a remote point or in a group by a manual or time switch. Some types have an individual clock and switch built in.

The pneumatic day-night thermostat uses a two-pressure air supply system, where changing the pressure at a central point from one value to the other actuates switching devices in the thermostat and indexes it.

The heating/cooling or summer/winter thermostat can have its action reversed and its setpoint changed in response to outdoor and comfort conditions. It is used to actuate controlled devices, such as valves or dampers, that regulate a heating source at one time and a cooling source at another. The pneumatic heating/cooling thermostat uses a two-pressure air supply similar to that described for occupied/unoccupied thermostats.

Multistage thermostats are arranged to operate two or more successive steps in sequence.

A submaster thermostat has its setpoint raised or lowered over a predetermined range, in accordance with variations in output from a master controller. The master controller can be a thermostat, manual switch, pressure controller or similar device. For example, a master thermostat measuring outdoor air temperature can be used to readjust the setpoint of a submaster thermostat that controls the water temperature in a heating system.

A wet-bulb thermostat is often used (in combination with a dry-bulb thermostat) for humidity control. Using a wick or another means of keeping the bulb wet with pure (distilled) water and rapid air motion to ensure a true wet-bulb measurement is essential. Wet-bulb thermostats are seldom used.

A dew-point thermostat is designed to control dew-point temperatures.

A dead-band thermostat has a wide differential over which the thermostat remains neutral, requiring neither heating nor cooling. This differential may be adjustable up to 10°F (-12°C) (see Section C403.2.4.2).

TIME SWITCH CONTROL. An automatic control device or system that controls lighting or other loads, including switching off, based on time schedules.

❖ Time switch controls are one type of lighting controls required in Sections C405.2 and C405.2.2. In areas of the building where the code doesn't require occupant sensor controls, time switch controls are required. The controls allow lights to be automatically turned off when people are not in an area, thus reducing energy use for lighting in such rooms. Occupant sensor controls, by comparison, trigger based on the presence of people in a space. Time switch controls allow lights to be turned on and off based on a programmed schedule.

***U*-FACTOR (THERMAL TRANSMITTANCE).** The coefficient of heat transmission (air to air) through a building component or assembly, equal to the time rate of heat flow per unit area and unit temperature difference between the warm side and cold side air films (Btu/h · ft^2 · °F) [W/(m^2 · K)].

❖ Thermal transmittance (*U*-factor) is a measure of how well a material or series of materials conducts heat. *U*-factors for window and door assemblies are the reciprocal of the assembly *R*-value:

$$U\text{-factor} = \frac{1}{R\text{-value}}$$

For other building assemblies, such as a wall or roof/ceiling, the *R*-value used in the above equation is the *R*-value of the entire assembly, not just the insulation. This distinction is important and reflects the provisions of Section C402.1.1. It also explains why there are differences between the comparable values of Tables C402.1.3 and C402.1.4.

VARIABLE REFRIGERANT FLOW SYSTEM. An engineered direct-expansion (DX) refrigerant system that incorporates a common condensing unit, at least one variable-capacity compressor, a distributed refrigerant piping network to multiple indoor fan heating and cooling units each capable of individual zone temperature control, through integral zone temperature control devices and a common communications network. Variable refrigerant flow utilizes three or more steps of control on common interconnecting piping.

❖ The code requires increased efficiency for variable refrigerant flow systems where the designer chooses compliance with Section C406.2 as the method to increase energy efficiency in a building design.

[M] VENTILATION. The natural or mechanical process of supplying conditioned or unconditioned air to, or removing such air from, any space.

❖ Ventilation is provided by the wind, naturally occurring pressure and temperature differences or mechanical systems typically consisting of fans and blowers.

[M] VENTILATION AIR. That portion of supply air that comes from outside (outdoors) plus any recirculated air that has been treated to maintain the desired quality of air within a designated space.

❖ Ventilation air can be used for comfort cooling, control of air contaminants, equipment cooling and replenishing oxygen levels (see commentary, "Ventilation").

VISIBLE TRANSMITTANCE [VT]. The ratio of visible light entering the space through the fenestration product assembly to the incident visible light. Visible transmittance includes the effects of glazing material and frame and is expressed as a number between 0 and 1.

❖ Use of daylight for energy conservation is now a fundamental part of the code. However, requirements for lower SHGC values for glazing can work against this, because glazing with lower SHGC values generally will allow less visible light through the glazing. Therefore, the code now must include a limit on VT.

WALK-IN COOLER. An enclosed storage space capable of being refrigerated to temperatures above 32°F (0°C) and less than 55°F (12.8°C) that can be walked into, has a ceiling height of not less than 7 feet (2134 mm) and has a total chilled storage area of less than 3,000 square feet (279 m^2).

❖ There are four related cooled spaces regulated by the code: Refrigerated warehouse cooler; Refrigerated warehouse freezer; Walk-in cooler and Walk-in freezer. The "walk-in spaces" are distinguished by being less than 3,000 square feet (279 m^2). The freezer spaces are distinguished from the coolers by having temperatures maintained at or below 32°F (0°C). The regulations for these four items are fairly consistent.

WALK-IN FREEZER. An enclosed storage space capable of being refrigerated to temperatures at or below 32°F (0°C) that can be walked into, has a ceiling height of not less than 7 feet (2134 mm) and has a total chilled storage area of less than 3,000 square feet (279 m^2).

❖ There are four related cooled spaces regulated by the code: Refrigerated warehouse cooler; Refrigerated warehouse freezer; Walk-in cooler and Walk-in freezer. The "walk-in spaces" are distinguished by being less than 3,000 square feet (279 m^2). The freezer spaces are distinguished from the coolers by having temperatures maintained at or below 32°F (0°C). The regulations for these four items are fairly consistent.

WALL, ABOVE-GRADE. A wall associated with the *building thermal envelope* that is more than 15 percent above grade and is on the exterior of the building or any wall that is associated with the *building thermal envelope* that is not on the exterior of the building.

❖ Walls are either above grade or below grade. The code is concerned with those walls that are part of the thermal envelop of a building—those separating conditioned spaces from unconditioned spaces and from the exterior. Walls that are not below grade are considered to be above-grade walls. Most of the walls covered by this definition are on the exterior of a building. In addition, the definition includes walls in the building that form part of the thermal envelope. Examples of such "interior" walls would be those separating conditioned space from an unconditioned parking garage and walls separating conditioned from unconditioned attic spaces. The location of grade has no impact on classification of an interior wall as an above-grade wall.

WALL, BELOW-GRADE. A wall associated with the basement or first story of the building that is part of the *building thermal envelope*, is not less than 85 percent below grade and is on the exterior of the building.

❖ Because below-grade walls are in contact with the ground and ground temperatures differ from air temperatures, the amount of energy transferred through a below-grade wall is different than the energy transferred through a wall predominantly above grade. Therefore, the code provides different thermal requirements for below-grade walls and above-grade walls. An individual wall enclosing conditioned space is classified as a below-grade wall where the gross wall area is 85 percent or more below grade and is bounded by soil; otherwise, the wall is classified as an above-grade wall (the residential provisions of the code define basement wall as 50 percent below grade). This definition is limited to walls on the exterior of the building, separating conditioned from nonconditioned space. This wall classification applies to the whole wall area, even if a portion of the individual wall is not below grade. Therefore, the above-grade portion of the wall is considered as part of the below-grade wall where the total wall is 85 percent or more below grade. Both sections of the wall (above grade and below grade) are then insulated as a "below-grade wall." Likewise, where an exterior wall is less than 15 percent below grade, the whole wall area is classified as an above-grade wall, including the portion underground.

The intent of this definition is to apply the provision to each wall enclosing the space. It is not intended to be applied to the aggregate of all of the walls of the basement or story. Therefore, this classification is done for each individual wall segment and not on an aggregate basis. The wall requirements apply only to the opaque basement wall area, excluding windows and doors. For purposes of meeting the code requirements, windows and doors in a below-grade wall are regulated as any other fenestration opening.

WATER HEATER. Any heating appliance or equipment that heats potable water and supplies such water to the potable hot water distribution system.

❖ A water heater is a closed pressure vessel or heat exchanger that is provided with a heat source and that supplies potable (drinkable) water to a building's hot water distribution system. Large commercial water heaters are sometimes referred to as "hot water supply boilers." Water heaters can be of the storage type with an integral storage vessel, circulating type for use with an external storage vessel, tankless instantaneous type without storage capacity and point-of-use type with or without storage capacity.

ZONE. A space or group of spaces within a building with heating or cooling requirements that are sufficiently similar so that desired conditions can be maintained throughout using a single controlling device.

❖ The simplest all-air system is a supply unit serving a single-temperature control zone. Ideally, this system responds completely to the space needs and well-designed control systems maintain temperature and humidity closely and efficiently. Single-zone systems can be shut down when not required without affecting the operation of adjacent areas. Thus, in a thorough discussion of the term "zone," the concept of "zoning" for temperature control requires consideration.

Exterior Zoning. Exterior zones are affected by varying weather conditions—wind, temperature and sun—and, depending on the geographic area and season, require both heating and cooling. This variation gives the designer considerable flexibility in choosing a system and results in the greatest advantages. The need for separate perimeter zone heating is determined by:

- Severity of the heating load (i.e., geographic location).
- Nature and orientation of the building envelope.
- Effects of downdraft at windows and the radiant effect of the cold glass surface (type of glass, area, height and *U*-factor).
- Type of occupancy (sedentary versus transient).

- Operating costs (in buildings such as offices and schools that are occupied for considerable periods). Fan operating costs can be reduced by heating with perimeter radiation during unoccupied periods rather than operating the main supply fans or local unit fans.

Interior Zoning. Conditions in interior spaces are relatively constant because they are isolated from external influences. Usually, interior spaces require cooling throughout the year. Interior spaces with a roof exposure, however, may require similar treatment to perimeter spaces requiring heat. To summarize, zone control is required when the conditions at the thermostat are not representative of all the rooms or the entire exposure. This situation will almost certainly occur if any of the following conditions exist:

- The building has more than one level.
- One or more spaces are used for entertaining large groups.
- One or more spaces have large glass areas.
- The building has an indoor swimming pool or hot tub.
- The building has a solarium or atrium. In addition, zoning may be required when several rooms or spaces are isolated from each other and from the thermostat.
- The building spreads out in many directions (wings).
- Some spaces are distinctly isolated from the rest of the building.
- The envelope only has one or two exposures.
- The building has a room or rooms in a basement.
- The building has a room or rooms in an attic space.
- The building has one or more rooms with a slab or exposed floor.
- Zone control can be achieved by insulation.
- There are discrete heating/cooling duct systems for each zone requiring control.

There are automatic zone damper systems in a single heating/cooling duct system.

Chapter 3 [CE]: General Requirements

General Comments

Chapter 3 [CE] specifies the climate zones that establish exterior design conditions and provides general requirements for interior design conditions, and materials, systems and equipment. In general, the climate zone provisions are determined simply by referring to the map (see Figure C301.1) or by looking at the tables [see Tables C301.1, C301.3(1) and C301.3(2)]. In addition, Section C302 provides the interior design conditions that are used for heating and cooling load calculations. Section C303 provides requirements for fenestration, identification of insulation and other basic general requirements for insulation materials.

Purpose

Climate has a major impact on the energy use of most commercial and residential buildings. The code establishes many requirements such as wall and roof insulation *R*-values, window and door thermal transmittance requirements (*U*-factors) and provisions that affect the mechanical systems based on the climate where the building is located. This chapter contains the information that will be used to properly assign the building location into the correct climate zone, which will then be used as the basis for establishing or eliminating requirements.

Materials and systems used to provide insulation and fenestration values, including *U*-factor and solar heat gain coefficient (SHGC) ratings, must be based on data used by appropriate tests. This establishes a level playing field for manufacturers of products.

Discussion and Development of the Climate Zone Map

The 2006 code made a dramatic shift in the classification of climate zones. While this change in the climate zone map was a part of the major revision to help simplify the code and make both compliance and enforcement easier, the climate zone revisions were a lengthy, very detailed and complicated process. Much of the new climate zone development was based on a paper titled "Climate Classification for Building Energy Codes and Standards." This paper was written by Robert S. Briggs, Robert G. Lucas and Z. Todd Taylor of the U.S. Department of Energy's Pacific Northwest National Laboratory (PNNL). Some aspects of this paper may help users better understand climate zones and also feel comfortable with these new classifications.

Climate zones were developed based on the following criteria:

1. Offer consistent climate materials for all compliance methods and code sections (including both commercial and residential).
2. Enable the code to be self-contained with respect to climate data.
3. Be technically sound.
4. Map to political boundaries.
5. Provide a long-term climate classification solution.
6. Be generic and neutral (i.e., not overly tailored to current code requirements).
7. Be useful in beyond-code and future-code contexts.
8. Offer a more concise set of climate zones and presentation formats.
9. Be acceptable to the American Society of Heating, Refrigerating and Air-Conditioning Engineers (ASHRAE), and usable in ASHRAE standards and guidelines.
10. Provide a basis for use outside of the United States.

The reasons that the authors cited for some of the less-obvious items include:

Item 4 – Mapping climate zones to easily recognizable political boundaries instead of to abstract climatic parameters facilitates code implementation. Users and jurisdictions are able to easily tell what requirements apply, which is not the case in some locations when climate parameters are used.

Item 7 – "Useful in future-code and beyond-code contexts" reflects the view that minimum acceptable practice codes and standards can provide an effective platform on which to build other efficiency programs. Beyond-code programs are likely to encourage features and technologies not included in current codes, many of which are likely to be more climate-sensitive than current requirements.

Item 9 – "Usable in ASHRAE standards and guidelines" is important because effective coordination of both content and formats used in the code and ASHRAE standards offers the potential to facilitate rapid migration of ASHRAE standards into model codes. Previous efforts to translate ASHRAE criteria into the simpler and more prescriptive forms most desired by the code enforcement community has, in some cases, added years to the process of getting updated criteria adopted and into widespread use.

The belief in developing the climate zones was that any new system needed to show substantial improvement over previous systems. In addition, any new classification must be at least roughly compatible with the previous climate-dependent requirements in order to allow for the conversion and inclusion of existing, generally accepted requirements. The intent was to develop a set of climate classifications that could support simple, approximate ways of prescribing energy-efficiency measures for buildings. It was not intending to develop a set of categories that could be used for all purposes.

The new climate zones were developed in an open process involving several standards committees of ASHRAE, the U.S. Department of Energy (DOE) staff and other interested parties.

Given the interest of the International Code Council (ICC)® and ASHRAE in producing documents that are capable of being used internationally, an effort was made to develop a system of climate zones that could work outside of the United States. The new climate definitions were developed using SI (The International System of Units, abbreviated SI from the French Le Systéme International d'Unités). By using the SI units and climate indices, which are widely available internationally, the climate zones and the development of building energy-efficiency provisions can be applied anywhere in the world. The boundaries between the various climate zones in Table C301.3(2) occur in multiples of 900 degree days Fahrenheit, which converts to 500 degree days Celsius. Distinguishing the climate zones at these numbers results in a clean and understandable division between the climate zones in either system of temperature measurement.

The developers of the climate zone map selected bands of 1,000 Heating Degree Days (HDD) 18°C (1800 HDD65°F) because they resulted in boundaries that align with boundaries established in ANSI/ASHRAE/IESNA 90.1-2001, plus they facilitate the use of both SI and inch-pound units, and were able to affect a significant reduction in the number of climate zones.

An objective for any effective classification is to maximize the differences between the selected criteria for each climate zone, while minimizing the variations that occur within the group. A large variation between the groups enables generalizations embodied in the code requirements to be better tailored to each climate zone. A small variation within each climate zone will ensure that the generalizations better fit the climate zones. It was the developers' feeling that the new classification better represents the climatic diversity, while defining more coherent climate zones than what the code previously used. It should be noted that mountainous regions defy clean geographic separation of clusters.

SECTION C301
CLIMATE ZONES

C301.1 General. Climate zones from Figure C301.1 or Table C301.1 shall be used in determining the applicable requirements from Chapter 4. Locations not in Table C301.1 (outside the United States) shall be assigned a climate zone based on Section C301.3.

❖ Climate involves temperature, moisture, wind and sun, and also includes both daily and seasonal patterns of variation of the parameters. To account for these variations, the code establishes climate zones that serve as the basis for the code provisions.

This section serves as the starting point for determining virtually all of the code requirements, especially under the prescriptive compliance paths. Because of their easy-to-understand graphic nature, maps have proven useful over the years as an effective way to enable code users to determine climate-dependent requirements. Therefore, for the United States, the climate zones are shown in the map in Figure C301.1. Because of the limited size of the map, the code also includes a listing of the climate zones by states and counties in Table C301.1. Table C301.1 will allow users to positively identify climate zone assignments in those few locations for which the map interpretation may be difficult. Whether the map or the county list is used, the climate classification for each area will be the same.

When dealing with the prescriptive compliance paths, the code user would simply look at the map or listing and select the proper climate zone based on the location of the building. When using a performance approach, additional climatic data may be needed.

Virtually every building energy code that has been developed for use in the United States has included a performance-based compliance path, which allows users to perform an energy analysis and demonstrate compliance based on equivalence with the prescriptive requirements. To perform these analyses, users must select appropriate weather data for their given project's location. The selection of appropriate weather data is straightforward for any project located in or around one of the various weather stations within the United States. For other locations, selecting the most appropriate weather site can be problematic. The codes themselves provide little guidance to help with this selection process. During the development of the new climate zones, the developers mapped every county in the United States to the most appropriate National Climatic Data Center "Solar and Meteorological Surface Observation Network" (SAMSON) station for each county as a whole.

This mapping is not included in the code but may be used in some compliance software. Designating an appropriate SAMSON station should not be considered to be the only climate data permitted for a given county. It could, however, be used in the absence of better information. Where local data better reflects regional or microclimatic conditions of an area, they would be appropriate to use. For example, elevation has a large impact on climate, and elevation can vary dramatically within individual counties, especially in the western United States. Where elevation differences are significant, code officials may require the use of sites that differ from the sites designated as being the most appropriate for the county. For additional information on this topic, review the paper "Climate Classification for Building Energy Codes and Standards," which is referenced in the commentary text that precedes Section C301.

The new climate classifications do not attempt to resolve the issue of what the appropriate treatment for elevation differences is. This aspect is left in the hands of the local code official.

C301.2 Warm humid counties. Warm humid counties are identified in Table C301.1 by an asterisk.

❖ Table C301.1 provides a listing of the counties within the southeastern United States that fall south and east of the white dashed line that appears in the map in Figure C301.1. The warm-humid climate designation includes parts of eight states and also covers all of Florida, Hawaii and the U.S. territories. Table C301.3(1) provides the details that were used to determine the classification of the warm-humid designation for the counties.

There currently are very few requirements in the code that are specifically tied to the warm-humid climate criteria. Although not tied directly to the warm-humid designation, many other code sections, such as those addressing moisture control and energy recovery ventilation systems, do take these climatic features into account.

C301.3 International climate zones. The climate zone for any location outside the United States shall be determined by applying Table C301.3(1) and then Table C301.3(2).

❖ Although the code and the climate zone classifications are predominantly used within the United States, they can be used in any location. Because the mapping and decisions that were made during the development of the new climate zones focused primarily on the United States, this section provides the details of how to properly classify the climate zones based on the thermal criteria [see Table C301.3(2)], the major climate types [see Table C301.3(1)] and warm-humid criteria (see Commentary Figure C301.3) for locations outside of the United States.

In developing the new climate zone designations, two climate zones were defined in the classification, but not thoroughly evaluated or actively applied because no sites in the United States or its territories required their use. The two climate zones are 1B [dry and > 5000 CDD10°C (9000 CDD50°F)], characterized as "very hot-dry," and 5°C [marine and 3000 < HDD18°C = 400 (5400, HDD65°F = 7200)], characterized as "cool marine." The marine (C) designation was not used for climate zones colder than Climate Zone 5 or hotter than Climate Zone 3, as marine climates are inherently neither very cold nor very hot. In addition, the humid (A) and dry (B) divisions were dropped for climate zones colder than Climate Zone 6 because they did not appear to be warranted based on differences in appropriate building design requirements. Reevaluation of these decisions might be warranted before applying the new climate classifications to locations outside of the United States.

C301.4 Tropical climate zone. The tropical climate zone shall be defined as:

1. Hawaii, Puerto Rico, Guam, American Samoa, U.S. Virgin Islands, Commonwealth of Northern Mariana Islands; and
2. Islands in the area between the Tropic of Cancer and the Tropic of Capricorn.

❖ The tropical climate zone was established as a subset of Climate Zone 1. The climate of tropical islands is uniquely constant, with moderate temperatures throughout the year. Similar to the development of the balance of the climate zones for the 2006 code, the tropical climate zone is also based on Koppen's classification of climates. Koppen divided the earth's climates into five major types, one of those being "tropical." According to Koppen, tropical climates are characterized by constant high temperature (at sea level and lower elevations) all 12 months of the year. Because of the constant nature of the climate, traditional construction methods and traditional HVAC installation found in buildings outside of the tropical environment may not be needed. For commercial buildings, the code does not provide different design or construction standards than provided for other Climate Zone 1 locations. Standards applicable to residential construction are provided in the residential provisions of the code.

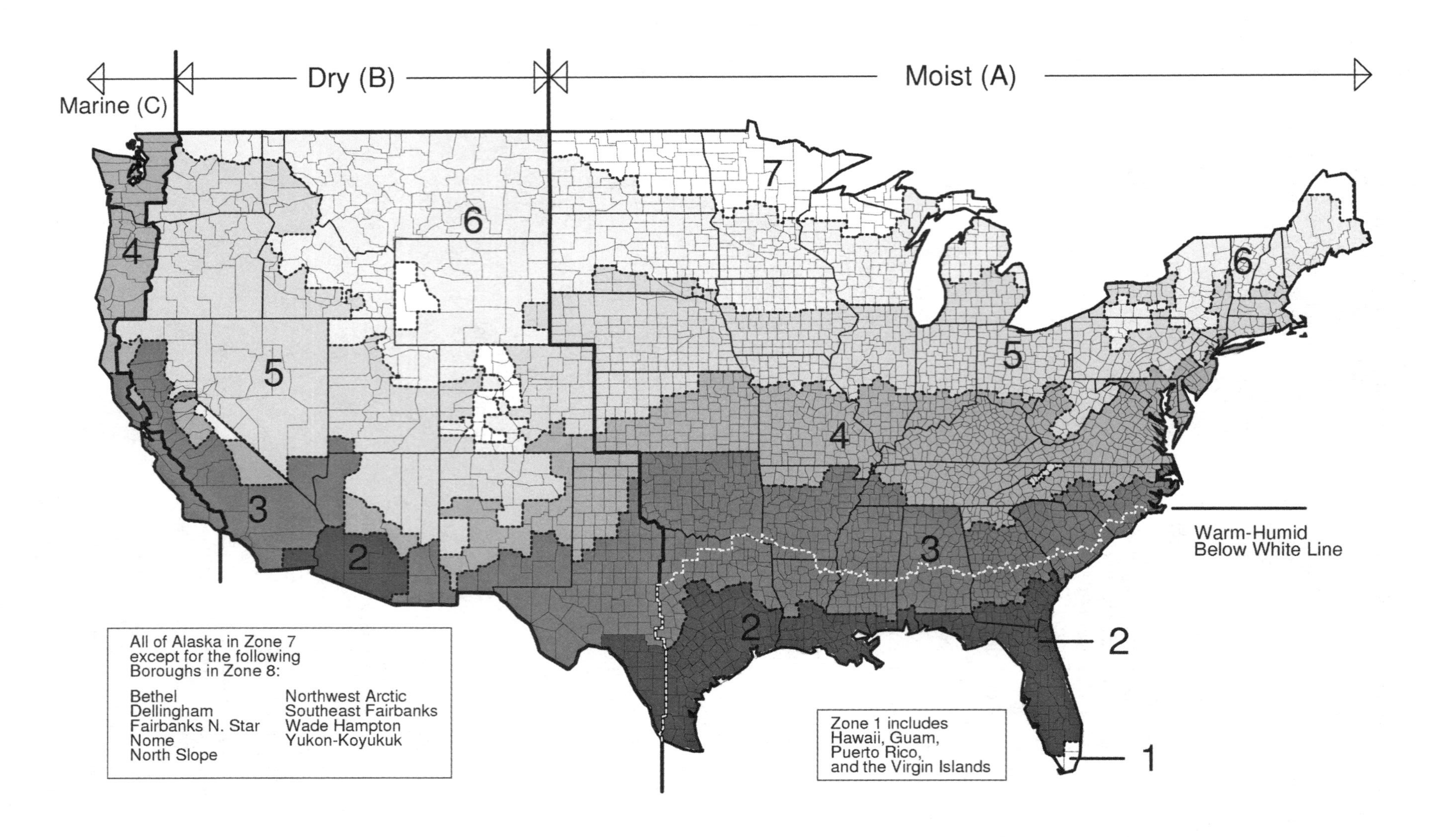

FIGURE C301.1
CLIMATE ZONES

ZONE NUMBER	CLIMATE ZONE NAME AND TYPE[2]	THERMAL CRITERIA [1, 3, 6]	REPRESENTATIVE U.S. CITY[4]
1A	Very Hot-Humid	9000 < CDD50°F	Miami, FL
1B[5]	Very Hot-Dry	9000 < CDD50°F	—
2A	Hot-Humid	6300 < CDD50°F ≤ 9000	Houston, TX
2B	Hot-Dry	6300 < CDD50°F ≤ 9000	Phoenix, AZ
3A	Warm-Humid	4500 < CDD50°F ≤ 6300	Memphis, TN
3B	Warm-Dry	4500 < CDD50°F ≤ 6300	El Paso, TX
3C	Warm-Marine	HDD65°F ≤ 3600	San Francisco, CA
4A	Mixed-Humid	CDD50°F ≤ 4500 AND HDD65°F ≤ 5400	Baltimore, MD
4B	Mixed-Dry	CDD50°F ≤ 4500 AND HDD65°F ≤ 5400	Albuquerque, NM
4C	Mixed-Marine	3600 < HDD65°F ≤ 5400	Salem, OR
5A	Cool-Humid	5400 < HDD65°F ≤ 7200	Chicago, IL
5B	Cool-Dry	5400 < HDD65°F ≤ 7200	Boise, ID
5C[5]	Cool-Marine	5400 < HDD65°F ≤ 7200	—
6A	Cool-Humid	7200 < HDD65°F ≤ 9000	Burlington, VT
6B	Cool-Dry	7200 < HDD65°F ≤ 9000	Helena, MT
7	Very Cold	9000 < HDD65°F ≤ 12600	Duluth, MN
8	Sub Arctic	12600 < HDD65°F	Fairbanks, AK

Notes:

1. Column 1 contains alphanumeric designations for each zone. These designations are intended for use when the climate zones are referenced in the code. The numeric part of the designation relates to the thermal properties of the zone. The letter part indicates the major climatic group to which the zone belongs; A indicates humid, B indicates dry, and C indicates marine. The climatic group designation was dropped for Zones 7 and 8 because the developers of the new climate zone classifications did not anticipate any building design criteria sensitive to the humid/dry/marine distinction in very cold climates. Zones 1B and 5C have been defined but are not used for the United States. Zone 6C (Marine and HDD18°C > 4000 (HDD65°F > 7200) might appear to be necessary for consistency. However, very few locations in the world are both as mild as is required by the marine zone definition and as cold as necessary to accumulate that many heating degree days. In addition, such sites do not appear climatically very different from sites in Zone 6A, which is where they are assigned in the absence of a Zone 6C.
2. Column 2 contains a descriptive name for each climate zone and the major climate type. The names can be used in place of the alphanumeric designations wherever a more descriptive designation is appropriate.
3. Column 3 contains definitions for the zone divisions based on degree day cooling and/or heating criteria. The humid/dry/marine divisions must be determined first before these criteria are applied. The definitions in Tables C301.3(1) and C301.3(2) contain logic capable of assigning a zone designation to any location with the necessary climate data anywhere in the world. However, the work to develop this classification focused on the 50 United States. Application of the classification to locations outside of the United States is untested.
4. Column 4 contains the name of a SAMSON station (National Climatic Data Center "Solar and Meteorological Surface Observation Network" station) found to best represent the climate zone as a whole. See the discussions at the beginning of this chapter regarding the development of the new climate zones for an explanation of how the representative cities were selected.
5. Zones 1B and 5C do not occur in the United States, and no representative cities were selected for these climate zones due to data limitations. Climates meeting the listed criteria do exist in such locations as Saudi Arabia; British Columbia, Canada; and Northern Europe.
6. SI to I-P Conversions:
   2500 CDD10°C = 4500 CDD50°F
   3000 HDD18°C = 5400 HDD65°F
   3500 CDD10°C = 6300 CDD50°F
   4000 HDD18°C = 7200 HDD65°F
   5000 CDD10°C = 9000 CDD50°F
   5000 HDD18°C = 9000 HDD65°F
   2000 HDD18°C = 3600 HDD65°F
   7000 HDD18°C = 12600 HDD65°F

Figure C301.3
CLIMATE ZONE DEFINITIONS

TABLE C301.3(1)
INTERNATIONAL CLIMATE ZONE DEFINITIONS

MAJOR CLIMATE TYPE DEFINITIONS
Marine (C) Definition—Locations meeting all four criteria: 1. Mean temperature of coldest month between -3°C (27°F) and 18°C (65°F). 2. Warmest month mean < 22°C (72°F). 3. At least four months with mean temperatures over 10°C (50°F). 4. Dry season in summer. The month with the heaviest precipitation in the cold season has at least three times as much precipitation as the month with the least precipitation in the rest of the year. The cold season is October through March in the Northern Hemisphere and April through September in the Southern Hemisphere.
Dry (B) Definition—Locations meeting the following criteria: Not marine and $P_{in} < 0.44 \times (TF - 19.5)$ [$P_{cm} < 2.0 \times (TC + 7)$ in SI units] where: P_{in} = Annual precipitation in inches (cm) T = Annual mean temperature in °F (°C)
Moist (A) Definition—Locations that are not marine and not dry.
Warm-humid Definition—Moist (A) locations where either of the following wet-bulb temperature conditions shall occur during the warmest six consecutive months of the year: 1. 67°F (19.4°C) or higher for 3,000 or more hours; or 2. 73°F (22.8°C) or higher for 1,500 or more hours.

For SI: °C = [(°F)-32]/1.8, 1 inch = 2.54 cm.

TABLE C301.3(2)
INTERNATIONAL CLIMATE ZONE DEFINITIONS

ZONE NUMBER	THERMAL CRITERIA	
	IP Units	SI Units
1	9000 < CDD50°F	5000 < CDD10°C
2	6300 < CDD50°F ≤ 9000	3500 < CDD10°C ≤ 5000
3A and 3B	4500 < CDD50°F ≤ 6300 AND HDD65°F ≤ 5400	2500 < CDD10°C ≤ 3500 AND HDD18°C ≤ 3000
4A and 4B	CDD50°F ≤ 4500 AND HDD65°F ≤ 5400	CDD10°C ≤ 2500 AND HDD18°C ≤ 3000
3C	HDD65°F ≤ 3600	HDD18°C ≤ 2000
4C	3600 < HDD65°F ≤ 5400	2000 < HDD18°C ≤ 3000
5	5400 < HDD65°F ≤ 7200	3000 < HDD18°C ≤ 4000
6	7200 < HDD65°F ≤ 9000	4000 < HDD18°C ≤ 5000
7	9000 < HDD65°F ≤ 12600	5000 < HDD18°C ≤ 7000
8	12600 < HDD65°F	7000 < HDD18°C

For SI: °C = [(°F)-32]/1.8.

TABLE C301.1
CLIMATE ZONES, MOISTURE REGIMES, AND WARM-HUMID DESIGNATIONS BY STATE, COUNTY AND TERRITORY

Key: A – Moist, B – Dry, C – Marine. Absence of moisture designation indicates moisture regime is irrelevant. Asterisk (*) indicates a warm-humid location.

US STATES

ALABAMA

3A Autauga*
2A Baldwin*
3A Barbour*
3A Bibb
3A Blount
3A Bullock*
3A Butler*
3A Calhoun
3A Chambers
3A Cherokee
3A Chilton
3A Choctaw*
3A Clarke*
3A Clay
3A Cleburne
3A Coffee*
3A Colbert
3A Conecuh*
3A Coosa
3A Covington*
3A Crenshaw*
3A Cullman
3A Dale*
3A Dallas*
3A DeKalb
3A Elmore*
3A Escambia*
3A Etowah
3A Fayette
3A Franklin
3A Geneva*
3A Greene
3A Hale
3A Henry*
3A Houston*
3A Jackson
3A Jefferson
3A Lamar
3A Lauderdale
3A Lawrence
3A Lee
3A Limestone
3A Lowndes*
3A Macon*
3A Madison
3A Marengo*
3A Marion
3A Marshall
2A Mobile*
3A Monroe*
3A Montgomery*
3A Morgan
3A Perry*
3A Pickens
3A Pike*
3A Randolph
3A Russell*
3A Shelby
3A St. Clair
3A Sumter
3A Talladega
3A Tallapoosa
3A Tuscaloosa
3A Walker
3A Washington*
3A Wilcox*
3A Winston

ALASKA

7 Aleutians East
7 Aleutians West
7 Anchorage
8 Bethel
7 Bristol Bay
7 Denali
8 Dillingham
8 Fairbanks North Star
7 Haines
7 Juneau
7 Kenai Peninsula
7 Ketchikan Gateway
7 Kodiak Island
7 Lake and Peninsula
7 Matanuska-Susitna
8 Nome
8 North Slope
8 Northwest Arctic
7 Prince of Wales Outer Ketchikan
7 Sitka
7 Skagway-Hoonah-Angoon
8 Southeast Fairbanks
7 Valdez-Cordova
8 Wade Hampton
7 Wrangell-Petersburg
7 Yakutat
8 Yukon-Koyukuk

ARIZONA

5B Apache
3B Cochise
5B Coconino
4B Gila
3B Graham
3B Greenlee
2B La Paz
2B Maricopa
3B Mohave
5B Navajo
2B Pima
2B Pinal
3B Santa Cruz
4B Yavapai
2B Yuma

ARKANSAS

3A Arkansas
3A Ashley
4A Baxter
4A Benton
4A Boone
3A Bradley
3A Calhoun
4A Carroll
3A Chicot
3A Clark
3A Clay
3A Cleburne
3A Cleveland
3A Columbia*
3A Conway
3A Craighead
3A Crawford
3A Crittenden
3A Cross
3A Dallas
3A Desha
3A Drew
3A Faulkner
3A Franklin
4A Fulton
3A Garland
3A Grant
3A Greene
3A Hempstead*
3A Hot Spring
3A Howard
3A Independence
4A Izard
3A Jackson
3A Jefferson
3A Johnson
3A Lafayette*
3A Lawrence
3A Lee
3A Lincoln
3A Little River*
3A Logan
3A Lonoke
4A Madison
4A Marion
3A Miller*
3A Mississippi
3A Monroe
3A Montgomery
3A Nevada
4A Newton
3A Ouachita
3A Perry
3A Phillips
3A Pike
3A Poinsett
3A Polk
3A Pope
3A Prairie
3A Pulaski
3A Randolph
3A Saline
3A Scott
4A Searcy
3A Sebastian
3A Sevier*
3A Sharp
3A St. Francis
4A Stone
3A Union*
3A Van Buren
4A Washington
3A White
3A Woodruff
3A Yell

CALIFORNIA

3C Alameda
6B Alpine
4B Amador
3B Butte
4B Calaveras
3B Colusa
3B Contra Costa
4C Del Norte
4B El Dorado
3B Fresno
3B Glenn
4C Humboldt
2B Imperial

(continued)

TABLE C301.1—continued
CLIMATE ZONES, MOISTURE REGIMES, AND WARM-HUMID DESIGNATIONS BY STATE, COUNTY AND TERRITORY

4B Inyo
3B Kern
3B Kings
4B Lake
5B Lassen
3B Los Angeles
3B Madera
3C Marin
4B Mariposa
3C Mendocino
3B Merced
5B Modoc
6B Mono
3C Monterey
3C Napa
5B Nevada
3B Orange
3B Placer
5B Plumas
3B Riverside
3B Sacramento
3C San Benito
3B San Bernardino
3B San Diego
3C San Francisco
3B San Joaquin
3C San Luis Obispo
3C San Mateo
3C Santa Barbara
3C Santa Clara
3C Santa Cruz
3B Shasta
5B Sierra
5B Siskiyou
3B Solano
3C Sonoma
3B Stanislaus
3B Sutter
3B Tehama
4B Trinity
3B Tulare
4B Tuolumne
3C Ventura
3B Yolo
3B Yuba

COLORADO

5B Adams
6B Alamosa
5B Arapahoe
6B Archuleta
4B Baca
5B Bent
5B Boulder
6B Chaffee
5B Cheyenne
7 Clear Creek
6B Conejos
6B Costilla
5B Crowley
6B Custer
5B Delta
5B Denver
6B Dolores
5B Douglas
6B Eagle
5B Elbert
5B El Paso
5B Fremont
5B Garfield
5B Gilpin
7 Grand
7 Gunnison
7 Hinsdale
5B Huerfano
7 Jackson
5B Jefferson
5B Kiowa
5B Kit Carson
7 Lake
5B La Plata
5B Larimer
4B Las Animas
5B Lincoln
5B Logan
5B Mesa
7 Mineral
6B Moffat
5B Montezuma
5B Montrose
5B Morgan
4B Otero
6B Ouray
7 Park
5B Phillips
7 Pitkin
5B Prowers
5B Pueblo
6B Rio Blanco
7 Rio Grande
7 Routt
6B Saguache
7 San Juan
6B San Miguel
5B Sedgwick
7 Summit
5B Teller
5B Washington
5B Weld
5B Yuma

CONNECTICUT

5A (all)

DELAWARE

4A (all)

DISTRICT OF COLUMBIA

4A (all)

FLORIDA

2A Alachua*
2A Baker*
2A Bay*
2A Bradford*
2A Brevard*
1A Broward*
2A Calhoun*
2A Charlotte*
2A Citrus*
2A Clay*
2A Collier*
2A Columbia*
2A DeSoto*
2A Dixie*
2A Duval*
2A Escambia*
2A Flagler*
2A Franklin*
2A Gadsden*
2A Gilchrist*
2A Glades*
2A Gulf*
2A Hamilton*
2A Hardee*
2A Hendry*
2A Hernando*
2A Highlands*
2A Hillsborough*
2A Holmes*
2A Indian River*
2A Jackson*
2A Jefferson*
2A Lafayette*
2A Lake*
2A Lee*
2A Leon*
2A Levy*
2A Liberty*
2A Madison*
2A Manatee*
2A Marion*
2A Martin*
1A Miami-Dade*
1A Monroe*
2A Nassau*
2A Okaloosa*
2A Okeechobee*
2A Orange*
2A Osceola*
2A Palm Beach*
2A Pasco*
2A Pinellas*
2A Polk*
2A Putnam*
2A Santa Rosa*
2A Sarasota*
2A Seminole*
2A St. Johns*
2A St. Lucie*
2A Sumter*
2A Suwannee*
2A Taylor*
2A Union*
2A Volusia*
2A Wakulla*
2A Walton*
2A Washington*

GEORGIA

2A Appling*
2A Atkinson*
2A Bacon*
2A Baker*
3A Baldwin
4A Banks
3A Barrow
3A Bartow
3A Ben Hill*
2A Berrien*
3A Bibb
3A Bleckley*
2A Brantley*
2A Brooks*
2A Bryan*
3A Bulloch*
3A Burke
3A Butts
3A Calhoun*
2A Camden*
3A Candler*
3A Carroll
4A Catoosa
2A Charlton*
2A Chatham*
3A Chattahoochee*
4A Chattooga
3A Cherokee
3A Clarke
3A Clay*
3A Clayton
2A Clinch*
3A Cobb
3A Coffee*
2A Colquitt*
3A Columbia
2A Cook*
3A Coweta

(continued)

TABLE C301.1—continued
CLIMATE ZONES, MOISTURE REGIMES, AND WARM-HUMID DESIGNATIONS BY STATE, COUNTY AND TERRITORY

3A Crawford
3A Crisp*
4A Dade
4A Dawson
2A Decatur*
3A DeKalb
3A Dodge*
3A Dooly*
3A Dougherty*
3A Douglas
3A Early*
2A Echols*
2A Effingham*
3A Elbert
3A Emanuel*
2A Evans*
4A Fannin
3A Fayette
4A Floyd
3A Forsyth
4A Franklin
3A Fulton
4A Gilmer
3A Glascock
2A Glynn*
4A Gordon
2A Grady*
3A Greene
3A Gwinnett
4A Habersham
4A Hall
3A Hancock
3A Haralson
3A Harris
3A Hart
3A Heard
3A Henry
3A Houston*
3A Irwin*
3A Jackson
3A Jasper
2A Jeff Davis*
3A Jefferson
3A Jenkins*
3A Johnson*
3A Jones
3A Lamar
2A Lanier*
3A Laurens*
3A Lee*
2A Liberty*
3A Lincoln
2A Long*
2A Lowndes*
4A Lumpkin
3A Macon*
3A Madison
3A Marion*
3A McDuffie
2A McIntosh*
3A Meriwether
2A Miller*
2A Mitchell*
3A Monroe
3A Montgomery*
3A Morgan
4A Murray
3A Muscogee
3A Newton
3A Oconee
3A Oglethorpe
3A Paulding
3A Peach*
4A Pickens
2A Pierce*
3A Pike
3A Polk
3A Pulaski*
3A Putnam
3A Quitman*
4A Rabun
3A Randolph*
3A Richmond
3A Rockdale
3A Schley*
3A Screven*
2A Seminole*
3A Spalding
4A Stephens
3A Stewart*
3A Sumter*
3A Talbot
3A Taliaferro
2A Tattnall*
3A Taylor*
3A Telfair*
3A Terrell*
2A Thomas*
3A Tift*
2A Toombs*
4A Towns
3A Treutlen*
3A Troup
3A Turner*
3A Twiggs*
4A Union
3A Upson
4A Walker
3A Walton
2A Ware*
3A Warren
3A Washington
2A Wayne*
3A Webster*
3A Wheeler*
4A White
4A Whitfield
3A Wilcox*
3A Wilkes
3A Wilkinson
3A Worth*

HAWAII

1A (all)*

IDAHO

5B Ada
6B Adams
6B Bannock
6B Bear Lake
5B Benewah
6B Bingham
6B Blaine
6B Boise
6B Bonner
6B Bonneville
6B Boundary
6B Butte
6B Camas
5B Canyon
6B Caribou
5B Cassia
6B Clark
5B Clearwater
6B Custer
5B Elmore
6B Franklin
6B Fremont
5B Gem
5B Gooding
5B Idaho
6B Jefferson
5B Jerome
5B Kootenai
5B Latah
6B Lemhi
5B Lewis
5B Lincoln
6B Madison
5B Minidoka
5B Nez Perce
6B Oneida
5B Owyhee
5B Payette
5B Power
5B Shoshone
6B Teton
5B Twin Falls
6B Valley
5B Washington

ILLINOIS

5A Adams
4A Alexander
4A Bond
5A Boone
5A Brown
5A Bureau
5A Calhoun
5A Carroll
5A Cass
5A Champaign
4A Christian
5A Clark
4A Clay
4A Clinton
5A Coles
5A Cook
4A Crawford
5A Cumberland
5A DeKalb
5A De Witt
5A Douglas
5A DuPage
5A Edgar
4A Edwards
4A Effingham
4A Fayette
5A Ford
4A Franklin
5A Fulton
4A Gallatin
5A Greene
5A Grundy
4A Hamilton
5A Hancock
4A Hardin
5A Henderson
5A Henry
5A Iroquois
4A Jackson
4A Jasper
4A Jefferson
5A Jersey
5A Jo Daviess
4A Johnson
5A Kane
5A Kankakee
5A Kendall
5A Knox
5A Lake
5A La Salle
4A Lawrence
5A Lee
5A Livingston
5A Logan
5A Macon
4A Macoupin
4A Madison
4A Marion
5A Marshall
5A Mason
4A Massac
5A McDonough
5A McHenry

(continued)

TABLE C301.1—continued
CLIMATE ZONES, MOISTURE REGIMES, AND WARM-HUMID DESIGNATIONS BY STATE, COUNTY AND TERRITORY

5A McLean
5A Menard
5A Mercer
4A Monroe
4A Montgomery
5A Morgan
5A Moultrie
5A Ogle
5A Peoria
4A Perry
5A Piatt
5A Pike
4A Pope
4A Pulaski
5A Putnam
4A Randolph
4A Richland
5A Rock Island
4A Saline
5A Sangamon
5A Schuyler
5A Scott
4A Shelby
5A Stark
4A St. Clair
5A Stephenson
5A Tazewell
4A Union
5A Vermilion
4A Wabash
5A Warren
4A Washington
4A Wayne
4A White
5A Whiteside
5A Will
4A Williamson
5A Winnebago
5A Woodford

INDIANA

5A Adams
5A Allen
5A Bartholomew
5A Benton
5A Blackford
5A Boone
4A Brown
5A Carroll
5A Cass
4A Clark
5A Clay
5A Clinton
4A Crawford
4A Daviess
4A Dearborn
5A Decatur
5A De Kalb
5A Delaware
4A Dubois
5A Elkhart
5A Fayette
4A Floyd
5A Fountain
5A Franklin
5A Fulton
4A Gibson
5A Grant
4A Greene
5A Hamilton
5A Hancock
4A Harrison
5A Hendricks
5A Henry
5A Howard
5A Huntington
4A Jackson
5A Jasper
5A Jay
4A Jefferson
4A Jennings
5A Johnson
4A Knox
5A Kosciusko
5A Lagrange
5A Lake
5A La Porte
4A Lawrence
5A Madison
5A Marion
5A Marshall
4A Martin
5A Miami
4A Monroe
5A Montgomery
5A Morgan
5A Newton
5A Noble
4A Ohio
4A Orange
5A Owen
5A Parke
4A Perry
4A Pike
5A Porter
4A Posey
5A Pulaski
5A Putnam
5A Randolph
4A Ripley
5A Rush
4A Scott
5A Shelby
4A Spencer
5A Starke
5A Steuben
5A St. Joseph
4A Sullivan
4A Switzerland
5A Tippecanoe
5A Tipton
5A Union
4A Vanderburgh
5A Vermillion
5A Vigo
5A Wabash
5A Warren
4A Warrick
4A Washington
5A Wayne
5A Wells
5A White
5A Whitley

IOWA

5A Adair
5A Adams
6A Allamakee
5A Appanoose
5A Audubon
5A Benton
6A Black Hawk
5A Boone
6A Bremer
6A Buchanan
6A Buena Vista
6A Butler
6A Calhoun
5A Carroll
5A Cass
5A Cedar
6A Cerro Gordo
6A Cherokee
6A Chickasaw
5A Clarke
6A Clay
6A Clayton
5A Clinton
5A Crawford
5A Dallas
5A Davis
5A Decatur
6A Delaware
5A Des Moines
6A Dickinson
5A Dubuque
6A Emmet
6A Fayette
6A Floyd
6A Franklin
5A Fremont
5A Greene
6A Grundy
5A Guthrie
6A Hamilton
6A Hancock
6A Hardin
5A Harrison
5A Henry
6A Howard
6A Humboldt
6A Ida
5A Iowa
5A Jackson
5A Jasper
5A Jefferson
5A Johnson
5A Jones
5A Keokuk
6A Kossuth
5A Lee
5A Linn
5A Louisa
5A Lucas
6A Lyon
5A Madison
5A Mahaska
5A Marion
5A Marshall
5A Mills
6A Mitchell
5A Monona
5A Monroe
5A Montgomery
5A Muscatine
6A O'Brien
6A Osceola
5A Page
6A Palo Alto
6A Plymouth
6A Pocahontas
5A Polk
5A Pottawattamie
5A Poweshiek
5A Ringgold
6A Sac
5A Scott
5A Shelby
6A Sioux
5A Story
5A Tama
5A Taylor
5A Union
5A Van Buren
5A Wapello
5A Warren
5A Washington
5A Wayne
6A Webster
6A Winnebago

(continued)

TABLE C301.1—continued
CLIMATE ZONES, MOISTURE REGIMES, AND WARM-HUMID DESIGNATIONS BY STATE, COUNTY AND TERRITORY

6A Winneshiek
5A Woodbury
6A Worth
6A Wright

KANSAS

4A Allen
4A Anderson
4A Atchison
4A Barber
4A Barton
4A Bourbon
4A Brown
4A Butler
4A Chase
4A Chautauqua
4A Cherokee
5A Cheyenne
4A Clark
4A Clay
5A Cloud
4A Coffey
4A Comanche
4A Cowley
4A Crawford
5A Decatur
4A Dickinson
4A Doniphan
4A Douglas
4A Edwards
4A Elk
5A Ellis
4A Ellsworth
4A Finney
4A Ford
4A Franklin
4A Geary
5A Gove
5A Graham
4A Grant
4A Gray
5A Greeley
4A Greenwood
5A Hamilton
4A Harper
4A Harvey
4A Haskell
4A Hodgeman
4A Jackson
4A Jefferson
5A Jewell
4A Johnson
4A Kearny
4A Kingman
4A Kiowa
4A Labette
5A Lane
4A Leavenworth
4A Lincoln
4A Linn
5A Logan
4A Lyon
4A Marion
4A Marshall
4A McPherson
4A Meade
4A Miami
5A Mitchell
4A Montgomery
4A Morris
4A Morton
4A Nemaha
4A Neosho
5A Ness
5A Norton
4A Osage
5A Osborne
4A Ottawa
4A Pawnee
5A Phillips
4A Pottawatomie
4A Pratt
5A Rawlins
4A Reno
5A Republic
4A Rice
4A Riley
5A Rooks
4A Rush
4A Russell
4A Saline
5A Scott
4A Sedgwick
4A Seward
4A Shawnee
5A Sheridan
5A Sherman
5A Smith
4A Stafford
4A Stanton
4A Stevens
4A Sumner
5A Thomas
5A Trego
4A Wabaunsee
5A Wallace
4A Washington
5A Wichita
4A Wilson
4A Woodson
4A Wyandotte

KENTUCKY

4A (all)

LOUISIANA

2A Acadia*
2A Allen*
2A Ascension*
2A Assumption*
2A Avoyelles*
2A Beauregard*
3A Bienville*
3A Bossier*
3A Caddo*
2A Calcasieu*
3A Caldwell*
2A Cameron*
3A Catahoula*
3A Claiborne*
3A Concordia*
3A De Soto*
2A East Baton Rouge*
3A East Carroll
2A East Feliciana*
2A Evangeline*
3A Franklin*
3A Grant*
2A Iberia*
2A Iberville*
3A Jackson*
2A Jefferson*
2A Jefferson Davis*
2A Lafayette*
2A Lafourche*
3A La Salle*
3A Lincoln*
2A Livingston*
3A Madison*
3A Morehouse
3A Natchitoches*
2A Orleans*
3A Ouachita*
2A Plaquemines*
2A Pointe Coupee*
2A Rapides*
3A Red River*
3A Richland*
3A Sabine*
2A St. Bernard*
2A St. Charles*
2A St. Helena*
2A St. James*
2A St. John the Baptist*
2A St. Landry*
2A St. Martin*
2A St. Mary*
2A St. Tammany*
2A Tangipahoa*
3A Tensas*
2A Terrebonne*
3A Union*
2A Vermilion*
3A Vernon*
2A Washington*
3A Webster*
2A West Baton Rouge*
3A West Carroll
2A West Feliciana*
3A Winn*

MAINE

6A Androscoggin
7 Aroostook
6A Cumberland
6A Franklin
6A Hancock
6A Kennebec
6A Knox
6A Lincoln
6A Oxford
6A Penobscot
6A Piscataquis
6A Sagadahoc
6A Somerset
6A Waldo
6A Washington
6A York

MARYLAND

4A Allegany
4A Anne Arundel
4A Baltimore
4A Baltimore (city)
4A Calvert
4A Caroline
4A Carroll
4A Cecil
4A Charles
4A Dorchester
4A Frederick
5A Garrett
4A Harford
4A Howard
4A Kent
4A Montgomery
4A Prince George's
4A Queen Anne's
4A Somerset
4A St. Mary's
4A Talbot
4A Washington
4A Wicomico
4A Worcester

MASSACHSETTS

5A (all)

MICHIGAN

6A Alcona
6A Alger

(continued)

TABLE C301.1—continued
CLIMATE ZONES, MOISTURE REGIMES, AND WARM-HUMID DESIGNATIONS BY STATE, COUNTY AND TERRITORY

5A Allegan
6A Alpena
6A Antrim
6A Arenac
7 Baraga
5A Barry
5A Bay
6A Benzie
5A Berrien
5A Branch
5A Calhoun
5A Cass
6A Charlevoix
6A Cheboygan
7 Chippewa
6A Clare
5A Clinton
6A Crawford
6A Delta
6A Dickinson
5A Eaton
6A Emmet
5A Genesee
6A Gladwin
7 Gogebic
6A Grand Traverse
5A Gratiot
5A Hillsdale
7 Houghton
6A Huron
5A Ingham
5A Ionia
6A Iosco
7 Iron
6A Isabella
5A Jackson
5A Kalamazoo
6A Kalkaska
5A Kent
7 Keweenaw
6A Lake
5A Lapeer
6A Leelanau
5A Lenawee
5A Livingston
7 Luce
7 Mackinac
5A Macomb
6A Manistee
6A Marquette
6A Mason
6A Mecosta
6A Menominee
5A Midland
6A Missaukee
5A Monroe
5A Montcalm
6A Montmorency
5A Muskegon
6A Newaygo
5A Oakland
6A Oceana
6A Ogemaw
7 Ontonagon
6A Osceola
6A Oscoda
6A Otsego
5A Ottawa
6A Presque Isle
6A Roscommon
5A Saginaw
6A Sanilac
7 Schoolcraft
5A Shiawassee
5A St. Clair
5A St. Joseph
5A Tuscola
5A Van Buren
5A Washtenaw
5A Wayne
6A Wexford

MINNESOTA

7 Aitkin
6A Anoka
7 Becker
7 Beltrami
6A Benton
6A Big Stone
6A Blue Earth
6A Brown
7 Carlton
6A Carver
7 Cass
6A Chippewa
6A Chisago
7 Clay
7 Clearwater
7 Cook
6A Cottonwood
7 Crow Wing
6A Dakota
6A Dodge
6A Douglas
6A Faribault
6A Fillmore
6A Freeborn
6A Goodhue
7 Grant
6A Hennepin
6A Houston
7 Hubbard
6A Isanti
7 Itasca
6A Jackson
7 Kanabec
6A Kandiyohi
7 Kittson
7 Koochiching
6A Lac qui Parle
7 Lake
7 Lake of the Woods
6A Le Sueur
6A Lincoln
6A Lyon
7 Mahnomen
7 Marshall
6A Martin
6A McLeod
6A Meeker
7 Mille Lacs
6A Morrison
6A Mower
6A Murray
6A Nicollet
6A Nobles
7 Norman
6A Olmsted
7 Otter Tail
7 Pennington
7 Pine
6A Pipestone
7 Polk
6A Pope
6A Ramsey
7 Red Lake
6A Redwood
6A Renville
6A Rice
6A Rock
7 Roseau
6A Scott
6A Sherburne
6A Sibley
6A Stearns
6A Steele
6A Stevens
7 St. Louis
6A Swift
6A Todd
6A Traverse
6A Wabasha
7 Wadena
6A Waseca
6A Washington
6A Watonwan
7 Wilkin
6A Winona
6A Wright
6A Yellow Medicine

MISSISSIPPI

3A Adams*
3A Alcorn
3A Amite*
3A Attala
3A Benton
3A Bolivar
3A Calhoun
3A Carroll
3A Chickasaw
3A Choctaw
3A Claiborne*
3A Clarke
3A Clay
3A Coahoma
3A Copiah*
3A Covington*
3A DeSoto
3A Forrest*
3A Franklin*
3A George*
3A Greene*
3A Grenada
2A Hancock*
2A Harrison*
3A Hinds*
3A Holmes
3A Humphreys
3A Issaquena
3A Itawamba
2A Jackson*
3A Jasper
3A Jefferson*
3A Jefferson Davis*
3A Jones*
3A Kemper
3A Lafayette
3A Lamar*
3A Lauderdale
3A Lawrence*
3A Leake
3A Lee
3A Leflore
3A Lincoln*
3A Lowndes
3A Madison
3A Marion*
3A Marshall
3A Monroe
3A Montgomery
3A Neshoba
3A Newton
3A Noxubee
3A Oktibbeha
3A Panola
2A Pearl River*
3A Perry*
3A Pike*
3A Pontotoc

(continued)

TABLE C301.1—continued
CLIMATE ZONES, MOISTURE REGIMES, AND WARM-HUMID DESIGNATIONS BY STATE, COUNTY AND TERRITORY

3A Prentiss
3A Quitman
3A Rankin*
3A Scott
3A Sharkey
3A Simpson*
3A Smith*
2A Stone*
3A Sunflower
3A Tallahatchie
3A Tate
3A Tippah
3A Tishomingo
3A Tunica
3A Union
3A Walthall*
3A Warren*
3A Washington
3A Wayne*
3A Webster
3A Wilkinson*
3A Winston
3A Yalobusha
3A Yazoo

MISSOURI

5A Adair
5A Andrew
5A Atchison
4A Audrain
4A Barry
4A Barton
4A Bates
4A Benton
4A Bollinger
4A Boone
5A Buchanan
4A Butler
5A Caldwell
4A Callaway
4A Camden
4A Cape Girardeau
4A Carroll
4A Carter
4A Cass
4A Cedar
5A Chariton
4A Christian
5A Clark
4A Clay
5A Clinton
4A Cole
4A Cooper
4A Crawford
4A Dade
4A Dallas
5A Daviess
5A DeKalb
4A Dent
4A Douglas
4A Dunklin
4A Franklin
4A Gasconade
5A Gentry
4A Greene
5A Grundy
5A Harrison
4A Henry
4A Hickory
5A Holt
4A Howard
4A Howell
4A Iron
4A Jackson
4A Jasper
4A Jefferson
4A Johnson
5A Knox
4A Laclede
4A Lafayette
4A Lawrence
5A Lewis
4A Lincoln
5A Linn
5A Livingston
5A Macon
4A Madison
4A Maries
5A Marion
4A McDonald
5A Mercer
4A Miller
4A Mississippi
4A Moniteau
4A Monroe
4A Montgomery
4A Morgan
4A New Madrid
4A Newton
5A Nodaway
4A Oregon
4A Osage
4A Ozark
4A Pemiscot
4A Perry
4A Pettis
4A Phelps
5A Pike
4A Platte
4A Polk
4A Pulaski
5A Putnam
5A Ralls
4A Randolph
4A Ray
4A Reynolds
4A Ripley
4A Saline
5A Schuyler
5A Scotland
4A Scott
4A Shannon
5A Shelby
4A St. Charles
4A St. Clair
4A Ste. Genevieve
4A St. Francois
4A St. Louis
4A St. Louis (city)
4A Stoddard
4A Stone
5A Sullivan
4A Taney
4A Texas
4A Vernon
4A Warren
4A Washington
4A Wayne
4A Webster
5A Worth
4A Wright

MONTANA

6B (all)

NEBRASKA

5A (all)

NEVADA

5B Carson City (city)
5B Churchill
3B Clark
5B Douglas
5B Elko
5B Esmeralda
5B Eureka
5B Humboldt
5B Lander
5B Lincoln
5B Lyon
5B Mineral
5B Nye
5B Pershing
5B Storey
5B Washoe
5B White Pine

NEW HAMPSHIRE

6A Belknap
6A Carroll
5A Cheshire
6A Coos
6A Grafton
5A Hillsborough
6A Merrimack
5A Rockingham
5A Strafford
6A Sullivan

NEW JERSEY

4A Atlantic
5A Bergen
4A Burlington
4A Camden
4A Cape May
4A Cumberland
4A Essex
4A Gloucester
4A Hudson
5A Hunterdon
5A Mercer
4A Middlesex
4A Monmouth
5A Morris
4A Ocean
5A Passaic
4A Salem
5A Somerset
5A Sussex
4A Union
5A Warren

NEW MEXICO

4B Bernalillo
5B Catron
3B Chaves
4B Cibola
5B Colfax
4B Curry
4B DeBaca
3B Dona Ana
3B Eddy
4B Grant
4B Guadalupe
5B Harding
3B Hidalgo
3B Lea
4B Lincoln
5B Los Alamos
3B Luna
5B McKinley
5B Mora
3B Otero
4B Quay
5B Rio Arriba
4B Roosevelt
5B Sandoval
5B San Juan
5B San Miguel
5B Santa Fe
4B Sierra
4B Socorro
5B Taos

(continued)

TABLE C301.1—continued
CLIMATE ZONES, MOISTURE REGIMES, AND WARM-HUMID DESIGNATIONS BY STATE, COUNTY AND TERRITORY

5B Torrance
4B Union
4B Valencia

NEW YORK

5A Albany
6A Allegany
4A Bronx
6A Broome
6A Cattaraugus
5A Cayuga
5A Chautauqua
5A Chemung
6A Chenango
6A Clinton
5A Columbia
5A Cortland
6A Delaware
5A Dutchess
5A Erie
6A Essex
6A Franklin
6A Fulton
5A Genesee
5A Greene
6A Hamilton
6A Herkimer
6A Jefferson
4A Kings
6A Lewis
5A Livingston
6A Madison
5A Monroe
6A Montgomery
4A Nassau
4A New York
5A Niagara
6A Oneida
5A Onondaga
5A Ontario
5A Orange
5A Orleans
5A Oswego
6A Otsego
5A Putnam
4A Queens
5A Rensselaer
4A Richmond
5A Rockland
5A Saratoga
5A Schenectady
6A Schoharie
6A Schuyler
5A Seneca
6A Steuben
6A St. Lawrence
4A Suffolk
6A Sullivan
5A Tioga
6A Tompkins
6A Ulster
6A Warren
5A Washington
5A Wayne
4A Westchester
6A Wyoming
5A Yates

NORTH CAROLINA

4A Alamance
4A Alexander
5A Alleghany
3A Anson
5A Ashe
5A Avery
3A Beaufort
4A Bertie
3A Bladen
3A Brunswick*
4A Buncombe
4A Burke
3A Cabarrus
4A Caldwell
3A Camden
3A Carteret*
4A Caswell
4A Catawba
4A Chatham
4A Cherokee
3A Chowan
4A Clay
4A Cleveland
3A Columbus*
3A Craven
3A Cumberland
3A Currituck
3A Dare
3A Davidson
4A Davie
3A Duplin
4A Durham
3A Edgecombe
4A Forsyth
4A Franklin
3A Gaston
4A Gates
4A Graham
4A Granville
3A Greene
4A Guilford
4A Halifax
4A Harnett
4A Haywood
4A Henderson
4A Hertford
3A Hok
3A Hyde
4A Iredell
4A Jackson
3A Johnston
3A Jones
4A Lee
3A Lenoir
4A Lincoln
4A Macon
4A Madison
3A Martin
4A McDowell
3A Mecklenburg
5A Mitchell
3A Montgomery
3A Moore
4A Nash
3A New Hanover*
4A Northampton
3A Onslow*
4A Orange
3A Pamlico
3A Pasquotank
3A Pender*
3A Perquimans
4A Person
3A Pitt
4A Polk
3A Randolph
3A Richmond
3A Robeson
4A Rockingham
3A Rowan
4A Rutherford
3A Sampson
3A Scotland
3A Stanly
4A Stokes
4A Surry
4A Swain
4A Transylvania
3A Tyrrell
3A Union
4A Vance
4A Wake
4A Warren
3A Washington
5A Watauga
3A Wayne
4A Wilkes
3A Wilson
4A Yadkin
5A Yancey

NORTH DAKOTA

6A Adams
7 Barnes
7 Benson
6A Billings
7 Bottineau
6A Bowman
7 Burke
6A Burleigh
7 Cass
7 Cavalier
6A Dickey
7 Divide
6A Dunn
7 Eddy
6A Emmons
7 Foster
6A Golden Valley
7 Grand Forks
6A Grant
7 Griggs
6A Hettinger
7 Kidder
6A LaMoure
6A Logan
7 McHenry
6A McIntosh
6A McKenzie
7 McLean
6A Mercer
6A Morton
7 Mountrail
7 Nelson
6A Oliver
7 Pembina
7 Pierce
7 Ramsey
6A Ransom
7 Renville
6A Richland
7 Rolette
6A Sargent
7 Sheridan
6A Sioux
6A Slope
6A Stark
7 Steele
7 Stutsman
7 Towner
7 Traill
7 Walsh
7 Ward
7 Wells
7 Williams

OHIO

4A Adams
5A Allen
5A Ashland

(continued)

TABLE C301.1—continued
CLIMATE ZONES, MOISTURE REGIMES, AND WARM-HUMID DESIGNATIONS BY STATE, COUNTY AND TERRITORY

5A Ashtabula
5A Athens
5A Auglaize
5A Belmont
4A Brown
5A Butler
5A Carroll
5A Champaign
5A Clark
4A Clermont
5A Clinton
5A Columbiana
5A Coshocton
5A Crawford
5A Cuyahoga
5A Darke
5A Defiance
5A Delaware
5A Erie
5A Fairfield
5A Fayette
5A Franklin
5A Fulton
4A Gallia
5A Geauga
5A Greene
5A Guernsey
4A Hamilton
5A Hancock
5A Hardin
5A Harrison
5A Henry
5A Highland
5A Hocking
5A Holmes
5A Huron
5A Jackson
5A Jefferson
5A Knox
5A Lake
4A Lawrence
5A Licking
5A Logan
5A Lorain
5A Lucas
5A Madison
5A Mahoning
5A Marion
5A Medina
5A Meigs
5A Mercer
5A Miami
5A Monroe
5A Montgomery
5A Morgan
5A Morrow
5A Muskingum
5A Noble
5A Ottawa
5A Paulding
5A Perry
5A Pickaway
4A Pike
5A Portage
5A Preble
5A Putnam
5A Richland
5A Ross
5A Sandusky
4A Scioto
5A Seneca
5A Shelby
5A Stark
5A Summit
5A Trumbull
5A Tuscarawas
5A Union
5A Van Wert
5A Vinton
5A Warren
4A Washington
5A Wayne
5A Williams
5A Wood
5A Wyandot

OKLAHOMA

3A Adair
3A Alfalfa
3A Atoka
4B Beaver
3A Beckham
3A Blaine
3A Bryan
3A Caddo
3A Canadian
3A Carter
3A Cherokee
3A Choctaw
4B Cimarron
3A Cleveland
3A Coal
3A Comanche
3A Cotton
3A Craig
3A Creek
3A Custer
3A Delaware
3A Dewey
3A Ellis
3A Garfield
3A Garvin
3A Grady
3A Grant
3A Greer
3A Harmon
3A Harper
3A Haskell
3A Hughes
3A Jackson
3A Jefferson
3A Johnston
3A Kay
3A Kingfisher
3A Kiowa
3A Latimer
3A Le Flore
3A Lincoln
3A Logan
3A Love
3A Major
3A Marshall
3A Mayes
3A McClain
3A McCurtain
3A McIntosh
3A Murray
3A Muskogee
3A Noble
3A Nowata
3A Okfuske
3A Oklahoma
3A Okmulgee
3A Osage
3A Ottawa
3A Pawnee
3A Payne
3A Pittsburg
3A Pontotoc
3A Pottawatomie
3A Pushmataha
3A Roger Mills
3A Rogers
3A Seminole
3A Sequoyah
3A Stephens
4B Texas
3A Tillman
3A Tulsa
3A Wagoner
3A Washington
3A Washita
3A Woods
3A Woodward

OREGON

5B Baker
4C Benton
4C Clackamas
4C Clatsop
4C Columbia
4C Coos
5B Crook
4C Curry
5B Deschutes
4C Douglas
5B Gilliam
5B Grant
5B Harney
5B Hood River
4C Jackson
5B Jefferson
4C Josephine
5B Klamath
5B Lake
4C Lane
4C Lincoln
4C Linn
5B Malheur
4C Marion
5B Morrow
4C Multnomah
4C Polk
5B Sherman
4C Tillamook
5B Umatilla
5B Union
5B Wallowa
5B Wasco
4C Washington
5B Wheeler
4C Yamhill

PENNSYLVANIA

5A Adams
5A Allegheny
5A Armstrong
5A Beaver
5A Bedford
5A Berks
5A Blair
5A Bradford
4A Bucks
5A Butler
5A Cambria
6A Cameron
5A Carbon
5A Centre
4A Chester
5A Clarion
6A Clearfield
5A Clinton
5A Columbia
5A Crawford
5A Cumberland
5A Dauphin
4A Delaware
6A Elk
5A Erie
5A Fayette
5A Forest
5A Franklin
5A Fulton
5A Greene
5A Huntingdon

(continued)

**TABLE C301.1—continued
CLIMATE ZONES, MOISTURE REGIMES, AND WARM-HUMID DESIGNATIONS BY STATE, COUNTY AND TERRITORY**

5A Indiana
5A Jefferson
5A Juniata
5A Lackawanna
5A Lancaster
5A Lawrence
5A Lebanon
5A Lehigh
5A Luzerne
5A Lycoming
6A McKean
5A Mercer
5A Mifflin
5A Monroe
4A Montgomery
5A Montour
5A Northampton
5A Northumberland
5A Perry
4A Philadelphia
5A Pike
6A Potter
5A Schuylkill
5A Snyder
5A Somerset
5A Sullivan
6A Susquehanna
6A Tioga
5A Union
5A Venango
5A Warren
5A Washington
6A Wayne
5A Westmoreland
5A Wyoming
4A York

RHODE ISLAND

5A (all)

SOUTH CAROLINA

3A Abbeville
3A Aiken
3A Allendale*
3A Anderson
3A Bamberg*
3A Barnwell*
3A Beaufort*
3A Berkeley*
3A Calhoun
3A Charleston*
3A Cherokee
3A Chester
3A Chesterfield
3A Clarendon
3A Colleton*
3A Darlington
3A Dillon
3A Dorchester*
3A Edgefield
3A Fairfield
3A Florence
3A Georgetown*
3A Greenville
3A Greenwood
3A Hampton*
3A Horry*
3A Jasper*
3A Kershaw
3A Lancaster
3A Laurens
3A Lee
3A Lexington
3A Marion
3A Marlboro
3A McCormick
3A Newberry
3A Oconee
3A Orangeburg
3A Pickens
3A Richland
3A Saluda
3A Spartanburg
3A Sumter
3A Union
3A Williamsburg
3A York

SOUTH DAKOTA

6A Aurora
6A Beadle
5A Bennett
5A Bon Homme
6A Brookings
6A Brown
6A Brule
6A Buffalo
6A Butte
6A Campbell
5A Charles Mix
6A Clark
5A Clay
6A Codington
6A Corson
6A Custer
6A Davison
6A Day
6A Deuel
6A Dewey
5A Douglas
6A Edmunds
6A Fall River
6A Faulk
6A Grant
5A Gregory
6A Haakon
6A Hamlin
6A Hand
6A Hanson
6A Harding
6A Hughes
5A Hutchinson
6A Hyde
5A Jackson
6A Jerauld
6A Jones
6A Kingsbury
6A Lake
6A Lawrence
6A Lincoln
6A Lyman
6A Marshall
6A McCook
6A McPherson
6A Meade
5A Mellette
6A Miner
6A Minnehaha
6A Moody
6A Pennington
6A Perkins
6A Potter
6A Roberts
6A Sanborn
6A Shannon
6A Spink
6A Stanley
6A Sully
5A Todd
5A Tripp
6A Turner
5A Union
6A Walworth
5A Yankton
6A Ziebach

TENNESSEE

4A Anderson
4A Bedford
4A Benton
4A Bledsoe
4A Blount
4A Bradley
4A Campbell
4A Cannon
4A Carroll
4A Carter
4A Cheatham
3A Chester
4A Claiborne
4A Clay
4A Cocke
4A Coffee
3A Crockett
4A Cumberland
4A Davidson
4A Decatur
4A DeKalb
4A Dickson
3A Dyer
3A Fayette
4A Fentress
4A Franklin
4A Gibson
4A Giles
4A Grainger
4A Greene
4A Grundy
4A Hamblen
4A Hamilton
4A Hancock
3A Hardeman
3A Hardin
4A Hawkins
3A Haywood
3A Henderson
4A Henry
4A Hickman
4A Houston
4A Humphreys
4A Jackson
4A Jefferson
4A Johnson
4A Knox
3A Lake
3A Lauderdale
4A Lawrence
4A Lewis
4A Lincoln
4A Loudon
4A Macon
3A Madison
4A Marion
4A Marshall
4A Maury
4A McMinn
3A McNairy
4A Meigs
4A Monroe
4A Montgomery
4A Moore
4A Morgan
4A Obion
4A Overton
4A Perry
4A Pickett
4A Polk
4A Putnam
4A Rhea
4A Roane

(continued)

TABLE C301.1—continued
CLIMATE ZONES, MOISTURE REGIMES, AND WARM-HUMID DESIGNATIONS BY STATE, COUNTY AND TERRITORY

4A Robertson
4A Rutherford
4A Scott
4A Sequatchie
4A Sevier
3A Shelby
4A Smith
4A Stewart
4A Sullivan
4A Sumner
3A Tipton
4A Trousdale
4A Unicoi
4A Union
4A Van Buren
4A Warren
4A Washington
4A Wayne
4A Weakley
4A White
4A Williamson
4A Wilson

TEXAS

2A Anderson*
3B Andrews
2A Angelina*
2A Aransas*
3A Archer
4B Armstrong
2A Atascosa*
2A Austin*
4B Bailey
2B Bandera*
2A Bastrop*
3B Baylor
2A Bee*
2A Bell*
2A Bexar*
3A Blanco*
3B Borden
2A Bosque*
3A Bowie*
2A Brazoria*
2A Brazos*
3B Brewster
4B Briscoe
2A Brooks*
3A Brown*
2A Burleson*
3A Burnet*
2A Caldwell*
2A Calhoun*
3B Callahan
2A Cameron*
3A Camp*
4B Carson
3A Cass*
4B Castro
2A Chambers*
2A Cherokee*
3B Childress
3A Clay
4B Cochran
3B Coke
3B Coleman
3A Collin*
3B Collingsworth
2A Colorado*
2A Comal*
3A Comanche*
3B Concho
3A Cooke
2A Coryell*
3B Cottle
3B Crane
3B Crockett
3B Crosby
3B Culberson
4B Dallam
3A Dallas*
3B Dawson
4B Deaf Smith
3A Delta
3A Denton*
2A DeWitt*
3B Dickens
2B Dimmit*
4B Donley
2A Duval*
3A Eastland
3B Ector
2B Edwards*
3A Ellis*
3B El Paso
3A Erath*
2A Falls*
3A Fannin
2A Fayette*
3B Fisher
4B Floyd
3B Foard
2A Fort Bend*
3A Franklin*
2A Freestone*
2B Frio*
3B Gaines
2A Galveston*
3B Garza
3A Gillespie*
3B Glasscock
2A Goliad*
2A Gonzales*
4B Gray
3A Grayson
3A Gregg*
2A Grimes*
2A Guadalupe*
4B Hale
3B Hall
3A Hamilton*
4B Hansford
3B Hardeman
2A Hardin*
2A Harris*
3A Harrison*
4B Hartley
3B Haskell
2A Hays*
3B Hemphill
3A Henderson*
2A Hidalgo*
2A Hill*
4B Hockley
3A Hood*
3A Hopkins*
2A Houston*
3B Howard
3B Hudspeth
3A Hunt*
4B Hutchinson
3B Irion
3A Jack
2A Jackson*
2A Jasper*
3B Jeff Davis
2A Jefferson*
2A Jim Hogg*
2A Jim Wells*
3A Johnson*
3B Jones
2A Karnes*
3A Kaufman*
3A Kendall*
2A Kenedy*
3B Kent
3B Kerr
3B Kimble
3B King
2B Kinney*
2A Kleberg*
3B Knox
3A Lamar*
4B Lamb
3A Lampasas*
2B La Salle*
2A Lavaca*
2A Lee*
2A Leon*
2A Liberty*
2A Limestone*
4B Lipscomb
2A Live Oak*
3A Llano*
3B Loving
3B Lubbock
3B Lynn
2A Madison*
3A Marion*
3B Martin
3B Mason
2A Matagorda*
2B Maverick*
3B McCulloch
2A McLennan*
2A McMullen*
2B Medina*
3B Menard
3B Midland
2A Milam*
3A Mills*
3B Mitchell
3A Montague
2A Montgomery*
4B Moore
3A Morris*
3B Motley
3A Nacogdoches*
3A Navarro*
2A Newton*
3B Nolan
2A Nueces*
4B Ochiltree
4B Oldham
2A Orange*
3A Palo Pinto*
3A Panola*
3A Parker*
4B Parmer
3B Pecos
2A Polk*
4B Potter
3B Presidio
3A Rains*
4B Randall
3B Reagan
2B Real*
3A Red River*
3B Reeves
2A Refugio*
4B Roberts
2A Robertson*
3A Rockwall*
3B Runnels
3A Rusk*
3A Sabine*
3A San Augustine*
2A San Jacinto*
2A San Patricio*
3A San Saba*

(continued)

TABLE C301.1—continued
CLIMATE ZONES, MOISTURE REGIMES, AND WARM-HUMID DESIGNATIONS BY STATE, COUNTY AND TERRITORY

3B Schleicher
3B Scurry
3B Shackelford
3A Shelby*
4B Sherman
3A Smith*
3A Somervell*
2A Starr*
3A Stephens
3B Sterling
3B Stonewall
3B Sutton
4B Swisher
3A Tarrant*
3B Taylor
3B Terrell
3B Terry
3B Throckmorton
3A Titus*
3B Tom Green
2A Travis*
2A Trinity*
2A Tyler*
3A Upshur*
3B Upton
2B Uvalde*
2B Val Verde*
3A Van Zandt*
2A Victoria*
2A Walker*
2A Waller*
3B Ward
2A Washington*
2B Webb*
2A Wharton*
3B Wheeler
3A Wichita
3B Wilbarger
2A Willacy*
2A Williamson*
2A Wilson*
3B Winkler
3A Wise
3A Wood*
4B Yoakum
3A Young
2B Zapata*
2B Zavala*

UTAH

5B Beaver
6B Box Elder
6B Cache
6B Carbon
6B Daggett
5B Davis
6B Duchesne
5B Emery
5B Garfield
5B Grand
5B Iron
5B Juab
5B Kane
5B Millard
6B Morgan
5B Piute
6B Rich
5B Salt Lake
5B San Juan
5B Sanpete
5B Sevier
6B Summit
5B Tooele
6B Uintah
5B Utah
6B Wasatch
3B Washington
5B Wayne
5B Weber

VERMONT

6A (all)

VIRGINIA

4A (all)

WASHINGTON

5B Adams
5B Asotin
5B Benton
5B Chelan
4C Clallam
4C Clark
5B Columbia
4C Cowlitz
5B Douglas
6B Ferry
5B Franklin
5B Garfield
5B Grant
4C Grays Harbor
4C Island
4C Jefferson
4C King
4C Kitsap
5B Kittitas
5B Klickitat
4C Lewis
5B Lincoln
4C Mason
6B Okanogan
4C Pacific
6B Pend Oreille
4C Pierce
4C San Juan
4C Skagit
5B Skamania
4C Snohomish
5B Spokane
6B Stevens
4C Thurston
4C Wahkiakum
5B Walla Walla
4C Whatcom
5B Whitman
5B Yakima

WEST VIRGINIA

5A Barbour
4A Berkeley
4A Boone
4A Braxton
5A Brooke
4A Cabell
4A Calhoun
4A Clay
5A Doddridge
5A Fayette
4A Gilmer
5A Grant
5A Greenbrier
5A Hampshire
5A Hancock
5A Hardy
5A Harrison
4A Jackson
4A Jefferson
4A Kanawha
5A Lewis
4A Lincoln
4A Logan
5A Marion
5A Marshall
4A Mason
4A McDowell
4A Mercer
5A Mineral
4A Mingo
5A Monongalia
4A Monroe
4A Morgan
5A Nicholas
5A Ohio
5A Pendleton
4A Pleasants
5A Pocahontas
5A Preston
4A Putnam
5A Raleigh
5A Randolph
4A Ritchie
4A Roane
5A Summers
5A Taylor
5A Tucker
4A Tyler
5A Upshur
4A Wayne
5A Webster
5A Wetzel
4A Wirt
4A Wood
4A Wyoming

WISCONSIN

6A Adams
7 Ashland
6A Barron
7 Bayfield
6A Brown
6A Buffalo
7 Burnett
6A Calumet
6A Chippewa
6A Clark
6A Columbia
6A Crawford
6A Dane
6A Dodge
6A Door
7 Douglas
6A Dunn
6A Eau Claire
7 Florence
6A Fond du Lac
7 Forest
6A Grant
6A Green
6A Green Lake
6A Iowa
7 Iron
6A Jackson
6A Jefferson
6A Juneau
6A Kenosha
6A Kewaunee
6A La Crosse
6A Lafayette
7 Langlade
7 Lincoln
6A Manitowoc
6A Marathon
6A Marinette
6A Marquette
6A Menominee
6A Milwaukee
6A Monroe
6A Oconto
7 Oneida
6A Outagamie
6A Ozaukee
6A Pepin
6A Pierce
6A Polk

(continued)

TABLE C301.1—continued
CLIMATE ZONES, MOISTURE REGIMES, AND WARM-HUMID DESIGNATIONS BY STATE, COUNTY AND TERRITORY

6A Portage
7 Price
6A Racine
6A Richland
6A Rock
6A Rusk
6A Sauk
7 Sawyer
6A Shawano
6A Sheboygan
6A St. Croix
7 Taylor
6A Trempealeau
6A Vernon
7 Vilas
6A Walworth
7 Washburn
6A Washington
6A Waukesha
6A Waupaca
6A Waushara
6A Winnebago
6A Wood

WYOMING

6B Albany
6B Big Horn
6B Campbell
6B Carbon
6B Converse
6B Crook
6B Fremont
5B Goshen
6B Hot Springs
6B Johnson
6B Laramie
7 Lincoln
6B Natrona
6B Niobrara
6B Park
5B Platte
6B Sheridan
7 Sublette
6B Sweetwater
7 Teton
6B Uinta
6B Washakie
6B Weston

US TERRITORIES

AMERICAN SAMOA

1A (all)*

GUAM

1A (all)*

NORTHERN MARIANA ISLANDS

1A (all)*

PUERTO RICO

1A (all)*

VIRGIN ISLANDS

1A (all)*

SECTION C302
DESIGN CONDITIONS

C302.1 Interior design conditions. The interior design temperatures used for heating and cooling load calculations shall be a maximum of 72°F (22°C) for heating and minimum of 75°F (24°C) for cooling.

❖ While the previous sections of Chapter 3 [CE] address outdoor design conditions, this section provides the interior conditions that will be used for properly sizing the mechanical equipment and the complete HVAC system. The proper sizing of mechanical equipment (see Sections C403.2.1 and C403.2.2) can vary depending on the selected design conditions. While the code does address oversizing equipment, it is not enforceable without establishing the exact design parameters. This section is included in the code only for system sizing, and it does not affect the interior design temperatures required by other codes such as Section 1204 of the *International Building Code®* (IBC®) or Section 602.2 of the *International Property Maintenance Code®* (IPMC®).

SECTION C303
MATERIALS, SYSTEMS AND EQUIPMENT

C303.1 Identification. Materials, systems and equipment shall be identified in a manner that will allow a determination of compliance with the applicable provisions of this code.

❖ The intent of this section is to make certain that sufficient information exists to determine compliance with the code during the plan review and field inspection phases. The permit applicant can submit the required equipment and materials information on the building plans, specification sheets or schedules; or in any other way that allows the code official to clearly identify which specifications apply to which portions of the building (i.e., which parts of the building are insulated to the levels listed). Materials information includes envelope insulation levels, glazing assembly *U*-factors and duct and piping insulation levels. Equipment information includes heating and cooling equipment and appliance efficiencies where high-efficiency equipment is claimed to meet code requirements.

This section contains specific material, equipment and system identification requirements for the approval and installation of the items required by the code. Although the means for permanent marking (tag, stencil, label, stamp, sticker, bar code, etc.) is often determined and applied by the manufacturer, if there is any uncertainty about the product, the mark is subject to the approval of the code official.

C303.1.1 Building thermal envelope insulation. An *R*-value identification mark shall be applied by the manufacturer to each piece of *building thermal envelope* insulation 12 inches (305 mm) or greater in width. Alternately, the insulation installers shall provide a certification listing the type, manufacturer and *R*-value of insulation installed in each element of the *building thermal envelope*. For blown or sprayed insulation (fiberglass and cellulose), the initial installed thickness, settled thickness, settled *R*-value, installed density, coverage area and number of bags installed shall be *listed* on the certification. For sprayed polyurethane foam (SPF) insulation, the installed thickness of the areas covered and *R*-value of installed thickness shall be *listed* on the certification. For insulated siding, the *R*-value shall be labeled on thew product's package and shall be *listed* on the certification. The insulation installer shall sign, date and post the certification in a conspicuous location on the job site.

❖ The thermal performance of insulation is rated in terms of *R*-value. For products lacking an *R*-value identification, the installer (or builder) must provide the insulation performance data. For example, some insulation materials, such as foamed-in-place urethane, can be installed in wall, floor and cathedral ceiling cavities. These products are not labeled, as is batt insulation, nor is it appropriate for them to be evaluated as required in the code for blown or sprayed insulation. However, the installer must certify the type, thickness and *R*-value of these materials.

The *R*-value of loose-fill insulation (blown or sprayed) is dependent on both the installed thickness and the installed density (number of bags used). Therefore, loose-fill insulation cannot be directly labeled by the manufacturer. Many blown insulation products carry a manufacturer's *R*-value guarantee when installed to a designated thickness, "inches = *R*-value." Blown insulation products lacking this manufacturer's guarantee can be subjected to special inspection and testing, what is referred to as "cookie cutting." Cookie cutting involves extracting a column of insulation with a cylinder to determine its density. The insulation depth and density must yield the specified *R*-value according to the manufacturer's bag label specification.

The code and Federal Trade Commission (FTC) Rule 460 require that installers of insulation in homes, apartments and manufactured housing units report this information to the authority having jurisdiction in the form of a certification posted in a conspicuous location (see Commentary Figure C303.1.1).

C303.1.1.1 Blown or sprayed roof/ceiling insulation. The thickness of blown-in or sprayed roof/ceiling insulation (fiberglass or cellulose) shall be written in inches (mm) on markers that are installed at least one for every 300 square feet (28 m^2) throughout the attic space. The markers shall be affixed to the trusses or joists and marked with the minimum initial installed thickness with numbers a minimum of 1 inch (25 mm) in height. Each marker shall face the attic access opening. Spray polyurethane foam thickness and installed *R*-value shall be *listed* on certification provided by the insulation installer.

❖ To help verify the installed *R*-value of blown-in or spray-applied insulation, the installer must certify the following information in a signed statement posted in a conspicuous place (see Section C303.1.1):

- The type of insulation used and manufacturer.

- The insulation's coverage per bag (the number of bags required to result in a given *R*-value for a given area), as well as the settled *R*-value.
- The initial and settled thickness.
- The number of bags installed.

Under circumstances where the insulation *R*-value is guaranteed, only the initial thickness is required on the certification.

This section helps demonstrate compliance and enforcement of the provisions found in Section C303.1.1. To assist with application and enforcement, loose-fill ceiling insulation also requires thickness markers that are attached to the framing and face the attic access. In a large space, markers placed evenly about every 17 feet (5182 mm) (with some markers at the edge of the space) will meet this requirement. For sprayed polyurethane, such markers are not effective. When using this product, the code requires that the measured thickness and *R*-value be recorded on the certificate.

C303.1.2 Insulation mark installation. Insulating materials shall be installed such that the manufacturer's *R*-value mark is readily observable upon inspection.

❖ For batt insulation, manufacturers' *R*-value designations and stripe codes are often printed directly on the insulation. Where possible, the insulation must be installed so these designations are readable. Backed floor batts can be installed with the designation against the underfloor, which means it would not be visible. In those cases, the *R*-value must be certified by the installer or be validated by some other means (see commentary, Section C303.1.1).

This Attic Has Been Insulated To

INSULATION CONTRACTORS ICAA ASSOCIATION OF AMERICA

R-

By A Professional Insulation Contractor

The insulation in this attic was installed by a qualified professional Contractor to the R-value stated above

CIMA

INSULATE TODAY SAVE TOMORROW

NAIMA
NORTH AMERICAN INSULATION MANUFACTURERS ASSOCIATION

Certificate of Insulation

BUILDING ADDRESS: ______________________

CONTRACTOR: ______________________

Installation Date ____________

License# ____________

Area Insulated	R-Value	Installed Thickness	Settled Thickness	Installed Density	No. Bags	Sq. Ft.
Attic						
Walls						
Floors						

I, ______________________, (print name) certify that this residence/building has been insulated to the stated R-value and that the installation is in conformance with all applicable codes, standards, regulations and specifications.

Authorized Signature ______________________ Date ________

Figure C303.1.1
SAMPLE CERTIFICATE OF INSULATION

(Logos courtesy of Cellulose Insulation Manufacturers Association, http:/cellulose.org, Insulation Contractors Association of America, www.insulate.org, and North American Insulation Manufacturers Association, www.NAIMA.org)

C303.1.3 Fenestration product rating. *U*-factors of fenestration products (windows, doors and skylights) shall be determined in accordance with NFRC 100.

> **Exception:** Where required, garage door *U*-factors shall be determined in accordance with either NFRC 100 or ANSI/DASMA 105.

U-factors shall be determined by an accredited, independent laboratory, and labeled and certified by the manufacturer.

Products lacking such a labeled *U*-factor shall be assigned a default *U*-factor from Table C303.1.3(1) or C303.1.3(2). The solar heat gain coefficient (SHGC) and *visible transmittance* (VT) of glazed fenestration products (windows, glazed doors and skylights) shall be determined in accordance with NFRC 200 by an accredited, independent laboratory, and labeled and certified by the manufacturer. Products lacking such a labeled SHGC or VT shall be assigned a default SHGC or VT from Table C303.1.3(3).

❖ Historically, the buyers of fenestration products received energy performance information in a variety of ways. Some manufacturers described performance by showing *R*-values of the glass. While the glass might have been a good performer, the rating did not include the effects of the frame. Other manufacturers touted the insulating value of different window components, but these, too, did not reflect the total window system performance. When manufacturers rated the entire product, some used test laboratory measurements and others used computer calculations. Even among those using test laboratory reports, the test laboratories often tested the products under different procedures, making an "apples-to-apples" comparison difficult. The different rating methods confused builders and consumers. They also created headaches for manufacturers trying to differentiate the performance of their products from the performance of their competitors' products.

The National Fenestration Rating Council, (NFRC) has developed a fenestration energy rating system based on whole-product performance. This accurately accounts for the energy-related effects of all the product's component parts and prevents information about a single component from being compared in a misleading way to other whole-product properties. With energy ratings based on whole-product performance, NFRC helps builders, designers and consumers directly compare products with different construction details and attributes.

The Door and Access Systems Manufacturers Association (DASMA) has developed a testing standard that provides equivalent test results with the NFRC 100 standard. DASMA 105 allows for testing of door assemblies in sizes not testable under NFRC 100.

Products that have been rated by NFRC-approved testing laboratories and certified by NFRC-accredited independent certification and inspection agencies carry a temporary and permanent label featuring the "NFRC-certified" mark. With this mark, the manufacturer stipulates that the energy performance of the product was determined according to NFRC rules and procedures.

By certifying and labeling their products, manufacturers are demonstrating their commitment to providing accurate energy and energy-related performance information. The code purposely sets the default values fairly high. This helps to encourage the use of products that have been tested and also ensures that products that have little energy savings value are not used inappropriately in the various climate zones. By setting the default value so high, it also will prevent someone from considering removing the label from a tested window and then using the default values. Therefore, the default values are most representative of the lower end of the energy-efficient products.

Products that are not NFRC or DASMA certified and do not exactly match the specifications in Tables C303.1.3(1) and C303.1.3(2) must use the tabular specification for the products they most closely resemble. In the absence of tested *U*-factors, the default *U*-factor for doors containing glazing can be a combination of the glazing and door *U*-factor as described in the definition for "*U*-factor" (see commentary, Section 202, "*U*-factor"). The NFRC procedures determine *U*-factor and SHGC ratings based on the whole fenestration assembly [untested fenestration products have default *U*-factors and SHGCs assigned in accordance with Tables C303.1.3(1) through C303.1.3(3)]. During construction inspection, the label on each glazing assembly should be checked for conformance to the *U*-factor specified on the approved plans. These labels must be left on the glazing until after the building has been inspected for compliance. A sample NFRC label is shown in Commentary Figure C303.1.3(1).

Products certified according to NFRC procedures are listed in the Certified Products Directory. The directory is published annually and contains energy performance information for over 1.4 million fenestration product options listed by over 450 manufacturers. When using the directory or shopping for NFRC-certified products, it is important to note:

1. A product is considered to be NFRC certified only if it carries the NFRC label. Simply being listed in this directory is not enough.
2. The NFRC-certified mark does not signify that the product meets any energy-efficiency standards or criteria.
3. NFRC sets no minimum performance standards nor does it mandate specific performance levels. Rather, NFRC ratings can be used to determine whether a product meets a state or local code, or other performance requirement, and to compare the energy performance of different products during plan review. For questions about NFRC and its rating and labeling system, more information is available on the organization's web site at www.nfrc.org. NFRC

adopted a new energy performance label in 2005. It lists the manufacturer, describes the product, provides a source for additional information and includes ratings for one or more energy performance characteristics.

The code offers an alternative to NFRC-certified glazed fenestration product *U*-factor ratings. In the absence of *U*-factors based on NFRC test procedures, the default *U*-factors in Table C303.1.3(1) must be used. When a composite of materials from two different product types is used, the code official should be consulted regarding how the product will be rated. Generally, the product must be assigned the higher *U*-factor, although an average based on the *U*-factors and areas may be acceptable in some cases.

The product cannot receive credit for a feature that cannot be seen. Because performance features such as argon fill and low-emissivity (low-E) coatings for glass are not visually verifiable, they do not receive credit in the default tables. Tested *U*-factors for these windows are often lower, so using tested *U*-factors is to the applicant's advantage. Commentary Figure C303.1.3(2) illustrates visually verifiable window characteristics, among other various window performance, function and cost considerations.

A single-glazed window with an installed storm window may be considered a double-glazed assembly and use the corresponding *U*-factor from the default table. For example, the *U*-factor 0.80 in Table C303.1.3(1) applies to a double-glazed, metal window without a thermal break (but with an installed storm window). If the storm window was not installed, the *U*-factor would be 1.20.

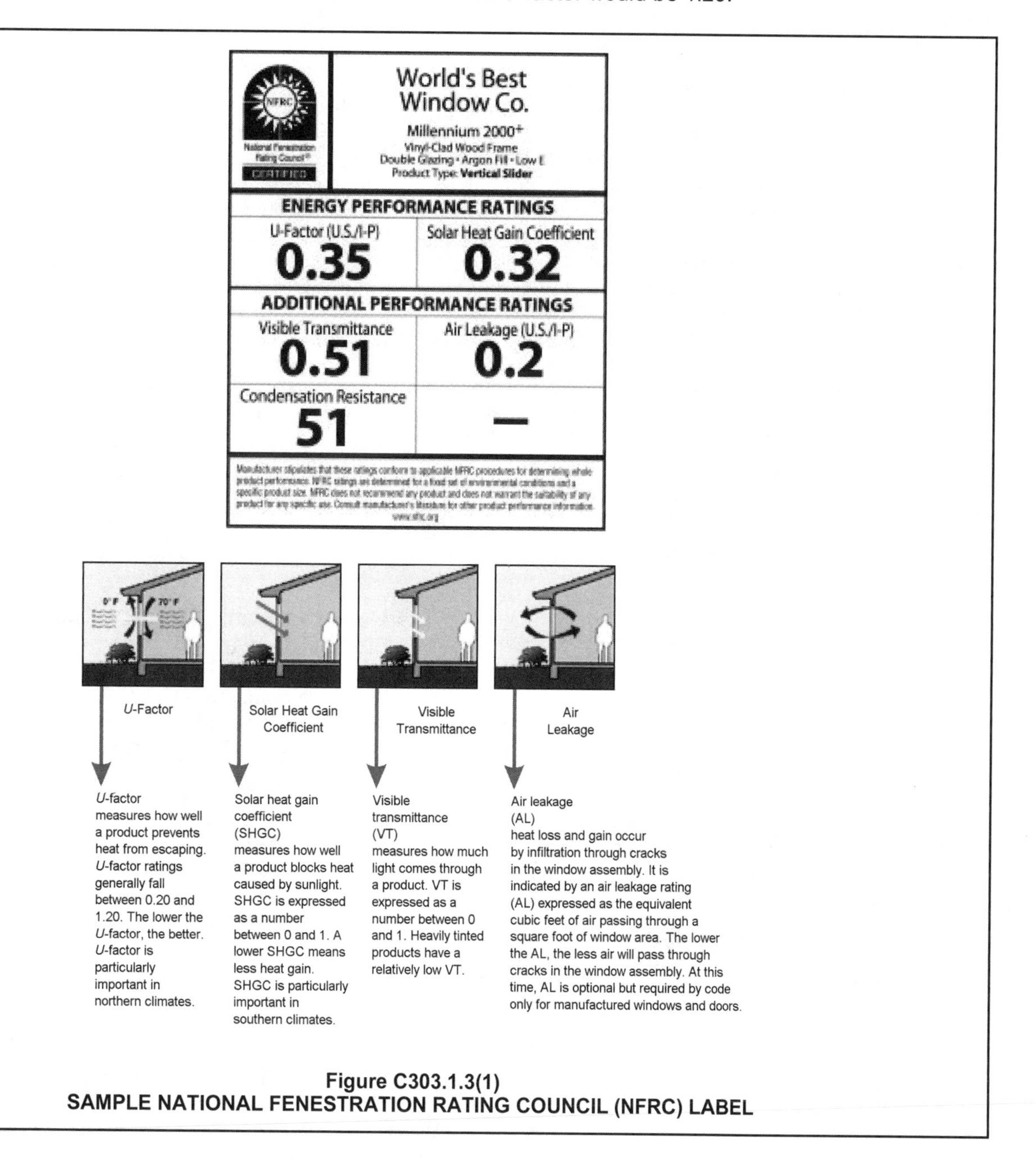

Figure C303.1.3(1)
SAMPLE NATIONAL FENESTRATION RATING COUNCIL (NFRC) LABEL

TABLE C303.1.3(1)
DEFAULT GLAZED FENESTRATION *U*-FACTORS

FRAME TYPE	SINGLE PANE	DOUBLE PANE	SKYLIGHT	
			Single	Double
Metal	1.20	0.80	2.00	1.30
Metal with Thermal Break	1.10	0.65	1.90	1.10
Nonmetal or Metal Clad	0.95	0.55	1.75	1.05
Glazed Block	0.60			

TABLE C303.1.3(2)
DEFAULT DOOR *U*-FACTORS

DOOR TYPE	*U*-FACTOR
Uninsulated Metal	1.20
Insulated Metal	0.60
Wood	0.50
Insulated, nonmetal edge, max 45% glazing, any glazing double pane	0.35

❖ Door *U*-factors in Table C303.1.3(2) should be used wherever NFRC-certified ratings are not available. There are a few other aspects to note about doors. Opaque door *U*-factors must include the effects of the door edge and the frame. Calculating *U*-factors based on a cross section through the insulated portion is not acceptable. To take credit for a thermal break, the door must have a thermal break in both the door slab and in the frame. The values in the table are founded on principles established in the 1997 ASHRAE *Handbook of Fundamentals*.

TABLE C303.1.3(3)
DEFAULT GLAZED FENESTRATION SHGC AND VT

	SINGLE GLAZED		DOUBLE GLAZED		GLAZED BLOCK
	Clear	Tinted	Clear	Tinted	
SHGC	0.8	0.7	0.7	0.6	0.6
VT	0.6	0.3	0.6	0.3	0.6

❖ This table offers an alternative to NFRC-certified SHGC and visible transmittance (VT) values based on visually verifiable characteristics of the fenestration product. The SHGC is the fraction of incident solar radiation absorbed and directly transmitted by the window area, then subsequently reradiated, conducted or convected inward. SHGC is a ratio, expressed as a number between 0 and 1. The lower a window's SHGC, the less solar heat it transmits. The VT is the ratio of visible light entering the space through the fenestration product assembly to the incident visible light. VT includes the effects of glazing material and frame, and is expressed as a number between 0 and 1.

An SHGC of 0.40 or less is recommended in cooling-dominated climates (Climate Zones 1–3). In heating-dominated climates, a high SHGC increases passive solar gain for the heating but reduces cooling-season performance. A low SHGC improves cooling-season performance but reduces passive solar gains for heating.

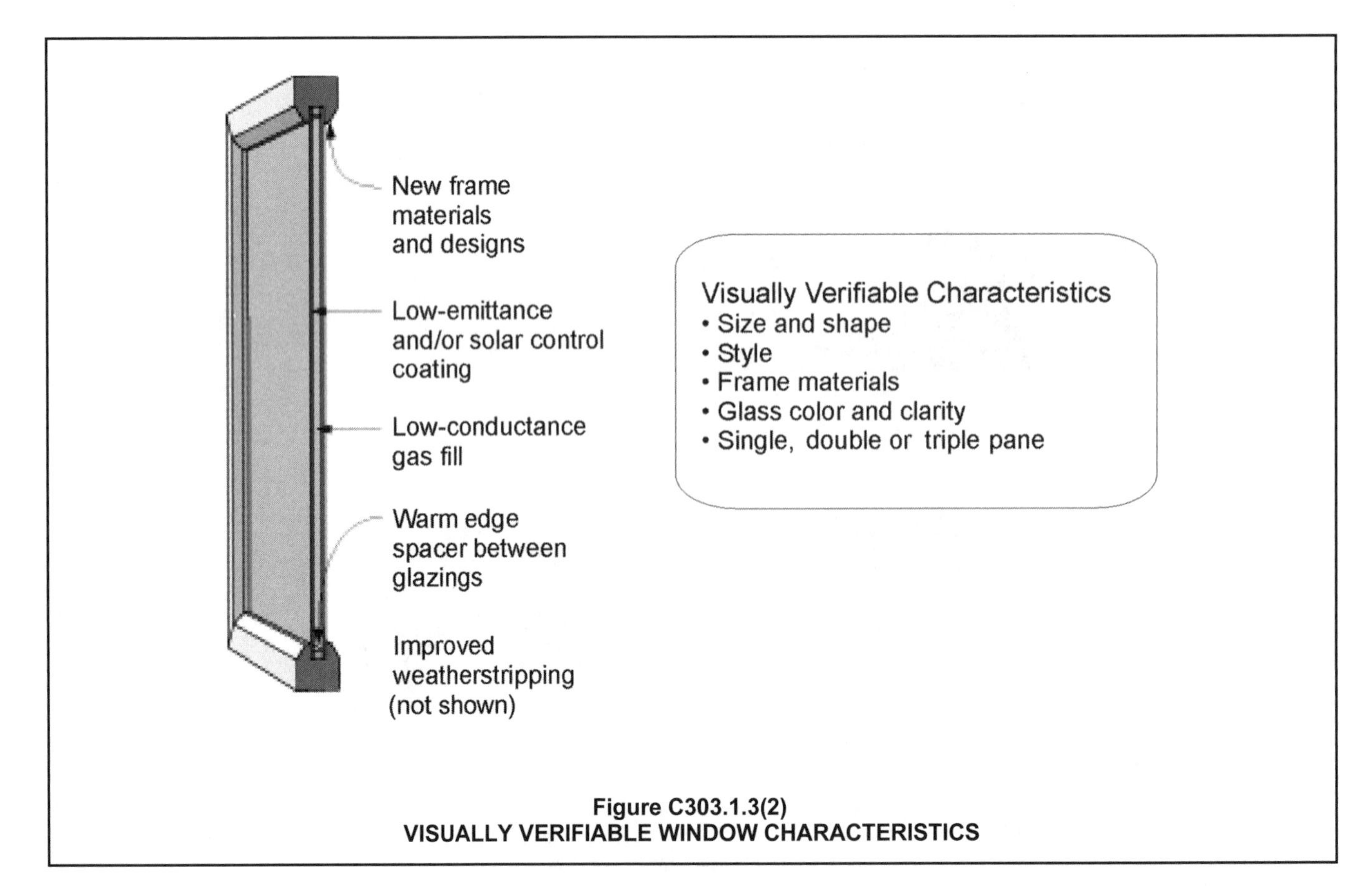

Figure C303.1.3(2)
VISUALLY VERIFIABLE WINDOW CHARACTERISTICS

C303.1.4 Insulation product rating. The thermal resistance (*R*-value) of insulation shall be determined in accordance with the U.S. Federal Trade Commission *R*-value rule (CFR Title 16, Part 460) in units of h × ft^2 × °F/Btu at a mean temperature of 75°F (24°C).

❖ This section brings two important requirements to the code.

First, the FTC *R*-value rule details specific test standards for insulation. The test standards are specific to the type of insulation and intended use. This clarifies any questions on the rating conditions to be used for insulation materials.

Second, the text above specifies the rating temperature to be used when evaluating the *R*-value of the product. Insulation products sometimes list several *R*-values based on different test temperatures. This eliminates any question as to which *R*-value to use. The temperature selected is a standard rating condition.

C303.1.4.1 Insulated siding. The thermal resistance (*R*-value) of insulated siding shall be determined in accordance with ASTM C 1363. Installation for testing shall be in accordance with the manufacturer's installation instructions.

❖ The ASTM C 1363 standard is the appropriate test method for determining the thermal resistance for insulated siding products. This standard, new to the 2015 edition of the code, will promote consistent compliance of this class of products.

C303.2 Installation. All materials, systems and equipment shall be installed in accordance with the manufacturer's installation instructions and the *International Building Code*.

❖ Manufacturers' installation instructions are thoroughly evaluated by the listing agency verifying that a safe installation is prescribed. When an appliance is tested to obtain a listing and label, the approval agency installs the appliance in accordance with the manufacturer's instructions. The appliance is tested under these conditions; thus, the installation instructions become an integral part of the labeling process. The listing agency can require that the manufacturer alter, delete or add information to the instructions as necessary to achieve compliance with applicable standards and code requirements.

Manufacturers' installation instructions are an enforceable extension of the code and must be in the hands of the code official when an inspection takes place. Inspectors must carefully and completely read and comprehend the manufacturer's instructions in order to properly perform an installation inspection. In some cases, the code will specifically address an installation requirement that is also addressed in the manufacturer's installation instructions. The code requirement may be the same or may exceed the requirement in the manufacturer's installation instructions. The manufacturer's installation instructions could contain requirements that exceed those in the code. In such cases, the more restrictive requirements would apply (see commentary, Section C106).

Even if an installation appears to be in compliance with the manufacturer's instructions, the installation cannot be completed or approved until all associated components, connections and systems that serve the appliance or equipment are also in compliance with the requirements of the applicable *International Codes*® (I-Codes®) of reference. For example, a gas-fired boiler installation must not be approved if the boiler is connected to a deteriorated, undersized or otherwise unsafe chimney or vent. Likewise, the same installation must not be approved if the existing gas piping has insufficient capacity to supply the boiler load or if the electrical supply circuit is inadequate or unsafe.

Manufacturers' installation instructions are often updated and changed for various reasons, such as changes in the appliance, equipment or material design; revisions to the product standards and as a result of field experiences related to existing installations. The code official should stay abreast of any changes by reviewing the manufacturer's instructions for every installation.

C303.2.1 Protection of exposed foundation insulation. Insulation applied to the exterior of basement walls, crawlspace walls and the perimeter of slab-on-grade floors shall have a rigid, opaque and weather-resistant protective covering to prevent the degradation of the insulation's thermal performance. The protective covering shall cover the exposed exterior insulation and extend a minimum of 6 inches (153 mm) below grade.

❖ The ultimate performance of insulation material is directly proportional to the workmanship involved in the material's initial installation, as well as the materials' integrity over the life of the structure. Accordingly, foundation wall and slab-edge insulation materials installed in the vicinity of the exterior grade line require protection from damage that could occur from contact by lawn-mowing and maintenance equipment, garden hoses, garden tools, perimeter landscape materials, etc. In addition, the long-term thermal performance of foam-plastic insulation materials is adversely affected by direct exposure to the sun. To protect the insulation from sunlight and physical damage, it must have a protective covering that is inflexible, puncture resistant, opaque and weather resistant.

C303.3 Maintenance information. Maintenance instructions shall be furnished for equipment and systems that require preventive maintenance. Required regular maintenance actions shall be clearly stated and incorporated on a readily accessible label. The label shall include the title or publication number for the operation and maintenance manual for that particular model and type of product.

❖ This section establishes an owner's responsibility for maintaining the building in accordance with the requirements of the code and other referenced standards. This section requires, among others, that mechanical and service water heating equipment and appliance maintenance information be made available to the owner/operator. This section does not require that labels be added to existing equipment; having the manufacturer's maintenance literature is usually sufficient to meet this requirement. During final occupancy inspection, the mechanical equipment and water heater should be inspected to verify that the information is taped to each unit or referenced on a label mounted in a conspicuous location on the units.

The code official has the authority to rule on the performance of maintenance work when equipment functions would be affected by such work. He or she also has the authority to require a building and its energy-using systems to be maintained in compliance with the public health and safety provisions required by other I-Codes.

Bibliography

The following resource materials were used in the preparation of the commentary for this chapter of the code.

ASHRAE-09, *Handbook of Fundamentals*. Atlanta, GA: American Society of Heating, Refrigerating and Air-conditioning Engineers, Inc., 2009.

Robert S. Briggs, Robert G. Lucas, and Z. Todd Taylor. *Climate Classification for Building Energy Codes and Standards*. Richland, WA: U.S. Department of Energy, Pacific Northwest National Laboratory (PNNL), 2002.

FTC 16 CFR, Part 460-05, *Labeling and Advertising of of Home Insulation*. Washington, DC: Federal Trade Commission, 2005.

Chapter 4 [CE]: Commercial Energy Efficiency

General Comments

Chapter 4 [CE] is used to demonstrate compliance for the design and construction of most types of commercial buildings. The definitions of residential and commercial buildings result in buildings of residential use that are greater than three stories in height above grade being classified as commercial buildings and regulated under this half of the code [see Commentary Figure 4C(1)]. Residential buildings including single-family homes, duplexes, townhouses and apartments three stories or less in height are covered in Chapter 4 [CE]. Commercial buildings will include Group R-1 occupancies (hotels and motels) of any height. Institutional occupancies such as hospitals and nursing homes are also considered to be commercial buildings.

In accordance with Sections C402.1.1 and C402.1.2, the thermal envelope requirements of the code do not apply to:

- Very low energy use buildings (less than 3.4 Btu per hour per square foot or 1 watt per square foot of floor area).
- Buildings or portions of buildings that are neither heated nor cooled.
- Greenhouses.
- Small equipment buildings not intended for human occupancy.

Some or all provisions of the code are exempted or modified for the following buildings or uses (see Chapter 5 [CE]):

- Continued use of existing buildings.
- Historic buildings.
- Additions, alterations, renovations or repairs.

Purpose

Chapter 4 [CE] contains several options to address the energy efficiency of commercial buildings. The three options are most clearly shown in Section C401.2, where compliance is required with either ANSI/ASHRAE/IESNA 90.1 or the remaining provisions of this chapter.

Besides adopting ANSI/ASHRAE/IESNA 90.1 by reference, Chapter 4 [CE] contains a set of requirements for the energy-efficient design of commercial buildings that provides an alternative way to show compliance for structures or systems based on ANSI/ASHRAE/IESNA

Airports	Indoor sporting facilities
Apartment buildings and condominiums (more than three stories)	Industrial work buildings
Assembly and conference areas	Laboratories
Banks	Libraries
Barber shops and beauty parlors	Museums and galleries
Bowling alleys	Nursing homes
Churches, synagogues and chapels	Offices
Commercial or industrial warehouses	Police and fire houses
Convention centers	Restaurants
Dormitories (more than three stories)	Retail, grocery and wholesale stores
Exhibit halls	Schools
Gymnasiums	Shopping malls
Health clubs	Shops
High-rise residential	Sporting arenas
Hospitals	Theaters and auditoriums
Hotels and motels	Warehouses and storage facilities

Note: This table includes only examples of building types covered by the code. It is not intended to be an exhaustive list. Other building types may be covered, even though they are not listed.

Figure 4C(1)
EXAMPLES OF BUILDING TYPES COVERED BY CHAPTER 4 [CE]

90.1 (see commentary, Section C401.2). The requirements specified in Sections C402 through C406 are reasonably equivalent to ANSI/ASHRAE/IESNA 90.1. Moreover, those portions of Chapter 4 [CE] are written in code language. The advice and recommendations contained in ANSI/ASHRAE/IESNA 90.1 are not found in the code.

The alternative methodology and prescriptive requirements in Chapter 4 [CE] were initially approved by the *International Energy Conservation Code*® (IECC®) Code Development Committee during the 1997 code development cycle to meet the needs of designers, builders and regulatory officials involved in the construction of commercial buildings who had requested a more easily understandable, usable and enforceable commercial energy conservation code.

This chapter is intended to promote the application of cost-effective design practices and technologies that minimize energy consumption without sacrificing either the comfort or productivity of occupants.

SECTION C401 GENERAL

C401.1 Scope. The provisions in this chapter are applicable to commercial *buildings* and their *building sites*.

❖ This chapter's requirements address the design of all building systems that affect the visual and thermal comfort of the occupants as well as building equipment that supports human use of a building, including:

- Wall, roof and floor insulation.
- Windows and skylights.
- Cooling equipment (air conditioners, chillers and cooling towers).
- Heating equipment (boilers, furnaces and heat pumps).
- Pumps, piping and liquid circulation systems.
- Heat rejection equipment (fan cooling towers, air-cooled condensers).
- Service water heating (kitchens, lavatories and pools).
- Electrical power and lighting systems (lighting types, densities, zones, and controls).
- Energy-using equipment (walk-in coolers and freezers, electric transformers and motors, elevators and escalators).

This chapter applies to new commercial buildings, including Group R occupancies, that contain residential uses but are outside of the code's definition of residential building. Chapter 4 [CE] does not apply to low-rise residential buildings such as single-family homes, duplexes and apartments three stories or less in height; however, these building types are covered by comparable provisions in this chapter.

Just as the code does not regulate or control the energy used by things such as office equipment and computers, the code does not limit or regulate the energy use intended primarily for manufacturing, or for commercial or industrial processing. Although the energy for manufacturing and processing is excluded, the envelope, mechanical systems, service water heating, and electrical power and lighting systems of these buildings are regulated. Chapter 4 [CE] includes a total building performance compliance evaluation, documentation and listing of software tools to determine the total building performance.

C401.2 Application. Commercial buildings shall comply with one of the following:

1. The requirements of ANSI/ASHRAE/IESNA 90.1.
2. The requirements of Sections C402 through C405. In addition, commercial buildings shall comply with Section C406 and tenant spaces shall comply with Section C406.1.1.
3. The requirements of Sections C402.5, C403.2, C404, C405.2, C405.3, C405.5, C405.6 and C407. The building energy cost shall be equal to or less than 85 percent of the standard reference design building.

❖ Because Chapter 4 [CE] applies to such a wide variety of buildings in various climate zones (see Tables C402.1.3 and C402.1.4), there must be flexibility in its application. The goal is to achieve energy-efficient building performance. When possible, this goal is achieved without requiring specific measures or products. Flexibility is offered at the prescriptive and systems levels, and at the whole-building level based on total building performance (TBP). The building designer must select the commercial building compliance path to be used to design the entire building from among the following:

1. ANSI/ASHRAE/IESNA 90.1.
2. IECC Chapter 4 [CE] prescriptive provisions (Sections C402, C403, C404, C405 and one of the six options in Section C406).
3. Total building performance provisions (Section C407).

Flexibility is achieved at the prescriptive level by specifying criteria in terms of component performance. The criterion for the roof component, for instance, is stated in terms of the *R*-value. Any construction assembly or method may be used as long as its *R*-value meets this criterion. For the overall building envelope, flexibility is offered through the component performance method of Section C402.1.5 through a *U*-factor-based method addressing each element of the building envelope including fenestration. The total building performance compliance path in Section C407 provides trade-offs among the envelope, lighting, mechanical and other systems.

Lighting control requirements can be satisfied with multiple occupant-sensor controls, time-switch controls, light reduction controls and daylight responsive controls but must be satisfied on a space-by-space basis. Lighting power requirements can be satisfied by any combination of equipment as long as the total connected interior lighting power is less than the total permitted lighting power from Section C405.4.1. The total exterior lighting power allowance for all exterior building applications is determined using the lighting zones information in Table C405.5.2(1) and the lighting power allowances in Table C405.5.2(2).

HVAC equipment and water heaters can be manufactured in a variety of ways as long as each piece of equipment meets the applicable overall criterion of energy efficiency: the energy efficiency ratio (EER). Averaging equipment efficiencies is not acceptable.

For the building envelope, Chapter 4 [CE] has prescriptive tables that give wall, floor and roof *R*-values; window *U*-factors; solar heat gain coefficient (SHGC); and opaque door values that satisfy the requirements for the specific climate zone. This is the easiest approach to meet the building envelope criteria. Where the prescriptive tables do not offer enough flexibility, the component performance method (Section C402.1.5) or an alternative method approved by the code official may be used. Generally, such approved alternatives allow consideration of window area and orientation; thermal mass; insulation position (on the inside or outside of the wall); daylighting; and internal gains from lights, equipment and people. Evaluation of such an alternative can be obtained by using a calculation software tool as listed in Section C407.6, work sheets and compliance manuals where approved by the code official (see commentary, Sections C101.3 and C101.5 or C103.1).

The greatest flexibility is achieved at the whole-building level under the total building performance approach of Section C407 or similar provisions in ANSI/ASHRAE/IESNA 90.1. Section C407 permits the comparative evaluation of annual energy costs to assess total building performance compliance. As long as the estimated annual energy cost of the proposed design is less than or equal to 85 percent of the estimated annual energy cost of the standard design, the proposed design meets the code. More information to determine the total building performance is in the commentary for Section C407. When this evaluation is complete, the code official may require that the results of the compliance calculations be certified by a registered design professional such as an architect or engineer.

C401.2.1 Application to replacement fenestration products. Where some or all of an existing *fenestration* unit is replaced with a new *fenestration* product, including sash and glazing, the replacement *fenestration* unit shall meet the applicable requirements for *U-factor* and *SHGC* in Table C402.4.

> **Exception:** An area-weighted average of the *U-factor* of replacement fenestration products being installed in the building for each fenestration product category listed in Table C402.4 shall be permitted to satisfy the *U-factor* requirements for each fenestration product category listed in Table C402.4. Individual fenestration products from different product categories listed in Table C402.4 shall not be combined in calculating the area-weighted average *U-factor*.

❖ For existing buildings where an existing fenestration unit is being replaced, the code requires that the new fenestration unit comply with the requirements for new construction found in Table 402.4. Many existing buildings were constructed under less stringent, or prior to, any energy conservation requirements. Upgrading windows at the time of replacement is an easy way to improve a building's efficiency. Chapter 5 [CE] provides additional requirements for existing buildings. The exceptions for windows in Chapter 5 address situations other than the replacement of a fenestration unit. The exception in this section provides flexibility for installing windows of each category. For example, while Table C402.4 would require operable fenestration products in a building in Climate Zone 3 to have a maximum *U*-factor of 0.60, windows of higher *U*-factors could be installed in some window openings provided other operable windows had a lower *U*-factor. The key is that the area of all the fenestration of that category (fixed vs. operable vs. entrance doors) average at or below the maximum *U*-factor for that zone.

SECTION C402 BUILDING ENVELOPE REQUIREMENTS

C402.1 General (Prescriptive). Building thermal envelope assemblies for buildings that are intended to comply with the code on a prescriptive basis, in accordance with the compliance path described in Item 2 of Section C401.2, shall comply with the following:

1. The opaque portions of the building thermal envelope shall comply with the specific insulation requirements of Section C402.2 and the thermal requirements of either the *R*-value-based method of Section C402.1.3; the *U*-, *C*- and *F*-factor-based method of Section C402.1.4; or the component performance alternative of Section C402.1.5.
2. Roof solar reflectance and thermal emittance shall comply with Section C402.3.

3. Fenestration in building envelope assemblies shall comply with Section C402.4.
4. Air leakage of building envelope assemblies shall comply with Section C402.5.

Alternatively, where buildings have a vertical fenestration area or skylight area exceeding that allowed in Section C402.4, the building and building thermal envelope shall comply with Section C401.2, Item 1 or Section C401.2, Item 3.

Walk-in coolers, walk-in freezers, refrigerated warehouse coolers and refrigerated warehouse freezers shall comply with Section C403.2.15 or C403.2.16.

❖ The building envelope is important to building energy efficiency. When it is cold outside, heat loss and air leakages through the building envelope add to the heating load. On hot days, solar gains through windows and infiltration of hot or humid air contribute to the air-conditioning (cooling) load. The building envelope requirements of Section C402 are intended to reduce heat gains and losses through the building envelope.

This section of the code provides acceptable methods for compliance evaluation for the insulation building envelope components including fenestration in that envelope [see Commentary Figure C402.1].

Determining the right amount and type of fenestration, and optimizing the levels of insulation are components of a delicate process that depends on climate, occupancy, schedules of operation, internal gains and other factors. The code sets minimum levels of thermal performance for all components of the building envelope and limits fenestration solar gain. While these limits ensure a minimum level of performance, they do not necessarily result in an optimum design. The designer is encouraged to use the code as a starting point; minimum compliance may not be the optimum solution.

The prescriptive building envelope requirements apply only to buildings that are conditioned (heating or cooling). Understanding the occupancy requirements for the commercial building is necessary in using the correct values for the *U*-factor, *C*-factor or *F*-factor of the building envelope requirements listed

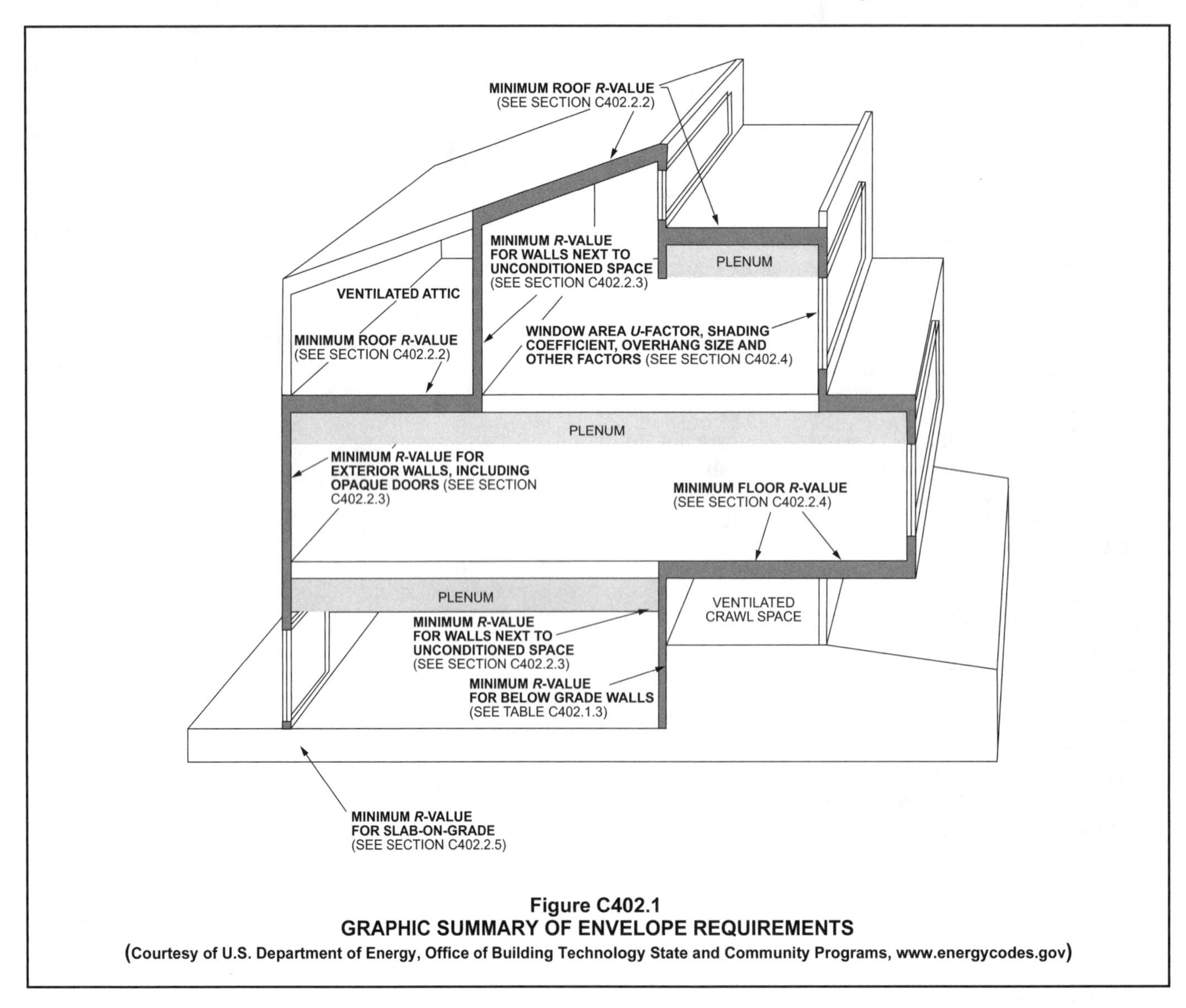

Figure C402.1
GRAPHIC SUMMARY OF ENVELOPE REQUIREMENTS
(Courtesy of U.S. Department of Energy, Office of Building Technology State and Community Programs, www.energycodes.gov)

in Tables C402.1.3 and C402.1.4. The definition of "Conditioned space" is important (see Chapter 2 [CE]). Notice that, based on Sections C402.1.1 and C402.1.2, certain buildings where there is no, or limited, energy used for space conditioning are exempt from the building envelope requirements.

Because the thermal envelope requirements do not apply to buildings that are neither heated nor cooled (Section C402.1.1), shell buildings may appear to be a special problem. Developers may try to obtain construction permits for a building before it is known how they will be used or even whether they will be heated or cooled. Although the code official has discretion in dealing with shell buildings, a common approach has been to postpone compliance until a subsequent permit application is filed where a use is established. For conditioned spaces, the envelope tables offer only two compliance options—Group R and all other. It is rare that a shell building is later made into a Group R occupancy; and since few buildings qualify as unconditioned, the prudent approach to shell buildings would be the installation of appropriate insulation. Further, it must be noted that Section C503.2 specifies that when a building is changed from nonconditioned to conditioned it must be brought into full compliance with the code. Again this suggests that the prudent course of action on a shell building is to assume it will be conditioned and therefore should be designed and built accordingly.

The section calls out the four key elements of successful building envelope design and construction: insulation, roofing, fenestration and air leakage (infiltration).

This section notes that if the designer wishes to exceed the fenestration limits of Section C402.4, then one of the other compliance paths identified in Section C401.2 must be used. Although not specifically stated, this is essentially the case where a designer wishes to exceed any limits of Section C402 beyond the alternatives provided in Section C402.

The final paragraph provides the code user a reference to the unique envelope provisions that apply to walk-in coolers, walk-in freezers, refrigerated warehouse coolers and refrigerated warehouse freezers.

C402.1.1 Low-energy buildings. The following low-energy buildings, or portions thereof separated from the remainder of the building by *building thermal envelope* assemblies complying with this section, shall be exempt from the *building thermal envelope* provisions of Section C402.

1. Those with a peak design rate of energy usage less than 3.4 Btu/h · ft^2 (10.7 W/m^2) or 1.0 watt per square foot (10.7 W/m^2) of floor area for space conditioning purposes.
2. Those that do not contain *conditioned space*.
3. Greenhouses.

❖ This section stipulates three conditions that permit a building or structure or portion of a building or structure to be exempt from the building envelope provisions of the code based on the marginal energy savings potential of such low-energy-use structures. These buildings are only exempted from the envelope provisions and must still be in compliance with the HVAC, service water heating, electrical and lighting systems provisions found in the balance of Chapter 4 [CE]. Section C402.1.2 provides another exception.

Item 1 exempts buildings and portions of buildings with peak rates of energy use below 3.4 Btu/h·ft^2 or 1.0 W/ft^2 (10.7 W/m^2). The phraseology, "a peak design rate of energy usage for space conditioning purposes," refers to the total peak primary energy used for space conditioning. There is no distinction in the code regarding where the energy is derived from, just a limit on usage. The usage is for space heating or cooling. There are no limits placed in such buildings for the energy used for lighting or service water heating other than those found in the balance of the chapter.

Few buildings designed for human occupancy will qualify for this exemption. If an exemption is claimed for a building, the permittee should provide enough supporting documentation to validate the claim. The peak rating of an appliance or piece of equipment can be determined by its nameplate rating or the manufacturer's literature.

A potential problem can exist when claiming this exemption. It is easy to modify the use of a structure after occupancy without obtaining an additional permit, thereby producing a noncomplying structure. Section C503.2 specifically requires where a "low-energy" building is changed to a fully conditioned building that the building must be brought into compliance. Although Section C503.2 will help regulate this problem, some building departments require a signed statement indicating that the permittee has claimed the exemption and that the structure will be brought into compliance with the code if its peak rate of energy use is raised above the maximum at any time thereafter.

Portions of buildings can also qualify for this exemption. Where a portion of a building meets the criteria for this exemption, that portion of the building is not required to comply with the envelope requirements. Other portions of the building, including the construction assemblies separating conditioned and unconditioned portions, define the limits of the building envelope that must meet the code requirements.

Item 2 indicates that the thermal envelope requirements of the code do not apply to buildings or portions of buildings that are neither heated nor cooled to create a "conditioned space." For a room or portion of a building to be considered neither heated nor cooled, the space must not contain any of the following:

1. A space-conditioning system designed to serve that space.

2. A space-conditioning register/diffuser or hydronic terminal unit serving the space.
3. An uninsulated duct or pipe where those items are normally required to be insulated.

In the past, the code only considered a space as being "conditioned" when it was being heated or cooled to keep the temperature within the human comfort range; however, based on the definition for "Conditioned space," even a space that is heated only to a level to prevent the freezing of the contents would still be considered a conditioned space and therefore unable to use Item 2. The space must also be physically separated from conditioned spaces by the building's thermal envelope.

Item 3 exempts greenhouses from the thermal envelope provisions of Chapter 4 [CE]. Greenhouses are specifically defined in Chapter 2 as structures or thermally isolated portions of a structure that maintain a specialized, sunlit environment exclusively used for, and essential to, the cultivation, protection and maintenance of plants. These buildings are not intended for human occupancy, but clearly people will be in such structures for "support" of the plants. Greenhouses are distinct from Items 1 and 2 in that there will be space conditioning, but that conditioning is aimed at the plants. The energy input of a greenhouse will likely exceed the limits of Item 1. Greenhouses are found in many areas for the cultivation of plants, including food plants. Greenhouses can be found at universities and other research facilities. Greenhouses are often found at botanical gardens where a controlled environment is needed to support plants that would not survive in the environment of the greenhouse's location. A greenhouse is not intended to be a sunroom in which you place a few potted plants.

C402.1.2 Equipment buildings. Buildings that comply with the following shall be exempt from the *building thermal envelope* provisions of this code:

1. Are separate buildings with floor area not more than 500 square feet (50 m^2).
2. Are intended to house electronic equipment with installed equipment power totaling not less than 7 watts per square foot (75 W/m^2) and not intended for human occupancy.
3. Have a heating system capacity not greater than (17,000 Btu/hr) (5 kW) and a heating thermostat set point that is restricted to not more than 50°F (10°C).
4. Have an average wall and roof *U*-factor less than 0.200 in *Climate Zones* 1 through 5 and less than 0.120 in *Climate Zones* 6 through 8.
5. Comply with the roof solar reflectance and thermal emittance provisions for *Climate Zone* 1.

❖ Section C402.1.2 is similar to Section C402.1.1 in that it allows buildings to be exempt from the standard envelope provisions of the code provided they meet the five criteria found here. Equipment buildings, shelters, or sheds are installed to protect electronic equipment from the weather and primarily provide cooling conditioning. Heating is installed for emergency backup operation and must be limited to 50°F or less by a setpoint. Due to the high density of electronic equipment installed, heat is rarely needed. The heat generated by the equipment results in cooling being the key issue; therefore, providing less insulation is actually desirable from an annual energy use standpoint.

This exemption is limited to stand-alone-equipment buildings no more than 500 square feet in area. Simplified insulation requirements that apply to an average of the roof and wall insulation are provided. This type of building is often made with 3 inches of concrete, internal foam insulation, and a plywood interior with similar construction for roof and walls. To reduce insulation requirements, the ASHRAE 90.1 option may be pursued as the building would qualify as a semiheated space. At around 7 watts per square foot of equipment load, the heat loss is offset by the equipment load, with the proposed insulation resulting in very little heating load.

It is important to note that this exemption applies only to the building thermal envelope provisions. Any HVAC, service water-heating or lighting systems in such buildings would still be required to meet the provisions of the code.

C402.1.3 Insulation component *R*-value-based method. *Building thermal envelope* opaque assemblies shall meet the requirements of Sections C402.2 and C402.4 based on the *climate zone* specified in Chapter 3. For opaque portions of the *building thermal envelope* intended to comply on an insulation component *R-value* basis, the *R*-values for insulation in framing cavities, where required, and for continuous insulation, where required, shall be not less than that specified in Table C402.1.3, based on the *climate zone* specified in Chapter 3. Commercial buildings or portions of commercial buildings enclosing Group R occupancies shall use the *R*-values from the "Group R" column of Table C402.1.3. Commercial buildings or portions of commercial buildings enclosing occupancies other than Group R shall use the *R*-values from the "All other" column of Table C402.1.3. The thermal resistance or *R*-value of the insulating material installed continuously within or on the below-grade exterior walls of the building envelope required in accordance with Table C402.1.3 shall extend to a depth of not less than 10 feet (3048 mm) below the outside finished ground level, or to the level of the lowest floor of the conditioned space enclosed by the below grade wall, whichever is less. Opaque swinging doors shall comply with Table C402.1.4 and opaque nonswinging doors shall comply with Table C402.1.3.

❖ The code provides three methods to determine compliance of the building thermal envelope. This section outlines the compliance method based on the provision of insulation with *R*-values not less than that specified in Table C402.1.3. The opaque portions of the envelope (walls, roofs, floors and opaque roll-up or sliding doors) must comply with the requirements of Section 402.2 and Table C402.1.3, while the fenestration (windows, glazed wall systems and skylights) must comply with Table C402.4. Commercial buildings or portions of commercial buildings enclosing Group R occupancies must use the *R*-values from the "Group R" column. Otherwise the "All other" column must be used for the *R*-values to determine the correct building thermal envelope. According to the *International Building Code*® (IBC®), Residential Group R includes the use of a building or structure, or a portion thereof, for sleeping purposes when not classified as an Institutional Group I or when not regulated by the *International Residential Code*® (IRC®). Group R includes the use of a building or structure, or a portion thereof, including Groups R-1, R-2, R-3 and R-4 as defined in the IBC.

Some of the envelope requirements are based on surface area. For instance, glazing area is limited to a percentage of the gross exterior wall area above grade and skylight area is limited to a percentage of the gross roof area. The rules for measuring surface areas are summarized in Commentary Figure C402.1.3. See Section C402.4 for additional information regarding fenestration.

The *R*-value of below-grade wall insulation must be equal to or greater than the criteria from Table C402.1.3, and must extend from the top of the foundation wall to a depth 10 feet (3048 mm) below the outside finished ground level or to the level of the basement floor, whichever is less [see Commentary Figure C402.1.3(1)]. This minimum level of insulation is required for the entire wall, including any portions that may be above grade. When the insulation is provided on the exterior side of the wall, the provisions of Section C303.2.1 should be reviewed.

C402.1.4 Assembly *U*-factor, *C*-factor or *F*-factor-based method. Building thermal envelope opaque assemblies intended to comply on an assembly *U*-, *C*- or *F*-factor basis shall have a *U*-, *C*- or *F*-factor not greater than that specified in Table C402.1.4. Commercial buildings or portions of commercial buildings enclosing Group R occupancies shall use the *U*-, *C*- or *F*-factor from the "Group R" column of Table C402.1.4. Commercial buildings or portions of commercial buildings enclosing occupancies other than Group R shall use the *U*-, *C*- or *F*-factor from the "All other" column of Table C402.1.4. The *C*-factor for the below-grade exterior walls of the building envelope, as required in accordance with Table C402.1.4, shall extend to a depth of 10 feet (3048 mm) below the outside finished ground level, or to the level of the lowest

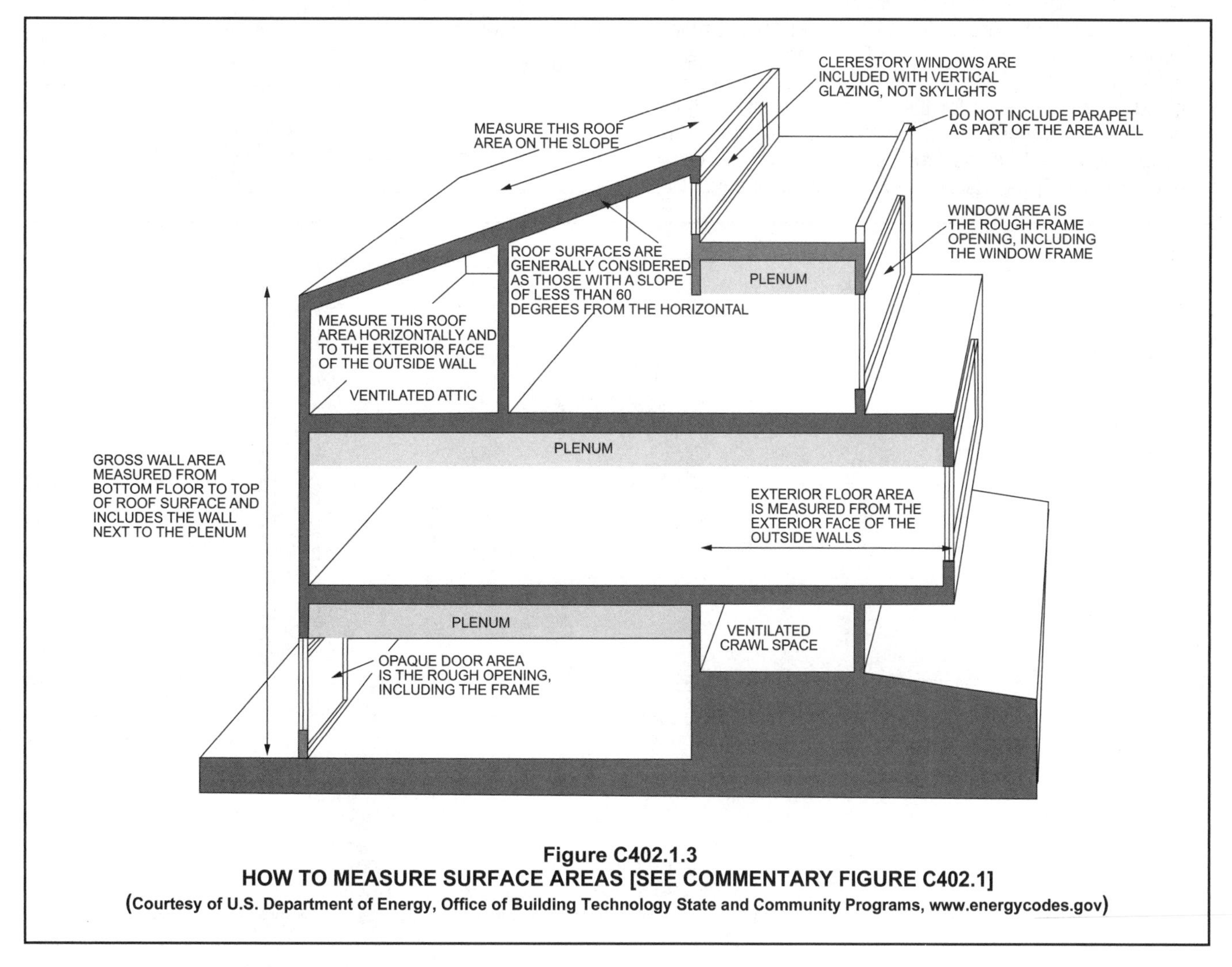

Figure C402.1.3
HOW TO MEASURE SURFACE AREAS [SEE COMMENTARY FIGURE C402.1]
(Courtesy of U.S. Department of Energy, Office of Building Technology State and Community Programs, www.energycodes.gov)

floor, whichever is less. Opaque swinging doors shall comply with Table C402.1.4 and opaque nonswinging doors shall comply with Table C402.1.3.

❖ This section allows an assembly with a *U*-factor, *C*-factor or *F*-factor equal to or less than that specified in Table C402.1.4 to be used in the design and construction instead of the *R*-values in Table C402.1.3. However, commercial buildings or portions of commercial buildings containing a Group R occupancy must use the *U*-factor, *C*-factor or *F*-factor from the "Group R" column. Group R includes the use of a building or structure, or a portion thereof, including Groups R-1, R-2, R-3 and R-4 as defined in the IBC. This will ensure the correct building thermal envelope will be determined.

Whereas the *R*-value table values refer to the *R*-values of the insulation materials alone, the *U*-factor table values refer to the *U*-values for the entire assembly, including air films and finish materials. Lower *U*-values are better in that they indicate reduced heat flow, while higher *R*-values are better in that they indicate reduced *resistance* to heat flow. Similarly, lower *C*-factors and *F*-factors indicate reduced heat flow through basement walls and the edges of slabs on grade, respectively.

Commentary Figure C402.1.4 provides an example of *U*-factor method. Shown in the figure are the *R*-values for the cavity and framing. The *U*-factor is 1/*R*-value. To calculate the *U*-factor for the combination of the cavity and framing combined, calculate the *U*-factor of the assembly based on the framing composing 15 percent of the assembly. An assembly with a *U*-factor equal to or less than that calculated in Table C402.1.4 is permitted in this method. The estimated framing is 15 percent of the wall area. The *U*-value may be calculated as follows:

$$U = \frac{0.15}{5.01} + \frac{1 - 0.15}{14.54} = 0.088$$

Section C402.1 also requires compliance with Section C402.2. Not every section of Section C402.2 directly addresses the *U*-factor approach to envelope compliance.

	CAVITY	FRAMING
	R-value	*R*-value
Outside air film	0.17	0.17
$^7/_8$-inch stucco	0.18	0.18
Building paper	0.06	0.06
Cavity insulation	13.00	
Framing		3.47
$^5/_8$-inch gypsum board	0.45	0.45
Inside air film	0.68	0.68
Sum of thermal resistance	14.54	5.01

Figure C402.1.4
EXAMPLE OF *U*-FACTOR METHOD

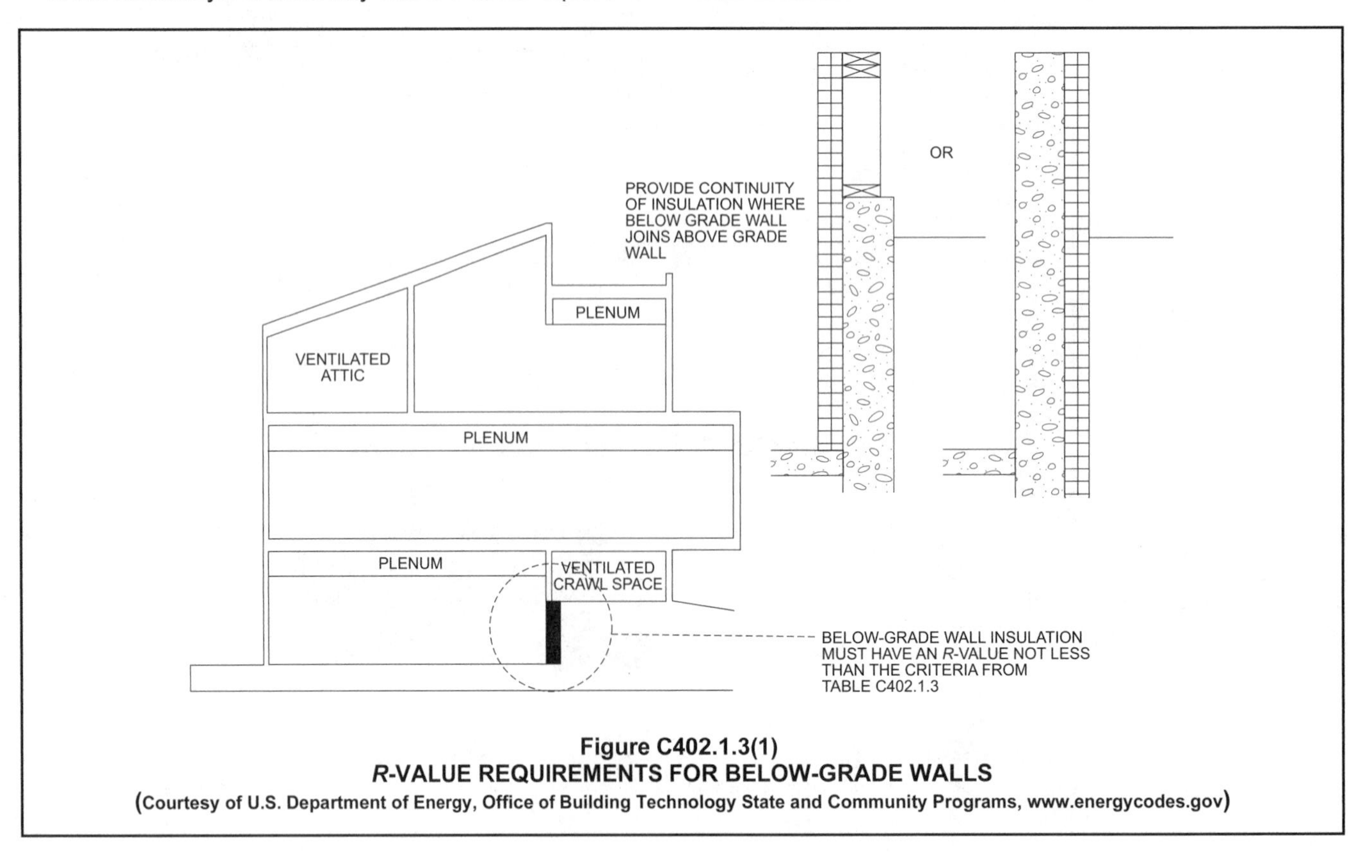

Figure C402.1.3(1)
***R*-VALUE REQUIREMENTS FOR BELOW-GRADE WALLS**
(Courtesy of U.S. Department of Energy, Office of Building Technology State and Community Programs, www.energycodes.gov)

TABLE C402.1.3
OPAQUE THERMAL ENVELOPE INSULATION COMPONENT MINIMUM REQUIREMENTS, *R*-VALUE METHOD[a]

CLIMATE ZONE	1		2		3		4 EXCEPT MARINE		5 AND MARINE 4		6		7		8	
	All other	Group R	All other	Group R	All other	Group R	All other	Group R	All other	Group R	All other	Group R	All other	Group R	All other	Group R
Roofs																
Insulation entirely above roof deck	R-20ci	R-25ci	R-25ci	R-25ci	R-25ci	R-25ci	R-30ci	R-30ci	R-30ci	R-30ci	R-30ci	R-30ci	R-35ci	R-35ci	R-35ci	R-35ci
Metal buildings[b]	R-19 + R-11 LS	R-19 + R-11 LS	R-19 + R11 LS	R-19 + R-11 LS	R-19 + R-11 LS	R-19 + R-11 LS	R-19 + R-11 LS	R-19 + R-11 LS	R-19 + R-11 LS	R-19 + R-11 LS	R-25 + R-11 LS	R-25 + R-11 LS	R-30 + R-11 LS	R-30 + R-11 LS	R-30 + R-11 LS	R-30 + R-11 LS
Attic and other	R-38	R-38	R-38	R-38	R-38	R-38	R-38	R-38	R-38	R-49	R-49	R-49	R-49	R-49	R-49	R-49
Walls, above grade																
Mass	R-5.7ci[c]	R-5.7ci[c]	R-5.7ci[c]	R-7.6ci	R-7.6ci	R-9.5ci	R-9.5ci	R-11.4ci	R-11.4ci	R-13.3ci	R-13.3ci	R-15.2ci	R-15.2ci	R-15.2ci	R-25ci	R-25ci
Metal building	R-13+ R-6.5ci	R-13 + R-6.5ci	R13 + R-6.5ci	R-13 + R-13ci	R-13 + R-6.5ci	R-13 + R-13ci	R-13 + R-13ci	R-13 + R-13ci	R-13 + R-13ci	R-13 + R-13ci	R-13 + R-13ci	R-13 + R-13ci	R-13 + R-13ci	R-13+ R-19.5ci	R-13 + R-13ci	R-13+ R-19.5ci
Metal framed	R-13 + R-5ci	R-13 + R-5ci	R-13 + R-5ci	R-13 + R-7.5ci	R-13 + R-7.5ci	R-13 + R-7.5ci	R-13 + R-7.5ci	R-13 + R-7.5ci	R-13 + R-7.5ci	R-13 + R-7.5ci	R-13 + R-7.5ci	R-13 + R-7.5ci	R-13 + R-7.5ci	R-13 + R-15.6ci	R-13 + R-7.5ci	R-13+ R17.5ci
Wood framed and other	R-13 + R-3.8ci or R-20	R-13 + R-3.8ci or R-20	R-13 + R-3.8ci or R-20	R-13 + R-3.8ci or R-20	R-13 + R-3.8ci or R-20	R-13 + R-3.8ci or R-20	R-13 + R-3.8ci or R-20	R-13 + R-3.8ci or R-20	R-13 + R-3.8ci or R-20	R-13 + R-7.5ci or R-20 + R-3.8ci	R-13 + R-7.5ci or R-20 + R-3.8ci	R-13 + R-7.5ci or R-20 + R-3.8ci	R-13 + R-7.5ci or R-20 + R-3.8ci	R-13 + R-7.5ci or R-20 + R-3.8ci	R13 + R-15.6ci or R-20 + R-10ci	R13 + R-15.6ci or R-20 + R-10ci
Walls, below grade																
Below-grade wall[d]	NR	NR	NR	NR	NR	NR	R-7.5ci	R-7.5ci	R-7.5ci	R-7.5ci	R-7.5ci	R-7.5ci	R-10ci	R-10ci	R-10ci	R-12.5ci
Floors																
Mass[e]	NR	NR	R-6.3ci	R-8.3ci	R-10ci	R-10ci	R-10ci	R-10.4ci	R-10ci	R-12.5ci	R-12.5ci	R-12.5ci	R-15ci	R-16.7ci	R-15ci	R-16.7ci
Joist/framing	NR	NR	R-30	R-30	R-30	R-30	R-30	R-30	R-30	R-30	R-30	R-30[f]	R-30[f]	R-30[f]	R-30[f]	R-30[f]
Slab-on-grade floors																
Unheated slabs	NR	NR	NR	NR	NR	NR	R-10 for 24″ below	R-10 for 24″ below	R-10 for 24″ below	R-10 for 24″ below	R-10 for 24″ below	R-15 for 24″ below	R-15 for 24″ below	R-15 for 24″ below	R-15 for 24″ below	R-20 for 24″ below
Heated slabs	R-7.5 for 12″ below	R-7.5 for 12″ below	R-7.5 for 12″ below	R-7.5 for 12″ below	R-10 for 24″ below	R-10 for 24″ below	R-15 for 24″ below	R-15 for 24″ below	R-15 for 36″ below	R-15 for 36″ below	R-15 for 36″ below	R-20 for 48″ below	R-20 for 24″ below	R-20 for 48″ below	R-20 for 48″ below	R-20 for 48″ below
Opaque doors																
Nonswinging	R-4.75	R-4.75	R-4.75	R-4.75	R-4.75	R-4.75	R-4.75	R-4.75	R-4.75	R-4.75	R-4.75	R-4.75	R-4.75	R-4.75	R-4.75	R-4.75

For SI: 1 inch = 25.4 mm, 1 pound per square foot = 4.88 kg/m^2, 1 pound per cubic foot = 16 kg/m^3.

ci = Continuous insulation, NR = No requirement, LS = Liner system.

a. Assembly descriptions can be found in ANSI/ASHRAE/IESNA Appendix A.

b. Where using *R*-value compliance method, a thermal spacer block shall be provided, otherwise use the *U*-factor compliance method in Table C402.1.4.

c. R-5.7ci is allowed to be substituted with concrete block walls complying with ASTM C 90, ungrouted or partially grouted at 32 inches or less on center vertically and 48 inches or less on center horizontally, with ungrouted cores filled with materials having a maximum thermal conductivity of 0.44 Btu-in/h-f^2 °F.

d. Where heated slabs are below grade, below-grade walls shall comply with the exterior insulation requirements for heated slabs.

e. "Mass floors" shall include floors weighing not less than:

1. 35 pounds per square foot of floor surface area; or
2. 25 pounds per square foot of floor surface area where the material weight is not more than 120 pounds per cubic foot.

f. Steel floor joist systems shall be insulated to R-38.

TABLE C402.1.4
OPAQUE THERMAL ENVELOPE ASSEMBLY MAXIMUM REQUIREMENTS, *U*-FACTOR METHOD[a, b]

CLIMATE ZONE	1		2		3		4 EXCEPT MARINE		5 AND MARINE 4		6		7		8	
	All other	Group R	All other	Group R	All other	Group R	All other	Group R	All other	Group R	All other	Group R	All other	Group R	All other	Group R
Roofs																
Insulation entirely above roof deck	U-0.048	U-0.039	U-0.039	U-0.039	U-0.039	U-0.039	U-0.032	U-0.032	U-0.032	U-0.032	U-0.032	U-0.032	U-0.028	U-0.028	U-0.028	U-0.028
Metal buildings	U-0.044	U-0.035	U-0.035	U-0.035	U-0.035	U-0.035	U-0.035	U-0.035	U-0.035	U-0.035	U-0.031	U-0.031	U-0.029	U-0.029	U-0.029	U-0.029
Attic and other	U-0.027	U-0.027	U-0.027	U-0.027	U-0.027	U-0.027	U-0.027	U-0.027	U-0.027	U-0.021	U-0.021	U-0.021	U-0.021	U-0.021	U-0.021	U-0.021
Walls, above grade																
Mass	U-0.151	U-0.151	U-0.151	U-0.123	U-0.123	U-0.104	U-0.104	U-0.090	U-0.090	U-0.080	U-0.080	U-0.071	U-0.071	U-0.061	U-0.061	U-0.061
Metal building	U-0.079	U-0.079	U-0.079	U-0.079	U-0.079	U-0.052	U-0.052	U-0.052	U-0.052	U-0.052	U-0.052	U-0.052	U-0.052	U-0.039	U-0.052	U-0.039
Metal framed	U-0.077	U-0.077	U-0.077	U-0.064	U-0.064	U-0.064	U-0.064	U-0.064	U-0.064	U-0.064	U-0.064	U-0.057	U-0.064	U-0.052	U-0.045	U-0.045
Wood framed and other[c]	U-0.064	U-0.064	U-0.064	U-0.064	U-0.064	U-0.064	U-0.064	U-0.064	U-0.064	U-0.064	U-0.051	U-0.051	U-0.051	U-0.051	U-0.036	U-0.036
Walls, below grade																
Below-grade wall[c]	C-1.140[e]	C-1.140[e]	C-1.140[e]	C-1.140[e]	C-1.140[e]	C-1.140[e]	C-0.119	C-0.119	C-0.119	C-0.119	C-0.119	C-0.119	C-0.092	C-0.092	C-0.092	C-0.092
Floors																
Mass[d]	U-0.322[e]	U-0.322[e]	U-0.107	U-0.087	U-0.076	U-0.076	U-0.076	U-0.074	U-0.074	U-0.064	U-0.064	U-0.057	U-0.055	U-0.051	U-0.055	U-0.051
Joist/framing	U-0.066[e]	U-0.066[e]	U-0.033	U-0.033	U-0.033	U-0.033	U-0.033	U-0.033	U-0.033	U-0.033	U-0.033	U-0.033	U-0.033	U-0.033	U-0.033	U-0.033
Slab-on-grade floors																
Unheated slabs	F-0.73[e]	F-0.73[e]	F-0.73[e]	F-0.73[e]	F-0.73[e]	F-0.73[e]	F-0.54	F-0.54	F-0.54	F-0.54	F-0.54	F-0.52	F-0.40	F-0.40	F-0.40	F-0.40
Heated slabs[f]	F-0.70	F-0.70	F-0.70	F-0.70	F-0.70	F-0.70	F-0.65	F-0.65	F-0.65	F-0.65	F-0.58	F-0.58	F-0.55	F-0.55	F-0.55	F-0.55
Opaque doors																
Swinging	U-0.61	U-0.61	U-0.61	U-0.61	U-0.61	U-0.61	U-0.61	U-0.61	U-0.37	U-0.37	U-0.37	U-0.37	U-0.37	U-0.37	U-0.37	U-0.37

For SI: 1 pound per square foot = 4.88 kg/m^2, 1 pound per cubic foot = 16 kg/m^3.

ci = Continuous insulation, NR = No requirement, LS = Liner system.

a. Use of Opaque assembly *U*-factors, *C*-factors, and *F*-factors from ANSI/ASHRAE/IESNA 90.1 Appendix A shall be permitted, provided the construction, excluding the cladding system on walls, complies with the appropriate construction details from ANSI/ASHRAE/ISNEA 90.1 Appendix A.

b. Opaque assembly *U*-factors based on designs tested in accordance with ASTM C 1363 shall be permitted. The *R*-value of continuous insulation shall be permitted to be added to or subtracted from the original tested design.

c. Where heated slabs are below grade, below-grade walls shall comply with the *F*-factor requirements for heated slabs.

d. "Mass floors" shall include floors weighing not less than:

1. 35 pounds per square foot of floor surface area; or
2. 25 pounds per square foot of floor surface area where the material weight is not more than 120 pounds per cubic foot.

e. These *C*-, *F*- and *U*-factors are based on assemblies that are not required to contain insulation.

f. Evidence of compliance with the *F*-factors indicated in the table for heated slabs shall be demonstrated by the application of the unheated slab *F*-factors and *R-values* derived from ASHRAE 90.1 Appendix A.

C402.1.4.1 Thermal resistance of cold-formed steel walls. U-factors of walls with cold-formed steel studs shall be permitted to be determined in accordance with Equation 4-1:

$$U = 1/[R_s + (ER)] \quad \textbf{(Equation 4-1)}$$

where:

R_s = The cumulative R-value of the wall components along the path of heat transfer, excluding the cavity insulation and steel studs.

ER = The effective R-value of the cavity insulation with steel studs.

❖ This section provides the calculation methodology as well as appropriate correction factors for determining the U-factors for steel-framed walls.

TABLE C402.1.4.1
EFFECTIVE *R*-VALUES FOR STEEL STUD WALL ASSEMBLIES

NOMINAL STUD DEPTH (inches)	SPACING OF FRAMING (inches)	CAVITY *R*-VALUE (insulation)	CORRECTION FACTOR (F_c)	EFFECTIVE *R*-VALUE (ER) (Cavity *R*-Value × F_c)
$3^1/_2$	16	13	0.46	5.98
		15	0.43	6.45
$3^1/_2$	24	13	0.55	7.15
		15	0.52	7.80
6	16	19	0.37	7.03
		21	0.35	7.35
6	24	19	0.45	8.55
		21	0.43	9.03
8	16	25	0.31	7.75
	24	25	0.38	9.50

❖ Table C402.1.4.1 provides a convenient means of estimating the "effective R-value" of steel studs combined with cavity insulation (the effective R-value would replace the "wood framing" and "cavity insulation" rows in the example U-value calculation above). This recognizes the fact that steel studs conduct a significant amount of heat energy through the insulation. As an example from the table, 6-inch steel studs spaced at 16 inches on center with R-21 insulation would result in an effective R-value of 7.35, a 65 percent reduction.

C402.1.5 Component performance alternative. Building envelope values and fenestration areas determined in accordance with Equation 4-2 shall be permitted in lieu of compliance with the U-, F- and C-factors in Tables C402.1.3 and C402.1.4 and the maximum allowable fenestration areas in Section C402.4.1.

$$A + B + C + D + E \leq \text{Zero} \quad \textbf{(Equation 4-2)}$$

where:

A = Sum of the (UA Dif) values for each distinct assembly type of the building thermal envelope, other than slabs on grade and below-grade walls.

UA Dif = UA Proposed - UA Table.

UA Proposed = Proposed U-value • Area.

UA Table = (U-factor from Table C402.1.3 or Table C402.1.4) • Area.

B = Sum of the (FL Dif) values for each distinct slab-on-grade perimeter condition of the building thermal envelope.

FL Dif = FL Proposed - FL Table.

FL Proposed = Proposed F-value • Perimeter length.

FL Table = (F-factor specified in Table C402.1.4) • Perimeter length.

C = Sum of the (CA Dif) values for each distinct below-grade wall assembly type of the building thermal envelope.

CA Dif = CA Proposed - CA Table

CA Proposed = Proposed C-value • Area.

CA Table = (Maximum allowable C-factor specified in Table C402.1.4) • Area.

Where the proposed vertical glazing area is less than or equal to the maximum vertical glazing area allowed by Section C402.4.1, the value of D (Excess Vertical Glazing Value) shall be zero. Otherwise:

D = (DA • UV) - (DA • U Wall), but not less than zero.

DA = (Proposed Vertical Glazing Area) - (Vertical Glazing Area allowed by Section C402.4.1).

UA Wall = Sum of the (UA Proposed) values for each opaque assembly of the exterior wall.

U Wall = Area-weighted average U-value of all above-grade wall assemblies.

UAV = Sum of the (UA Proposed) values for each vertical glazing assembly.

UV = UAV/total vertical glazing area.

Where the proposed skylight area is less than or equal to the skylight area allowed by Section C402.4.1, the value of E (Excess Skylight Value) shall be zero. Otherwise:

E = (EA • US) - (EA • U Roof), but not less than zero.

EA = (Proposed Skylight Area) - (Allowable Skylight Area as specified in Section C402.4.1).

U Roof = Area-weighted average U-value of all roof assemblies.

UAS = Sum of the (UA Proposed) values for each skylight assembly.

US = UAS/total skylight area.

❖ The component performance alternative provides an envelope U-value trade-off methodology for commercial buildings. It is similar to the "Total UA Alternative" for residential buildings but accounts for slab edge F-factors, basement wall C-factors, and any fenestration areas in excess of the code limits. It allows envelope heat loss to be calculated using a simple spreadsheet (see Commentary Figure C402.1.5) for projects that are not using COMcheck or a full-blown total building performance analysis.

For each envelope component (roof, mass wall,

framed wall, window, etc.), the user first multiplies the area by the proposed *U*-value, then multiplies the area by the table *U*-value, and finally subtracts the second value from the first. If the sum of all those "UA Dif" values is less than zero, the heat loss through the envelope will be no greater than the heat loss through a "code minimum" envelope. This provides design flexibility while maintaining limits on heat loss. For instance, larger window areas can be traded off against better roof insulation, or thinner slab edge insulation can be traded against thicker frame wall insulation.

Of the five factors in the formula—A, B, C, D and E—all projects will use factor A, which includes the *U*-values of most above-grade envelope components. They will likely also use either factor B or C, depending on whether they use slabs on grade or basement walls. Factors D and E will only be necessary for projects that have greater vertical glazing area or skylight area than allowed by code. Commentary Figure C402.1.5 demonstrates the calculations under the component performance alternative for an example building.

C402.2 Specific building thermal envelope insulation requirements (Prescriptive). Insulation in building thermal envelope opaque assemblies shall comply with Sections C402.2.1 through C402.2.6 and Table C402.1.3.

❖ Item 1 of Section C401.2 requires that all three building envelope compliance methods must include compliance with the requirements of Section C402.2. Many of the subsections to Section C402.2 speak specifically to the *R*-value method and therefore are not readily applicable to either the *U*-factor or component performance methods. Clearly for the *U*-factor approach (Section C402.1.4) or the component performance approach (Section C402.1.5) the use of the *R*-value (Table C402.1.3) would not be appropriate.

C402.2.1 Multiple layers of continuous insulation board. Where two or more layers of continuous insulation board are used in a construction assembly, the continuous insulation boards shall be installed in accordance with Section C303.2. Where the continuous insulation board manufacturer's instructions do not address installation of two or more layers, the edge joints between each layer of continuous insulation boards shall be staggered.

❖ Whether continuous insulation is used in specific compliance with Table C402.1.3 or in order to achieve the *U*-factors specified in Table C402.1.4 or the component performance alternative, this section applies to all multiple layer installations. Staggering helps the assembly resist air leakage.

C402.2.2 Roof assembly. The minimum thermal resistance (*R*-value) of the insulating material installed either between the roof framing or continuously on the roof assembly shall be as specified in Table C402.1.3, based on construction materials used in the roof assembly. Skylight curbs shall be insulated to the level of roofs with insulation entirely above deck or R-5, whichever is less.

Exceptions:

1. Continuously insulated roof assemblies where the thickness of insulation varies 1 inch (25 mm) or less and where the area-weighted *U*-factor is equivalent to the same assembly with the *R*-value specified in Table C402.1.3.
2. Where tapered insulation is used with insulation entirely above deck, the *R*-value where the insulation thickness varies 1 inch (25 mm) or less from the minimum thickness of tapered insulation shall comply with the *R*-value specified in Table C402.1.3.
3. Unit skylight curbs included as a component of a skylight listed and labeled in accordance with NFRC 100 shall not be required to be insulated.

Insulation installed on a suspended ceiling with removable ceiling tiles shall not be considered part of the minimum thermal resistance of the roof insulation.

❖ As discussed in the definition for "Roof assembly," roof surfaces are generally viewed as having an angle less than 60 degrees (1.1 rad) from the horizontal. The gross area of the roof includes all roof surfaces exposed to outside air or unconditioned (ventilated) space. Roof surfaces are measured from the exterior faces of exterior walls and the centerline of walls separating buildings, and should include all roof or ceiling components through which heat can flow, with the exception of service openings (see Commentary Figure C402.1.3). Service openings include roof hatches, duct penetrations and pipe penetrations. Skylights are not considered service openings and are addressed as a part of the fenestration items in Section C402.4.

The roof *R*-value must be greater than or equal to the value specified in Table C402.1.3 based on the insulation position and the construction materials. If the building has more than one type of roof construction, the *R*-value for each type of construction must be determined individually for comparison to the tabular criteria (see Commentary Figure C402.2.2). With the exception of skylights (see commentary, Section C402.4.1), service openings and roof penetrations need not be considered.

Because of the likelihood that the insulation would be displaced or not reinstalled correctly, the code prohibits considering the insulation on the upper side of removable ceiling tiles as a portion of the required insulation. Therefore, on suspended ceiling systems with removable tiles, any insulation installed at this location is not considered as contributing to the minimum required building thermal envelope provisions. This provision does not prohibit such insulation from being installed but simply does not give any credit for the insulation.

Exception 1 permits a roof that is "continuously

insulated" to have areas that do not meet the required *R*-values, provided that the area-weighted values are equivalent to the specified insulation values. Therefore, while one section may have less insulation due to this slope, other portions of the roof would be above the values required. Therefore, in this situation the weighted average of the insulation would meet the required values even though some portions may be less than that specified in Table C402.1.3. When applying the exception, it is important to notice that

	Example building: 2-story building with 10,000 SF each floor, 10,000 exterior wall area, 5,000 SF floor over parking, no basement walls, and 40% vertical glazing (instead of code max 30%). In this case, the extra glazing area is accommodated in the design by use of a triple-glazed curtain wall with a very low *U*-value.							
	Formula: (A + B + C + D + E ≤ Zero)							
		Area	**Proposed *U*-value**	**Proposed UA (U x Area)**	**Table *U*-factor**	**Table UA (U x Area)**	**UA Dif (Proposed UA - Table UA)**	**Totals**
	roof - insul above deck	10000	0.03	300	0.034	340	-40	
	wall 1 - mass wall	6000	0.09	540	0.078	468	72	
	wall 2 - steel stud	4000	0.055	220	0.055	220	0	
	floor - framed	5000	0.029	145	0.029	145	0	
	skylight	100	0.5	50	0.5	50	0	
	VG 1 - alum curtain wall	3000	0.22	660	0.38	1140	-480	
	VG 2 - wood framed	1000	0.3	300	0.3	300	0	
A	**Sum of the (UA Dif) values for envelope assemblies**						-448	**-448**
		Length of slab edge	**Proposed *F*-value**	**Proposed F x Length**	**Table *F*-factor**	**Table F x Length**	**FL Dif**	
	slab edge - perimeter	200	0.54	108	0.528	105.6	2.4	
	slab edge - at garage	100	0.62	62	0.528	52.8	9.2	
B	**Sum of the (FL Dif) values for both slab-on-grade perimeter conditions**						11.6	**11.6**
C	**(no basement walls in this design)**							**0**
	Uwall	0.076	= Area-weighted avg *U*-value of above-grade wall assemblies					
	UAV	960	= Sum of the (UA Proposed) values for each vertical glazing assembly					
	UV	0.24	= UAV / total vertical glazing area					
	DA	1000	= (Proposed VG Area) – (VG Area allowed by Section C402.3.1)					
	VGA	4000	= Proposed Vertical Glazing Area					
	Allow VG Area	3000	= 30% max from Section C402.3.1					
	Wall Area	10000	= Gross wall area					
	UA Wall	760	= Uwall x Wall Area)					
D	**Excess vert glazing area**	164	(DA x UVG) – (DA x UWall) - Zero if ≤ zero					**164**
E	**Excess skylight area**	(Proposed skylight area is less than allowable area, so value is zero)						**0**
Component Performance: (A + B + C + D + E) - OK since less than zero.								**-272**

Figure C402.1.5
COMPONENT PERFORMANCE EXAMPLE

the variation in insulation thickness is limited to 1 inch (25 mm). This limitation on the thickness variation will help ensure more consistent insulation coverage and also reduce the number of roofs that qualify to use this exception. This 1-inch (25 mm) limitation does not prevent the provisions from being applied to roofs that have a greater variation; it simply does not allow additional thickness to be factored into the average insulation values. Where the variation exceeds 1 inch (25 mm), it would be permissible to go to the thinnest spot and measure the *R*-value at that point (for the example, call this Point "a"). Then go to a point that is 1 inch (25 mm) thicker than Point "a" and measure the *R*-value there (for the example, call this Point "b"). The remaining portions of the roof that are thicker than that additional 1-inch (25 mm) portion (Point "b") would simply be assumed to have the same *R*-value that Point "b" had. All portions of the roof that meet or exceed the Point "b" *R*-value would simply use the Point "b" *R*-value when determining the area-weighted *U*-factor for the roof.

Exception 2 addresses another common design where the insulation is tapered in order to provide a slope for drainage. In the case of this exception, the taper is limited to a maximum of 1 inch and the required *R*-value is determined at the minimum thickness of the insulation.

Exception 3 exempts the curbs of unit skylights from the insulation requirements of this section provided the skylight is listed and labeled under the NFRC 100 standard.

C402.2.3 Thermal resistance of above-grade walls. The minimum thermal resistance (*R*-value) of materials installed in the wall cavity between framing members and continuously on the walls shall be as specified in Table C402.1.3, based on framing type and construction materials used in the wall assembly. The *R*-value of integral insulation installed in concrete masonry units shall not be used in determining compliance with Table C402.1.3.

"Mass walls" shall include walls:

1. Weighing not less than 35 psf (170 kg/m^2) of wall surface area.
2. Weighing not less than 25 psf (120 kg/m^2) of wall surface area where the material weight is not more than 120 pcf (1900 kg/m^3).
3. Having a heat capacity exceeding 7 Btu/ft^2 · °F (144 kJ/m^2 · K).
4. Having a heat capacity exceeding 5 Btu/ft^2 · °F (103 kJ/m^2 · K), where the material weight is not more than 120 pcf (1900 kg/m^3).

❖ Table C402.1.3 contains the prescriptive criteria for walls based on framing type, insulation position and construction materials used in the wall assembly.

Above-grade walls, as defined in Chapter 2, include all opaque walls exposed to the outside air or forming the building thermal envelope (see Commentary Figure C402.2.3). Walls adjacent to attic spaces and crawl spaces need to be evaluated.

This section also provides two other requirements that play an important part in determining the appropriate insulation requirements for the walls forming

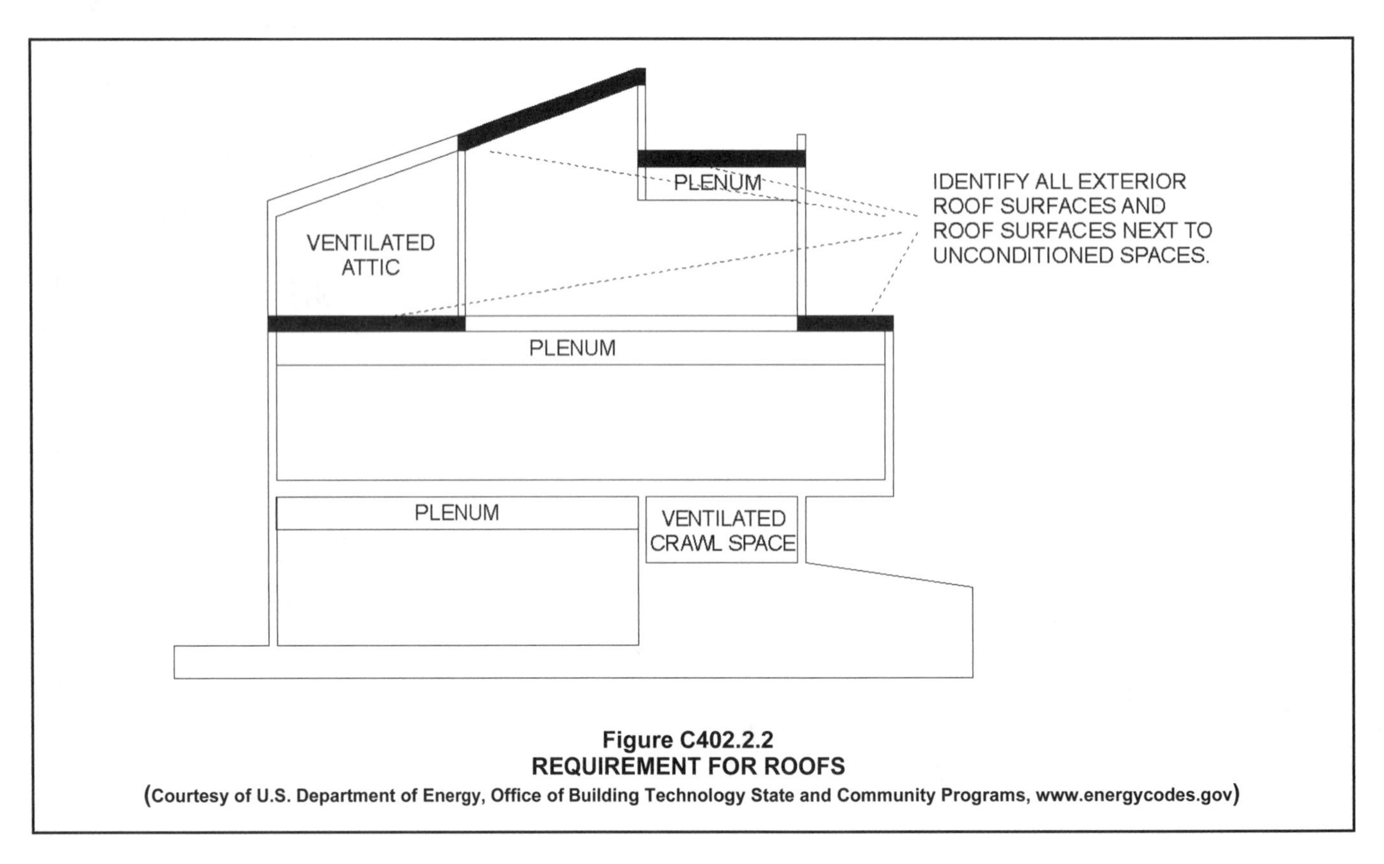

Figure C402.2.2
REQUIREMENT FOR ROOFS
(Courtesy of U.S. Department of Energy, Office of Building Technology State and Community Programs, www.energycodes.gov)

the thermal barrier envelope. The first is the requirement that insulation within the cells of concrete masonry units (CMUs) are not permitted to be considered when applying the prescriptive insulation requirements of Table C402.1.3. Therefore, a CMU wall would either need to have the level of insulation required in the "wood framed and other" row of the table, or it would need to meet the "mass" wall provisions if appropriate. This requirement essentially mandates that the insulation be provided either on the exterior or interior side of the CMU wall or, in the case of a multiwythe wall, between the wythes of masonry material. The second issue is that the provisions designate what is considered a "mass" wall. Because of the thermal lag that exists with a mass wall and the fact that the code requires continuous insulation (ci), the insulation level of a mass wall generally will be reduced from the level of insulation required for other above-grade walls in moderate climate zones.

The area of the above-grade wall is used when applying the fenestration limits of Section C402.4.1 and Table C402.4. The gross area of above-grade walls is measured on the exterior and includes between-floor spandrels, peripheral edges of flooring, window areas (including sash) and doors.

C402.2.4 Floors. The thermal properties (component *R*-values or assembly *U*-, *C*- or *F*-factors) of floor assemblies over outdoor air or unconditioned space shall be as specified in Table C402.1.3 or C402.1.4 based on the construction materials used in the floor assembly. Floor framing cavity insulation or structural slab insulation shall be installed to maintain permanent contact with the underside of the subfloor decking or structural slabs.

Exceptions:

1. The floor framing cavity insulation or structural slab insulation shall be permitted to be in contact with the top side of sheathing or continuous insulation installed on the bottom side of floor assemblies where combined with insulation that meets or exceeds the minimum *R*-value in Table C402.1.3 for "Metal framed" or "Wood framed and other" values for "Walls, Above Grade" and extends from the bottom to the top of all perimeter floor framing or floor assembly members.
2. Insulation applied to the underside of concrete floor slabs shall be permitted an airspace of not more than 1 inch (25 mm) where it turns up and is in contact with the underside of the floor under walls associated with the *building thermal envelope*.

❖ This section requires floor insulation to meet the *R*-value or *U*-value requirements of the respective tables. Floor insulation is required to be in contact with the floor slab or subfloor decking above unless it meets the requirements of one of the exceptions noted. The exceptions were added to accommodate an airspace between the insulation and the floor above, which results in a warmer floor. The first exception applies to framed floors. It allows a gap above cavity insulation, provided that the bottom of the cavity insulation is in contact with continuous insulation or some form of "sheathing," often gypsum board. At the perimeter walls, the gap above the cav-

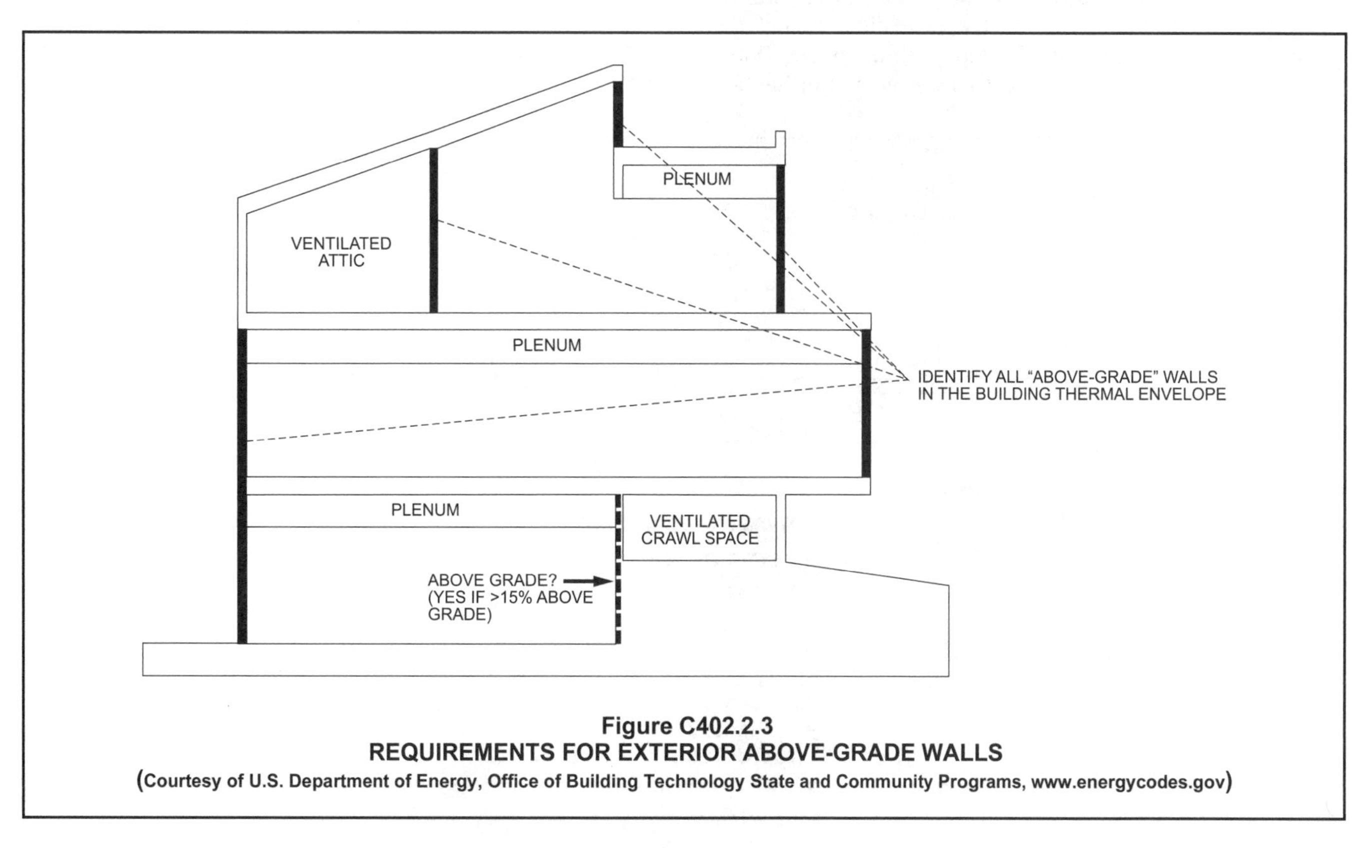

Figure C402.2.3
REQUIREMENTS FOR EXTERIOR ABOVE-GRADE WALLS
(Courtesy of U.S. Department of Energy, Office of Building Technology State and Community Programs, www.energycodes.gov)

ity insulation must be insulated as required for exterior walls. The second exception applies to concrete slab floors and similarly allows an air gap above the floor insulation, provided that the gap is insulated at the perimeter as required for exterior walls.

Tables C402.1.3 and C402.1.4 differentiate for the climate zones "All other" and "Group R" categories. As a result, the *U*-factor, *F*-factor and *R*-values differ in the tables and should be used accordingly to determine the correct building thermal envelope.

The *R*-value for floors over outdoor air and unconditioned space must be greater than or equal to the criteria in Table C402.1.3. The criterion applies to all floors over unconditioned spaces, including floors over unconditioned garages and crawl spaces (see Commentary Figure C402.2.4).

The footnotes to Tables C402.1.3 and C402.1.4 designate what is considered a "mass" floor so that the application of the tables is clear. Because of the thermal lag that exists with a mass floor and the fact that the code requires continuous insulation, the insulation level of a mass floor generally will be reduced from the level of insulation required for other floor assemblies.

The gross floor area over outdoor air or unconditioned space typically is measured from the exterior face of exterior walls and the centerline of walls separating buildings. The floor area must include all floor components through which heat may flow.

The exceptions provide alternate designs and installation criteria for floor insulation.

C402.2.5 Slabs-on-grade perimeter insulation. Where the slab on grade is in contact with the ground, the minimum thermal resistance (*R*-value) of the insulation around the perimeter of unheated or heated slab-on-grade floors designed in accordance with the *R*-value method of Section C402.1.3 shall be as specified in Table C402.1.3. The insulation shall be placed on the outside of the foundation or on the inside of the foundation wall. The insulation shall extend downward from the top of the slab for a minimum distance as shown in the table or to the top of the footing, whichever is less, or downward to at least the bottom of the slab and then horizontally to the interior or exterior for the total distance shown in the table. Insulation extending away from the building shall be protected by pavement or by not less than of 10 inches (254 mm) of soil.

Exception: Where the slab-on-grade floor is greater than 24 inches (61 mm) below the finished exterior grade, perimeter insulation is not required.

❖ Section C402.2.5 and Table C402.1.3 give the criteria for insulating slabs on grade serving a conditioned space. The installed *R*-value must be equal to or greater than the value specified in Table C402.1.3, which shows values for both unheated and heated slabs. Unheated slabs will only require insulation in the most severe climate zone. Commentary Figure C402.2.5 illustrates the requirements and shows acceptable installation methods.

When the insulation is installed inside the foundation slab wall, both horizontal and vertical dimensions may be added to achieve the total developed length required by the code. Critical to the installation is the thermal break required by the code, evidenced by installing the insulation from the top of the slab downward. The thermal break minimizes conductive losses through edge effects. The protection of exposed foundation insulation is required by Section C303.2.1.

C402.2.6 Insulation of radiant heating systems. *Radiant heating system* panels, and their associated components that are installed in interior or exterior assemblies shall be insulated with a minimum of R-3.5 (0.62 $m^2/K \cdot W$) on all sur-

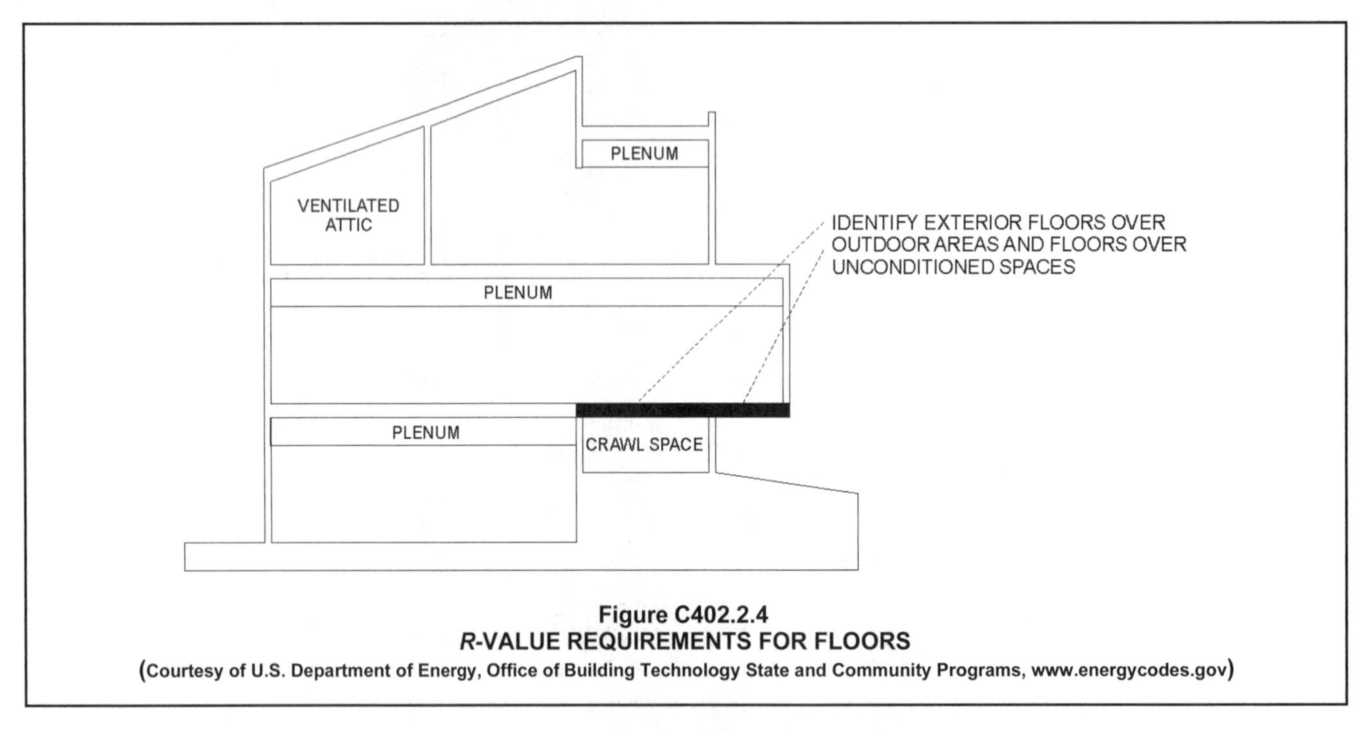

Figure C402.2.4
***R*-VALUE REQUIREMENTS FOR FLOORS**
(Courtesy of U.S. Department of Energy, Office of Building Technology State and Community Programs, www.energycodes.gov)

faces not facing the space being heated. *Radiant heating system* panels that are installed in the *building thermal envelope* shall be separated from the exterior of the building or unconditioned or exempt spaces by not less than the *R*-value of insulation installed in the opaque assembly in which they are installed or the assembly shall comply with Section C402.1.4.

Exception: Heated slabs on grade insulated in accordance with Section C402.2.5.

❖ This section requires radiant panels, and associated U-bends and headers, designed for heating an indoor space to be insulated with a minimum of R-3.5 (0.62 $m^2/K \cdot W$). Radiant panels should direct the heat where it is needed. For example, radiant panels in unconditioned warehouses are used to provide spot heating for workers below; however, any heat that radiates from the top of the panel toward the ceiling above is wasted. The code now clarifies that where such panels are installed within an assembly that is part of the building's thermal envelope, there must be insulation meeting the thermal envelope requirements behind the radiant panel.

C402.3 Roof solar reflectance and thermal emittance. Low-sloped roofs directly above cooled conditioned spaces in *Climate Zones* 1, 2 and 3 shall comply with one or more of the options in Table C402.3.

Exceptions: The following roofs and portions of roofs are exempt from the requirements of Table C402.3:

1. Portions of the roof that include or are covered by the following:
   1.1. Photovoltaic systems or components.
   1.2. Solar air or water-heating systems or components.
   1.3. Roof gardens or landscaped roofs.
   1.4. Above-roof decks or walkways.
   1.5. Skylights.
   1.6. HVAC systems and components, and other opaque objects mounted above the roof.
2. Portions of the roof shaded during the peak sun angle on the summer solstice by permanent features of the building or by permanent features of adjacent buildings.
3. Portions of roofs that are ballasted with a minimum stone ballast of 17 pounds per square foot [74 kg/m^2] or 23 psf [117 kg/m^2] pavers.

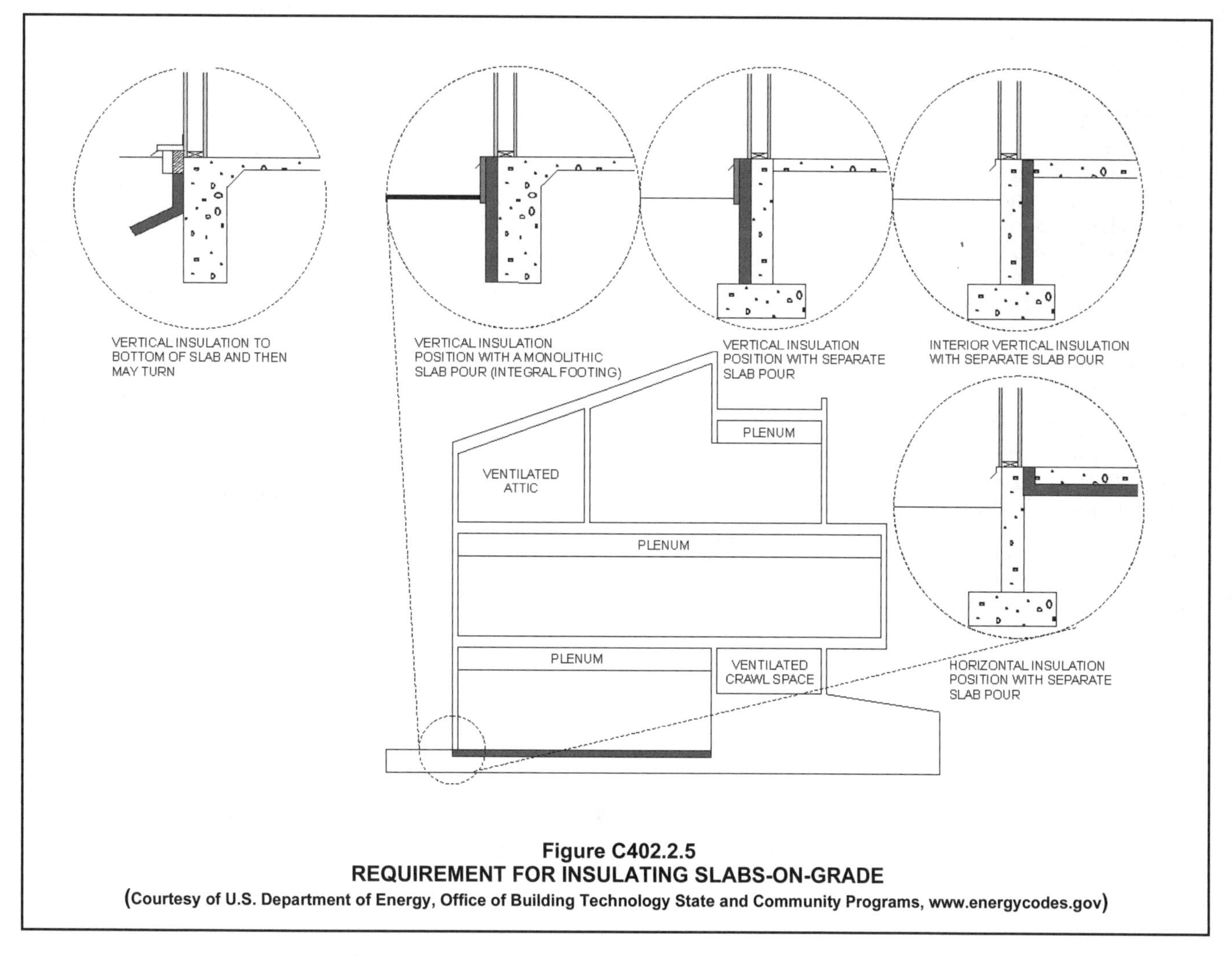

Figure C402.2.5
REQUIREMENT FOR INSULATING SLABS-ON-GRADE
(Courtesy of U.S. Department of Energy, Office of Building Technology State and Community Programs, www.energycodes.gov)

4. Roofs where not less than 75 percent of the roof area complies with one or more of the exceptions to this section.

❖ Cool roofs are important in reducing the cooling load in a building with low-slope roofs in Climate Zones 1, 2 and 3, where temperatures are often hot. A cool roof is a roofing system that can deliver high solar reflectance (the ability to reflect the visible, infrared and ultraviolet wavelengths of the sun, reducing the heat transfer to the building) and high thermal emittance (the ability to radiate absorbed or nonreflected solar energy). Cool roofs can reduce the building heat gain and cooling loads. They can help reduce the urban heat island effect. Cool roofs offer both immediate and long-term savings in building energy costs by enhancing the life expectancy of the roof membrane and building's cooling equipment by reducing the building's heat gain.

This section requires low-sloped roofs that are located directly above cooled, conditioned spaces in Climate Zones 1, 2 and 3 to meet one or more of the roof reflectance and emittance options in Table C402.3. See the definition of "Low-sloped roof" in Chapter 2 [CE]. This section provides exceptions to the requirement of solar reflectant materials. Exception 1 exempts those portions of roofs that have photovoltaic systems or components, solar air or water-heating systems, gardens or landscaping, skylights, HVAC system components and walkways and decks mounted on the roof. Exception 2 exempts portions of the roof shaded during the peak sun angle on the summer solstice by permanent features of the building, or by permanent features of adjacent buildings. Exception 3 exempts portions of roofs that are ballasted or have pavers of minimum densities. These materials have a similar impact on the building as do low-reflectance roofing materials. Exception 4 looks at the total roof area and exempts the total roof if 75 percent of the roof area is covered by the materials and/or installations found in Exceptions 1 through 3.

TABLE C402.3. See below.

❖ Section C402.3 requires roofs to comply with one of the options of Table C402.3. The options are either in compliance with the first line of the table or with the second line of the table. Each option specifies through the appropriate footnotes how products need to be tested to demonstrate the solar reflectance, thermal emittance and solar reflectance index, respectively. The footnotes provide options with respect to the testing standard. Footnote a provides default values if roofing products have not been tested for 3 years. Roofing products are to be labeled with the results of their testing.

C402.3.1 Aged roof solar reflectance. Where an aged solar reflectance required by Section C402.3 is not available, it shall be determined in accordance with Equation 4-3.

$$R_{aged} = [0.2+0.7(R_{initial}-0.2)] \quad \textbf{(Equation 4-3)}$$

where:

R_{aged} = The aged solar reflectance.

$R_{initial}$ = The initial solar reflectance determined in accordance with CRRC-1 Standard.

❖ Previous editions of the code allowed compliance to be based on either the initial product or an aged product. The code now requires compliance based on a 3-year-aged analysis of the products. As the testing process is obviously time consuming—not all products have had a completed 3-year evaluation—this section provides a formula to provide a default 3-year aged, factor-based testing of the initial product. Experience with testing products for the 3-year time period indicates Equation 4-3 provides a conservative value for products that have not been fully tested.

C402.4 Fenestration (Prescriptive). Fenestration shall comply with Sections C402.4 through C402.4.4 and Table C402.4. Daylight responsive controls shall comply with this section and Section C405.2.3.1.

❖ "Fenestration" basically includes all of the items in the building thermal envelope that are not regulated under Section C402.2. In general, windows, doors and skylights are considered fenestration (see definition and commentary for "Fenestration" in Chapter 2 [CE]). It is important to notice when applying the fenestration provisions to commercial structures that "opaque doors" (Section C402.4.4) are not regulated by Section C402.4 (see commentary, Sections C402.4.4, C402.1.3 and C402.1.4). Therefore, if a door has less than 50-percent glass area, it is not required to comply with Section C402.4 or Table C402.4. All other types of fenestration, such as skylights, windows, doors with 50 percent or more glass area and glazed block, are regulated by this section.

The amount of vertical fenestration in walls and skylights in roofs is dependent on the provision of

TABLE C402.3
MINIMUM ROOF REFLECTANCE AND EMITTANCE OPTIONS[a]

Options
Three-year aged solar reflectance[b] of 0.55 and 3-year aged thermal emittance[c] of 0.75
Three-year-aged solar reflectance index[d] of 64

a. The use of area-weighted averages to comply with these requirements shall be permitted. Materials lacking 3-year-aged tested values for either solar reflectance or thermal emittance shall be assigned both a 3-year-aged solar reflectance in accordance with Section C402.3.1 and a 3-year-aged thermal emittance of 0.90.

b. Aged solar reflectance tested in accordance with ASTM C 1549, ASTM E 903 or ASTM E 1918 or CRRC-1 Standard.

c. Aged thermal emittance tested in accordance with ASTM C 1371 or ASTM E 408 or CRRC-1 Standard.

d. Solar reflectance index (SRI) shall be determined in accordance with ASTM E 1980 using a convection coefficient of 2.1 Btu/h • ft^2 •°F (12W/m^2 • K). Calculation of aged SRI shall be based on aged tested values of solar reflectance and thermal emittance.

daylight responsive controls. The details for such controls are found in the lighting section of the code.

C402.4.1 Maximum area. The vertical fenestration area (not including opaque doors and opaque spandrel panels) shall not be greater than 30 percent of the gross above-grade wall area. The skylight area shall not be greater than 3 percent of the gross roof area.

❖ This section establishes that a maximum of 30 percent of the gross above-grade wall area of a building can be vertical fenestration and that a maximum of 3 percent of the gross roof area can be skylights. Because the thermal performance of fenestrations is lower than that required for the wall and roof under Sections C402.1 and C402.2, the code limits the amount of wall and roof that may be fenestration in order to minimize the overall reduced performance of the wall.

Daylight responsive controls are not required for the daylight areas associated with either the vertical fenestration or the skylights allowed under this section. Subsequent sections that allow for increases in these maximum areas do require daylight responsive controls. In addition, Section C402.4.2 sets a minimum amount of skylights above certain uses. Section C402.4.1.1 provides a method to allow a greater amount of vertical fenestration in some climate zones. Section C402.4.1.2 provides a method to allow a greater amount of skylights. Sections C402.4.3.1 and C402.4.3.2 allow increased SHGC and *U*-factors, respectively, in skylights. To use any of these greater allowances in a design, the code requires the installation of daylight responsive controls complying with Section 405.2.

Even with the expanded limits found in Sections C402.4.1.1 and C402.4.1.2, a designer may wish to provide larger areas of fenestration. Under the component performance alternative found in Section C402.1.5, an evaluation of the building envelope elements can be provided to see whether increased efficiency in other elements can be used to balance the greater window/skylight area. But where Section C402.1.5 doesn't provide a solution and the building design exceeds these area limitations, the prescriptive provisions cannot be used. The designer will not need to pursue either the total-building-performance requirements of Section C407 or the ANSI/ASHRAE/IESNA 90.1 to demonstrate compliance with the building envelope requirements (see Sections C401.2 and C402.1).

The gross area of exterior walls is the normal projection of all exterior walls, including the area of all windows and doors installed therein. The gross wall area is the area of the wall measured on the exterior face from the top of the lowest floor to the bottom of the roof. The gross wall area does not include semi-exterior walls or interior partitions. Generally, the gross wall area is based on the area of above-grade walls; therefore, the above-grade portion of below-grade walls and the area of fenestration in those portions typically are excluded when calculating the percentage of openings.

When determining the gross wall area, it is important to remember that the area should include not only the exterior exposed walls but also the areas of gable and dormer walls, band joists between floors, between-floor spandrels, peripheral edges of floors, roof and attic knee walls, skylight shafts and any other wall that forms the boundary of conditioned spaces (the building's thermal envelope). The wall between an unconditioned garage and a conditioned interior space should not be overlooked in calculating the gross wall area as it would also be part of the thermal envelope.

It is also important to remember that opaque doors (doors having less than 50 percent glass area) are regulated by Section C402.4.4. Opaque doors would not be included within the 30-percent limitation for fenestration.

TABLE C402.4
BUILDING ENVELOPE FENESTRATION MAXIMUM *U*-FACTOR AND SHGC REQUIREMENTS

CLIMATE ZONE	1		2		3		4 EXCEPT MARINE		5 AND MARINE 4		6		7		8	
Vertical fenestration																
***U*-factor**																
Fixed fenestration	0.50		0.50		0.46		0.38		0.38		0.36		0.29		0.29	
Operable fenestration	0.65		0.65		0.60		0.45		0.45		0.43		0.37		0.37	
Entrance doors	1.10		0.83		0.77		0.77		0.77		0.77		0.77		0.77	
SHGC																
Orientation[a]	SEW	N	SEW	N	SEW	N	SEW	N	SEW	N	SEW	N	SEW	N	SEW	N
PF < 0.2	0.25	0.33	0.25	0.33	0.25	0.33	0.40	0.53	0.40	0.53	0.40	0.53	0.45	NR	0.45	NR
0.2 ≤ PF < 0.5	0.30	0.37	0.30	0.37	0.30	0.37	0.48	0.58	0.48	0.58	0.48	0.58	NR	NR	NR	NR
PF ≥ 0.5	0.40	0.40	0.40	0.40	0.40	0.40	0.64	0.64	0.64	0.64	0.64	0.64	NR	NR	NR	NR
Skylights																
U-factor	0.75		0.65		0.55		0.50		0.50		0.50		0.50		0.50	
SHGC	0.35		0.35		0.35		0.40		0.40		0.40		NR		NR	

NR = No requirement, PF = Projection factor.

a. "N" indicates vertical fenestration oriented within 45 degrees of true north. "SEW" indicates orientations other than "N." For buildings in the southern hemisphere, reverse south and north. Buildings located at less than 23.5 degrees latitude shall use SEW for all orientations.

The fenestration area is the entire window and glazed door area, including the frame. For premanufactured windows and doors, this area must be considered as the rough frame opening.

Windows located in interior walls adjacent to unconditioned space would need to be included when determining the percentage of fenestration. Although these openings do not have a solar heat gain coefficient (SHGC) concern, they would still need to comply with the appropriate *U*-factor from Table C402.4 and would represent a reduction from the wall insulation levels required in Tables C402.1.3 and C402.1.4. Therefore, it is appropriate that their size be included in the 30-percent limit on the amount of fenestration openings.

C402.4.1.1 Increased vertical fenestration area with daylight responsive controls. In *Climate Zones* 1 through 6, not more than 40 percent of the gross above-grade wall area shall be permitted to be vertical fenestration, provided all of the following requirements are met:

1. In buildings not greater than two stories above grade, not less than 50 percent of the net floor area is within a *daylight zone*.
2. In buildings three or more stories above grade, not less than 25 percent of the net floor area is within a *daylight zone*.
3. *Daylight responsive controls* complying with Section C405.2.3.1 are installed in *daylight zones*.
4. Visible transmittance (VT) of vertical fenestration is not less than 1.1 times solar heat gain coefficient (SHGC).

Exception: Fenestration that is outside the scope of NFRC 200 is not required to comply with Item 4.

❖ This section establishes that a maximum of 40 percent of the gross wall area of a building can be vertical fenestration. Because the thermal performance of fenestrations is lower than that required for the wall and roof under Sections C402.1 and C402.2, the code limits the amount of wall and roof that may be fenestration in order to minimize the reduced efficiency. This section allows the increase where a minimum portion of the net floor area is in daylight zones. See the definition of "Floor area, net" in Chapter 2 [CE]. See the commentary to Section C402.4.1 regarding calculation of wall and fenestration area. Also see the same commentary for designs that would exceed even the 40 percent allowed by this section.

Section 405.2.3 provides the requirements for daylight responsive controls and lays out the method by which daylight areas are determined.

Table C402.4 provides both the *U*-factor and SHGC requirements for all types of fenestration and skylights.

C402.4.1.2 Increased skylight area with daylight responsive controls. The skylight area shall be permitted to be not more than 5 percent of the roof area provided *daylight responsive controls* complying with Section C405.2.3.1 are installed in *daylight zones* under skylights.

❖ Daylight responsive controls can save substantial energy. With daylighting, a larger window/skylight area actually can become an energy-saving measure as it reduces both the need for electric lighting and cooling loads, more than making up for thermal losses through the increased fenestration areas. Daylight responsive control system requirements are specified in Section C405.2.3.1.

C402.4.2 Minimum skylight fenestration area. In an enclosed space greater than 2,500 square feet (232 m^2) in floor area, directly under a roof with not less than 75 percent of the ceiling area with a ceiling height greater than 15 feet (4572 mm), and used as an office, lobby, atrium, concourse, corridor, storage space, gymnasium/exercise center, convention center, automotive service area, space where manufacturing occurs, nonrefrigerated warehouse, retail store, distribution/sorting area, transportation depot or workshop, the total *daylight zone* under skylights shall be not less than half the floor area and shall provide one of the following:

1. A minimum skylight area to *daylight zone* under skylights of not less than 3 percent where all skylights have a VT of at least 0.40 as determined in accordance with Section C303.1.3.
2. A minimum skylight effective aperture of at least 1 percent, determined in accordance with Equation 4-4.

$$\text{Skylight Effective Aperture} = \frac{0.85 \cdot \text{Skylight Area} \cdot \text{Skylight VT} \cdot \text{WF}}{\text{Daylight zone under skylight}}$$

(Equation 4-4)

where:

Skylight area	=	Total fenestration area of skylights.
Skylight VT	=	Area weighted average visible transmittance of skylights.
WF	=	Area weighted average well factor, where well factor is 0.9 if light well depth is less than 2 feet (610 mm), or 0.7 if light well depth is 2 feet (610 mm) or greater.
Light well depth	=	Measure vertically from the underside of the lowest point of the skylight glazing to the ceiling plane under the skylight.

Exception: Skylights above *daylight zones* of enclosed spaces are not required in:

1. Buildings in *Climate Zones* 6 through 8.
2. Spaces where the designed *general lighting* power densities are less than 0.5 W/ft^2 (5.4 W/m^2).
3. Areas where it is documented that existing structures or natural objects block direct beam sunlight

on at least half of the roof over the enclosed area for more than 1,500 daytime hours per year between 8 a.m. and 4 p.m.

4. Spaces where the *daylight zone* under rooftop monitors is greater than 50 percent of the enclosed space floor area.

5. Spaces where the total area minus the area of *daylight zones* adjacent to vertical fenestration is less than 2,500 square feet (232 m^2), and where the lighting is controlled according to Section C405.2.3.

❖ Daylighting can be a major energy-efficiency asset, provided electric lighting is reduced when daylight is available. Skylight area was adjusted based on substantial energy savings available from correctly utilized top lighting. With higher daylight transmittance, more electric lighting can be displaced and result in a greater net energy savings.

The intent of this section is to recognize the significant energy savings that can be achieved through natural lighting of a space by combining light transmitting fenestration with automatic lighting controls. The benefit of this combination has been demonstrated through modeling studies by the American Architectural Manufacturers Association (AAMA) Skylight Council and research conducted by other parties, including http://www.h-m-g.com/ASHRAE_Daylighting, as a reference. This section is based on studies done regarding lowering the *U*-factor for vertical fenestration in a number of different climate zones and product categories, permitting skylights to constitute a greater percentage of the roof when combined with daylight responsive controls and allowing skylights to have a higher SHGC in combination with higher VT.

This section applies to individual spaces that meet the three key thresholds (minimum area 2,500 square feet; minimum ceiling height 15 feet; located directly under the roof) and contain one of the listed uses. If the space is divided by walls to areas less than 2,500 square feet, this section does not apply. For example, a top floor under a roof could be occupied by a series of offices. An open office floor plan with a space exceeding 2,500 square feet would need to be designed to comply; however, were that space divided into multiple separate offices with floor-to-ceiling partitions, each office being less than 2,500 square feet, then this section doesn't apply.

There are five additional exceptions where a minimum of skylights would not be required. Exception 1 exempts buildings in the most northerly climate zones (in the Northern Hemisphere) based on lower light levels and greater need for the roof portion of the thermal envelope to be a significant barrier to heat loss. Exceptions 2, 3 and 4 apply to situations in the building design where the gains in energy savings from daylighting would not outweigh the added losses from heating or cooling. Exception 2 exempts spaces where the lighting levels are already low and additional daylight does not result in an overall energy savings. Exception 3 addresses buildings where adjacent buildings or natural features limit the amount of available light on the roof. The designer must submit appropriate documentation demonstrating the extent of shading of the roof area in order to justify the exception. Exception 4 exempts areas that already will have significant daylighting from rooftop monitors (see definition, "Rooftop monitor," Chapter 2 [CE]. Finally, Exception 5 exempts spaces where the area of daylight zones next to vertical fenestration already play a significant role in the space.

C402.4.2.1 Lighting controls in daylight zones under skylights. *Daylight responsive controls* complying with Section C405.2.3.1 shall be provided to control all electric lights with *daylight zones* under skylights.

❖ In order to achieve energy savings in daylight zones, the code requires daylight responsive controls in these areas. This section is a subsection to Section C402.4.2, which requires skylights over some spaces that contain specific uses. The requirement applies to those spaces and does not require daylight responsive controls for all areas beneath skylights. For example, where a building doesn't have any spaces that qualify under Section C402.4.2 and skylights are limited to 3 percent of the roof, no daylight responsive controls are necessary.

C402.4.2.2 Haze factor. Skylights in office, storage, automotive service, manufacturing, nonrefrigerated warehouse, retail store and distribution/sorting area spaces shall have a glazing material or diffuser with a haze factor greater than 90 percent when tested in accordance with ASTM D 1003.

Exception: Skylights designed and installed to exclude direct sunlight entering the occupied space by the use of fixed or automated baffles or the geometry of skylight and light well.

❖ In order to not limit the design of the building, the haze factor for skylights has been limited to certain uses where more of a diffuse skylight is shown to be more effective and where glare from direct sunlight can be an issue. The exception provides a skylight shading option for these spaces. The uses listed in this section do not include all of the uses specified in Section C402.4.2 where a minimum amount of skylight area is required. Again this section is a subsection of Section C402.4.2, which requires a minimum amount of skylights above certain spaces. Where Section C402.4.2 applies to a building, this section also applies.

C402.4.3 Maximum *U*-factor and SHGC. The maximum *U*-factor and solar heat gain coefficient (SHGC) for fenestration shall be as specified in Table C402.4.

The window projection factor shall be determined in accordance with Equation 4-5.

$$PF = A/B \qquad \textbf{(Equation 4-5)}$$

where:

PF = Projection factor (decimal).

A = Distance measured horizontally from the furthest continuous extremity of any overhang, eave or permanently attached shading device to the vertical surface of the glazing.

B = Distance measured vertically from the bottom of the glazing to the underside of the overhang, eave or permanently attached shading device.

Where different windows or glass doors have different *PF* values, they shall each be evaluated separately.

❖ Fenestration performance includes not only the *U*-factor or thermal transmittance, but also information about its light- and heat-transmitting capabilities. SHGC is a measure of all solar radiation, both transmitted and absorbed, that contributes to the cooling load (or offsets the heating load) in buildings. It is expressed as a number between 0 and 1. The lower a window's SHGC, the less solar heat it transmits, and the greater its shading ability. SHGC is affected by the composition of the glass; coatings that may be applied to the surfaces of glass; and internal shading devices, such as draperies or blinds. Solar heat gain and daylighting are also affected by permanently mounted overhangs and other shading technologies, such as electrochromic windows. Default values for *U*-factor and SHGC can be obtained from the tables in Section C303.1.3.

Table C402.4 provides both the *U*-factor and SHGC requirements for all types of fenestration and skylights. Adjustment factors, which in previous codes had to be mathematically applied, are not built into the values in Table C402.4.

The PF is determined by Equation 4-5. Commentary Figure C402.4.3 shows how the projection factor is determined.

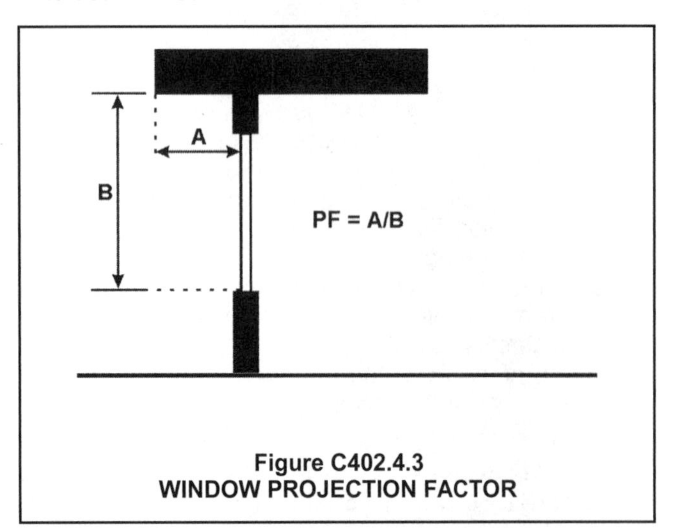

Figure C402.4.3
WINDOW PROJECTION FACTOR

C402.4.3.1 Increased skylight SHGC. In *Climate Zones* 1 through 6, skylights shall be permitted a maximum SHGC of 0.60 where located above *daylight zones* provided with *daylight responsive controls*.

❖ The key to this section is that the increased SHGC is only permitted where daylight areas are provided with daylight responsive controls that will reduce the overall lighting draw in correspondence with adequate lighting being provided by daylighting.

This section allows skylights in Climate Zones 1 through 6 a maximum SHGC of 0.60 where located above daylight zones provided with daylight responsive controls. This section and Section C402.4.3.2 allow for a broader range of skylights, including plastic dome skylights, used for daylighting purposes. Energy analyses have shown that the daylighting savings when used with automatic daylight responsive controls outweigh any energy impacts from these changes in the *U*-factor and SHGC for the noted climate zones.

C402.4.3.2 Increased skylight *U*-factor. Where skylights are installed above *daylight zones* provided with *daylight responsive controls*, a maximum *U*-factor of 0.9 shall be permitted in *Climate Zones* 1 through 3 and a maximum *U*-factor of 0.75 shall be permitted in *Climate Zones* 4 through 8.

❖ This is to allow more use of daylighting in lieu of electric lighting. There is a better likelihood that this will take place due to the requirement for daylight responsive controls (see commentary, Section C402.4.3.1).

C402.4.3.3 Dynamic glazing. Where *dynamic glazing* is intended to satisfy the SHGC and VT requirements of Table C402.4, the ratio of the higher to lower labeled SHGC shall be greater than or equal to 2.4, and the *dynamic glazing* shall be automatically controlled to modulate the amount of solar gain into the space in multiple steps. Dynamic glazing shall be considered separately from other fenestration, and area-weighted averaging with other fenestration that is not dynamic glazing shall not be permitted.

Exception: Dynamic glazing is not required to comply with this section where both the lower and higher labeled SHGC already comply with the requirements of Table C402.4.

❖ When calculated in accordance with National Fenestration Rating Council (NFRC) procedures, dynamic glazing has two values each listed for *U*-factor, SHGC and VT based on the range over which the product can modify its properties. This section clarifies which values to use for the determination of compliance with the code and that the upper and lower SHGC values must not exceed a ratio of 2.4.

For virtually all fenestration products, the NFRC ratings specify a single SHGC and a single VT, all determined at a standard rating condition. However, dynamic glazing has the ability to change properties based on sunlight or electric charges and so the NFRC provides two ratings for SHGC and VT: one rating is for the maximum value, and the other is for the minimum value. This section specifies which rating is to be used for code compliance.

C402.4.3.4 Area-weighted *U*-factor. An area-weighted average shall be permitted to satisfy the *U*-factor requirements for each fenestration product category listed in Table C402.4. Individual fenestration products from different fenestration product categories listed in Table C402.4 shall not be combined in calculating area-weighted average *U*-factor.

❖ Area-weighted *U*-factor averages may be used to comply with the *U*-factor requirements in Table C402.4, which is similar to what is allowed in Chapter 4 [CE] and ANSI/ASHRAE/IESNA 90.1. There is a large diversity of fenestration products in commercial construction, and enforcement issues can arise where there are a small number of products that do not meet the prescriptive requirements, yet the overall performance of all fenestration assemblies is well below the requirement. Area-weighted averaging would allow the use of these products. Enforcement by the code official should not be a problem in that Section C103.2 already requires area-weighted *U*-factor calculations to be provided on construction documents.

Using the *U*-factor requirement of 0.36 for Climate Zone 6 for fixed fenestration, this section provides two options for compliance. The simpler option is to ensure that all of the windows and doors have labeled NFRC values of 0.36 or less. This approach also is more likely to ensure adequate performance and comfort throughout the home. Alternatively, a weighted average may be taken of the values from all windows and doors to see if the "weighted average" is less than or equal to 0.36. Another option is the following example. Assume 100 square feet (9.3 m^2) of 0.32 windows, 100 square feet (9.3 m^2) of 0.36 windows and one 20-square-foot (1.8 m^2) door with a *U*-factor of 0.40: [(100 x 0.32) + (100 x 0.36) + (20 x 0.40)]/(100 + 100 + 20) = 0.345 (weighted average *U*-factor). Therefore, because the weighted average *U*-factor is less than the required 0.36, the fenestration in this example would be in compliance with the code.

C402.4.4 Doors. *Opaque doors* shall comply with the applicable requirements for doors as specified in Tables C402.1.3 and C402.1.4 and be considered part of the gross area of above-grade walls that are part of the building *thermal envelope*. Other doors shall comply with the provisions of Section C402.4.3 for vertical fenestration.

❖ All doors, whether glazed, opaque or partially glazed, must meet the applicable requirements as either an opaque assembly (Section C402.1.3 or C402.1.4) or as fenestration (Section C402.4 and Table C402.4). "Opaque" doors are defined in Chapter 2 [CE] as doors that have less than 50 percent of their area as glass. This distinction is important because opaque doors are regulated under the building envelope insulation requirements of Sections C402.1.3 and C402.1.4 and, therefore, must comply with the provisions in the respective tables. Doors with a greater amount of glass must comply with the "fenestration" provisions of Section C402.4. This section merely emphasizes certain requirements for doors in the building thermal envelope.

C402.5 Air leakage—thermal envelope (Mandatory). The *thermal envelope* of buildings shall comply with Sections C402.5.1 through C402.5.8, or the building *thermal envelope* shall be tested in accordance with ASTM E 779 at a pressure differential of 0.3 inch water gauge (75 Pa) or an equivalent method approved by the code official and deemed to comply with the provisions of this section when the tested air leakage rate of the building thermal envelope is not greater than 0.40 cfm/ft^2 (0.2 L/s · m^2). Where compliance is based on such testing, the building shall also comply with Sections C402.5.5, C402.5.6 and C402.5.7.

❖ Key to the success of any building thermal envelope is keeping the conditioned air inside the building and the nonconditioned air outside. The assessment of that success is tied to limiting air leakage through the envelope (infiltration from the outside and exfiltration to the outside). This section provides two main options to provide limited air leakage in a building's thermal envelope. The first avenue of compliance is to design and construct the building in compliance with the prescriptive standards of Sections C402.5.1 through C402.5.8. The second avenue of compliance is to have the "tightness" of the envelope tested in accordance with ASTM E 779 standard. The second avenue also requires prescriptive compliance with three of the envelope construction requirements.

This testing option for thermal envelope compliance requires the completed building to be tested in accordance with ASTM E 779 or an equivalent method approved by the code official. ASTM E 779 has a test method consisting of mechanical pressurization or depressurization of a building and measurements of the resulting airflow rates at given indoor-outdoor static pressure differences. From the relationship between the airflow rates and pressure differences, the air leakage characteristics of a building envelope are determined. For the purpose of this test method, many multizone buildings can be treated as single-zone buildings by opening interior doors or by inducing equal pressures in adjacent zones. The test in ASTM E 779 must be done to the completed building to ensure the airtightness of the air barriers.

Window, door and service openings in the building thermal envelope can be sources of considerable infiltration unless the requirements for air leakage in Section C402.5 are addressed.

C402.5.1 Air barriers. A continuous air barrier shall be provided throughout the building thermal envelope. The air barriers shall be permitted to be located on the inside or outside of the building envelope, located within the assemblies composing the envelope, or any combination thereof. The air barrier shall comply with Sections C402.5.1.1 and C402.5.1.2.

Exception: Air barriers are not required in buildings located in *Climate Zone* 2B.

❖ Air barriers are well-known technologies for achieving airtightness for the building envelope. According to the Department of Energy (DOE) and National Response Coordination Center (NRCC), uncontrolled air movement through the building envelope (infiltration and exfiltration) can account for up to 50 percent of heating loads and a significant part of cooling loads, representing as much as 30 percent of a building's annual HVAC costs.

Air barriers are not required in buildings located in Climate Zone 2B because they are not needed for the insulating value to reduce the HVAC energy costs. In the United States, Climate Zone 2B is limited to five counties in southern Arizona, one county in southern California and 14 counties in southern Texas. The exception is only to the air barrier requirement. The requirements found in Sections C402.5.2, C402.5.4, C402.5.5, C402.5.6 and C402.5.8 will still apply in these areas. Sections C402.5.3 and C402.5.7 also provide exceptions for Climate Zone 2.

C402.5.1.1 Air barrier construction. The *continuous air barrier* shall be constructed to comply with the following:

1. The air barrier shall be continuous for all assemblies that are the thermal envelope of the building and across the joints and assemblies.
2. Air barrier joints and seams shall be sealed, including sealing transitions in places and changes in materials. The joints and seals shall be securely installed in or on the joint for its entire length so as not to dislodge, loosen or otherwise impair its ability to resist positive and negative pressure from wind, stack effect and mechanical ventilation.
3. Penetrations of the air barrier shall be caulked, gasketed or otherwise sealed in a manner compatible with the construction materials and location. Joints and seals associated with penetrations shall be sealed in the same manner or taped or covered with moisture vapor-permeable wrapping material. Sealing materials shall be appropriate to the construction materials being sealed and shall be securely installed around the penetration so as not to dislodge, loosen or otherwise impair the penetrations' ability to resist positive and negative pressure from wind, stack effect and mechanical ventilation. Sealing of concealed fire sprinklers, where required, shall be in a manner that is recommended by the manufacturer. Caulking or other adhesive sealants shall not be used to fill voids between fire sprinkler cover plates and walls or ceilings.
4. Recessed lighting fixtures shall comply with Section C402.5.8. Where similar objects are installed that penetrate the air barrier, provisions shall be made to maintain the integrity of the air barrier.

❖ This section prescribes the construction of continuous air barriers. The air barrier shall be continuous for all assemblies that are the thermal envelope of the building. Air barrier joints and seams shall be sealed. Sealing the building envelope is critical to obtain good thermal performance for the building. The seal will prevent warm, conditioned air from leaking out around doors, windows and other cracks during the heating season, thereby reducing the cost of heating the building. During the hot summer months, a proper seal will stop hot air from entering the building, helping to reduce the air-conditioning load on the building. Any penetration in the building envelope must be completely sealed during the construction process, including holes made for the installation of plumbing, electrical, lighting and heating and cooling systems (see Commentary Figure C402.5.1.1).

C402.5.1.2 Air barrier compliance options. A continuous air barrier for the opaque building envelope shall comply with Section C402.5.1.2.1 or C402.5.1.2.2.

❖ A continuous air barrier for the opaque building envelope must comply with one of the two options regarding the assemblies of materials and components. This section is not required if under Section C402.5 the option to test the air barrier was selected.

C402.5.1.2.1 Materials. Materials with an air permeability not greater than 0.004 cfm/ft^2 (0.02 L/s · m^2) under a pressure differential of 0.3 inch water gauge (75 Pa) when tested in accordance with ASTM E 2178 shall comply with this section. Materials in Items 1 through 16 shall be deemed to comply with this section, provided joints are sealed and materials are installed as air barriers in accordance with the manufacturer's instructions.

1. Plywood with a thickness of not less than $^3/_8$ inch (10 mm).
2. Oriented strand board having a thickness of not less than $^3/_8$ inch (10 mm).
3. Extruded polystyrene insulation board having a thickness of not less than $^1/_2$ inch (12.7 mm).
4. Foil-back polyisocyanurate insulation board having a thickness of not less than $^1/_2$ inch (12.7 mm).
5. Closed-cell spray foam a minimum density of 1.5 pcf (2.4 kg/m^3) having a thickness of not less than 1$^1/_2$ inches (38 mm).
6. Open-cell spray foam with a density between 0.4 and 1.5 pcf (0.6 and 2.4 kg/m^3) and having a thickness of not less than 4.5 inches (113 mm).
7. Exterior or interior gypsum board having a thickness of not less than $^1/_2$ inch (12.7 mm).
8. Cement board having a thickness of not less than $^1/_2$ inch (12.7 mm).
9. Built-up roofing membrane.
10. Modified bituminous roof membrane.
11. Fully adhered single-ply roof membrane.
12. A Portland cement/sand parge, or gypsum plaster having a thickness of not less than $^5/_8$ inch (15.9 mm).
13. Cast-in-place and precast concrete.

14. Fully grouted concrete block masonry.
15. Sheet steel or aluminum.
16. Solid or hollow masonry constructed of clay or shale masonry units.

❖ The air barrier materials must have a very low air-leakage rate according to the leakage criteria in the code. The *National Building Code of Canada* and the *Massachusetts Building Code* consider 0.004 cfm/ft^2 (0.02 L/s × m^2) under a pressure differential of 0.3-inch w.g. (75 Pa) [the air permeance of $^1/_2$-inch (12 mm) unpainted gypsum board] as the maximum air-leakage rate for the air barrier material as part of the opaque envelope. In order to achieve a reasonable whole-building airtightness, the basic materials selected for the air barrier must be resistant to air leakage. The 16 materials specified in this section are materials that have been tested and meet the criteria or are presumed to pass the test.

C402.5.1.2.2 Assemblies. Assemblies of materials and components with an average air leakage not greater than 0.04 cfm/ft^2 (0.2 L/s · m^2) under a pressure differential of 0.3 inch of water gauge (w.g.)(75 Pa) when tested in accordance with ASTM E 2357, ASTM E 1677 or ASTM E 283 shall comply with this section. Assemblies listed in Items 1 through 3 shall be deemed to comply, provided joints are sealed and the requirements of Section C402.5.1.1 are met.

1. Concrete masonry walls coated with either one application of block filler or two applications of a paint or sealer coating.
2. Masonry walls constructed of clay or shale masonry units with a nominal width of 4 inches (102 mm) or more.
3. A Portland cement/sand parge, stucco or plaster not less than $^1/_2$ inch (12.7 mm) in thickness.

❖ Assemblies are materials and components that are listed in Tables C402.1.3 and C402.1.4. When essentially air-tight materials are assembled together by appropriately including sealing and taping, the assembly will have a higher leakage than the original air barrier material, primarily due to higher leakage at the joints.

C402.5.2 Air leakage of fenestration. The air leakage of fenestration assemblies shall meet the provisions of Table C402.5.2. Testing shall be in accordance with the applicable reference test standard in Table C402.5.2 by an accredited,

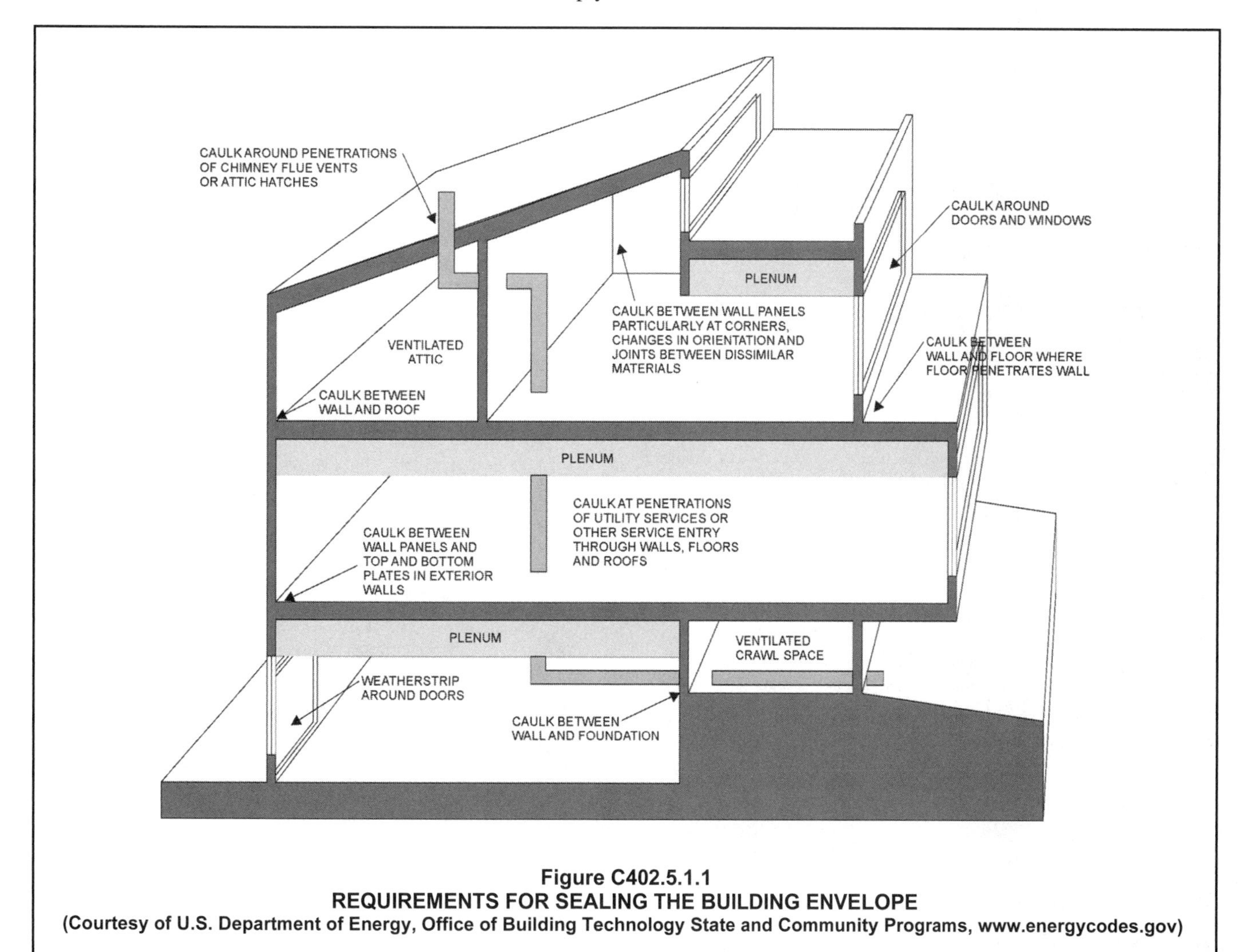

Figure C402.5.1.1
REQUIREMENTS FOR SEALING THE BUILDING ENVELOPE
(Courtesy of U.S. Department of Energy, Office of Building Technology State and Community Programs, www.energycodes.gov)

independent testing laboratory and *labeled* by the manufacturer.

Exceptions:

1. Field-fabricated fenestration assemblies that are sealed in accordance with Section C402.5.1.
2. Fenestration in buildings that comply with the testing alternative of Section C402.5 are not required to meet the air leakage requirements in Table C402.5.2.

❖ Manufactured door and window assemblies must meet requirements for air leakage in AAMA/WDMA/CSA 101/I.S.2/A440. Manufacturers typically will have an independent laboratory or testing agency perform the necessary testing to verify compliance with the requirements and provide documentation to their customers. To ensure that complying products are installed, designers should include notes on the plans or language in the construction specifications. Window and door schedules on the drawings are the appropriate places to reference the standards and test procedures of AAMA/WDMA/CSA 101/I.S.2/A440. The code also will accept testing to NFRC 400 since that standard also uses ASTM E 283 as a basis and provides comparable results. These requirements do not apply to site-constructed windows and doors sealed in accordance with this section. Where the designer chooses to have the whole building tested in accordance with the option provided in Section C402.5, then individual fenestration products do not have to meet the requirements of the table.

TABLE C402.5.2
MAXIMUM AIR LEAKAGE RATE
FOR FENESTRATION ASSEMBLIES

FENESTRATION ASSEMBLY	MAXIMUM RATE (CFM/FT²)	TEST PROCEDURE
Windows	0.20[a]	AAMA/WDMA/ CSA101/I.S.2/A440 or NFRC 400
Sliding doors	0.20[a]	
Swinging doors	0.20[a]	
Skylights – with condensation weepage openings	0.30	
Skylights – all other	0.20[a]	
Curtain walls	0.06	NFRC 400 or ASTM E 283 at 1.57 psf (75 Pa)
Storefront glazing	0.06	
Commercial glazed swinging entrance doors	1.00	
Revolving doors	1.00	
Garage doors	0.40	ANSI/DASMA 105, NFRC 400, or ASTM E 283 at 1.57 psf (75 Pa)
Rolling doors	1.00	
High-speed doors	1.30	

For SI: 1 cubic foot per minute = 0.47 L/s, 1 square foot = 0.093 m^2.

a. The maximum rate for windows, sliding and swinging doors, and skylights is permitted to be 0.3 cfm per square foot of fenestration or door area when tested in accordance with AAMA/WDMA/CSA101/I.S.2/A440 at 6.24 psf (300 Pa).

C402.5.3 Rooms containing fuel-burning appliances. In *Climate Zones* 3 through 8, where open combustion air ducts provide combustion air to open combustion space conditioning fuel-burning appliances, the appliances and combustion air openings shall be located outside of the *building thermal envelope* or enclosed in a room isolated from inside the thermal envelope. Such rooms shall be sealed and insulated in accordance with the envelope requirements of Table C402.1.3 or C402.1.4, where the walls, floors and ceilings shall meet the minimum of the below-grade wall *R*-value requirement. The door into the room shall be fully gasketed, and any water lines and ducts in the room insulated in accordance with Section C403. The combustion air duct shall be insulated, where it passes through conditioned space, to a minimum of R-8.

Exceptions:

1. Direct vent appliances with both intake and exhaust pipes installed continuous to the outside.
2. Fireplaces and stoves complying with Sections 901 through 905 of the *International Mechanical Code*, and Section 2111.13 of the *International Building Code*.

❖ Space conditioning fuel-burning appliances require a supply of air for combustion. If that air is being supplied from outside of the building, the result can be a hole in a building's thermal envelope. Such is not the case for direct vent appliances (Exception 1) where the combustion air is piped from the outside and delivered into the appliance. But where the air is "supplied" by an opening in the envelope, this section provides two options.

The first option is simply to put the appliance outside the building envelope. Such an installation would deliver conditioned air into the conditioned space through ductwork. Where such ductwork penetrated the envelope, the code requires appropriate sealing.

The second option is to create an enclosed room within the thermal envelope, but isolated from the balance of the conditioned space through insulated and sealed assemblies. The walls, floor and ceiling of the enclosed room must be insulated to the equivalent level of that required for below-grade walls appropriate to the climate zone. It should be noted that for Climate Zone 3 there is no insulation required. Other measures must be included to isolate this room from the balance of the fully conditioned space, including gasketing the door and insulating water lines and ducts passing through the room. If there is a duct bringing outside air into this room, it too needs to be insulated where it passes through any conditioned space.

C402.5.4 Doors and access openings to shafts, chutes, stairways and elevator lobbies. Doors and access openings from conditioned space to shafts, chutes stairways and elevator lobbies not within the scope of the fenestration assemblies covered by Section C402.5.2 shall be gasketed, weatherstripped or sealed.

Exceptions:

1. Door openings required to comply with Section 716 or 716.5 of the *International Building Code.*
2. Doors and door openings required to comply with UL 1784 by the *International Building Code.*

❖ This section requires doors and access openings to shafts, chutes, stairways and elevator lobbies to meet the requirements of Section C402.4.3; or to be gasketed, weatherstripped or sealed to minimize warm, conditioned air from leaking out around doors and access openings to shafts, chutes, stairways and elevator lobbies. This will lower the HVAC load and reduce the energy usage for the building. The exceptions are for door openings required to comply with Section 716 or 716.5 of the IBC; or doors and door openings required by the IBC to comply with UL 1784.

Sections 716 and 716.5 of the IBC contain requirements for fire protective openings. The seals and gaskets necessary to reduce air leakage in doors are not compatible with the fire-protected doors; therefore, given that these doors are usually of tight construction, the code exempts them from the air-leakage requirements. Similarly, the code exempts any door that is required to meet the tightness criteria of UL 1784. Again, the air leakage on these doors is expected to be low.

C402.5.5 Air intakes, exhaust openings, stairways and shafts. Stairway enclosures, elevator shaft vents and other outdoor air intakes and exhaust openings integral to the building envelope shall be provided with dampers in accordance with Section C403.2.4.3.

❖ Air intakes, exhaust openings, stairways and shafts have been recognized as uncontrolled sources of building air leakage that can create a stack effect in large buildings and significantly depressurize lower floors in the building. This can cause unintended air and moisture migration into the building and waste energy.

C402.5.6 Loading dock weatherseals. Cargo doors and loading dock doors shall be equipped with weatherseals to restrict infiltration when vehicles are parked in the doorway.

❖ Weather seals help to reduce the amount of infiltration that occurs when a trailer pulls up to unload at an open loading dock door.

C402.5.7 Vestibules. Building entrances shall be protected with an enclosed vestibule, with all doors opening into and out of the vestibule equipped with self-closing devices. Vestibules shall be designed so that in passing through the vestibule it is not necessary for the interior and exterior doors to open at the same time. The installation of one or more revolving doors in the building entrance shall not eliminate the requirement that a vestibule be provided on any doors adjacent to revolving doors.

Exceptions: Vestibules are not required for the following:

1. Buildings in *Climate Zones* 1 and 2.
2. Doors not intended to be used by the public, such as doors to mechanical or electrical equipment rooms, or intended solely for employee use.
3. Doors opening directly from a *sleeping unit* or dwelling unit.
4. Doors that open directly from a space less than 3,000 square feet (298 m^2) in area.
5. Revolving doors.
6. Doors used primarily to facilitate vehicular movement or material handling and adjacent personnel doors
7. Doors that have an air curtain with a minimum velocity of 6.56 feet per second (2 m/s) at the floor that have been tested in accordance with ANSI/AMCA 220, and installed in accordance with the manufacturer's instructions. Manual or automatic controls shall be provided that will operate the air curtain with the opening and closing of the door. Air curtains and their controls shall comply with Section C408.2.3.

❖ The use of vestibules helps to reduce the loss of conditioned air (either heated or cooled) when exterior doors are open. Exception 1 waives the requirement for some of the milder climate zones.

Although the code does not include any minimum dimensions for the vestibule, it does specify that they should be sized "so that in passing through the vestibule it is not necessary for the interior and exterior doors to open at the same time." Although not required by the code, the IBC and ICC A117.1 do contain requirements that would affect the vestibules and the doors, especially doors into buildings that are required to be accessible. Section 1203.5.1.1 of the IBC establishes the size and area of openings that must exist to allow one space to be ventilated through another. Therefore, a code official may determine that where the openings are less than that required by Section 1203.5.1 of the IBC, the spaces may be considered as separate. ICC A117.1 requires that "two doors in series," such as those in a vestibule, must have a minimum of 48 inches (1219 mm) plus the width of any door swinging into the space. In addition, ICC A117.1 requires the space to permit a minimum 60-inch (1524 mm) turning circle within it. Commentary Figure C402.5.7 shows examples of the vestibule requirements in ICC A117.1.

Vestibules are not required in buildings in Climate Zones 1 and 2, and doors not intended to be used as a building entrance door, such as: doors to mechanical or electrical equipment rooms; doors opening directly from a sleeping unit or dwelling unit; doors that open directly from a space less than 3,000 square feet (298 m^2) in area; and revolving doors where there is a minimal heating and cooling load change from conditioned space from the exterior. Doors used for vehicular movement or material handling in the dock area of a warehouse or a manufacturing plant do not require vestibules, as the heating and cooling load are minimal to none.

Finally the code allows an air curtain tested in accordance with the AMCA 220 standard. An air curtain with a minimum velocity of 6.56 feet per second provides an effective barrier at a door to minimize the loss of conditioned air to the outside or the intrusion of unconditioned air into the building.

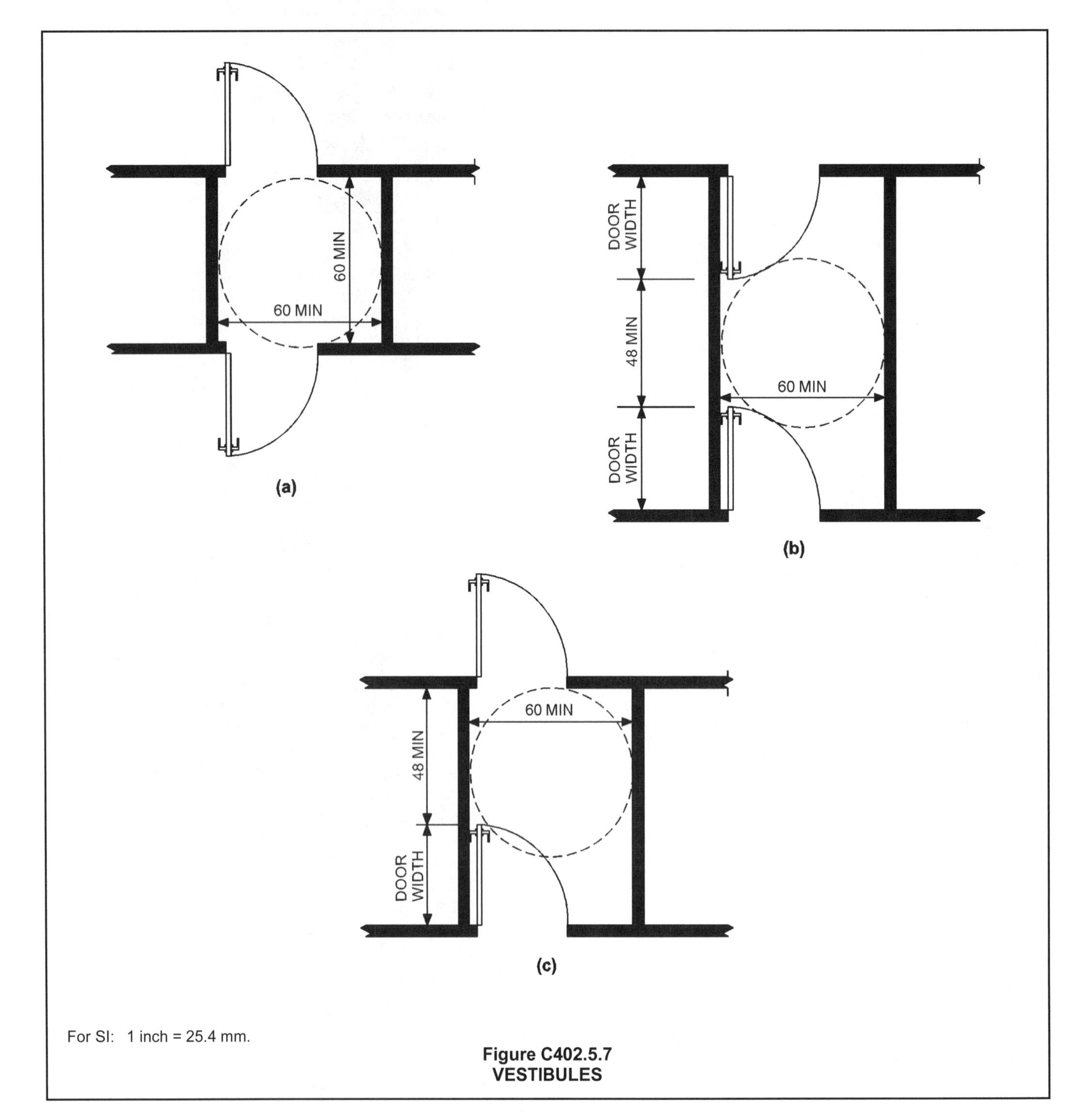

Figure C402.5.7
VESTIBULES

C402.5.8 Recessed lighting. Recessed luminaires installed in the *building thermal envelope* shall be all of the following:

1. IC-rated.
2. Labeled as having an air leakage rate of not more 2.0 cfm (0.944 L/s) when tested in accordance with ASTM E 283 at a 1.57 psf (75 Pa) pressure differential.
3. Sealed with a gasket or caulk between the housing and interior wall or ceiling covering.

❖ The installation of recessed lighting fixtures in the building envelope can not only lead to an increased loss of energy through the envelope, but also may create a fire hazard if incorrectly covered with insulation. These fixtures also act as chimneys, transferring heat loss and moisture through the building envelope into attic spaces. The heat loss resulting from improperly insulated recessed lighting fixtures can be significant. The prescriptive requirements found here make the commercial structures consistent with the requirement for residential uses found in Chapter 4 [RE] (see commentary, Section R402.4.5).

Recessed lighting fixtures must be insulation contact (IC) rated. IC-rated lights are typically double-can fixtures, with one can inside another (see Commentary Figure C402.5.8). The outer can (in contact with insulation) is tested to make sure it remains cool enough to avoid a fire hazard. An IC-rated fixture should have the IC rating stamped on the fixture or printed on an attached label.

All recessed lights must be tightly sealed or gasketed to prevent air leakage through the fixture into the ceiling cavity.

SECTION C403
BUILDING MECHANICAL SYSTEMS

C403.1 General. Mechanical systems and equipment serving the building heating, cooling or ventilating needs shall comply with Section C403.2 and shall comply with Sections C403.3 and C403.4 based on the equipment and systems provided.

Walk-in coolers, walk-in freezers, refrigerated warehouse coolers and refrigerated warehouse freezers shall comply with Section C403.2.15 or C403.2.16.

❖ Section C403 is divided into five parts. Sections C403.1 and C403.2 address general issues that are applicable to all systems. Section C403.3 addresses where economizers are required and the applicable design standards. Section C403.4 is applicable to hydronic and multizone HVAC systems controls and equipment. The provisions of Section C403.4 apply only when the equipment regulated by the section is included in the design of an HVAC system. Section C403.5 covers refrigeration equipment.

Section C403.3 applies to buildings served by unitary or packaged HVAC equipment as listed in Tables C403.2.3(1) through C403.2.3(5), economizers as listed in Table C403.3(1) and hydronic system controls. Section C403.4 applies to buildings served by HVAC equipment, including economizers, meeting the minimum efficiency requirements of Table C403.2.3(1) or C403.2.3(2) and having variable air volume (VAV) fan controls, hydronic system controls, hydronic heat pump systems, heat rejection equipment and fan speed controls. Unitary equipment is self-contained, packaged equipment from a single manufacturer, while hydronic systems consist of separate components, such as chillers, cooling towers and air handlers. Single-zone systems are those that serve just one zone and are controlled by a single thermostat. Multiple-zone systems are capable of simultaneously satisfying the comfort needs of a number of zones.

C403.2 Provisions applicable to all mechanical systems (Mandatory). Mechanical systems and equipment serving the building heating, cooling or ventilating needs shall comply with Sections C403.2.1 through C403.2.16.

❖ As stated in Section C403.1, the provisions of this section apply to all HVAC systems regardless of the

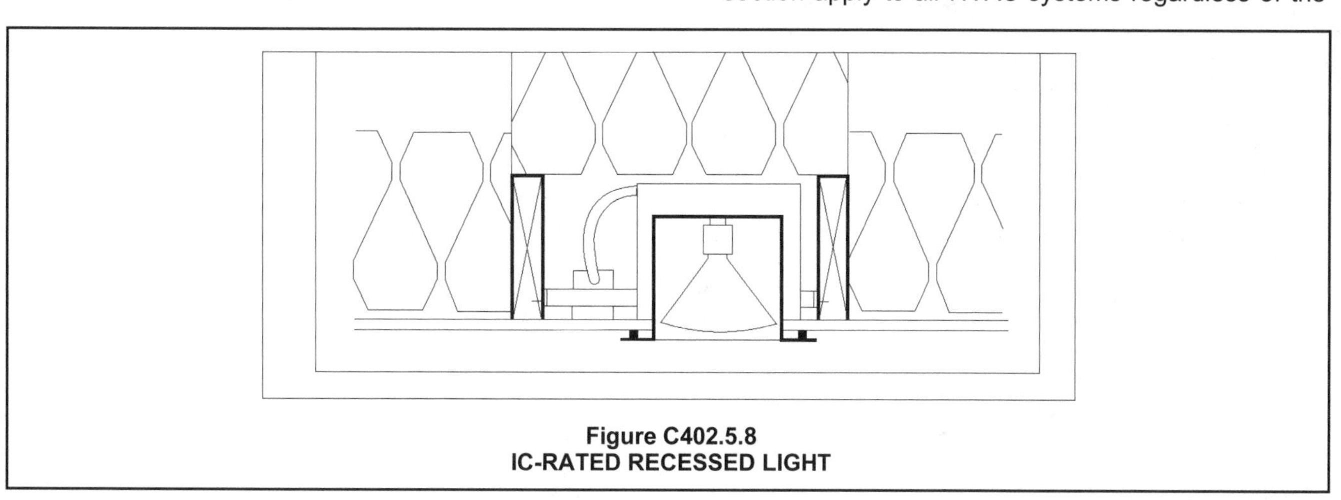

Figure C402.5.8
IC-RATED RECESSED LIGHT

complexity of the system and apply to the equipment specified herein. In general, the requirements of Section C403.2 deal with equipment efficiencies, proper system design/sizing, controls and installation. These elements provide for effective energy performance regardless of the type of systems used.

C403.2.1 Calculation of heating and cooling loads. Design loads associated with heating, ventilating and air conditioning of the building shall be determined in accordance with ANSI/ASHRAE/ACCA Standard 183 or by an *approved* equivalent computational procedure using the design parameters specified in Chapter 3. Heating and cooling loads shall be adjusted to account for load reductions that are achieved where energy recovery systems are utilized in the HVAC system in accordance with the ASHRAE *HVAC Systems and Equipment Handbook* by an approved equivalent computational procedure.

❖ The heating and cooling loads must be determined in accordance with the procedures described in ANSI/ASHRAE/ACCA 183. The designer must calculate the heating and cooling loads before selecting or sizing the HVAC equipment. This requirement should lead to the selection of equipment that is neither oversized nor undersized for the intended application. Oversized equipment not only increases owner cost, but usually operates less efficiently than properly sized equipment. It also can result in reduced comfort because of, for example, a lack of humidity control in cooling systems and fluctuating temperatures from short cycling. Undersizing obviously will result in poor temperature control in extreme weather, but also can increase energy usage at other times. For example, an undersized heating system may have to be operated 24 hours a day because it has an insufficient capacity to warm up the building each morning in a timely manner.

Accurate calculations of expected heating and cooling loads begin with a reliable simulation tool. The code requires that calculation procedures be in accordance with the methods espoused in ANSI/ASHRAE/ACCA 183, which standardize the methodology in the ASHRAE *Handbook of Fundamentals*. ANSI/ASHRAE/ACCA 183 sets the minimum requirements for the methods and procedures used to perform peak cooling and heating load calculations. Energy recovery systems can recover substantial amounts of energy and significantly reduce the heating and cooling loads that the system actually sees. If these heating and cooling load reductions are not factored into sizing the system, it will be oversized, less efficient and less able to control humidity in the cooling mode. This would defeat the purpose and value of the energy recovery system.

The code does not specify the limit design parameters. For instance, there is no "correct" inside space temperature for all applications; what is appropriate for a nursing home may not be appropriate for a detention center. Likewise, there are no "absolute" or "correct" equipment loads; one office space may have personal computers and printers on every desk in a 10-foot by 10-foot (3048-mm by 3048-mm) cubicle, while another may have many private offices with only small office equipment loads. The sheer number of variables and the wide range of reasonable values for each of them makes regulation of the details of load calculations impractical. The proper selection of design parameters is left up to the professional judgment of the designer, but requires prudent cross-examination by the code official.

While there is no one "correct" inside temperature for all applications, the code does specify a heating and cooling temperature that must be met in the design process. Section C302.1 provides interior design conditions that are used when sizing equipment. By providing these design conditions, it helps designers select the HVAC equipment and controls that will be energy efficient and best serve the needs of the building's owner and/or occupants.

C403.2.2 Equipment sizing. The output capacity of heating and cooling equipment shall be not greater than the loads calculated in accordance with Section C403.2.1. A single piece of equipment providing both heating and cooling shall satisfy this provision for one function with the capacity for the other function as small as possible, within available equipment options.

Exceptions:

1. Required standby equipment and systems provided with controls and devices that allow such systems or equipment to operate automatically only when the primary equipment is not operating.
2. Multiple units of the same equipment type with combined capacities exceeding the design load and provided with controls that have the capability to sequence the operation of each unit based on load.

❖ Once load calculations have been completed, using them for selecting equipment is nearly self-regulating because of market-driven incentives. For example, it is unlikely that the designer would select a 12-ton (10 886.2 kg) unit to do the job of a 6-ton (5443.1 kg) unit because of the additional first costs. Nevertheless, a sizing restriction remains in the code to ensure that each major piece of equipment is properly sized.

Most HVAC system components are less efficient at part load than they are at full load. Section C403.2.2, therefore, requires that HVAC systems and equipment be sized to provide not more than the cooling and heating design loads calculated in accordance with Section C403.2.1.

The intent of this restriction is often misunderstood because the term "oversizing" is commonly used in two different ways. As it applies here, "oversizing" means selecting equipment that has a higher output capability than required by the design load. But the term also is used to describe the sizing of equipment using design parameters that are more conservative than is typical for the industry, such as sizing a duct for a 0.05 inch/100-foot pressure drop, using a 49-inch (1245 mm) fan when a 44-inch (1118 mm) fan would work; sizing a coil or filter bank for 300 feet per minute (fpm) versus 450 fpm; or selecting a cooling tower to deliver 75°F (24°C) condenser water. These designs are not intended to increase the capacity of the system and are not considered "oversizing" in the context of this code. On the other hand, sizing a chiller for 200 tons (181 437 kg) when the load is 150 tons (136 078 kg), or sizing a fan for 10,000 cfm (5 m^3/s) when the load requires only 8,000 cfm (4 m^3/s) is considered oversizing. This type of oversizing will almost always result in reduced energy efficiency. The important distinction between the two is that "oversizing" in this context means specifying equipment having a design capacity exceeding the required design load.

For a single piece of equipment that has both heating and cooling capability, only one function, either the heating or the cooling, need meet the oversizing restriction. Capacity for the other function must be the smallest size, within the available equipment options, that meets the load requirement. This primarily is intended to apply to unitary equipment, such as heat pumps or gas/electric units. Such equipment generally is selected to meet the space-cooling load, which often leaves them oversized for heating because of the nature of the equipment (such as heat pumps where their capacity in the heating mode will be determined by their capacity in the cooling mode because the same compressor and heat exchangers are used) or by limited heating options (such as gas/electric units for which there typically are only one or two gas furnace options, the smallest of which may be larger than needed). Another example is direct gas-fired absorption chillers/heaters that have both heating and cooling capability.

Section C403.2.1 gives designers flexibility in the calculation of heating and cooling loads. This prevents the strict enforcement of the sizing restriction in this section. The designer may simply adjust the design parameters in load calculations until an oversized piece of equipment is justified by the calculation. Nevertheless, a sizing restriction remains in the code to ensure that load calculations are done for each major piece of equipment (chiller, boiler, air-conditioning unit, etc.) and that they are used to help in the selection of that equipment. The accuracy of the calculations and the extent to which they are used in the equipment selection process is left up to the aptitude and professionalism of the designer.

There are two exceptions to the system-sizing equipment. First, equipment capacity may exceed design loads if the equipment is intended for standby use only, and controls are installed that allow its operation only when the primary equipment is not in operation. Second, multiple pieces of the same equipment type, such as multiple chillers or boilers, may have a total capacity exceeding the design load if controls are installed to sequence or otherwise optimally control the equipment as a function of the operating load. This is an ideal design technique to allow for additional load for future expansion, without reducing energy efficiency. It also provides redundancy to reduce the impact of equipment failure.

C403.2.3 HVAC equipment performance requirements. Equipment shall meet the minimum efficiency requirements of Tables C403.2.3(1), C403.2.3(2), C403.2.3(3), C403.2.3(4), C403.2.3(5), C403.2.3(6), C403.2.3(7), C403.2.3(8) and C403.2.3(9) when tested and rated in accordance with the applicable test procedure. Plate-type liquid-to-liquid heat exchangers shall meet the minimum requirements of Table C403.2.3(10). The efficiency shall be verified through certification under an *approved* certification program or, where a certification program does not exist, the equipment efficiency ratings shall be supported by data furnished by the manufacturer. Where multiple rating conditions or performance requirements are provided, the equipment shall satisfy all stated requirements. Where components, such as indoor or outdoor coils, from different manufacturers are used, calculations and supporting data shall be furnished by the designer that demonstrates that the combined efficiency of the specified components meets the requirements herein.

❖ HVAC equipment must meet or exceed the energy efficiencies shown in Tables C403.2.3(1) through C403.2.3(10) if the equipment is to be used in a building or an HVAC system falling within the scope of the code. Each table has reference notes that must be adhered to. Efficiencies must be measured in accordance with the rating procedures specified in the tables. Equipment ratings may be self-certified by the manufacturer or rated under a nationally recognized certification program. Field tests are not required. Where more than one requirement applies to a given piece of equipment, all must be met. Please note that Tables C403.2.3(1), (2) and (7) have different efficiency requirements based on the date on which the system is approved for permit. Installations of new equipment in existing buildings must also comply with these efficiency standards.

Tables C403.2.3(1) through C403.2.3(10) cover the majority of equipment used in the design of contemporary HVAC systems; however, there are equipment types that are not covered. Although these pieces of equipment are not required to meet any particular efficiency level to comply with the code, omission from the tables does not imply such equipment cannot be used.

Some of the equipment efficiency requirements of this section are also requirements of the National Appliance Energy Conservation Act (NAECA). The NAECA requirements apply to new equipment installed in existing buildings, as well as new buildings, and are established by the federal government. This is referenced in Tables C403.2.3(1), C403.2.3(2) and C403.2.3(4).

Product development for water-cooled chillers in recent years has focused largely on improving off-design and part load performance. In particular, variable speed drives (VSDs) have gone through significant technology advancements and are now finding widespread application in water-cooled chillers. The use of VSDs has resulted in a significant improvement in the off-design and part-load improvement of the chiller's performance. Improvements of not greater than 30 percent in integrated part load value (IPLV) are possible. Partially offsetting the part-load performance improvement is a small decrease in full-load efficiency at design conditions, nominally not greater than 4 percent. The decrease in full-load efficiency is due to inherent electronic drive losses and power line filters.

The code updates the current minimum efficiency requirements for water chilling and reflects new AHRI 550/590 fouling factors and IPLV equations. IPLV is defined in the commentary to Tables C403.2.3(1) through C403.2.3(7). Second, effective January 1, 2010, an additional path of compliance for water-cooled chillers was required. Both paths have been maintained in the current edition of the code. Path A is intended for applications where significant operating time is expected at full-load conditions. Path B is an alternative set of efficiency levels for water-cooled chillers intended for applications where significant time is expected at part load. All Path B chillers are required to be equipped with demand-limiting controls. Compliance with the code can be achieved by meeting the requirements of either Path A or B. However, both full-load and IPLV levels must be met to fulfill the requirements of Path A or B.

Also, the code combines all water-cooled positive displacement chillers into one category and adds a new size category for centrifugal chillers greater than or equal to 600 tons (544 metric tons). The air-cooled chiller without the condenser equipment-type category has been eliminated. All air-cooled chillers without the condensers must now be rated with matching condensers. The minimum efficiencies of air-cooled chillers have been updated with the code change. Efficiencies are now expressed in energy efficiency ratios (EER) for air-cooled chillers, kW/ton for water-cooled chillers and coefficient of performance (COP) for absorption chillers to reflect current industry practices.

The user should exercise discretion when applying the standard efficiency ratings established in this section to compare different types of HVAC equipment. The performance efficiency ratings are sometimes established using different standardized rating conditions, such as ground-water-source heat pumps, which are rated at 70°F (21°C) compared to water-source heat pumps, which are rated at 85°F (29°C). More importantly, the ratings may not include the energy used to operate auxiliary systems that may be required for the primary equipment to function as a complete system. For example:

- The efficiency ratings for water-cooled equipment cannot be directly compared to those for air-cooled equipment. Water-cooled equipment ratings do not include the energy used by the condenser water pumps and cooling tower fans, while air-cooled package ratings include condenser fan energy.
- The ratings for condensing units cannot be directly compared to ratings for packaged or split-system air conditioners. Condensing unit ratings do not include the energy used by indoor air-handling fans.
- Efficiency ratings for different types of furnaces account only for gas usage, but do not include the energy used by combustion air fans and indoor air-handler fans that vary between product categories.
- The efficiency of a chilled-water system cannot be compared to a unitary direct-expansion system using standard ratings. Chilled-water-system efficiency does not include the energy used by chilled-water pumps and air-handler fans.

Different types of equipment may be applied in many different ways, and the manufacturer can test the performance of equipment as it exists only when it leaves the factory. A fair comparison between different types of equipment, such as water- versus air-cooled equipment, requires knowledge of the auxiliary equipment needed for the application as a complete system and the energy they use at both full and part load. Often an energy analysis using a simulation tool [see Table C407.5.1(1)] is the only way to make the best possible comparison.

Example of Multiple Requirements—Unitary Heat Pump

What are the efficiency requirements for a 10-ton (9072 kg) unitary air conditioner and condensing unit? (Hint: 10 tons is equal to 120,000 Btu/h.)

Table C403.2.3(2) contains the requirements for unitary conditioner and condensing units. The heat pump must meet the high temperature COP requirements for heating (COP of 3.3). It also must meet the full-load EER requirements for cooling (11.0 EER).

Example of Multifunction Equipment Requirements—Space- and Water-heating Boiler

A 2,600,000-Btu/h (754 kw/h) input gas-fired water-tube boiler is to provide domestic water heating in the form of an external storage tank and also will provide space heating. The boiler has two-stage gas valves. What efficiency requirements must be met?

The boiler serves as a "nonstorage" (instantaneous) water heater and, therefore, must meet the Table C404.2 requirement for a full-load thermal efficiency, E_t, of not less than 80 percent. Because it provides space heating as well, the boiler also must meet Table C403.2.3(5) requirements for a thermal efficiency, E_t, of not less than 82 percent at its minimum firing rate (low fire). In this case, the Table C403.2.3(5) requirement is more stringent.

TABLES C403.2.3(1) through C403.2.3(10). See pages C4-34 through C4-43

❖ These tables provide the minimum efficiencies and appropriate test standards for the various types of mechanical equipment used to condition a building. Many of the minimum efficiency standards were increased in the 2015 code. Because the equipment generally is tested and labeled, the inspector will not need to be greatly aware of the standards but would only need to confirm the equipment's labeled efficiency meets the appropriate minimum requirements. Designers would need to be aware of these minimum efficiencies in order to specify equipment that is able to meet the energy conservation requirements of the code.

Historically, energy codes limited only the full-load equipment efficiency (i.e., the efficiency of the equipment at standard rating conditions representative of typical peak design or full-load conditions). Even though equipment full-load efficiency is important, particularly in areas with high utility demand charges, performance at part load is usually more relevant in determining annual equipment energy costs. In recognition of this issue, the code requires that most equipment be tested and meet efficiency limits at part-load as well as full-load conditions.

New for the 2015 edition of the code is Table C403.2.3(9), which provides minimum efficiency requirements for equipment serving the unique environment of computer equipment rooms. Computer rooms are defined in Section C202 to be rooms with a primary function of housing equipment for processing and storage of electronic data. This type of equipment generates a significant level of internal heat. The HVAC system must address this heat (remove it) in order to maintain the functionality of the computer equipment. The cooling equipment for these spaces operates differently than equipment where the heating load is related to the human occupants of the space. The test procedure for this equipment is found in ASHRAE Standard 127.

Actual equipment-load profiles will vary by system type, building architecture, occupancy patterns and climate. Therefore, it is possible to determine performance only at representative part-load conditions. Although the results will not predict actual performance precisely, they can be used to compare different manufacturers and different equipment types in much the same way that the Environmental Protection Agency's (EPA) mileage ratings allow comparison of different cars, even though actual mileage will vary. Three types of part-load descriptors are used in Tables C403.2.3(1) through C403.2.3(7).

1. Heating seasonal performance factor (HSPF), annual fuel utilization efficiency (AFUE), seasonal energy efficiency ratio (SEER). These are seasonal performance descriptors for gas furnaces and air conditioners, respectively, determined in accordance with the corresponding test procedures listed. These ratings are based on a typical weather profile and include the effects of equipment cycling. All products covered by the NAECA must be rated using these descriptors.

2. IPLV. IPLV is a measure of part-load performance for some Air Conditioning, Heating and Refrigeration Institute (AHRI)-rated equipment with unloading capability. The units of IPLV are the same as those for the corresponding full-load descriptor; for instance, an IPLV for an air conditioner has the same units as the EER. IPLV is a weighted average of the steady state equipment performance at several load conditions. An IPLV may contain up to three part-load efficiencies, in addition to the full-load efficiency. Because measurements are steady-state, IPLV does not include the effects of equipment cycling; equipment is rated only at the part loading allowed by the equipment and its controls.

3. Low- or High-temperature ratings and COP. Some equipment is rated at an off-design condition, in addition to the standard full-load condition. Although not actually a part-load condition, because the equipment is tested at full capacity, the off-design rating can give an indication of equipment performance at different temperature conditions. Examples include high [47°F (8°C)] and low [17°F (-8°C)] outside air temperature COP ratings for heat pumps, and high- [85°F (29°C)] and low- [75°F (24°C)] temperature ratings for hydronic heat pumps. The ratings are steady state and do not include the effect of equipment cycling.

For additional commentary related to the tables, see the commentary to Section C403.2.3.

TABLE C403.2.3(1)
MINIMUM EFFICIENCY REQUIREMENTS:
ELECTRICALLY OPERATED UNITARY AIR CONDITIONERS AND CONDENSING UNITS

EQUIPMENT TYPE	SIZE CATEGORY	HEATING SECTION TYPE	SUBCATEGORY OR RATING CONDITION	MINIMUM EFFICIENCY		TEST PROCEDURE[a]
				Before 1/1/2016	As of 1/1/2016	
Air conditioners, air cooled	< 65,000 Btu/h[b]	All	Split System	13.0 SEER	13.0 SEER	AHRI 210/240
			Single Package	13.0 SEER	14.0 SEER[c]	
Through-the-wall (air cooled)	≤ 30,000 Btu/h[b]	All	Split system	12.0 SEER	12.0 SEER	
			Single Package	12.0 SEER	12.0 SEER	
Small-duct high-velocity (air cooled)	< 65,000 Btu/h[b]	All	Split System	11.0 SEER	11.0 SEER	
Air conditioners, air cooled	≥ 65,000 Btu/h and < 135,000 Btu/h	Electric Resistance (or None)	Split System and Single Package	11.2 EER 11.4 IEER	11.2 EER 12.8 IEER	AHRI 340/360
		All other	Split System and Single Package	11.0 EER 11.2 IEER	11.0 EER 12.6 IEER	
	≥ 135,000 Btu/h and < 240,000 Btu/h	Electric Resistance (or None)	Split System and Single Package	11.0 EER 11.2 IEER	11.0 EER 12.4 IEER	
		All other	Split System and Single Package	10.8 EER 11.0 IEER	10.8 EER 12.2 IEER	
	≥ 240,000 Btu/h and < 760,000 Btu/h	Electric Resistance (or None)	Split System and Single Package	10.0 EER 10.1 IEER	10.0 EER 11.6 IEER	
		All other	Split System and Single Package	9.8 EER 9.9 IEER	9.8 EER 11.4 IEER	
	≥ 760,000 Btu/h	Electric Resistance (or None)	Split System and Single Package	9.7 EER 9.8 IEER	9.7 EER 11.2 IEER	
		All other	Split System and Single Package	9.5 EER 9.6 IEER	9.5 EER 11.0 IEER	
Air conditioners, water cooled	< 65,000 Btu/h[b]	All	Split System and Single Package	12.1 EER 12.3 IEER	12.1 EER 12.3 IEER	AHRI 210/240
	≥ 65,000 Btu/h and < 135,000 Btu/h	Electric Resistance (or None)	Split System and Single Package	12.1 EER 12.3 IEER	12.1 EER 13.9 IEER	AHRI 340/360
		All other	Split System and Single Package	11.9 EER 12.1 IEER	11.9 EER 13.7 IEER	
	≥ 135,000 Btu/h and < 240,000 Btu/h	Electric Resistance (or None)	Split System and Single Package	12.5 EER 12.5 IEER	12.5 EER 13.9 IEER	
		All other	Split System and Single Package	12.3 EER 12.5 IEER	12.3 EER 13.7 IEER	
	≥ 240,000 Btu/h and < 760,000 Btu/h	Electric Resistance (or None)	Split System and Single Package	12.4 EER 12.6 IEER	12.4 EER 13.6 IEER	
		All other	Split System and Single Package	12.2 EER 12.4 IEER	12.2 EER 13.4 IEER	
	≥ 760,000 Btu/h	Electric Resistance (or None)	Split System and Single Package	12.2 EER 12.4 IEER	12.2 EER 13.5 IEER	
		All other	Split System and Single Package	12.0 EER 12.2 IEER	12.0 EER 13.3 IEER	

(continued)

TABLE C403.2.3(1)—continued
MINIMUM EFFICIENCY REQUIREMENTS:
ELECTRICALLY OPERATED UNITARY AIR CONDITIONERS AND CONDENSING UNITS

EQUIPMENT TYPE	SIZE CATEGORY	HEATING SECTION TYPE	SUB-CATEGORY OR RATING CONDITION	MINIMUM EFFICIENCY		TEST PROCEDURE[a]
				Before 1/1/2016	As of 1/1/2016	
Air conditioners, evaporatively cooled	< 65,000 Btu/h[b]	All	Split System and Single Package	12.1 EER 12.3 IEER	12.1 EER 12.3 IEER	AHRI 210/240
	≥ 65,000 Btu/h and < 135,000 Btu/h	Electric Resistance (or None)	Split System and Single Package	12.1 EER 12.3 IEER	12.1 EER 12.3 IEER	AHRI 340/360
		All other	Split System and Single Package	11.9 EER 12.1 IEER	11.9 EER 12.1 IEER	
	≥ 135,000 Btu/h and < 240,000 Btu/h	Electric Resistance (or None)	Split System and Single Package	12.0 EER 12.2 IEER	12.0 EER 12.2 IEER	
		All other	Split System and Single Package	11.8 EER 12.0 IEER	11.8 EER 12.0 IEER	
	≥ 240,000 Btu/h and < 760,000 Btu/h	Electric Resistance (or None)	Split System and Single Package	11.9 EER 12.1 IEER	11.9 EER 12.1 IEER	
		All other	Split System and Single Package	11.7 EER 11.9 IEER	11.7 EER 11.9 IEER	
	≥ 760,000 Btu/h	Electric Resistance (or None)	Split System and Single Package	11.7 EER 11.9 IEER	11.7 EER 11.9 IEER	
		All other	Split System and Single Package	11.5 EER 11.7 IEER	11.5 EER 11.7 IEER	
Condensing units, air cooled	≥ 135,000 Btu/h			10.5 EER 11.8 IEER	10.5 EER 11.8 IEER	AHRI 365
Condensing units, water cooled	≥ 135,000 Btu/h			13.5 EER 14.0 IEER	13.5 EER 14.0 IEER	
Condensing units, evaporatively cooled	≥ 135,000 Btu/h			13.5 EER 14.0 IEER	13.5 EER 14.0 IEER	

For SI: 1 British thermal unit per hour = 0.2931 W.

a. Chapter 6 contains a complete specification of the referenced test procedure, including the reference year version of the test procedure.

b. Single-phase, air-cooled air conditioners less than 65,000 Btu/h are regulated by NAECA. SEER values are those set by NAECA.

c. Minimum efficiency as of January 1, 2015.

TABLE C403.2.3(2)
MINIMUM EFFICIENCY REQUIREMENTS:
ELECTRICALLY OPERATED UNITARY AND APPLIED HEAT PUMPS

EQUIPMENT TYPE	SIZE CATEGORY	HEATING SECTION TYPE	SUBCATEGORY OR RATING CONDITION	MINIMUM EFFICIENCY		TEST PROCEDURE[a]
				Before 1/1/2016	As of 1/1/2016	
Air cooled (cooling mode)	< 65,000 Btu/h[b]	All	Split System	13.0 SEER[c]	14.0 SEER[c]	AHRI 210/240
			Single Package	13.0 SEER[c]	14.0 SEER[c]	
Through-the-wall, air cooled	≤ 30,000 Btu/h[b]	All	Split System	12.0 SEER	12.0 SEER	
			Single Package	12.0 SEER	12.0 SEER	
Single-duct high-velocity air cooled	< 65,000 Btu/h[b]	All	Split System	11.0 SEER	11.0 SEER	
Air cooled (cooling mode)	≥ 65,000 Btu/h and < 135,000 Btu/h	Electric Resistance (or None)	Split System and Single Package	11.0 EER 11.2 IEER	11.0 EER 12.0 IEER	AHRI 340/360
		All other	Split System and Single Package	10.8 EER 11.0 IEER	10.8 EER 11.8 IEER	
	≥ 135,000 Btu/h and < 240,000 Btu/h	Electric Resistance (or None)	Split System and Single Package	10.6 EER 10.7 IEER	10.6 EER 11.6 IEER	
		All other	Split System and Single Package	10.4 EER 10.5 IEER	10.4 EER 11.4 IEER	
	≥ 240,000 Btu/h	Electric Resistance (or None)	Split System and Single Package	9.5 EER 9.6 IEER	9.5 EER 10.6 IEER	
		All other	Split System and Single Package	9.3 EER 9.4 IEER	9.3 EER 9.4 IEER	
Water to Air: Water Loop (cooling mode)	< 17,000 Btu/h	All	86°F entering water	12.2 EER	12.2 EER	ISO 13256-1
	≥ 17,000 Btu/h and < 65,000 Btu/h	All	86°F entering water	13.0 EER	13.0 EER	
	≥ 65,000 Btu/h and < 135,000 Btu/h	All	86°F entering water	13.0 EER	13.0 EER	
Water to Air: Ground Water (cooling mode)	< 135,000 Btu/h	All	59°F entering water	18.0 EER	18.0 EER	ISO 13256-1
Brine to Air: Ground Loop (cooling mode)	< 135,000 Btu/h	All	77°F entering water	14.1 EER	14.1 EER	ISO 13256-1
Water to Water: Water Loop (cooling mode)	< 135,000 Btu/h	All	86°F entering water	10.6 EER	10.6 EER	ISO 13256-2
Water to Water: Ground Water (cooling mode)	< 135,000 Btu/h	All	59°F entering water	16.3 EER	16.3 EER	
Brine to Water: Ground Loop (cooling mode)	< 135,000 Btu/h	All	77°F entering fluid	12.1 EER	12.1 EER	

(continued)

TABLE C403.2.3(2)—continued
MINIMUM EFFICIENCY REQUIREMENTS:
ELECTRICALLY OPERATED UNITARY AND APPLIED HEAT PUMPS

EQUIPMENT TYPE	SIZE CATEGORY	HEATING SECTION TYPE	SUBCATEGORY OR RATING CONDITION	MINIMUM EFFICIENCY		TEST PROCEDURE[a]
				Before 1/1/2016	As of 1/1/2016	
Air cooled (heating mode)	< 65,000 Btu/h[b]	—	Split System	7.7 HSPF[c]	8.2 HSPF[c]	AHRI 210/240
		—	Single Package	7.7 HSPF[c]	8.0 HSPF[c]	
Through-the-wall, (air cooled, heating mode)	≤ 30,000 Btu/h[b] (cooling capacity)	—	Split System	7.4 HSPF	7.4 HSPF	
		—	Single Package	7.4 HSPF	7.4 HSPF	
Small-duct high velocity (air cooled, heating mode)	< 65,000 Btu/h[b]	—	Split System	6.8 HSPF	6.8 HSPF	
Air cooled (heating mode)	≥ 65,000 Btu/h and < 135,000 Btu/h (cooling capacity)	—	47°F db/43°F wb outdoor air	3.3 COP	3.3 COP	AHRI 340/360
			17°F db/15°F wb outdoor air	2.25 COP	2.25 COP	
	≥ 135,000 Btu/h (cooling capacity)	—	47°F db/43°F wb outdoor air	3.2 COP	3.2 COP	
			17°F db/15°F wb outdoor air	2.05 COP	2.05 COP	
Water to Air: Water Loop (heating mode)	< 135,000 Btu/h (cooling capacity)	—	68°F entering water	4.3 COP	4.3 COP	ISO 13256-1
Water to Air: Ground Water (heating mode)	< 135,000 Btu/h (cooling capacity)	—	50°F entering water	3.7 COP	3.7 COP	
Brine to Air: Ground Loop (heating mode)	< 135,000 Btu/h (cooling capacity)	—	32°F entering fluid	3.2 COP	3.2 COP	
Water to Water: Water Loop (heating mode)	< 135,000 Btu/h (cooling capacity)	—	68°F entering water	3.7 COP	3.7 COP	ISO 13256-2
Water to Water: Ground Water (heating mode)	< 135,000 Btu/h (cooling capacity)	—	50°F entering water	3.1 COP	3.1 COP	
Brine to Water: Ground Loop (heating mode)	< 135,000 Btu/h (cooling capacity)	—	32°F entering fluid	2.5 COP	2.5 COP	

For SI: 1 British thermal unit per hour = 0.2931 W, °C = [(°F) - 32]/1.8.

a. Chapter 6 contains a complete specification of the referenced test procedure, including the reference year version of the test procedure.

b. Single-phase, air-cooled air conditioners less than 65,000 Btu/h are regulated by NAECA. SEER values are those set by NAECA.

c. Minimum efficiency as of January 1, 2015.

TABLE C403.2.3(3)
MINIMUM EFFICIENCY REQUIREMENTS:
ELECTRICALLY OPERATED PACKAGED TERMINAL AIR CONDITIONERS,
PACKAGED TERMINAL HEAT PUMPS, SINGLE-PACKAGE VERTICAL AIR CONDITIONERS,
SINGLE VERTICAL HEAT PUMPS, ROOM AIR CONDITIONERS AND ROOM AIR-CONDITIONER HEAT PUMPS

EQUIPMENT TYPE	SIZE CATEGORY (INPUT)	SUBCATEGORY OR RATING CONDITION	MINIMUM EFFICIENCY	TEST PROCEDURE[a]
PTAC (cooling mode) new construction	All Capacities	95°F db outdoor air	14.0 – (0.300 × Cap/1000) EER[c]	AHRI 310/380
PTAC (cooling mode) replacements[b]	All Capacities	95°F db outdoor air	10.9 - (0.213 × Cap/1000) EER	
PTHP (cooling mode) new construction	All Capacities	95°F db outdoor air	14.0 - (0.300 × Cap/1000) EER	
PTHP (cooling mode) replacements[b]	All Capacities	95°F db outdoor air	10.8 - (0.213 × Cap/1000) EER	
PTHP (heating mode) new construction	All Capacities	—	3.2 - (0.026 × Cap/1000) COP	
PTHP (heating mode) replacements[b]	All Capacities	—	2.9 - (0.026 × Cap/1000) COP	
SPVAC (cooling mode)	< 65,000 Btu/h	95°F db/ 75°F wb outdoor air	9.0 EER	AHRI 390
	≥ 65,000 Btu/h and < 135,000 Btu/h	95°F db/ 75°F wb outdoor air	8.9 EER	
	≥ 135,000 Btu/h and < 240,000 Btu/h	95°F db/ 75°F wb outdoor air	8.6 EER	
SPVHP (cooling mode)	< 65,000 Btu/h	95°F db/ 75°F wb outdoor air	9.0 EER	
	≥ 65,000 Btu/h and < 135,000 Btu/h	95°F db/ 75°F wb outdoor air	8.9 EER	
	≥ 135,000 Btu/h and < 240,000 Btu/h	95°F db/ 75°F wb outdoor air	8.6 EER	
SPVHP (heating mode)	< 65,000 Btu/h	47°F db/ 43°F wb outdoor air	3.0 COP	AHRI 390
	≥ 65,000 Btu/h and < 135,000 Btu/h	47°F db/ 43°F wb outdoor air	3.0 COP	
	≥ 135,000 Btu/h and < 240,000 Btu/h	47°F db/ 75°F wb outdoor air	2.9 COP	
Room air conditioners, with louvered sides	< 6,000 Btu/h	—	9.7 SEER	ANSI/ AHAM RAC-1
	≥ 6,000 Btu/h and < 8,000 Btu/h	—	9.7 EER	
	≥ 8,000 Btu/h and < 14,000 Btu/h	—	9.8 EER	
	≥ 14,000 Btu/h and < 20,000 Btu/h	—	9.7 SEER	
	≥ 20,000 Btu/h	—	8.5 EER	
Room air conditioners, without louvered sides	< 8,000 Btu/h	—	9.0 EER	
	≥ 8,000 Btu/h and < 20,000 Btu/h	—	8.5 EER	
	≥ 20,000 Btu/h	—	8.5 EER	
Room air-conditioner heat pumps with louvered sides	< 20,000 Btu/h	—	9.0 EER	
	≥ 20,000 Btu/h	—	8.5 EER	
Room air-conditioner heat pumps without louvered sides	< 14,000 Btu/h	—	8.5 EER	
	≥ 14,000 Btu/h	—	8.0 EER	

(continued)

TABLE C403.2.3(3)—continued
MINIMUM EFFICIENCY REQUIREMENTS: ELECTRICALLY OPERATED PACKAGED TERMINAL AIR CONDITIONERS, PACKAGED TERMINAL HEAT PUMPS, SINGLE-PACKAGE VERTICAL AIR CONDITIONERS, SINGLE VERTICAL HEAT PUMPS, ROOM AIR CONDITIONERS AND ROOM AIR-CONDITIONER HEAT PUMPS

EQUIPMENT TYPE	SIZE CATEGORY (INPUT)	SUBCATEGORY OR RATING CONDITION	MINIMUM EFFICIENCY	TEST PROCEDURE[a]
Room air conditioner casement only	All capacities	—	8.7 EER	ANSI/ AHAM RAC-1
Room air conditioner casement-slider	All capacities	—	9.5 EER	

For SI: 1 British thermal unit per hour = 0.2931 W, °C = [(°F) - 32]/1.8, wb = wet bulb, db = dry bulb.

"Cap" = The rated cooling capacity of the project in Btu/h. Where the unit's capacity is less than 7000 Btu/h, use 7000 Btu/h in the calculation. Where the unit's capacity is greater than 15,000 Btu/h, use 15,000 Btu/h in the calculations.

a. Chapter 6 contains a complete specification of the referenced test procedure, including the referenced year version of the test procedure.

b. Replacement unit shall be factory labeled as follows: "MANUFACTURED FOR REPLACEMENT APPLICATIONS ONLY: NOT TO BE INSTALLED IN NEW CONSTRUCTION PROJECTS." Replacement efficiencies apply only to units with existing sleeves less than 16 inches (406 mm) in height and less than 42 inches (1067 mm) in width.

c. Before January 1, 2015 the minimum efficiency shall be 13.8 - (0.300 x Cap/1000) EER.

TABLE 403.2.3(4)
WARM-AIR FURNACES AND COMBINATION WARM-AIR FURNACES/AIR-CONDITIONING UNITS, WARM-AIR DUCT FURNACES AND UNIT HEATERS, MINIMUM EFFICIENCY REQUIREMENTS

EQUIPMENT TYPE	SIZE CATEGORY (INPUT)	SUBCATEGORY OR RATING CONDITION	MINIMUM EFFICIENCY[d, e]	TEST PROCEDURE[a]
Warm-air furnaces, gas fired	< 225,000 Btu/h	—	78% AFUE or 80%E_t[c]	DOE 10 CFR Part 430 or ANSI Z21.47
	≥ 225,000 Btu/h	Maximum capacity[c]	80%E_t[f]	ANSI Z21.47
Warm-air furnaces, oil fired	< 225,000 Btu/h	—	78% AFUE or 80%E_t[c]	DOE 10 CFR Part 430 or UL 727
	≥ 225,000 Btu/h	Maximum capacity[b]	81%E_t[g]	UL 727
Warm-air duct furnaces, gas fired	All capacities	Maximum capacity[b]	80%E_c	ANSI Z83.8
Warm-air unit heaters, gas fired	All capacities	Maximum capacity[b]	80%E_c	ANSI Z83.8
Warm-air unit heaters, oil fired	All capacities	Maximum capacity[b]	80%E_c	UL 731

For SI: 1 British thermal unit per hour = 0.2931 W.

a. Chapter 6 contains a complete specification of the referenced test procedure, including the referenced year version of the test procedure.

b. Minimum and maximum ratings as provided for and allowed by the unit's controls.

c. Combination units not covered by the National Appliance Energy Conservation Act of 1987 (NAECA) (3-phase power or cooling capacity greater than or equal to 65,000 Btu/h [19 kW]) shall comply with either rating.

d. E_t = Thermal efficiency. See test procedure for detailed discussion.

e. E_c = Combustion efficiency (100% less flue losses). See test procedure for detailed discussion.

f. E_c = Combustion efficiency. Units shall also include an IID, have jackets not exceeding 0.75 percent of the input rating, and have either power venting or a flue damper. A vent damper is an acceptable alternative to a flue damper for those furnaces where combustion air is drawn from the conditioned space.

g. E_t = Thermal efficiency. Units shall also include an IID, have jacket losses not exceeding 0.75 percent of the input rating, and have either power venting or a flue damper. A vent damper is an acceptable alternative to a flue damper for those furnaces where combustion air is drawn from the conditioned space.

TABLE C403.2.3(5)
MINIMUM EFFICIENCY REQUIREMENTS: GAS- AND OIL-FIRED BOILERS

EQUIPMENT TYPE[a]	SUBCATEGORY OR RATING CONDITION	SIZE CATEGORY (INPUT)	MINIMUM EFFICIENCY[d, e]	TEST PROCEDURE
Boilers, hot water	Gas-fired	< 300,000 Btu/h	80% AFUE	10 CFR Part 430
		≥ 300,000 Btu/h and ≤ 2,500,000 Btu/h[b]	80% E_t	10 CFR Part 431
		> 2,500,000 Btu/h[a]	82% E_c	
	Oil-fired[c]	< 300,000 Btu/h	80% AFUE	10 CFR Part 430
		≥ 300,000 Btu/h and ≤ 2,500,000 Btu/h[b]	82% E_t	10 CFR Part 431
		> 2,500,000 Btu/h[a]	84% E_c	
Boilers, steam	Gas-fired	< 300,000 Btu/h	75% AFUE	10 CFR Part 430
	Gas-fired- all, except natural draft	≥ 300,000 Btu/h and ≤ 2,500,000 Btu/h[b]	79% E_t	10 CFR Part 431
		> 2,500,000 Btu/h[a]	79% E_t	
	Gas-fired-natural draft	≥ 300,000 Btu/h and ≤ 2,500,000 Btu/h[b]	77% E_t	
		> 2,500,000 Btu/h[a]	77% E_t	
	Oil-fired[c]	< 300,000 Btu/h	80% AFUE	10 CFR Part 430
		≥ 300,000 Btu/h and ≤ 2,500,000 Btu/h[b]	81% E_t	10 CFR Part 431
		> 2,500,000 Btu/h[a]	81% E_t	

For SI: 1 British thermal unit per hour = 0.2931 W.

a. These requirements apply to boilers with rated input of 8,000,000 Btu/h or less that are not packaged boilers and to all packaged boilers. Minimum efficiency requirements for boilers cover all capacities of packaged boilers.

b. Maximum capacity – minimum and maximum ratings as provided for and allowed by the unit's controls.

c. Includes oil-fired (residual).

d. E_c = Combustion efficiency (100 percent less flue losses).

e. E_t = Thermal efficiency. See referenced standard for detailed information.

TABLE C403.2.3(6)
MINIMUM EFFICIENCY REQUIREMENTS:
CONDENSING UNITS, ELECTRICALLY OPERATED

EQUIPMENT TYPE	SIZE CATEGORY	MINIMUM EFFICIENCY[b]	TEST PROCEDURE[a]
Condensing units, air cooled	≥ 135,000 Btu/h	10.1 EER 11.2 IPLV	AHRI 365
Condensing units, water or evaporatively cooled	≥ 135,000 Btu/h	13.1 EER 13.1 IPLV	

For SI: 1 British thermal unit per hour = 0.2931 W.

a. Chapter 6 contains a complete specification of the referenced test procedure, including the referenced year version of the test procedure.

b. IPLVs are only applicable to equipment with capacity modulation.

TABLE C403.2.3(7) WATER CHILLING PACKAGES – EFFICIENCY REQUIREMENTS[a, b, d]

EQUIPMENT TYPE	SIZE CATEGORY	UNITS	BEFORE 1/1/2015		AS OF 1/1/2015		TEST PROCEDURE[c]
			Path A	Path B	Path A	Path B	
Air-cooled chillers	< 150 Tons	EER (Btu/W)	≥ 9.562 FL	NAc	≥ 10.100 FL	≥ 9.700 FL	AHRI 550/ 590
			≥ 12.500 IPLV		≥ 13.700 IPLV	≥ 15,800 IPLV	
	≥ 150 Tons		≥ 9.562 FL	NAc	≥ 10.100 FL	≥ 9.700 FL	
			≥ 12.500 IPLV		≥ 14.000 IPLV	≥ 16.100 IPLV	
Air cooled without condenser, electrically operated	All capacities	EER (Btu/W)	Air-cooled chillers without condenser shall be rated with matching condensers and complying with air-cooled chiller efficiency requirements.				
Water cooled, electrically operated positive displacement	< 75 Tons	kW/ton	≤ 0.780 FL	≤ 0.800 FL	≤ 0.750 FL	≤ 0.780 FL	
			≤ 0.630 IPLV	≤ 0.600 IPLV	≤ 0.600 IPLV	≤ 0.500 IPLV	
	≥ 75 tons and < 150 tons		≤ 0.775 FL	≤ 0.790 FL	≤ 0.720 FL	≤ 0.750 FL	
			≤ 0.615 IPLV	≤ 0.586 IPLV	≤ 0.560 IPLV	≤ 0.490 IPLV	
	≥ 150 tons and < 300 tons		≤ 0.680 FL	≤ 0.718 FL	≤ 0.660 FL	≤ 0.680 FL	
			≤ 0.580 IPLV	≤ 0.540 IPLV	≤ 0.540 IPLV	≤ 0.440 IPLV	
	≥ 300 tons and < 600 tons		≤ 0.620 FL	≤ 0.639 FL	≤ 0.610 FL	≤ 0.625 FL	
			≤ 0.540 IPLV	≤ 0.490 IPLV	≤ 0.520 IPLV	≤ 0.410 IPLV	
	≥ 600 tons		≤ 0.620 FL	≤ 0.639 FL	≤ 0.560 FL	≤ 0.585 FL	
			≤ 0.540 IPLV	≤ 0.490 IPLV	≤ 0.500 IPLV	≤ 0.380 IPLV	
Water cooled, electrically operated centrifugal	< 150 Tons	kW/ton	≤ 0.634 FL	≤ 0.639 FL	≤ 0.610 FL	≤ 0.695 FL	
			≤ 0.596 IPLV	≤ 0.450 IPLV	≤ 0.550 IPLV	≤ 0.440 IPLV	
	≥ 150 tons and < 300 tons		≤ 0.634 FL	≤ 0.639 FL	≤ 0.610 FL	≤ 0.635 FL	
			≤ 0.596 IPLV	≤ 0.450 IPLV	≤ 0.550 IPLV	≤ 0.400 IPLV	
	≥ 300 tons and < 400 tons		≤ 0.576 FL	≤ 0.600 FL	≤ 0.560 FL	≤ 0.595 FL	
			≤ 0.549 IPLV	≤ 0.400 IPLV	≤ 0.520 IPLV	≤ 0.390 IPLV	
	≥ 400 tons and < 600 tons		≤ 0.576 FL	≤ 0.600 FL	≤ 0.560 FL	≤ 0.585 FL	
			≤ 0.549 IPLV	≤ 0.400 IPLV	≤ 0.500 IPLV	≤ 0.380 IPLV	
	≥ 600 Tons		≤ 0.570 FL	≤ 0.590 FL	≤ 0.560 FL	≤ 0.585 FL	
			≤ 0.539 IPLV	≤ 0.400 IPLV	≤ 0.500 IPLV	≤ 0.380 IPLV	
Air cooled, absorption, single effect	All capacities	COP	≥ 0.600 FL	NA[c]	≥ 0.600 FL	NA[c]	AHRI 560
Water cooled absorption, single effect	All capacities	COP	≥ 0.700 FL	NA[c]	≥ 0.700 FL	NA[c]	
Absorption, double effect, indirect fired	All capacities	COP	≥ 1.000 FL	NA[c]	≥ 1.000 FL	NA[c]	
			≥ 1.050 IPLV		≥ 1.050 IPLV		
Absorption double effect direct fired	All capacities	COP	≥ 1.000 FL	NA[c]	≥ 1.000 FL	NA[c]	
			≥ 1.000 IPLV		≥ 1.050 IPLV		

a. The requirements for centrifugal chiller shall be adjusted for nonstandard rating conditions in accordance with Section C403.2.3.1 and are only applicable for the range of conditions listed in Section C403.2.3.1. The requirements for air-cooled, water-cooled positive displacement and absorption chillers are at standard rating conditions defined in the reference test procedure.

b. Both the full-load and IPLV requirements shall be met or exceeded to comply with this standard. Where there is a Path B, compliance can be with either Path A or Path B for any application.

c. NA means the requirements are not applicable for Path B and only Path A can be used for compliance.

d. FL represents the full-load performance requirements and IPLV the part-load performance requirements.

TABLE C403.2.3(8)
MINIMUM EFFICIENCY REQUIREMENTS:
HEAT REJECTION EQUIPMENT

EQUIPMENT TYPE[a]	TOTAL SYSTEM HEAT REJECTION CAPACITY AT RATED CONDITIONS	SUBCATEGORY OR RATING CONDITION[i]	PERFORMANCE REQUIRED[b, c, d, g, h]	TEST PROCEDURE[e, f]
Propeller or axial fan open-circuit cooling towers	All	95°F entering water 85°F leaving water 75°F entering wb	≥ 40.2 gpm/hp	CTI ATC-105 and CTI STD-201
Centrifugal fan open-circuit cooling towers	All	95°F entering water 85°F leaving water 75°F entering wb	≥ 20.0 gpm/hp	CTI ATC-105 and CTI STD-201
Propeller or axial fan closed-circuit cooling towers	All	102°F entering water 90°F leaving water 75°F entering wb	≥ 14.0 gpm/hp	CTI ATC-105S and CTI STD-201
Centrifugal fan closed-circuit cooling towers	All	102°F entering water 90°F leaving water 75°F entering wb	≥ 7.0 gpm/hp	CTI ATC-105S and CTI STD-201
Propeller or axial fan evaporative condensers	All	Ammonia Test Fluid 140°F entering gas temperature 96.3°F condensing temperature 75°F entering wb	≥ 134,000 Btu/h·hp	CTI ATC-106
Centrifugal fan evaporative condensers	All	Ammonia Test Fluid 140°F entering gas temperature 96.3°F condensing temperature 75°F entering wb	≥ 110,000 Btu/h·hp	CTI ATC-106
Propeller or axial fan evaporative condensers	All	R-507A Test Fluid 165°F entering gas temperature 105°F condensing temperature 75°F entering wb	≥ 157,000 Btu/h·hp	CTI ATC-106
Centrifugal fan evaporative condensers	All	R-507A Test Fluid 165°F entering gas temperature 105°F condensing temperature 75°F entering wb	≥ 135,000 Btu/h·hp	CTI ATC-106
Air-cooled condensers	All	125°F Condensing Temperature 190°F Entering Gas Temperature 15°F subcooling 95°F entering db	≥ 176,000 Btu/h·hp	AHRI 460

For SI: °C = [(°F)-32]/1.8, L/s · kW = (gpm/hp)/(11.83), COP = (Btu/h · hp)/(2550.7),
db = dry bulb temperature, °F, wb = wet bulb temperature, °F.

a. The efficiencies and test procedures for both open- and closed-circuit cooling towers are not applicable to hybrid cooling towers that contain a combination of wet and dry heat exchange sections.

b. For purposes of this table, open circuit cooling tower performance is defined as the water flow rating of the tower at the thermal rating condition listed in Table 403.2.3(8) divided by the fan nameplate-rated motor power.

c. For purposes of this table, closed-circuit cooling tower performance is defined as the water flow rating of the tower at the thermal rating condition listed in Table 403.2.3(8) divided by the sum of the fan nameplate-rated motor power and the spray pump nameplate-rated motor power.

d. For purposes of this table, air-cooled condenser performance is defined as the heat rejected from the refrigerant divided by the fan nameplate-rated motor power.

e. Chapter 6 contains a complete specification of the referenced test procedure, including the referenced year version of the test procedure. The certification requirements do not apply to field-erected cooling towers.

f. Where a certification program exists for a covered product and it includes provisions for verification and challenge of equipment efficiency ratings, then the product shall be listed in the certification program; or, where a certification program exists for a covered product, and it includes provisions for verification and challenge of equipment efficiency ratings, but the product is not listed in the existing certification program, the ratings shall be verified by an independent laboratory test report.

g. Cooling towers shall comply with the minimum efficiency listed in the table for that specific type of tower with the capacity effect of any project-specific accessories and/or options included in the capacity of the cooling tower

h. For purposes of this table, evaporative condenser performance is defined as the heat rejected at the specified rating condition in the table divided by the sum of the fan motor nameplate power and the integral spray pump nameplate power

i. Requirements for evaporative condensers are listed with ammonia (R-717) and R-507A as test fluids in the table. Evaporative condensers intended for use with halocarbon refrigerants other than R-507A shall meet the minimum efficiency requirements listed in this table with R-507A as the test fluid.

TABLE C403.2.3(9)
MINIMUM EFFICIENCY AIR CONDITIONERS AND CONDENSING UNITS SERVING COMPUTER ROOMS

EQUIPMENT TYPE	NET SENSIBLE COOLING CAPACITY[a]	MINIMUM SCOP-127[b] EFFICIENCY DOWNFLOW UNITS / UPFLOW UNITS	TEST PROCEDURE
Air conditioners, air cooled	< 65,000 Btu/h	2.20 / 2.09	ANSI/ASHRAE 127
	≥ 65,000 Btu/h and < 240,000 Btu/h	2.10 / 1.99	
	≥ 240,000 Btu/h	1.90 / 1.79	
Air conditioners, water cooled	< 65,000 Btu/h	2.60 / 2.49	
	≥ 65,000 Btu/h and < 240,000 Btu/h	2.50 / 2.39	
	≥ 240,000 Btu/h	2.40 /2.29	
Air conditioners, water cooled with fluid economizer	< 65,000 Btu/h	2.55 /2.44	
	≥ 65,000 Btu/h and < 240,000 Btu/h	2.45 / 2.34	
	≥ 240,000 Btu/h	2.35 / 2.24	
Air conditioners, glycol cooled (rated at 40% propylene glycol)	< 65,000 Btu/h	2.50 / 2.39	
	≥ 65,000 Btu/h and < 240,000 Btu/h	2.15 / 2.04	
	≥ 240,000 Btu/h	2.10 / 1.99	
Air conditioners, glycol cooled (rated at 40% propylene glycol) with fluid economizer	< 65,000 Btu/h	2.45 / 2.34	
	≥ 65,000 Btu/h and < 240,000 Btu/h	2.10 / 1.99	
	≥ 240,000 Btu/h	2.05 / 1.94	

For SI: 1 British thermal unit per hour = 0.2931 W.

a. Net sensible cooling capacity: the total gross cooling capacity less the latent cooling less the energy to the air movement system. (Total Gross – latent – Fan Power).

b. Sensible coefficient of performance (SCOP-127): a ratio calculated by dividing the net sensible cooling capacity in watts by the total power input in watts (excluding reheaters and humidifiers) at conditions defined in ASHRAE Standard 127. The net sensible cooling capacity is the gross sensible capacity minus the energy dissipated into the cooled space by the fan system.

TABLE C403.2.3(10)
HEAT TRANSFER EQUIPMENT

EQUIPMENT TYPE	SUBCATEGORY	MINIMUM EFFICIENCY	TEST PROCEDURE[a]
Liquid-to-liquid heat exchangers	Plate type	NR	AHRI 400

NR = No Requirement.

a. Chapter 6 contains a complete specification of the referenced test procedure, including the referenced year version of the test procedure.

C403.2.3.1 Water-cooled centrifugal chilling packages. Equipment not designed for operation at AHRI Standard 550/590 test conditions of 44°F (7°C) leaving chilled-water temperature and 2.4 gpm/ton evaporator fluid flow and 85°F (29°C) entering condenser water temperature with 3 gpm/ton (0.054 I/s • kW) condenser water flow shall have maximum full-load kW/ton (FL) and part-load ratings requirements adjusted using Equations 4-6 and 4-7.

$$FL_{adj} = FL/K_{adj} \quad \textbf{(Equation 4-6)}$$

$$PL_{Vadj} = IPLV/K_{adj} \quad \textbf{(Equation 4-7)}$$

where:

K_{adj} = $A \times B$

FL = Full-load kW/ton value as specified in Table C403.2.3(7).

FL_{adj} = Maximum full-load kW/ton rating, adjusted for nonstandard conditions.

$IPLV$ = Value as specified in Table C403.2.3(7).

PLV_{adj} = Maximum $NPLV$ rating, adjusted for nonstandard conditions.

A = $0.00000014592 \cdot (LIFT)^4 - 0.0000346496 \cdot (LIFT)^3 + 0.00314196 \cdot (LIFT)^2 - 0.147199 \cdot (LIFT) + 3.9302$

B = $0.0015 \cdot L_{vg}E_{vap} + 0.934$

$LIFT$ = $L_{vg}Cond - L_{vg}E_{vap}$

$L_{vg}C_{ond}$ = Full-load condenser leaving fluid temperature (°F).

$L_{vg}E_{vap}$ = Full-load evaporator leaving temperature (°F).

The FL_{adj} and PLV_{adj} values are only applicable for centrifugal chillers meeting all of the following full-load design ranges:

1. Minimum evaporator leaving temperature: 36°F.
2. Maximum condenser leaving temperature: 115°F.
3. 20°F ≤ $LIFT$ ≤ 80°F.

❖ This section is somewhat self-explanatory. The equipment list in Table C403.2.3(7) is limited to equipment that is designed in the range of AHRI 550/590 test conditions. The code, however, intends that equipment outside of those test conditions also meet the operational criteria. This section provides requirements for that equipment.

The intent of this section is that water-cooled centrifugal chilling packages would comply with water as the tested fluid at covered temperature and flow combinations. AHRI 550/590 does not allow for testing with secondary coolants and it is impractical to require it in manufacturer's test facilities used for certification and performance testing.

C403.2.3.2 Positive displacement (air- and water-cooled) chilling packages. Equipment with a leaving fluid temperature higher than 32°F (0°C) and water-cooled positive displacement chilling packages with a condenser leaving fluid temperature below 115°F (46°C) shall meet the requirements of Table C403.2.3(7) when tested or certified with water at standard rating conditions, in accordance with the referenced test procedure.

❖ Positive displacement (both air- and water-cooled) chillers, with glycol added for freeze protection when the unit is off or for winter operation, would likely have used a secondary coolant with a freeze point below 27°F (-0.8°C). However, if the positive displacement chiller was designed to create a cooling temperature greater than 32°F (0°C), machine changes might hinder its ability to meet the requirements. Table C403.2.3(7) addresses positive displacement (air- and water-cooled) chilling packages with a leaving fluid temperature greater than 32°F (0°C) and water-cooled positive displacement chilling packages with a condenser leaving fluid temperature below 115°F (46°C) when tested or certified with water at standard rating conditions, in accordance with the referenced test procedure (see commentary, Section C403.2.3.1).

C403.2.4 HVAC system controls. Each heating and cooling system shall be provided with thermostatic controls as specified in Section C403.2.4.1, C403.2.4.1.3, C403.2.4.2, C403.2.4.3, C403.3.1, C403.4, C403.4.1 or C403.4.4.

❖ Thermostatic controls are required for each heating and cooling system. Sections C403.2.4.1 through C403.2.4.4 cover the related requirements.

C403.2.4.1 Thermostatic controls. The supply of heating and cooling energy to each *zone* shall be controlled by individual thermostatic controls capable of responding to temperature within the *zone*. Where humidification or dehumidification or both is provided, at least one humidity control device shall be provided for each humidity control system.

Exception: Independent perimeter systems that are designed to offset only building envelope heat losses, gains or both serving one or more perimeter *zones* also served by an interior system provided:

1. The perimeter system includes at least one thermostatic control *zone* for each building exposure having exterior walls facing only one orientation (within +/-45 degrees) (0.8 rad) for more than 50 contiguous feet (15 240 mm); and
2. The perimeter system heating and cooling supply is controlled by thermostats located within the *zones* served by the system.

❖ A thermostat is required to control heating or cooling in each zone (see Commentary Figure C403.2.4.1). Thermostats must have the capability to automatically set back or shut down heating and cooling systems when appropriate. Thermostats also must have an accessible override so occupants can operate the system during off hours. Heat pumps with supplementary electric-resistance heaters must have thermostats specifically designed for heat pump operation (see commentary, Section C403.2.4.1.1).

A programmable thermostat must be used to meet these requirements. These thermostats are available for heating only, cooling only, heating and cooling, and heat pump systems. These thermostats can set back or shut down the system during nights and weekends. In addition, occupants can temporarily override the thermostat and it will return to the original schedule without reprogramming.

Thermostats that control the temperature in areas where heating or cooling systems must operate continuously, or areas with a small load demand and manual shutoff switch, do not require a setback or shutoff control (see Section C403.2.4.2).

An HVAC thermostatic control "zone" is defined as a space or group of spaces with load characteristics sufficiently similar such that the desired space conditions can be maintained throughout with a single controlling device. This section requires that the supply of heating and cooling to each such zone be controlled by an individual temperature controller that responds to the temperature within the zone.

To meet this requirement, spaces must be grouped into proper control zones. For instance, spaces with exterior wall and glass exposures typically cannot be zoned with interior spaces. Similarly, spaces with

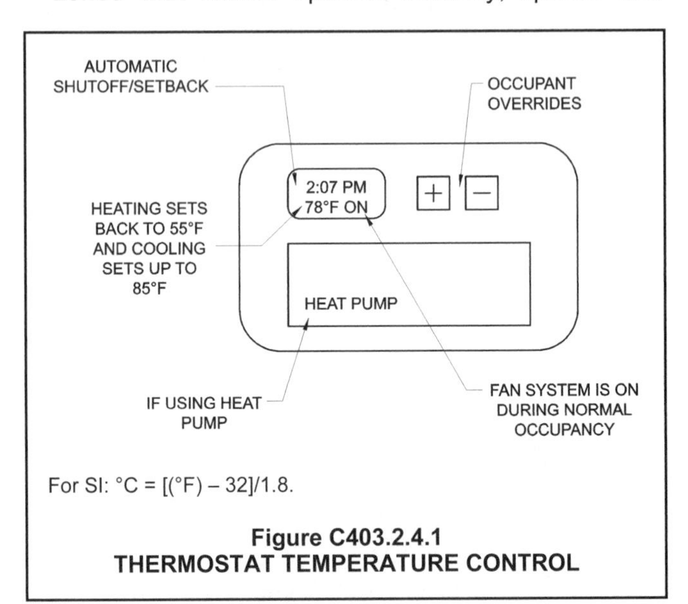

For SI: °C = [(°F) – 32]/1.8.

Figure C403.2.4.1
THERMOSTAT TEMPERATURE CONTROL

windows facing one direction should not be zoned with spaces having windows facing another orientation, unless the spaces are sufficiently open to one another and have reasonably uniform temperatures. Zoning in this manner generally does not apply to dwelling units or sleeping rooms (guest rooms), such as in high-rise apartment or hotel occupancies, since these types of uses are normally considered a single control zone or grouped together in a control zone with similar exposure characteristics.

The independent perimeter system exception is mandatory only as it applies to zones that are served by two independent HVAC systems: the perimeter system and the interior system. The perimeter system is designed to offset only skin loads—those loads that result from energy transfer through the building envelope. Typically, these systems are designed for heating only. The interior system is designed to handle cooling loads from light, equipment and people. This system also may be designed to handle skin-cooling loads if the perimeter system is heating only.

Commentary Figure C403.2.4.1(1) shows an example of this HVAC system design. The perimeter system consists of a heating-only fan coil, one for each building exposure. The interior system consists of a cooling-only VAV system serving the entire floor, including all exposures, as well as interior zones. If the perimeter system is capable of supplying heating to several zones at once, this design would not strictly meet the thermostatic control requirements of this section.

The heating fan coil shown in Commentary Figure C403.2.4.1(2) serves four zones of the VAV system. Therefore, heating energy from the fan coil is capable of being controlled by individual thermostats in each zone as required. For this reason, the interior system and the perimeter system will likely fight each other, with the perimeter system overheating some spaces and the interior system overcooling them to compensate. This is obviously energy wasteful. But the system can be designed to address this inefficiency, and the exception requires this to be the case as follows:

- The perimeter system must have not less than one zone for each major exposure, defined as an exterior wall that faces 50 contiguous feet (15 240 mm) or more in one direction. For example,

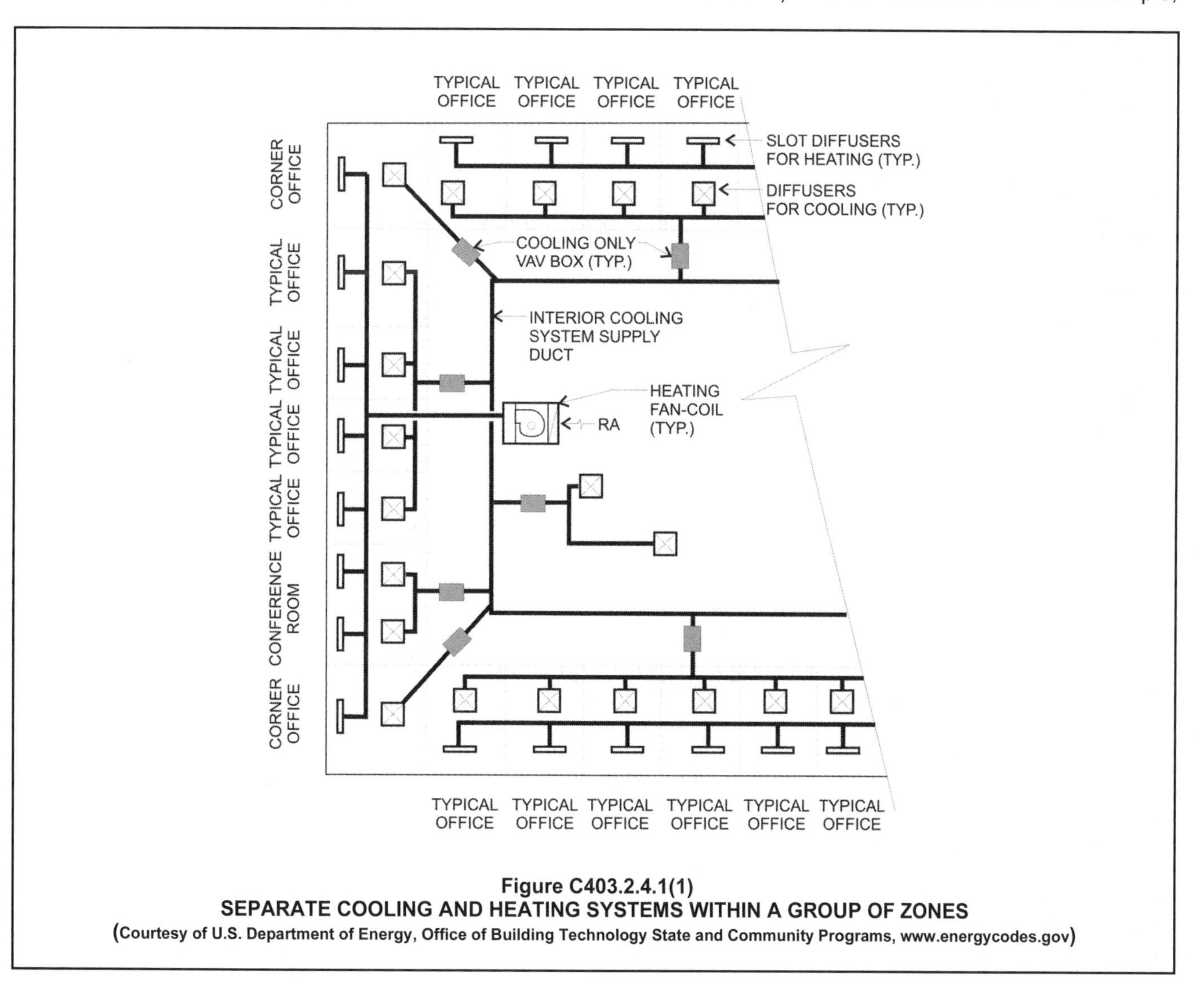

Figure C403.2.4.1(1)
SEPARATE COOLING AND HEATING SYSTEMS WITHIN A GROUP OF ZONES
(Courtesy of U.S. Department of Energy, Office of Building Technology State and Community Programs, www.energycodes.gov)

in Commentary Figure C403.2.4.1(2), a zone must be designated for each of the exposures that exceed 50 feet (15 240 mm) in length, but the shorter exposures on the serrated side of the building need not have individual zones. The shorter exposures could be served by a nearby zone with a reasonably similar load or exposure profile.

- Each perimeter system zone must be controlled by one or more thermostats located in the zones served. In the past, perimeter systems often were controlled by outside air sensors that would reset the output of the system proportional to the outside air temperature. But since solar loads can offset some of the heat loss from a space, this type of control inevitably causes overheating by the perimeter system when the sun is shining and subsequently fighting with the cooling system. Even when this control is improved by solar compensation, it still can result in wasteful fighting between interior and perimeter systems as a result of varying internal loads. Therefore, only controls that respond to the temperature within the zones served are allowed.

In Commentary Figure C403.2.4.1(2), this requirement might be met by controlling the perimeter fan coil off one of the thermostats controlling one of the four interior system VAV zones on the exposure. Alternatively, all four thermostat signals could be monitored along with the one requiring the most heat used to control the fan coil. Finally, a completely independent thermostat could be installed in one of the rooms on the exposure to control the fan coil, set to a setpoint that was below those controlling the VAV boxes and interlocked to other control points accordingly. A control schematic and narrative, as required for complex systems by Section C403.4.4, could be used as a basis for evaluating compliance (see commentary, Section C403.4.4).

As with thermostats, the code requires that humidity control systems be connected to humidistats or other such devices that would be capable of controlling and regulating their operation.

C403.2.4.1.1 Heat pump supplementary heat. Heat pumps having supplementary electric resistance heat shall have controls

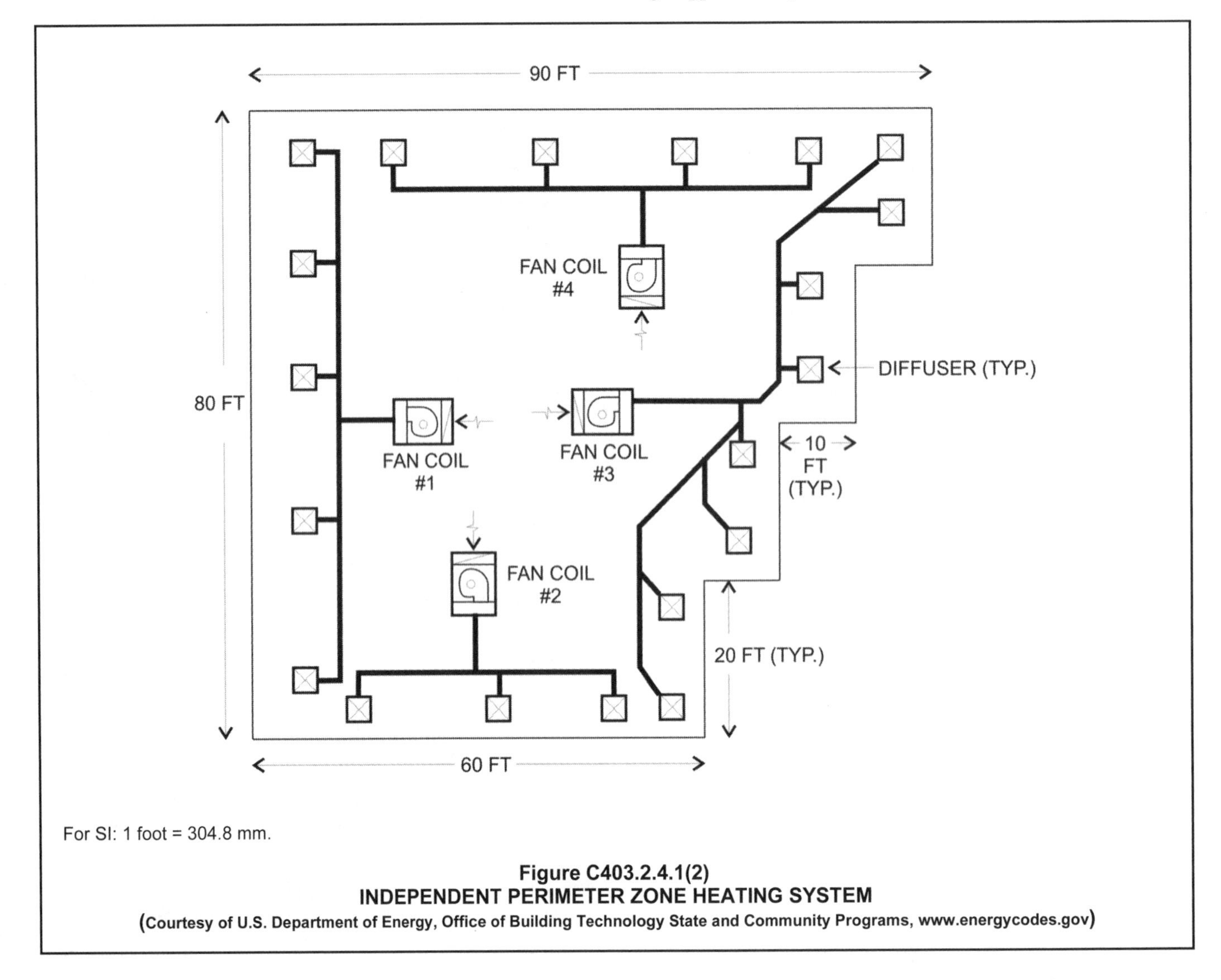

For SI: 1 foot = 304.8 mm.

Figure C403.2.4.1(2)
INDEPENDENT PERIMETER ZONE HEATING SYSTEM
(Courtesy of U.S. Department of Energy, Office of Building Technology State and Community Programs, www.energycodes.gov)

that, except during defrost, prevent supplementary heat operation where the heat pump can provide the heating load.

❖ Heat pumps with supplementary electric-resistance heaters must have thermostats designed specifically for heat pump operation.

Heat pump installations must include a thermostat that prevents supplementary heat from coming on when the heating requirements can be met by the heat pump alone. The intent is to use only the higher efficiency heat source in most conditions.

A two-stage thermostat with setpoints that control supplementary heating on the second stage (lower setpoint) and compression heating on the first stage (higher setpoint) meets this requirement. An optional energy conservation measure is to install an outside thermostat wired in series with the second stage. The outside thermostat can lock out the supplemental heat source while the outside temperature is high enough for the heat pump to meet the load.

The heat pump control system must be designed and adjusted so that under normal operating conditions, the refrigeration cycle (compression heating) will start first (before the supplementary heating) and shut off last (after the supplementary heating). This control requirement is intended to maximize energy savings by minimizing the use of the less efficient, electric-resistance supplementary heater. Standard heat pump thermostats supplied by heat pump manufacturers meet this requirement.

The heating capacity of air-source heat pumps will decrease as outside air temperatures fall. To make up for this deficiency, auxiliary heaters often are installed to augment the heat output for the heat pump. For heat pumps equipped with electric-resistance auxiliary heaters (with a COP of 1), the efficiency of the system is significantly reduced compared to the heat pump operating alone (with a COP typically greater than 2). The code, therefore, requires that controls be installed to prevent the auxiliary electric heater operation when the heating load can be met by the heat pump alone [at temperatures above 40°F (4°C)].

Of primary concern is the morning warm-up when the space may be well below the setpoint, even during relatively mild weather. The heat pump could warm the space sufficiently and quickly by itself, but typical thermostatic controls would cause the auxiliary heater to operate as well, wasting energy.

The best way to resolve this problem is to use an electronic thermostat designed specifically for use with heat pumps. This thermostat can sense whether the heat pump is raising space temperature during warm-up at a sufficient rate, or maintaining space temperature during normal operation and energizing the auxiliary heat only as required. More traditional electric controls also can be used, as demonstrated by the following two examples on two-stage thermostats.

Example of Heat Pump Auxiliary Heat Control—Two-stage Thermostat

Will a simple, two-stage thermostat, wired to bring on the auxiliary heat as the second stage, meet the requirements of this section?

No, because it will still cause auxiliary heat to be brought on during warm-up even when outdoor temperatures are mild and the heat pump has adequate capacity by itself. It can be used in conjunction with an outside air thermostat.

Example of Heat Pump Auxiliary Heat Control—Two-stage Thermostat with Outside Air Lock Out

Will an outdoor thermostat, wired to lock out auxiliary heat operation during mild weather, meet the requirements of this section?

Yes, but only when used in conjunction with a two-stage thermostat and only if wired properly. Occasionally, manufacturers' installation diagrams show outdoor thermostats wired to provide an additional thermostat stage while using only a single-stage thermostat. It is wired so that electric heat operates with the heat pump when outdoor temperatures are cold (below the outdoor thermostat setpoint). This may cause the auxiliary heat to operate when it is not required because the heat pump may be able to meet the load even during cold weather. The suggested system using conventional controls includes both of the following:

- A two-stage thermostat with the first stage wired to energize the heat pump and the second stage wired to bring on the auxiliary heat.
- An outdoor thermostat wired in series with the second stage so that the auxiliary heat will operate only if both the second stage of heat is required and the outdoor air is cold.

The outdoor thermostat setpoint must be set to the temperature at which the heat pump capacity will be insufficient to warm up the space in a reasonable time during warm-up [e.g., 40°F (4°C)].

C403.2.4.1.2 Deadband. Where used to control both heating and cooling, *zone* thermostatic controls shall be capable of providing a temperature range or deadband of at least 5°F (2.8°C) within which the supply of heating and cooling energy to the *zone* is capable of being shut off or reduced to a minimum.

Exceptions:

1. Thermostats requiring manual changeover between heating and cooling modes.
2. Occupancies or applications requiring precision in indoor temperature control as *approved* by the *code official*.

❖ Zone thermostatic controls that control both space heating and cooling must have a temperature differential range, or deadband, of not less than 5°F

(2.8°C) within which the supply of heating and cooling to the space is shut off or reduced to a minimum.

Commentary Figure C403.2.4.1.2 shows a proportional control scheme that meets this requirement. This might apply to a VAV zone where the cooling source is cold supply air and heating is provided by reheat or perhaps an independent perimeter heating system. The point from which the cooling supply is shut off or reduced to its minimum position to where the heating is turned on is called the deadband and must be not less than 5°F (2.8°C).

The deadband requirement typically is met using dual setpoint thermostats, which essentially are two thermostats built into the same enclosure. One thermostat controls heating and one controls cooling. The deadband can be achieved by setting the two setpoints not less than 5°F (2.8°C) apart. For proportional controls, such as pneumatic controls, that are calibrated so that the thermostat setpoint is at the midpoint of the control band, the setpoints would have to be set apart by at least 5°F (2.8°C) plus one throttling range. For instance, in Commentary Figure C403.2.4.1.2, the throttling range indicated is 2°F (1.1°C), so the deadband would be maintained by a heating setpoint of 69°F (21°C) and a cooling setpoint of 76°F (24°C).] Where two thermostats are used, Section C403.2.4.1.3 requires a mechanism and/or programming to make sure that the deadband setpoints are maintained.

Another type of thermostat that would meet the requirement is a deadband or hesitation thermostat. This thermostat is designed to have a temperature range within which its output signal is neutral, calling for neither heating nor cooling.

There are two exceptions. Under Exception 1, deadband controls are not necessary for thermostats that require a manual changeover between heating and cooling. This is typical of many residential thermostats. Occupants generally will allow the space temperature to swing considerably before changing the heating/cooling mode, thereby causing an effective deadband.

Exception 2 addresses thermostats in spaces that have special occupancies where precise space temperature control is required and therefore do not need deadband control. Examples include areas housing temperature-sensitive equipment or processes, such as specialty manufacturing areas and hospital operating rooms; or sensitive materials, such as a museum or art gallery. Generally, the temperature deadband will not be a problem for people, but it could affect certain processes or materials. Other examples where deadband control may not be appropriate include patient care areas for people who may be sensitive to wide temperature swings. Consult with the code official because many jurisdictions may already have an established policy regarding deadband control.

Buildings where deadband controls are appropriate include office buildings, retail stores, schools and hotels.

Example of Deadband Requirement—DDC Systems

A direct digital control (DDC) system using a space sensor and a "smart" controller is to be used to control a VAV box with hot water reheat. Does it have to meet the deadband requirement?

Yes. This system qualifies as a "zone thermostat," although it uses a space sensor and computer rather than a conventional thermostat to control space temperature. The software in the "smart" controller would have to support two separate control loops with individual setpoints, one for heating and one for cooling, each with separate output signals connecting

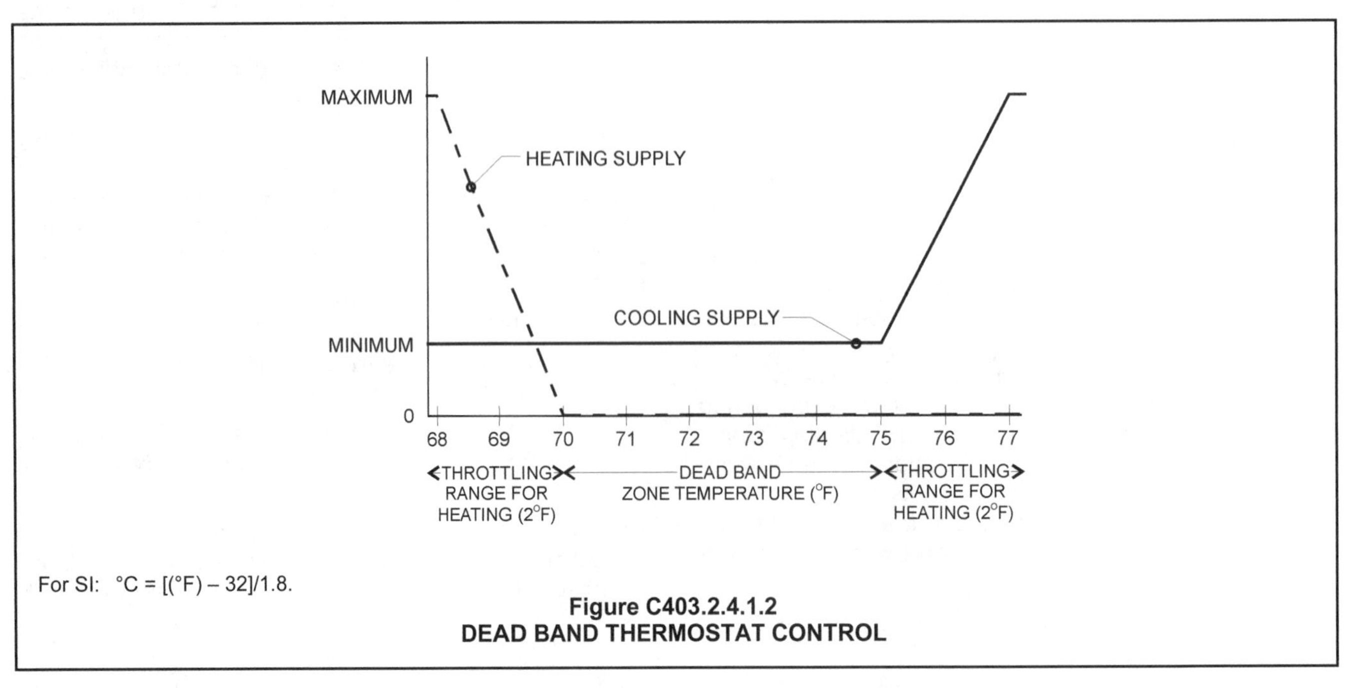

Figure C403.2.4.1.2
DEAD BAND THERMOSTAT CONTROL

to the VAV damper and reheat control valve, respectively.

Example of Deadband Requirements—Single-setpoint Thermostat

A single-setpoint thermostat is proposed to control a VAV box with hot water reheat. Because the thermostat can be adjusted in the winter to setpoints appropriate for heating, then changed in the summer to a setpoint 5°F (2.8°C) degrees higher, does it meet the deadband requirement?

No. This does not meet the intent of this section. The deadband must be continuous and automatic, unless the changeover requires a manual operation to move from heating to cooling mode. The system, as proposed, would only have a single setpoint that would allow the system to cycle from heating to cooling within a single degree during either winter or summer operation.

C403.2.4.1.3 Set point overlap restriction. Where a *zone* has a separate heating and a separate cooling thermostatic control located within the *zone*, a limit switch, mechanical stop or direct digital control system with software programming shall be provided with the capability to prevent the heating set point from exceeding the cooling set point and to maintain a deadband in accordance with Section C403.2.4.1.2.

❖ This section is new to the 2015 code and makes explicit what should have been understood in previous editions of the code. The deadband is intended to prevent simultaneous operation of the heating and cooling system within a zone. Where two different thermostats are provided for the heating and cooling systems, they must be linked in a manner of the designer's choosing to prevent overlap.

C403.2.4.2 Off-hour controls. Each *zone* shall be provided with thermostatic setback controls that are controlled by either an automatic time clock or programmable control system.

Exceptions:

1. *Zones* that will be operated continuously.
2. *Zones* with a full HVAC load demand not exceeding 6,800 Btu/h (2 kW) and having a readily accessible manual shutoff switch.

❖ Most HVAC unitary systems serve spaces that are occupied intermittently but in a fairly predictable cycle. To reduce system energy use during off hours, the code requires that zones and the HVAC systems that serve them be equipped with automatic controls that will shut off the system or setback setpoints. Examples of acceptable automatic controls are time clocks, programmable time switches, energy management systems, DDC systems, wind-up bypass timers and occupant sensors.

Because the term "HVAC system" applies to all equipment that provides any or all of the heating, ventilation or cooling functions serving a zone, this section requires time controls on systems ranging from simple ventilation fans to large chiller plants.

Exception 1 recognizes that under certain circumstances, setback or shutdown controls may not be necessary. For instance, time controls are not practical for systems serving spaces that are expected to be in continuous operation. Examples include hospitals, police stations and detention facilities, central computer rooms and some 24-hour retail establishments. Consult with the code official because many jurisdictions may already have an established policy regarding setback and shutdown controls for spaces expected to be in continuous operation.

Exception 2 acknowledges that small equipment, having a full-load demand of 2 kW (6,800 Btu/h) or less, may have readily accessible manual on-off controls instead of automatic controls. This exception is intended to apply to small, independent systems, such as conference room exhaust fans or small toilet room exhaust fans. The intent is that all energy associated with the operation of the equipment be at 2 kW (6,800 Btu/h) or less. For instance, a fan coil that would use chilled or hot water, requiring the operation of a remote chiller or boiler, may use in excess of 2 kW (6,800 Btu/h) of energy when it operates, even though the fan of the fan coil unit itself may use less than 2 kW (6,800 Btu/h). In this case, the fan coil would have to be automatically controlled, although many fan coils may be interlocked to the same time clock.

Historically, heat pump systems with electric-resistance heat were considered to be less efficient when operated intermittently because of the increased use of the resistance heat during warm-up. But this increase is reduced by the use of proper controls that lock out the auxiliary heat when the heat pump can handle the load (controls that are required by Section C403.2.4.1.1), and in most cases, the savings resulting from reduced fan energy use and reduced heat gains and losses during setback will offset the increased costs from the resistance heater.

Example of Time Controls—Hotel Guest Rooms

Does a hotel sleeping unit (guest room) fan coil require individual time-clock controls?

No. Hotel guest rooms, although not actually continuously occupied, are continuously available for occupancy and, thus, fall under Exception 1. It should be noted that systems that automatically setback or setup thermostat setpoints when rooms are not rented may be cost effective. Because the code does not specifically list the types of uses that are covered by Exception 1, it is always a good idea to discuss this issue with the code official during the design and permit application stages rather than wait until after a noncomplying system is installed.

Example of Time Controls—Equipment Room Cooling Unit

An air conditioner serves an elevator equipment room in an office building, and is controlled by a thermostat that cycles the indoor supply fan and the compressor when there is no need for cooling. Does this unit need time controls so that it shuts off when the building is unoccupied at night?

No. The equipment must be maintained at a given temperature at all times, and thus, this unit qualifies for Exception 1. Furthermore, when the elevators are inactive at night, the air conditioner will automatically shut off because there is no load in the space.

C403.2.4.2.1 Thermostatic setback capabilities. Thermostatic setback controls shall have the capability to set back or temporarily operate the system to maintain *zone* temperatures down to 55°F (13°C) or up to 85°F (29°C).

❖ Zone thermostatic controls used to control space heating must be capable of being set to 55°F (13°C) or lower. Those controlling space cooling must be capable of being set to 85°F (29°C) or higher. A single thermostat controlling both heating and cooling must meet both requirements.

Note that the code does not require that a single thermostat be used to meet the entire range of setpoints. For instance, one thermostat with a range from 65°F (18°C) to 80°F (27°C) may be used to control heating and cooling during normal hours, while one or two other thermostats that can be set down to 55°F (13°C) and up to 85°F (29°C) can be installed for night setback and setup.

Thermostatic controls may be set either locally (with adjustment buttons, switches, knobs, etc.) or remotely (such as by a building automation system). Setpoints also may be changed by replacing elements for thermostats whose setpoints are functions of their sensing elements.

It is important to note that the code does not mandate that the setback features are used. It strictly requires that the installed thermostat has the capability to set back or temporarily operate the system at the alternative temperatures. The code also does not require or restrict the use of locking thermostat covers or other means to restrict adjustment by occupants. The decision to limit or restrict occupant access to the thermostat or their ability to make adjustments to it is up to the building owner.

C403.2.4.2.2 Automatic setback and shutdown capabilities. Automatic time clock or programmable controls shall be capable of starting and stopping the system for seven different daily schedules per week and retaining their programming and time setting during a loss of power for at least 10 hours. Additionally, the controls shall have a manual override that allows temporary operation of the system for up to 2 hours; a manually operated timer capable of being adjusted to operate the system for up to 2 hours; or an occupancy sensor.

❖ Off-hour controls of the automatic time clock and programmable types must be able to control system starts and stops over seven different daily schedules per week, and must retain that programming for not less than 10 hours if power is lost. The true 7-day-programmable thermostat is not your typical weekday/weekend or weekday/Saturday/Sunday thermostat most often used in residential applications. The 7-day-programmable controls also must be equipped with a manual override, such as a windup timer or factory push-button presets that will control comfort heating or cooling for periods not greater than 2 hours. The standard sharing capabilities of most energy management systems generally will meet this criterion. An occupant-sensor-based strategy for off-hour HVAC control can be used instead of the 2-hour temporary operation limitation because this control methodology responds directly to the presence or absence of building occupants.

C403.2.4.2.3 Automatic start capabilities. Automatic start controls shall be provided for each HVAC system. The controls shall be capable of automatically adjusting the daily start time of the HVAC system in order to bring each space to the desired occupied temperature immediately prior to scheduled occupancy.

❖ This section requires each HVAC system to have automatic start controls. Automatic start controls are simply another available option on HVAC systems for off-hour controls. Allowing the occupants to set a time for the HVAC system to operate based on the normal occupancy times of the building saves energy by enabling complete shut down during off hours.

C403.2.4.3 Shutoff dampers. Outdoor air intake and exhaust openings and stairway and shaft vents shall be provided with Class I motorized dampers. The dampers shall have an air leakage rate not greater than 4 cfm/ft^2 (20.3 L/s · m^2) of damper surface area at 1.0 inch water gauge (249 Pa) and shall be labeled by an approved agency when tested in accordance with AMCA 500D for such purpose.

Outdoor air intake and exhaust dampers shall be installed with automatic controls configured to close when the systems or spaces served are not in use or during unoccupied period warm-up and setback operation, unless the systems served require outdoor or exhaust air in accordance with the *International Mechanical Code* or the dampers are opened to provide intentional economizer cooling.

Stairway and shaft vent dampers shall be installed with automatic controls configured to open upon the activation of any fire alarm initiating device of the building's fire alarm system or the interruption of power to the damper.

Exception: Gravity (nonmotorized) dampers shall be permitted to be used as follows:

1. In buildings less than three stories in height above grade plane.
2. In buildings of any height located in *Climate Zones* 1, 2 or 3.
3. Where the design exhaust capacity is not greater than 300 cfm (142 L/s).

Gravity (nonmotorized) dampers shall have an air leakage rate not greater than 20 cfm/ft^2 (101.6 L/s • m^2) where not less than 24 inches (610 mm) in either dimension and 40 cfm/ft^2 (203.2 L/s • m^2) where less than 24 inches (610 mm) in either dimension. The rate of air leakage shall be determined at 1.0 inch water gauge (249 Pa) when tested in accordance with AMCA 500D for such purpose. The dampers shall be labeled by an approved agency.

❖ Regardless of system size, fans that either bring outside air into the building or exhaust air to the outside must have dampers that automatically close when the fan is shut off. These dampers are required to be motorized unless one of three exceptions applies.

This requirement is intended to reduce infiltration into the building when ventilation systems are off. Infiltration will speed up the natural cooling or warming of the space during off hours, thereby increasing the energy required to maintain setback temperatures and possibly increasing the energy used to bring the space back to normal occupied temperatures.

The three exceptions provide options where gravity dampers may be used in lieu of the generally required motorized dampers. These exceptions address situations where the level of energy loss and, therefore, the cost effectiveness of the motorized dampers is not as great.

The exceptions allow the use of less-expensive gravity dampers in buildings where stack effects are less pronounced (less than three stories in height), in buildings in warm climates (Climate Zones 1, 2 and 3) where temperature differences between indoor and outdoor are not as pronounced, and on small exhaust systems where motorized dampers may be cost prohibitive.

Damper performance is specified for both mechanical and gravity dampers. Maximum air leakages is set for both categories. The dampers must be tested in accordance with the AMCA 500D standard.

C403.2.4.4 Zone isolation. HVAC systems serving *zones* that are over 25,000 square feet (2323 m^2) in floor area or that span more than one floor and are designed to operate or be occupied nonsimultaneously shall be divided into isolation areas. Each isolation area shall be equipped with isolation devices and controls configured to automatically shut off the supply of conditioned air and outdoor air to and exhaust air from the isolation area. Each isolation area shall be controlled independently by a device meeting the requirements of Section C403.2.4.2.2. Central systems and plants shall be provided with controls and devices that will allow system and equipment operation for any length of time while serving only the smallest isolation area served by the system or plant.

Exceptions:

1. Exhaust air and outdoor air connections to isolation areas where the fan system to which they connect is not greater than 5,000 cfm (2360 L/s).
2. Exhaust airflow from a single isolation area of less than 10 percent of the design airflow of the exhaust system to which it connects.
3. Isolation areas intended to operate continuously or intended to be inoperative only when all other isolation areas in a *zone* are inoperative.

❖ In large buildings it is often the case that portions of those buildings operate on different schedules, particularly where there are distinct tenant spaces in a building. This section requires that, where a large central system is designed to serve a large floor area having portions operating on different schedules, the system include mechanisms by which distinctly operating areas can be isolated. The purpose should be obvious: if portions of the building are unoccupied, it is a waste of energy that they be conditioned simply because another portion of the zone is occupied. Isolation areas can be as small as one zone, but more practically zones will be grouped together in a single isolation area. Each distinct area so isolated needs to have isolation devices and controls that allow each zone to be shut off or set back individually. Essentially each isolation area must have individual automatic shutdown controls meeting the requirements of Section C403.2.4.2 as if each area were a separate HVAC system. This allows each isolation area to automatically operate on different time schedules.

Isolation devices and controls are not required for the following areas: In Exception 1, where the fan system to the area is 5,000 cfm (2360 L/s) or smaller. In other words, for exhaust fans or outdoor air ventilation fans 5,000 cfm (2360 L/s) or smaller, no isolation devices (such as dampers) or controls need be installed at the fan or at any exhaust or outdoor air supply to any isolation area served by the system. Exception 2 exempts providing a distinct exhaust system to an isolation area where the exhaust airflow is less than 10 pecent of the flow of the total exhaust system. Exception 3 reemphasizes the intent of the requirement of isolation areas. If the whole zone is expected to be operated continuously—or on the same schedule as the balance of the area—no isolation is needed.

C403.2.4.5 Snow- and ice-melt system controls. Snow- and ice-melting systems shall include automatic controls capable of shutting off the system when the pavement temperature is above 50°F (10°C) and no precipitation is falling and an automatic or manual control that will allow shutoff when the outdoor temperature is above 40°F (4°C).

❖ Snow- and ice-melting systems, regardless of the source of the energy for the system, are required to include automatic controls capable of shutting off the system when the pavement temperature is greater than 50°F (10°C), no precipitation is falling and the outdoor temperature is greater than 40°F (4°C). The automatic shutoff prevents loss of energy often occurring when building occupants simply forget to turn the system off.

C403.2.4.6 Freeze protection system controls. Freeze protection systems, such as heat tracing of outdoor piping and heat exchangers, including self-regulating heat tracing, shall include automatic controls configured to shut off the systems when outdoor air temperatures are above 40°F (4°C) or when the conditions of the protected fluid will prevent freezing.

❖ Where piping and equipment are located outdoors or in unconditioned spaces, freeze protection heating systems are often installed to prevent freezing in the winter. A common type of such systems is electric-resistance heat tracing wound around piping through which a current is run. This action heats the piping in order to keep it about above 32°F (0°C). Clearly, when outdoor temperatures reach a level where the liquid in the system will no longer freeze, putting any energy into such a freeze protection system is wasteful. The code simply requires automatic controls be provided to shut off the system as prescribed.

C403.2.4.7 Economizer fault detection and diagnostics (FDD). Air-cooled unitary direct-expansion units listed in Tables C403.2.3(1) through C403.2.3(3) and variable refrigerant flow (VRF) units that are equipped with an economizer in accordance with Section C403.3 shall include a fault detection and diagnostics (FDD) system complying with the following:

1. The following temperature sensors shall be permanently installed to monitor system operation:
   1.1. Outside air.
   1.2. Supply air.
   1.3. Return air.
2. Temperature sensors shall have an accuracy of ±2°F (1.1°C) over the range of 40°F to 80°F (4°C to 26.7°C).
3. Refrigerant pressure sensors, where used, shall have an accuracy of ±3 percent of full scale.
4. The unit controller shall be capable of providing system status by indicating the following:
   4.1. Free cooling available.
   4.2. Economizer enabled.
   4.3. Compressor enabled.
   4.4. Heating enabled.
   4.5. Mixed air low limit cycle active.
   4.6. The current value of each sensor.
5. The unit controller shall be capable of manually initiating each operating mode so that the operation of compressors, economizers, fans and the heating system can be independently tested and verified.
6. The unit shall be capable of reporting faults to a fault management application accessible by day-to-day operating or service personnel, or annunciated locally on zone thermostats.
7. The FDD system shall be capable of detecting the following faults:
   7.1. Air temperature sensor failure/fault.
   7.2. Not economizing when the unit should be economizing.
   7.3. Economizing when the unit should not be economizing.
   7.4. Damper not modulating.
   7.5. Excess outdoor air.

❖ Commercial HVAC systems have been shown to have problems with economizer function, control, and performance in field studies and utility-sponsored maintenance programs. This results in reduced energy efficiency and potential energy savings from the economizer with fan-only operation. These fault detection and diagnostics (FDD) specifications have been standardized in California Title 24. Manufacturers were involved in the development of these criteria and are already manufacturing to these standards. Adding such systems will provide building owners key information regarding the operation of their HVAC systems.

C403.2.5 Hot water boiler outdoor temperature setback control. Hot water boilers that supply heat to the building through one- or two-pipe heating systems shall have an outdoor setback control that lowers the boiler water temperature based on the outdoor temperature.

❖ This additional control provides a simple method of further energy savings.

C403.2.6 Ventilation. Ventilation, either natural or mechanical, shall be provided in accordance with Chapter 4 of the *International Mechanical Code*. Where mechanical ventilation is provided, the system shall provide the capability to reduce the outdoor air supply to the minimum required by Chapter 4 of the *International Mechanical Code*.

❖ HVAC systems must be capable of supplying outside air at the minimum rate required by Chapter 4 of the IMC. The minimum volume of outside air required may be higher to make up for process exhaust or other special requirements, or if energy costs are subsequently reduced, as occurs with economizer controls.

Note that the IMC does not limit outside air intake to the minimum levels prescribed; rather, it requires systems to be designed for those levels and have the capability to operate at those levels. This results from many experts feeling that higher rates are required for adequate indoor air quality.

Even though adequate indoor air quality should take precedence over energy conservation, the energy consequence of achieving this goal by dilution with outdoor air must not be overlooked. Conditioning outside air is extremely energy intensive in many climates and the designer will ideally economize outside air intake for ventilation.

A detailed discussion of ways to achieve acceptable indoor air quality at minimal energy usage is beyond the scope of this commentary. The amount of ouside air required can be minimized by use of any of the following:

- Source control (eliminate or minimize internal pollutant sources).
- Source containment (use of local exhaust systems or isolation of contaminants into special rooms or areas).
- Demand controls (use of sensors to adjust outside air intake as a function of indoor contaminant levels.
- Use of transfer air (diluting pollutants in one room by transferring abundant ventilation air from a less polluted or overventilated space rather than using outside air).
- Proper design and selection of air distribution systems to improve ventilation effectiveness.

Where outside air rates must be high, the use of heat recovery systems, such as plane heat exchangers, run-around loops, sensible and total energy wheels, and heat pipe heat exchangers, can reduce the energy impact.

C403.2.6.1 Demand controlled ventilation. Demand control ventilation (DCV) shall be provided for spaces larger than 500 square feet (46.5 m^2) and with an average occupant load of 25 people per 1,000 square feet (93 m^2) of floor area (as established in Table 403.3.1.1 of the *International Mechanical Code*) and served by systems with one or more of the following:

1. An air-side economizer.
2. Automatic modulating control of the outdoor air damper.
3. A design outdoor airflow greater than 3,000 cfm (1416 L/s).

Exception: Demand control ventilation is not required for systems and spaces as follows:

1. Systems with energy recovery complying with Section C403.2.7.
2. Multiple-*zone* systems without direct digital control of individual *zones* communicating with a central control panel.
3. Systems with a design outdoor airflow less than 1,200 cfm (566 L/s).
4. Spaces where the supply airflow rate minus any makeup or outgoing transfer air requirement is less than 1,200 cfm (566 L/s).
5. Ventilation provided for process loads only.

❖ Demand control ventilation (DCV) is a control method that links the amount of fresh air in the ventilation at occupancy of a building. DCV optimizes the volume of fresh air introduced into a building by resetting the minimum economizer damper opening based on the actual occupancy as determined by the level of carbon dioxide (CO_2) in the space. Energy consumption is reduced because less outside air needs to be heated or cooled. This was added to the code for spaces greater than 500 square feet (46 m^2) with an occupant load of 25 occupants per 1,000 square feet (93 m^2). While the code states an absolute occupant load average of 25 per 1,000 square feet, in practice the provision is applied where the average occupant load is 25 per 1,000 square feet or greater. DCV is required where served by one or more of the following systems: an air-side economizer, automatic modulating control of outdoor air damper or outdoor design airflow of greater than 3,000 cfm (1400 m^3/s). Exceptions are for energy recovery ventilation (ERV) systems complying with Section C403.2.7. Also exempt are multiple-zone systems without direct digital control (DDC) of individual zones, communicating with a central control panel. Systems with a design outdoor airflow less than 1,200 cfm (600 m^3/s) and spaces where the supply airflow rate minus any makeup or outgoing transfer air requirement is less than 1,200 cfm (600 m^3/s) are exempt. Another exception is ventilation provided for process loads, for example, on a conveyor in a manufacturing facility. These do not need fresh air.

C403.2.6.2 Enclosed parking garage ventilation controls. Enclosed parking garages used for storing or handling automobiles operating under their own power shall employ contamination-sensing devices and automatic controls configured to stage fans or modulate fan average airflow rates to 50 percent or less of design capacity, or intermittently operate fans less than 20 percent of the occupied time or as required to maintain acceptable contaminant levels in accordance with *International Mechanical Code* provisions. Failure of contamination sensing devices shall cause the exhaust fans to operate continuously at design airflow.

Exceptions:

1. Garages with a total exhaust capacity less than 22,500 cfm (10 620 L/s) with ventilation systems that do not utilize heating or mechanical cooling.
2. Garages that have a garage area to ventilation system motor nameplate power ratio that exceeds 1125 cfm/hp (710 L/s/kW) and do not utilize heating or mechanical cooling.

❖ This new provision allows reduction of garage ventilation rates, but the systems must result in compliance with the IMC. They must have sensors trigger system operation when contaminant levels exceed allowed values. The system must fail in the operating mode if the contamination sensors fail.

C403.2.7 Energy recovery ventilation systems. Where the supply airflow rate of a fan system exceeds the values specified in Tables C403.2.7(1) and C403.2.7(2), the system shall include an energy recovery system. The energy recovery system shall have the capability to provide a change in the enthalpy of the outdoor air supply of not less than 50 percent of the difference between the outdoor air and return air enthalpies, at design conditions. Where an air economizer is required, the energy recovery system shall include a bypass or controls which permit operation of the economizer as required by Section C403.3.

Exception: An energy recovery ventilation system shall not be required in any of the following conditions:

1. Where energy recovery systems are prohibited by the *International Mechanical Code*.
2. Laboratory fume hood systems that include at least one of the following features:
   - 2.1. Variable-air-volume hood exhaust and room supply systems capable of reducing exhaust and makeup air volume to 50 percent or less of design values.
   - 2.2. Direct makeup (auxiliary) air supply equal to at least 75 percent of the exhaust rate, heated not warmer than 2°F (1.1°C) above room setpoint, cooled to not cooler than 3°F (1.7°C) below room setpoint, no humidification added, and no simultaneous heating and cooling used for dehumidification control.
3. Systems serving spaces that are heated to less than 60°F (15.5°C) and are not cooled.
4. Where more than 60 percent of the outdoor heating energy is provided from site-recovered or site solar energy.
5. Heating energy recovery in *Climate Zones* 1 and 2.
6. Cooling energy recovery in *Climate Zones* 3C, 4C, 5B, 5C, 6B, 7 and 8.
7. Systems requiring dehumidification that employ energy recovery in series with the cooling coil.
8. Where the largest source of air exhausted at a single location at the building exterior is less than 75 percent of the design *outdoor air* flow rate.
9. Systems expected to operate less than 20 hours per week at the *outdoor air* percentage covered by Table C403.2.7(1).
10. Systems exhausting toxic, flammable, paint or corrosive fumes or dust.
11. Commercial kitchen hoods used for collecting and removing grease vapors and smoke.

❖ This section provides a threshold requirement for the mandatory use of ERV systems. The code requires the energy recovery system to have the capability to provide a change in the enthalpy of the outdoor air supply of not less than 50 percent of the difference between the outdoor-air and return-air enthalpies, at design conditions. Permeable and nonpermeable plate-type heat exchangers, heat pipes, run-around loops and enthalpy wheels are used to meet this requirement. In the 2015 edition of the standard, the requirement was expanded to cover buildings where systems are run on a more continuous basis [see Table C403.2.7(2)] and to buildings in specified climate zones that are operating with ventilation rates less than 30 percent. There are a number of exceptions to this requirement.

TABLE C403.2.7(1)
ENERGY RECOVERY REQUIREMENT
(Ventilation systems operating less than 8,000 hours per year)

CLIMATE ZONE	PERCENT (%) OUTDOOR AIR AT FULL DESIGN AIRFLOW RATE							
	≥ 10% and < 20%	≥ 20% and < 30%	≥ 30% and < 40%	≥ 40% and < 50%	≥ 50% and < 60%	≥ 60% and < 70%	≥ 70% and < 80%	≥ 80%
	DESIGN SUPPLY FAN AIRFLOW RATE (cfm)							
3B, 3C, 4B, 4C, 5B	NR	NR	NR	NR	NR	NR	NR	NR
1B, 2B, 5C	NR	NR	NR	NR	≥ 26,000	≥ 12,000	≥ 5,000	≥ 4,000
6B	≥ 28,000	≥ 26,5000	≥ 11,000	≥ 5,500	≥ 4,500	≥ 3,500	≥ 2,500	≥ 1,500
1A, 2A, 3A, 4A, 5A, 6A	≥ 26,000	≥ 16,000	≥ 5,500	≥ 4,500	≥ 3,500	≥ 2,000	≥ 1,000	> 0
7, 8	≥ 4,500	≥ 4,000	≥ 2,500	≥ 1,000	> 0	> 0	> 0	> 0

For SI: 1 cfm = 0.4719 L/s.
NR = Not Required.

TABLE C403.2.7(2)
ENERGY RECOVERY REQUIREMENT
(Ventilation systems operating not less than 8,000 hours per year)

CLIMATE ZONE	PERCENT (%) OUTDOOR AIR AT FULL DESIGN AIRFLOW RATE							
	≥ 10% and < 20%	≥ 20% and < 30%	≥ 30% and < 40%	≥ 40% and < 50%	≥ 50% and < 60%	≥ 60% and < 70%	≥ 70% and < 80%	≥ 80%
	Design Supply Fan Airflow Rate (cfm)							
3C	NR	NR	NR	NR	NR	NR	NR	NR
1B, 2B, 3B, 4C, 5C	NR	≥ 19,500	≥ 9,000	≥ 5,000	≥ 4,000	≥ 3,000	≥ 1,500	> 0
1A, 2A, 3A, 4B, 5B	≥ 2,500	≥ 2,000	≥ 1,000	≥ 500	> 0	> 0	> 0	> 0
4A, 5A, 6A, 6B, 7, 8	> 0	> 0	> 0	> 0	> 0	> 0	> 0	> 0

For SI: 1 cfm = 0.4719 L/s.
NR = Not required

Requirements for outdoor ventilation air rates in ASHRAE 62.1, and subsequently the model codes, have increased over recent years. This has placed new demands on HVAC equipment and operating budgets for buildings. New ozone-friendly refrigerants also have reduced equipment capacity. These conditions have increased the value of energy recovery in ventilation systems. ERV reduces the load on the system caused by outdoor air by taking advantage of the work that has been done to heat, cool, humidify or dehumidify a space. Instead of losing that energy to the atmosphere when air is exhausted, a portion of the energy is recovered and used to pretreat incoming outdoor air, thus reducing the loads on the HVAC system. ERV systems are particularly effective in this application. Studies indicate this would result in a 15-percent reduction of the total energy used in commercial heating, refrigeration, ventilation and air conditioning.

The provisions and the exceptions define the boundary conditions within which mandatory requirements for the application of ERV technology are justified. There are several different types of ERV systems that are provided by a variety of manufacturers and suppliers. The provisions, however, are generic in that they treat all ERV systems the same without being specific or limiting as to type. Not all ERV system types are equally efficient, but under the conditions and limitations in the code, all system types will accomplish the objective of substantial energy efficiency.

Exception 1, referencing the IMC, correlates with prohibitions found in IMC Section 514. This will prohibit the use of ERV systems in hazardous exhaust systems and commercial kitchen hoods. The code must not require ERV in those exhaust systems in order to avoid a conflict between the IMC and the code. Not all laboratory fume hoods are hazardous exhaust systems. Exception 2 for laboratory fume hoods is, therefore, appropriate and does not pose any conflict with Exception 1 since any laboratory fume hood that is a hazardous exhaust system would be prohibited by Exception 1 and the IMC from having an ERV system.

Many of the exceptions address conditions where the input of energy is limited and where the installation of a system to recover ventilation energy would be of limited benefit. Exceptions 3 through 8 each exemplify where the benefit is limited. Exception 9 similarly exempts systems that are expected to operate less than 20 hours per week at the outdoor ventilation percentage covered by Table C403.2.7(1).

Exceptions 10 and 11 are provided for consistency with the IMC, which prohibits heat recovery ventilation systems in such installations.

C403.2.8 Kitchen exhaust systems. Replacement air introduced directly into the exhaust hood cavity shall not be greater than 10 percent of the hood exhaust airflow rate. Conditioned supply air delivered to any space shall not exceed the greater of the following:

1. The ventilation rate required to meet the space heating or cooling load.
2. The hood exhaust flow minus the available transfer air from adjacent space where available transfer air is considered that portion of outdoor ventilation air not required to satisfy other exhaust needs, such as restrooms, and not required to maintain pressurization of adjacent spaces.

Where total kitchen hood exhaust airflow rate is greater than 5,000 cfm (2360 L/s), each hood shall be a factory-built commercial exhaust hood listed by a nationally recognized testing laboratory in compliance with UL 710. Each hood shall have a maximum exhaust rate as specified in Table C403.2.8 and shall comply with one of the following:

1. Not less than 50 percent of all replacement air shall be transfer air that would otherwise be exhausted.
2. Demand ventilation systems on not less than 75 percent of the exhaust air that are capable of not less than a 50-percent reduction in exhaust and replacement air system airflow rates, including controls necessary to modulate airflow in response to appliance operation and to maintain full capture and containment of smoke, effluent and combustion products during cooking and idle.
3. Listed energy recovery devices with a sensible heat recovery effectiveness of not less than 40 percent on not less than 50 percent of the total exhaust airflow.

TABLE C403.2.8
MAXIMUM NET EXHAUST FLOW RATE, CFM PER LINEAR FOOT OF HOOD LENGTH

TYPE OF HOOD	LIGHT-DUTY EQUIPMENT	MEDIUM-DUTY EQUIPMENT	HEAVY-DUTY EQUIPMENT	EXTRA-HEAVY-DUTY EQUIPMENT
Wall-mounted canopy	140	210	280	385
Single island	280	350	420	490
Double island (per side)	175	210	280	385
Eyebrow	175	175	NA	NA
Backshelf/Pass-over	210	210	280	NA

For SI:1 cfm = 0.4719 L/s; 1 foot = 305 mm.
NA = Not Allowed.

Where a single hood, or hood section, is installed over appliances with different duty ratings, the maximum allowable flow rate for the hood or hood section shall be based on the requirements for the highest appliance duty rating under the hood or hood section.

Exception: Where not less than 75 percent of all the replacement air is transfer air that would otherwise be exhausted.

❖ The intent is to conserve energy through the use of engineered hoods or performance-based hoods that have been validated based on consensus standard test methods. It should be noted that ASHRAE research has not demonstrated that exhaust rate reductions substantially beyond 30 percent can or should be recommended at this time. These new provisions basically outlaw "short-circuit" hoods. Research and the California Energy Commission have shown that direct supply of makeup air, in excess of 10 percent of hood exhaust airflow, into the hood cavity significantly deteriorates the Capture and Containment (C&C) performance of hoods. This research has also demonstrated that short-circuit hoods waste energy and degrade kitchen environment and hygiene. Engineers are often in the habit of simply providing makeup air units in kitchens to provide makeup air equal to the exhaust flow rate even when "free" transfer air is available from adjacent spaces. Adding makeup air when transfer air is available is a wasteful design practice and should be prohibited. Using available transfer air saves energy and reduces the first cost of the makeup unit and exhaust system in the adjacent spaces. It simply requires some engineering and coordination to provide a path for the transfer air.

C403.2.9 Duct and plenum insulation and sealing. Supply and return air ducts and plenums shall be insulated with a minimum of R-6 insulation where located in unconditioned spaces and where located outside the building with a minimum of R-8 insulation in *Climate Zones* 1 through 4 and a minimum of R-12 insulation in *Climate Zones* 5 through 8. Where located within a building envelope assembly, the duct or plenum shall be separated from the building exterior or unconditioned or exempt spaces by a minimum of R-8 insulation in *Climate Zones* 1 through 4 and a minimum of R-12 insulation in *Climate Zones* 5 through 8.

Exceptions:

1. Where located within equipment.
2. Where the design temperature difference between the interior and exterior of the duct or plenum is not greater than 15°F (8°C).

Ducts, air handlers and filter boxes shall be sealed. Joints and seams shall comply with Section 603.9 of the *International Mechanical Code*.

❖ "Duct systems" are defined as a continuous passageway for the transmission of air that includes ducts, duct fittings, dampers, plenums, fans, and accessory air-handling equipment and appliances. Ducts and plenums must be constructed to meet the construction standards established in the IMC or equivalent. In addition to the requirements of the IMC, duct systems also must be insulated for compliance with the code. When the duct is in the thermal envelope, the code requires a minimum of R-8 insulation in Climate Zones 1 through 4 and a minimum of R-12 in Climate Zones 5 through 8. Ducts are not required by the code to be insulated when located in the conditioned space. The exception with the temperature differential is an exception to the required insulation of ducts in unconditioned spaces. There still may be a need for some lesser amount of insulation on the ducts to prevent condensation problems. Section 603.12 of the IMC states that "provisions shall be made to prevent the formation of condensation on the exterior of any duct."

All supply and return ductwork and air plenums must be thermally insulated as required by this section. Insulation requirements are divided into primary categories, one for ducts that are outdoors and exposed to weather and one for ducts that are either in unconditioned spaces (enclosed but not in the building envelope) or a building envelope assembly.

This section is for the insulation material as installed, excluding air film resistance, but including the effect of the compression of duct wraps as they are typically installed. Duct wraps are generally assumed to be compressed not greater than 75 percent of their nominal thickness when they are installed. Material conductivity is often tested as measured in accordance with ASTM C 518 at 75°F (24°C) mean temperature.

Common duct insulation materials that meet the *R*-value levels in this section are shown in Commentary Figure C403.2.9.

The following duct and plenum applications need not meet the insulation requirements of this section:

1. Ducts, plenums and casings that are factory installed and an integral part of equipment rated in accordance with Section C403.2.3. As for factory-installed piping insulation, this exception is intended to apply to casings around tested equipment. The energy losses through these casings are accounted for in the energy performance tests and ratings. Although they are not always included in performance testing, optional casings such as economizer sections also should be exempted from insulation requirements if they are insulated to the same extent as the equipment to which they are attached. This is a practical requirement because designers seldom have the option to specify casing insulation levels.
2. Exhaust air ducts, and ducts or plenums located anywhere the design temperature difference between the air in the duct and the air surrounding the duct is less than 15°F (8.3°C). Simply stated, this includes ducts or plenums located in the conditioned space they serve. The designer should keep in mind that certain

minimum insulation levels still may be required to prevent condensation.

The following discussion is provided for the benefit of the user in identifying specific duct insulation applications. Commentary Figure C403.2.9 identifies the primary duct configurations that have requirements from this section. The insulation requirements for each duct location identified in Commentary Figure C403.2.9 are described in the following and numbered as shown in the figure. In each section, the default insulation requirement is identified.

1. Unit Casings and plenums—not required (N/R). Factory-installed insulation that is part of the HVAC equipment is covered through the minimum equipment efficiencies of Section C403.2.3 and does not have requirements in this section (see Exception 1). This includes equipment casings and plenums. This exception does not apply to built-up systems with field-fabricated plenums. These systems must be insulated in accordance with the requirements for ductwork in this section. Perimeter mechanical rooms that are used as mixing plenums must be insulated as if they are on the exterior of the building or to the requirements of Section C403.2.3, whichever is more stringent.
2. Exhaust ductwork—N/R. Except to prevent condensation, exhaust ductwork also does not need to meet the requirements of this section because it applies only to supply and return ductwork.
3. Supply and return ducts in vented attic—R-6. These ducts are located in an attic, which is vented to the outside. Crawl spaces and attics,

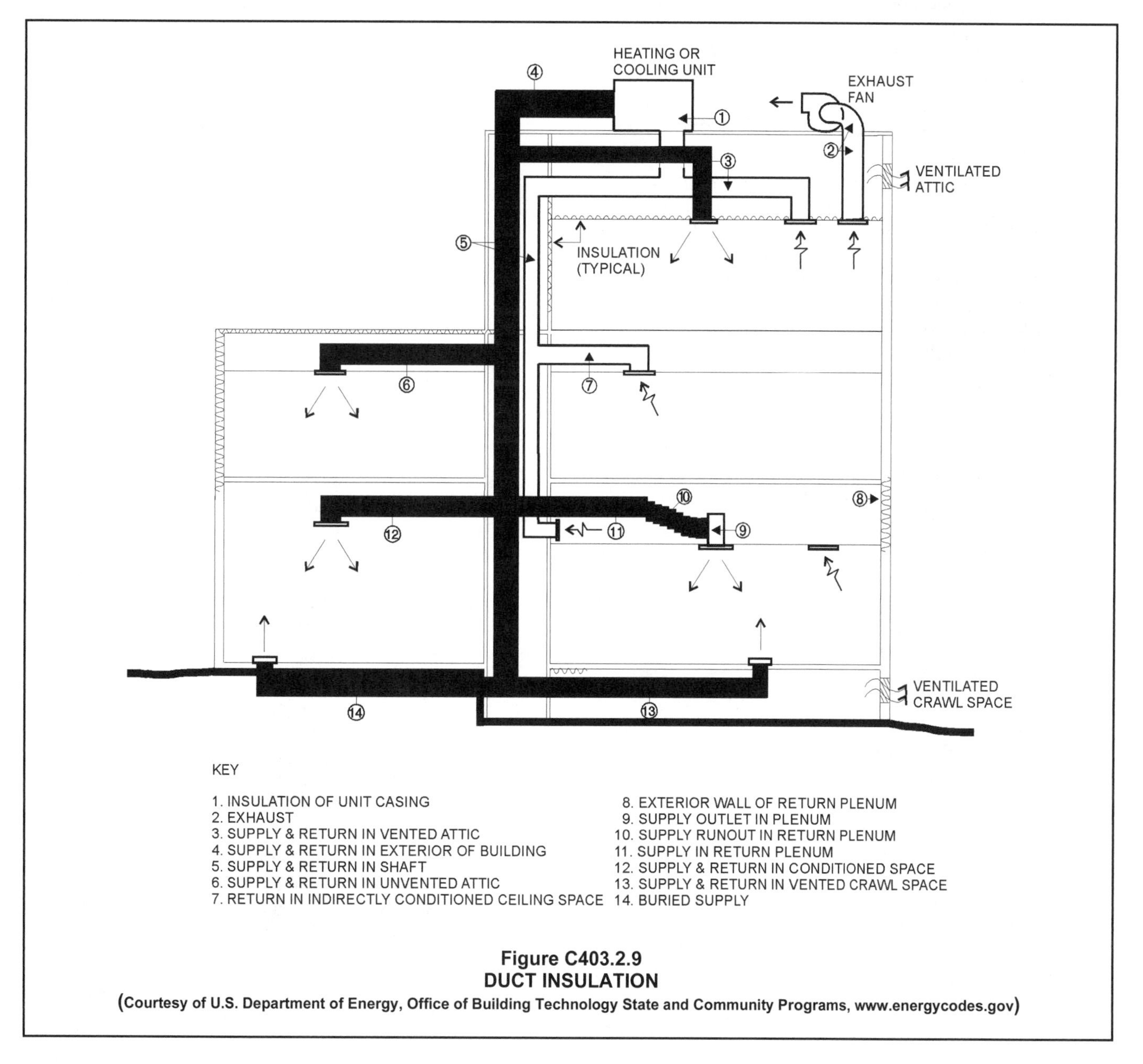

Figure C403.2.9
DUCT INSULATION
(Courtesy of U.S. Department of Energy, Office of Building Technology State and Community Programs, www.energycodes.gov)

vented or otherwise open to the outdoors, are considered unconditioned spaces for the purpose of determining duct insulation requirements. A vented attic generally will run 5°F to 20°F (2.8°C to 11.1°C) warmer than the outside air temperature during peak cooling, and be close to the outside air temperature during peak heating.

4. Supply and return on exterior of the building—R-8 or R-12. Exposed ductwork must have minimum insulation installed. The minimum in Climate Zones 1 through 4 is R-8; the minimum in Climate Zones 5 through 8 is R-12.
5. Supply and return ducts in exposed shaft—R-6. These ducts are located in unconditioned space similar to Item 3. However, this wall cavity should not experience the extreme temperatures of the attic because the space is not vented to the outdoors and has minimal roof exposure. Where design temperatures for the shaft space are not available from load calculations, assume that at design conditions the shaft temperature is near outside conditions at both cooling and heating peaks.
6. Supply (and return) ducts in unvented attic—R-6. Here the ducts are located in an unventilated attic. The temperature in the attic typically is determined through load calculations. The attic temperature often will be higher than space temperatures because of solar gains on the roof and heat from recessed light fixtures. An unvented attic generally will run 15°F to 30°F (8.3°C to 16.7°C) warmer than the outside air temperature during peak cooling, and be close to the outside air temperature during peak heating.

   Therefore, in most cases, supply ducts in unventilated attics must be insulated to a minimum of R-6 as a result of exceeding the 15°F (8.3°C) design temperature difference.
7. Return ducts in indirectly conditioned ceiling spaces—N/R. This requirement is similar to Item 6, but the ceiling-space temperatures will be essentially equal to space temperatures for heating and slightly higher than space temperatures for cooling because of heat gains from recessed light fixtures. In general, no insulation will be required on return ducts. See Item 11 for supply ducts in indirectly conditioned spaces.
8. Exterior wall of return plenum—(see the following discussion). In this case, the ceiling space is being used as a return plenum. The exterior walls of the space are effectively return duct walls exposed to the outside air. This wall may be insulated to the requirements of either Section C402.2.3 for above-grade exterior walls or the requirements for ducts located in the exterior of the building from this section. Where the wall is insulated to the lesser requirement (often the duct insulation requirement), a performance-based path to envelope compliance must be pursued (such as that found in Section C407), or an approved simulation tool must be used.
9. Supply outlet in return plenum—N/R. The sheet metal plenum surrounding the air outlet is part of the supply duct system and, therefore, must be insulated the same as the supply ducts (see Item 11).
10. Supply runout in return plenum—N/R. Typically, runouts of up to 10 feet (3048 mm) to a terminal device (supply outlet or VAV box) are constructed using flexible duct having up to 1 inch (25 mm) of insulation to address sound attenuation and air-sealing issues. However, the duct remains part of the supply duct system and, therefore, must be insulated the same as supply ducts (see Item 11).
11. Supply ducts in return plenum—N/R. For supply ducts located within a return plenum, the ΔT is the difference between the supply and return temperatures. In general, the temperature in a return plenum will be 2°F to 5°F (1.1°C to 2.8°C) warmer than the space temperature because of heat from the lights. Unless the supply is that serving a hybrid or cold-air distribution system, from Exception 2, supply ducts in indirectly conditioned spaces, such as ceiling plenums, do not require insulation.
12. Supply and return ducts in conditioned space—N/R. In accordance with Exception 2, supply and return ducts, or plenums, located in the conditioned space they serve do not require insulation. From a practical standpoint, insulation may be desirable for cooling ducts to prevent condensation if the duct passes near local areas of high humidity, as might occur in a kitchen or natatorium. For typical spaces, condensation generally will not occur, even at very low supply temperatures, because the space's relative humidity will be lowered correspondingly by the dry air being supplied.
13. Supply and return ducts in vented crawl space—R-6. This is similar to the case of ducts in a vented attic (see Item 3). Because the space is vented and unlikely to experience strong solar loads, crawl space temperatures will be very close to the outdoor temperature.
14. Supply and return ducts below grade—R-6. The amount of insulation required for ducts located below grade depends on the local ground temperature. Although the ground temperature does not track the temperature extremes of an unconditioned space and is technically located outside

the building envelope, the ductwork is not exposed to the same solar and temperature loads as exposed ductwork. Practically speaking, this situation is much closer to ductwork in an unconditioned space than ductwork on the roof.

Sealing Requirements for Ductwork. All ducts, air handlers and filter boxes must be sealed. Although ducts in conditioned spaces do not need to be insulated, they must be sealed. Sealing helps control the infiltration of unconditioned air into the ducts and the exfiltration of conditioned air from the ducts.

Longitudinal seams are those parallel to the direction of airflow; spiral joints do not require sealing in any seal class. All other duct connections are considered transverse joints.

The code accepts a variety of methods and materials that may be used for sealing. However, the code does prohibit the use of unlisted pressure-sensitive tapes, which generally tend to peel with time and are, therefore, not a reliable sealant in the long run. The type of sealing material used must be appropriate for the ductwork materials and the anticipated operating pressures, and not simply be a short-term solution that does not remain capable of sealing the joints.

It should be noted that the code's requirements for sealing apply to "ductwork" and not to the insulation that surrounds the ductwork. The purpose for this sealing is to limit the amount of air leakage from the duct. Although it is not a code requirement, the sealing of the insulation materials would be a good item to consider. Depending on the location and environment of the duct, it would be possible for joints in the insulation to permit the movement of air and moisture into the insulation and, therefore, permit condensation to form and possibly reduce the effectiveness of the insulation. The insulation manufacturer's instructions should be reviewed to determine its recommendations and appropriate materials for effectively sealing the insulation.

C403.2.9.1 Duct construction. Ductwork shall be constructed and erected in accordance with the *International Mechanical Code*.

❖ The provisions and standards listed in this section apply primarily to duct construction rather than energy-conservation issues. Although they are general issues, they do also contribute to energy conservation. Ductwork must be constructed and erected in accordance with the IMC.

C403.2.9.1.1 Low-pressure duct systems. Longitudinal and transverse joints, seams and connections of supply and return ducts operating at a static pressure less than or equal to 2 inches water gauge (w.g.) (498 Pa) shall be securely fastened and sealed with welds, gaskets, mastics (adhesives), mastic-plus-embedded-fabric systems or tapes installed in accordance with the manufacturer's instructions. Pressure classifications specific to the duct system shall be clearly indicated on the construction documents in accordance with the *International Mechanical Code*.

Exception: Locking-type longitudinal joints and seams, other than the snap-lock and button-lock types, need not be sealed as specified in this section.

❖ Supply and return ducts must be made air tight by sealing longitudinal and transverse joints and connections. Sealing helps control the infiltration of unconditioned air into the ducts and the exfiltration of conditioned air from the ducts. Leaky ductwork allows the conditioned air to escape into wall and floor cavities or attic areas where the duct is located instead of being distributed to the registers and spaces where it is needed.

If not sealed with welds or gaskets, ducts must be sealed using mastic with fibrous backing tape or an equivalent approved by the code official as a long-term seal, and installed according to the manufacturer's instructions. For fibrous ducts, listed pressure-sensitive tape may be used when it is installed in accordance with NAIMA AH 116. Determining the appropriate requirements and enforcing them is made easier by requiring the duct system pressures to be indicated on the construction documents.

C403.2.9.1.2 Medium-pressure duct systems. Ducts and plenums designed to operate at a static pressure greater than 2 inches water gauge (w.g.) (498 Pa) but less than 3 inches w.g. (747 Pa) shall be insulated and sealed in accordance with Section C403.2.9. Pressure classifications specific to the duct system shall be clearly indicated on the construction documents in accordance with the *International Mechanical Code*.

❖ As discussed in the commentary to Section C403.2.9.1.1, ductwork must be properly constructed and sealed. This section provides the operating pressures that classify a duct as a medium-pressure duct and make the connection to the requirements of the IMC. Sealing of the ducts is done with the proper materials as designated in Section C403.2.9.

C403.2.9.1.3 High-pressure duct systems. Ducts and plenums designed to operate at static pressures greater than 3 inches water gauge (747 Pa) shall be insulated and sealed in accordance with Section C403.2.9. In addition, ducts and plenums shall be leak tested in accordance with the SMACNA *HVAC Air Duct Leakage Test Manual* and shown to have a rate of air leakage (CL) less than or equal to 4.0 as determined in accordance with Equation 4-8.

$$CL = F/P^{0.65} \quad \textbf{(Equation 4-8)}$$

where:

F = The measured leakage rate in cfm per 100 square feet of duct surface.

P = The static pressure of the test.

Documentation shall be furnished by the designer demonstrating that representative sections totaling at least 25 percent of the duct area have been tested and that all tested sections comply with the requirements of this section.

❖ As discussed in the commentary to Section C403.2.9.1.1, ductwork must be properly constructed and sealed. This section provides the operating pressures that classify a duct as a high-pressure duct (3 inches water gauge and greater) and make the connection to the requirements of the IMC. Sealing of the ducts is done with the proper materials as designated in Section C403.2.9.

In addition to the requirements of the IMC and Section C403.2.9, leak testing is required when evaluating high-pressure commercial duct systems for compliance with the code (see commentary, Section C403.2.9).

Commercial ductwork designed to operate at static pressures in excess of 3-inch w.g. (747 kPa) must meet leakage limitations as follows:

- Ducts must be tested in accordance with the procedures outlined in Section 5 of the Sheet Metal and Air Conditioning Contractors National Association's (SMACNA) *HVAC Duct Leakage Test Manual* with tests reported using forms equivalent to those outlined in Section 6 of that manual.
- To reduce costs, the entire duct system need not be tested. Representative sections can be tested if these sections represent not less than 25 percent of the total installed duct area for the tested pressure class.
- Duct leakage at a test pressure equal to the design duct pressure must result in a rate of air leakage not greater than 4.0 as determined in accordance with Equation 4-8. The rate of air leakage according to this formula must be less than 4.0 when the test pressure is equal to the design duct pressure class. The pressure classification of ducts must be shown on the construction documents, and be equal to or exceed the design pressure of the system required by the IMC.

C403.2.10 Piping insulation. Piping serving as part of a heating or cooling system shall be thermally insulated in accordance with Table C403.2.10.

Exceptions:

1. Factory-installed piping within HVAC equipment tested and rated in accordance with a test procedure referenced by this code.
2. Factory-installed piping within room fan-coils and unit ventilators tested and rated according to AHRI 440 (except that the sampling and variation provisions of Section 6.5 shall not apply) and AHRI 840, respectively.
3. Piping that conveys fluids that have a design operating temperature range between 60°F (15°C) and 105°F (41°C).
4. Piping that conveys fluids that have not been heated or cooled through the use of fossil fuels or electric power.
5. Strainers, control valves, and balancing valves associated with piping 1 inch (25 mm) or less in diameter.
6. Direct buried piping that conveys fluids at or below 60°F (15°C).

❖ The values in Table C403.2.10 are minimum thicknesses of insulation having a conductivity of 0.27 Btu/in/(h x ft^2 x °F) (0.00008 kw) (R-3.7 per inch) for each fluid category. This conductivity is typical of fiberglass and most cellular and phenolic foam insulations. The thermal insulation conductivity increases with temperature. For insulation with a thermal conductivity not equal to 0.27 Btu/in/(h x ft^2 x °F) (0.00008 kw) at a mean temperature of 75°F (25°C), the minimum required pipe thickness is adjusted using this equation.

Insulation is not regulated in the following cases:

1. Piping that is factory installed in equipment that is tested and rated in accordance with Section C403.2.3. The intent here is to exempt piping in equipment where the energy performance is tested and piping losses are ostensibly accounted for in the ratings.
2. The code exempts room fan coils and unit ventilators that are tested and rated according to AHRI 440 and AHRI 840 from the piping insulation requirements. The code did not accept the sampling and variation provisions in Section 6.5 of AHRI 440 because the standard's language was ambiguous and permissive as to testing a representative sample by the manufacturer. This representative sample would have been the standard model of this particular size having the lowest air delivery and, therefore, was not accepted by the code.
3. Piping conveying fluids that have design operating temperatures between 60°F and 105°F (15°C and 41°C), such as in typical condenser water piping.
4. Piping that conveys fluids that have not been heated or cooled using fossil fuels or electricity. This exception is intended to cover gas piping, cold domestic water piping, waste and vent piping, rain water piping, solar-heated water piping, etc., that may carry fluids with operating temperatures outside the 55°F to 105°F (13°C to 41°C) range but for which no energy was consumed to bring the fluids to these temperatures. Even though insulating such piping will have no energy impact and is not required, it may be desirable in some cases, and possibly required by some municipalities, to prevent condensation.

5. Insulation for strainers and valves associated with small pipe, simply because this represents a negligible energy savings.

6. Direct buried piping that conveys fluids of less than 60°F (15°C), because this is fluid for which little energy was consumed to bring it to that temperature.

Insulation also is not required where heat gain or heat loss from the piping will not increase building energy costs. An example of piping falling into this exception might be condensate drains.

TABLE C403.2.10. See below.

❖ All piping associated with HVAC systems must be thermally insulated in accordance with this table.

C403.2.10.1 Protection of piping insulation. Piping insulation exposed to weather shall be protected from damage, including that due to sunlight, moisture, equipment maintenance and wind, and shall provide shielding from solar radiation that can cause degradation of the material. Adhesive tape shall not be permitted.

❖ This section requires protection for piping insulation exposed in exterior installations. The piping insulation should be protected from sunlight, moisture, wind and solar radiation, but also from personnel who may step on it, run into it with equipment, etc., and cause it to be damaged. Adhesive tape is not permitted to be used because it will limit maintenance and damage the insulation's permeability characteristics. Removing the adhesive tape can damage the integrity of the original insulation, especially if the insulation has reached thermoset state. Protection also can keep silted pipe insulation from commonly separating, thus saving additional energy cost. Repair torn or damaged insulation with duct tape listed and labeled to UL 181B.

C403.2.11 Mechanical systems commissioning and completion requirements. Mechanical systems shall be commissioned and completed in accordance with Section C408.2.

❖ Further examination of the reference to the IMC in Section C408.2 reveals that it requires proper test and balancing procedures to promote stable temperature control without wide fluctuations in occupant comfort. The supply system balancing requirements in the code are limited to providing a means for balancing. For air systems, this requirement can be met by installing manual dampers at each branch of ductwork or by installing adjustable registers, grilles or diffusers that can be used to regulate airflow to the space.

The IMC as an extension of the code requires initial and supplementary balancing for commissioning of the building's ventilation system. Because many systems function differently with seasonal change and after construction, this preparatory work is intended to confirm code compliance with required outdoor airflow rates.

Effective testing, adjusting and balancing of air systems requires a planned procedure by first adjusting the fan speed to meet the design flow conditions and then modulating dampers to minimize the throttling losses. Damper throttling is discouraged as the sole means for balancing fans.

Air and hydronic system balancing procedures may be obtained in the National Environmental Balancing Bureau's (NEBB) Procedural Standards or the Asso-

TABLE C403.2.10
MINIMUM PIPE INSULATION THICKNESS (in inches)[a, c]

FLUID OPERATING TEMPERATURE RANGE AND USAGE (°F)	INSULATION CONDUCTIVITY		NOMINAL PIPE OR TUBE SIZE (inches)				
	Conductivity Btu · in./(h · ft² · °F)[b]	Mean Rating Temperature, °F	< 1	1 to < $1^1/_2$	$1^1/_2$ to < 4	4 to < 8	≥ 8
> 350	0.32 – 0.34	250	4.5	5.0	5.0	5.0	5.0
251 – 350	0.29 – 0.32	200	3.0	4.0	4.5	4.5	4.5
201 – 250	0.27 – 0.30	150	2.5	2.5	2.5	3.0	3.0
141 – 200	0.25 – 0.29	125	1.5	1.5	2.0	2.0	2.0
105 – 140	0.21 – 0.28	100	1.0	1.0	1.5	1.5	1.5
40 – 60	0.21 – 0.27	75	0.5	0.5	1.0	1.0	1.0
< 40	0.20 – 0.26	50	0.5	1.0	1.0	1.0	1.5

For SI: 1 inch = 25.4 mm, °C = [(°F) - 32]/1.8.

a. For piping smaller than $1^1/_2$ inches and located in partitions within conditioned spaces, reduction of these thicknesses by 1 inch shall be permitted (before thickness adjustment required in footnote b) but not to a thickness less than 1 inch.

b. For insulation outside the stated conductivity range, the minimum thickness (T) shall be determined as follows:

$T = r\{(1 + t/r)^{K/k} - 1\}$

where:

T = minimum insulation thickness,

r = actual outside radius of pipe,

t = insulation thickness listed in the table for applicable fluid temperature and pipe size,

K = conductivity of alternate material at mean rating temperature indicated for the applicable fluid temperature (Btu • in/h • ft² • °F) and

k = the upper value of the conductivity range listed in the table for the applicable fluid temperature.

c. For direct-buried heating and hot water system piping, reduction of these thicknesses by $1^1/_2$ inches (38 mm) shall be permitted (before thickness adjustment required in footnote b but not to thicknesses less than 1 inch (25 mm).

ciated Air Balance Council's (AABC) National Standards (see commentary, Section C408.2.2).

Overflow in hydronic systems increases pumping cost, and any decrease in flow reduces effective heating and cooling at design conditions. Although a variety of techniques exist for balancing hydronic systems, the code neither requires hydronic systems to actually be balanced nor specifies any particular procedure to achieve that balance.

The procedure most often used is based on a water-temperature measurement between the supply and the return at the terminal unit. This method, as well as others, requires attention to detail and keeping accurate field measurements to successfully balance the system.

This requires water systems to have a means for balancing and pressure test connections to achieve hydronic balance for terminal units, branch lines, zones, risers and mains. The means for balancing is intentionally left unspecified, but often incorporates balancing valves, proportional flow circuits, variable speed pumping, trimming of pump impellers and combinations of these items.

C403.2.12 Air system design and control. Each HVAC system having a total fan system motor nameplate horsepower (hp) exceeding 5 hp (3.7 kW) shall comply with the provisions of Sections C403.2.12.1 through C403.2.12.3.

❖ This section adds the requirements of Sections C403.2.12.1 through C403.2.12.3 for the design of HVAC fan systems and regulates the energy use for these systems.

C403.2.12.1 Allowable fan motor horsepower. Each HVAC system at fan system design conditions shall not exceed the allowable *fan system motor nameplate hp* (Option 1) or *fan system bhp* (Option 2) as shown in Table C403.2.12.1(1). This includes supply fans, exhaust fans, return/relief fans, and fan-powered terminal units associated with systems providing heating or cooling capability. Single-*zone* variable air volume systems shall comply with the constant volume fan power limitation.

Exceptions:

1. Hospital, vivarium and laboratory systems that utilize flow control devices on exhaust or return to maintain space pressure relationships necessary for occupant health and safety or environmental control shall be permitted to use variable volume fan power limitation.
2. Individual exhaust fans with motor nameplate horsepower of 1 hp (0.746 kW) or less are exempt from the allowable fan horsepower requirement.

❖ This section specifies requirements for the design of HVAC fan systems and regulates the motor energy use for these systems. Option 1 provides a simple option based on nameplate motor horsepower (hp) and can be used by most projects. Option 2 is a more complex option based on brake horsepower (bhp) for more complex systems that might require additional fan power allowances to overcome the pressure drop of certain required components, such as specialized filtration, heat recovery devices and sound attenuators.

TABLE C403.2.12.1(2). See page C4-63.

❖ NC35 specified in the sound attenuation section refers to Noise Criterion (NC) rating. Noise criteria are established to address the rating of indoor noise including noise from air-conditioning equipment.

C403.2.12.2 Motor nameplate horsepower. For each fan, the fan brake horsepower shall be indicated on the construction documents and the selected motor shall be not larger than the first available motor size greater than the following:

1. For fans less than 6 bhp (4413 W), 1.5 times the fan brake horsepower.
2. For fans 6 bhp (4413 W) and larger, 1.3 times the fan brake horsepower.
3. Systems complying with Section C403.2.12.1 *fan system motor nameplate hp* (Option 1).

❖ The purpose of this section is to limit the size of the fan selected for the design requirements to prevent using oversized equipment, which would waste energy.

C403.2.12.3 Fan efficiency. Fans shall have a fan efficiency grade (FEG) of not less than 67 when determined in accordance with AMCA 205 by an *approved*, independent testing laboratory and labeled by the manufacturer. The total efficiency of the fan at the design point of operation shall be

TABLE C403.2.12.1(1)
FAN POWER LIMITATION

	LIMIT	CONSTANT VOLUME	VARIABLE VOLUME
Option 1: Fan system motor nameplate hp	Allowable nameplate motor hp	$hp \leq CFM_S \cdot 0.0011$	$hp \leq CFM_S \cdot 0.0015$
Option 2: Fan system bhp	Allowable fan system bhp	$bhp \leq CFM_S \cdot 0.00094 + A$	$bhp \leq CFM_S \cdot 0.0013 + A$

For SI: 1 bhp = 735.5 W, 1 hp = 745.5 W, 1 cfm = 0.4719 L/s.

where:

CFM_S = The maximum design supply airflow rate to conditioned spaces served by the system in cubic feet per minute.

hp = The maximum combined motor nameplate horsepower.

Bhp = The maximum combined fan brake horsepower.

A = Sum of $[PD \times CFM_D / 4131]$

where:

PD = Each applicable pressure drop adjustment from Table C403.2.12.1(2) in. w.c.

CFM_D = The design airflow through each applicable device from Table C403.2.12.1(2) in cubic feet per minute.

within 15 percentage points of the maximum total efficiency of the fan.

Exception: The following fans are not required to have a fan efficiency grade:

1. Fans of 5 hp (3.7 kW) or less as follows:
   1.1. Single fan with a motor nameplate horsepower of 5 hp (3.7 kW) or less, unless Exception 1.2 applies.
   1.2. Multiple fans in series or parallel that have a combined motor nameplate horsepower of 5 hp (3.7 kW) or less and are operated as the functional equivalent of a single fan.
2. Fans that are part of equipment covered under Section C403.2.3.
3. Fans included in an equipment package certified by an *approved agency* for air or energy performance.
4. Powered wall/roof ventilators.
5. Fans outside the scope of AMCA 205.
6. Fans that are intended to operate only during emergency conditions.

❖ This section adds requirements for minimum fan efficiency by means of a Fan Efficiency Grade (FEG) when tested to AMCA 205 by an independent third-party laboratory. The code provisions also require that fans be labeled as such by the manufacturer. AMCA 205 defines the FEG metric and how it is to be calculated by manufacturers when rating their fans.

Regulating the FEG level of the fan alone is inadequate for reducing fan energy consumption; right-sizing fans is also important. As such, both AMCA 205 and the code require a "right-sizing" provision that stipulates fans be sized/selected to operate within 15 percentage points of the fan's rated peak total efficiency. That the "right-sized" fan is used must be verified during plan review. Engineers/specifiers will have to indicate compliance by documenting the fan's peak total efficiency, which is provided by the manufacturer as part of the fan data. Because the code requires an *approved*, independent laboratory to test fans to AMCA 205, thereby confirming the rated peak total efficiency, this data will be easily available. The engineer/designer needs to "sign off" or "affirm" that the selected fan's operating total efficiency will be less than the rated peak total efficiency. It might also be checked by commissioning providers and/or test and balance contractors.

C403.2.13 Heating outside a building. Systems installed to provide heat outside a building shall be radiant systems.

TABLE C403.2.12.1(2)
FAN POWER LIMITATION PRESSURE DROP ADJUSTMENT

DEVICE	ADJUSTMENT
Credits	
Fully ducted return and/or exhaust air systems	0.5 inch w.c. (2.15 in w.c. for laboratory and vivarium systems)
Return and/or exhaust airflow control devices	0.5 inch w.c.
Exhaust filters, scrubbers or other exhaust treatment	The pressure drop of device calculated at fan system design condition
Particulate filtration credit: MERV 9 thru 12	0.5 inch w.c.
Particulate filtration credit: MERV 13 thru 15	0.9 inch. w.c.
Particulate filtration credit: MERV 16 and greater and electronically enhanced filters	Pressure drop calculated at 2x clean filter pressure drop at fan system design condition.
Carbon and other gas-phase air cleaners	Clean filter pressure drop at fan system design condition.
Biosafety cabinet	Pressure drop of device at fan system design condition.
Energy recovery device, other than coil runaround loop	(2.2 × energy recovery effectiveness) – 0.5 inch w.c. for each airstream.
Coil runaround loop	0.6 inch w.c. for each airstream.
Evaporative humidifier/cooler in series with another cooling coil	Pressure drop of device at fan system design conditions.
Sound attenuation section (fans serving spaces with design background noise goals below NC35)	0.15 inch w.c.
Exhaust system serving fume hoods	0.35 inch w.c.
Laboratory and vivarium exhaust systems in high-rise buildings	0.25 inch w.c./100 feet of vertical duct exceeding 75 feet.
Deductions	
Systems without central cooling device	- 0.6 in. w.c.
Systems without central heating device	- 0.3 in. w.c.
Systems with central electric resistance heat	- 0.2 in. w.c.

For SI: 1 inch w.c. = 249 Pa, 1 inch = 25.4 mm.

w.c. = water column, NC = Noise criterion.

Such heating systems shall be controlled by an occupancy sensing device or a timer switch, so that the system is automatically deenergized when no occupants are present.

❖ This section requires occupant sensors or a time-switch control to limit a radiant system installed to provide heat outside a building. It easily can double as an HVAC off-hour control by adding an interlock wired as an input to the DDC zone controller. This avoids wasting energy during unoccupied hours, while still allowing a system that is shut off during off-hours to automatically restart and temporarily operate in order to maintain the space at a setback or setup temperature setpoint.

C403.2.14 Refrigeration equipment performance. Refrigeration equipment shall have an energy use in kWh/day not greater than the values of Tables C403.2.14(1) and C403.2.14(2) when tested and rated in accordance with AHRI Standard 1200. The energy use shall be verified through certification under an approved certification program or, where a certification program does not exist, the energy use shall be supported by data furnished by the equipment manufacturer.

❖ Manufactured refrigeration equipment can be a significant user of energy. With the development of AHRI Standard 1200, there is now a method by which efficiency of this equipment can be measured. The equipment should be labeled with its tested efficiency. Where equipment hasn't been tested, the designer must be able to supply energy-use data. Such data will most likely come from the manufacturer. Sections C403.2.15 and C403.2.16 separately address larger equipment often described as walk-in. Section C403.2.17 specifically addresses refrigerated display cases.

C403.2.15 Walk-in coolers, walk-in freezers, refrigerated warehouse coolers and refrigerated warehouse freezers. *Refrigerated warehouse coolers* and *refrigerated warehouse freezers* shall comply with this section. *Walk-in coolers* and *walk-in freezers* that are not either site assembled or site constructed shall comply with the following:

1. Be equipped with automatic door-closers that firmly close walk-in doors that have been closed to within 1 inch (25 mm) of full closure.

   Exception: Automatic closers are not required for doors more than 45 inches (1143 mm) in width or more than 7 feet (2134 mm) in height.

2. Doorways shall have strip doors, curtains, spring-hinged doors or other method of minimizing infiltration when doors are open.

3. *Walk-in coolers* and *refrigerated warehouse coolers* shall contain wall, ceiling, and door insulation of not less than R-25 and *walk-in freezers* and *refrigerated warehouse freezers* shall contain wall, ceiling and door insulation of not less than R-32.

   Exception: Glazed portions of doors or structural members need not be insulated.

4. *Walk-in freezers* shall contain floor insulation of not less than R-28.

5. Transparent reach-in doors for *walk-in freezers* and windows in *walk-in freezer* doors shall be of triple-pane glass, either filled with inert gas or with heat-reflective treated glass.

6. Windows and transparent reach-in doors for *walk-in coolers* doors shall be of double-pane or triple-pane, inert gas-filled, heat-reflective treated glass.

7. Evaporator fan motors that are less than 1 hp (0.746 kW) and less than 460 volts shall use electronically commutated motors, brushless direct- current motors, or 3-phase motors.

8. Condenser fan motors that are less than 1 hp (0.746 kW) shall use electronically commutated motors, permanent split capacitor-type motors or 3-phase motors.

9. Where antisweat heaters without antisweat heater controls are provided, they shall have a total door rail, glass and frame heater power draw of not more than 7.1 W/ft^2 (76 W/m^2) of door opening for *walk-in freezers* and 3.0 W/ft^2 (32 W/m^2) of door opening for *walk-in coolers.*

10. Where antisweat heater controls are provided, they shall reduce the energy use of the antisweat heater as a function of the relative humidity in the air outside the door or to the condensation on the inner glass pane.

11. Lights in *walk-in coolers*, *walk-in freezers*, *refrigerated warehouse coolers* and *refrigerated warehouse freezers* shall either use light sources with an efficacy of not less than 40 lumens per watt, including ballast losses, or shall use light sources with an efficacy of not less than 40 lumens per watt, including ballast losses, in conjunction with a device that turns off the lights within 15 minutes when the space is not occupied.

❖ Sections C403.2.15 and C403.2.16 address walk-in coolers, walk-in freezers, refrigerated warehouse coolers and refrigerated warehouse freezers. The criteria for these features are nearly identical. Section C403.2.16 addresses walk-in coolers and walk-in freezers that are site assembled or site constructed. All other coolers and freezers must comply with Section C403.2.15. Definitions of these four categories are found in Chapter 2 [CE]. A key difference is the walk-in coolers and freezers are under 3,000 square feet. Where they are 3,000 square feet or larger, they fall into the warehouse categories. In this section, Item 11 addresses lights in these spaces. Two options appear to both address where efficacy is not less than 40 lumens per watt and allow an option with or without automatic shut-off controls. Under Section C403.2.16, the option is high-efficacy lighting or an automated sensor shut-off.

TABLE C403.2.14(1)
MINIMUM EFFICIENCY REQUIREMENTS: COMMERCIAL REFRIGERATION

EQUIPMENT TYPE	APPLICATION	ENERGY USE LIMITS (kWh per day)[a]	TEST PROCEDURE
Refrigerator with solid doors	Holding Temperature	0.10 • V + 2.04	AHRI 1200
Refrigerator with transparent doors		0.12 • V + 3.34	
Freezers with solid doors		0.40 • V + 1.38	
Freezers with transparent doors		0.75 • V + 4.10	
Refrigerators/freezers with solid doors		the greater of 0.12 • V + 3.34 or 0.70	
Commercial refrigerators	Pulldown	0.126 • V + 3.51	

a. V = volume of the chiller or frozen compartment as defined in AHAM-HRF-1.

TABLE C403.2.14(2)
MINIMUM EFFICIENCY REQUIREMENTS: COMMERCIAL REFRIGERATORS AND FREEZERS

EQUIPMENT TYPE				ENERGY USE LIMITS (kWh/day)[a, b]	TEST PROCEDURE
Equipment Class[c]	Family Code	Operating Mode	Rating Temperature		
VOP.RC.M	Vertical open	Remote condensing	Medium	0.82 • TDA + 4.07	AHRI 1200
SVO.RC.M	Semivertical open	Remote condensing	Medium	0.83 • TDA + 3.18	
HZO.RC.M	Horizontal open	Remote condensing	Medium	0.35 • TDA + 2.88	
VOP.RC.L	Vertical open	Remote condensing	Low	2.27 • TDA + 6.85	
HZO.RC.L	Horizontal open	Remote condensing	Low	0.57 • TDA + 6.88	
VCT.RC.M	Vertical transparent door	Remote condensing	Medium	0.22 TDA + 1.95	
VCT.RC.L	Vertical transparent door	Remote condensing	Low	0.56 • TDA + 2.61	
SOC.RC.M	Service over counter	Remote condensing	Medium	0.51 • TDA + 0.11	
VOP.SC.M	Vertical open	Self-contained	Medium	1.74 • TDA + 4.71	
SVO.SC.M	Semivertical open	Self-contained	Medium	1.73 • TDA + 4.59	
HZO.SC.M	Horizontal open	Self-contained	Medium	0.77 • TDA + 5.55	
HZO.SC.L	Horizontal open	Self-contained	Low	1.92 • TDA + 7.08	
VCT.SC.I	Vertical transparent door	Self-contained	Ice cream	0.67 • TDA + 3.29	
VCS.SC.I	Vertical solid door	Self-contained	Ice cream	0.38 • V + 0.88	
HCT.SC.I	Horizontal transparent door	Self-contained	Ice cream	0.56 • TDA + 0.43	
SVO.RC.L	Semivertical open	Remote condensing	Low	2.27 • TDA + 6.85	
VOP.RC.I	Vertical open	Remote condensing	Ice cream	2.89 • TDA + 8.7	
SVO.RC.I	Semivertical open	Remote condensing	Ice cream	2.89 • TDA + 8.7	
HZO.RC.I	Horizontal open	Remote condensing	Ice cream	0.72 • TDA + 8.74	
VCT.RC.I	Vertical transparent door	Remote condensing	Ice cream	0.66 • TDA + 3.05	
HCT.RC.M	Horizontal transparent door	Remote condensing	Medium	0.16 • TDA + 0.13	

(continued)

TABLE C403.2.14(2)—continued
MINIMUM EFFICIENCY REQUIREMENTS: COMMERCIAL REFRIGERATORS AND FREEZERS

EQUIPMENT TYPE				ENERGY USE LIMITS (kWh/day)[a, b]	TEST PROCEDURE
Equipment Class[c]	Family Code	Operating Mode	Rating Temperature		
HCT.RC.L	Horizontal transparent door	Remote condensing	Low	0.34 • TDA + 0.26	AHRI 1200
HCT.RC.I	Horizontal transparent door	Remote condensing	Ice cream	0.4 • TDA + 0.31	
VCS.RC.M	Vertical solid door	Remote condensing	Medium	0.11 • V + 0.26	
VCS.RC.L	Vertical solid door	Remote condensing	Low	0.23 • V + 0.54	
VCS.RC.I	Vertical solid door	Remote condensing	Ice cream	0.27 • V + 0.63	
HCS.RC.M	Horizontal solid door	Remote condensing	Medium	0.11 • V + 0.26	
HCS.RC.L	Horizontal solid door	Remote condensing	Low	0.23 • V + 0.54	
HCS.RC.I	Horizontal solid door	Remote condensing	Ice cream	0.27 • V + 0.63	
HCS.RC.I	Horizontal solid door	Remote condensing	Ice cream	0.27 • V + 0.63	
SOC.RC.L	Service over counter	Remote condensing	Low	1.08 • TDA + 0.22	
SOC.RC.I	Service over counter	Remote condensing	Ice cream	1.26 • TDA + 0.26	
VOP.SC.L	Vertical open	Self-contained	Low	4.37 • TDA + 11.82	
VOP.SC.I	Vertical open	Self-contained	Ice cream	5.55 • TDA + 15.02	
SVO.SC.L	Semivertical open	Self-contained	Low	4.34 • TDA + 11.51	
SVO.SC.I	Semivertical open	Self-contained	Ice cream	5.52 • TDA + 14.63	
HZO.SC.I	Horizontal open	Self-contained	Ice cream	2.44 • TDA + 9.0	
SOC.SC.I	Service over counter	Self-contained	Ice cream	1.76 • TDA + 0.36	
HCS.SC.I	Horizontal solid door	Self-contained	Ice cream	0.38 • V + 0.88	

a. V = Volume of the case, as measured in accordance with Appendix C of AHRI 1200.

b. TDA = Total display area of the case, as measured in accordance with Appendix D of AHRI 1200.

c. Equipment class designations consist of a combination [(in sequential order separated by periods (AAA).(BB).(C))] of:

(AAA) An equipment family code where:
- VOP = vertical open
- SVO = semivertical open
- HZO = horizontal open
- VCT = vertical transparent doors
- VCS = vertical solid doors
- HCT = horizontal transparent doors
- HCS = horizontal solid doors
- SOC = service over counter

(BB) An operating mode code:
- RC = remote condensing
- SC = self-contained

(C) A rating temperature code:
- M = medium temperature (38°F)
- L = low temperature (0°F)
- I = ice-cream temperature (15°F)

For example, ''VOP.RC.M'' refers to the "vertical-open, remote-condensing, medium-temperature" equipment class.

C403.2.16 Walk-in coolers and walk-in freezers. Site-assembled or site-constructed *walk-in coolers* and *walk-in freezers* shall comply with the following:

1. Automatic door closers shall be provided that fully close walk-in doors that have been closed to within 1 inch (25 mm) of full closure.

   Exception: Closers are not required for doors more than 45 inches (1143 mm) in width or more than 7 feet (2134 mm) in height.

2. Doorways shall be provided with strip doors, curtains, spring-hinged doors or other method of minimizing infiltration when the doors are open.
3. Walls shall be provided with insulation having a thermal resistance of not less than R-25, ceilings shall be provided with insulation having a thermal resistance of not less than R-25 and doors of *walk-in coolers* and *walk-in freezers* shall be provided with insulation having a thermal resistance of not less than R-32.

   Exception: Insulation is not required for glazed portions of doors or at structural members associated with the walls, ceiling or door frame.

4. The floor of *walk-in freezers* shall be provided with insulation having a thermal resistance of not less than R-28.
5. Transparent reach-in doors for and windows in opaque *walk-in freezer* doors shall be provided with triple-pane glass having the interstitial spaces filled with inert gas or provided with heat-reflective treated glass.
6. Transparent reach-in doors for and windows in opaque *walk-in cooler* doors shall be double-pane heat-reflective treated glass having the interstitial space gas filled.
7. Evaporator fan motors that are less than 1 hp (0.746 kW) and less than 460 volts shall be electronically commutated motors or 3-phase motors.
8. Condenser fan motors that are less than 1 hp (0.746 kW) in capacity shall be of the electronically commutated or permanent split capacitor-type or shall be 3-phase motors.

   Exception: Fan motors in *walk-in coolers* and *walk-in freezers* combined in a single enclosure greater than 3,000 square feet (279 m^2) in floor area are exempt.

9. Antisweat heaters that are not provided with antisweat heater controls shall have a total door rail, glass and frame heater power draw not greater than 7.1 W/ft2 (76 W/m^2) of door opening for *walk-in freezers*, and not greater than 3.0 W/ft^2 (32 W/m^2) of door opening for *walk-in coolers*.
10. Antisweat heater controls shall be capable of reducing the energy use of the antisweat heater as a function of the relative humidity in the air outside the door or to the condensation on the inner glass pane.
11. Light sources shall have an efficacy of not less than 40 lumens per Watt, including any ballast losses, or shall be provided with a device that automatically turns off the lights within 15 minutes of when the *walk-in cooler* or *walk-in freezer* was last occupied.

❖ See the commentary to Section C403.2.15.

C403.2.17 Refrigerated display cases. Site-assembled or site-constructed refrigerated display cases shall comply with the following:

1. Lighting and glass doors in refrigerated display cases shall be controlled by one of the following:
   1.1. Time switch controls to turn off lights during nonbusiness hours. Timed overrides for display cases shall turn the lights on for up to 1 hour and shall automatically time out to turn the lights off.

   1.2. Motion sensor controls on each display case section that reduce lighting power by at least 50 percent within 3 minutes after the area within the sensor range is vacated.
2. Low-temperature display cases shall incorporate temperature-based defrost termination control with a time-limit default. The defrost cycle shall terminate first on an upper temperature limit breach and second upon a time limit breach.
3. Antisweat heater controls shall reduce the energy use of the antisweat heater as a function of the relative humidity in the air outside the door or to the condensation on the inner glass pane.

❖ Refrigerated display cases are used to display products that must be refrigerated. Compared to the features addressed in Sections C403.2.15 and C403.2.16, these units are factory assembled rather than site assembled. Many energy-efficiency requirements are similar to walk-in units.

C403.3 Economizers (Prescriptive). Each cooling system shall include either an air or water economizer complying with Sections C403.3.1 through C403.3.4.

Exceptions: Economizers are not required for the systems listed below.

1. In cooling systems for buildings located in *Climate Zones* 1A and 1B.
2. In *climate zones* other than 1A and 1B, where individual fan cooling units have a capacity of less than 54,000 Btu/h (15.8 kW) and meet one of the following:
   2.1. Have direct expansion cooling coils.

   2.2. The total chilled water system capacity less the capacity of fan units with air economizers is less than the minimum specified in Table C403.3(1).

   The total supply capacity of all fan-cooling units not provided with economizers shall not exceed 20 percent of the total supply capacity of all fan-cooling units in the building or 300,000 Btu/h (88 kW), whichever is greater.

3. Where more than 25 percent of the air designed to be supplied by the system is to spaces that are designed to be humidified above 35°F (1.7°C) dew-point temperature to satisfy process needs.
4. Systems that serve *residential* spaces where the system capacity is less than five times the requirement listed in Table C403.3(1).
5. Systems expected to operate less than 20 hours per week.
6. Where the use of *outdoor air* for cooling will affect supermarket open refrigerated casework systems.
7. Where the cooling *efficiency* meets or exceeds the *efficiency* requirements in Table C403.3(2).
8. Chilled-water cooling systems that are passive (without a fan) or use induction where the total chilled water system capacity less the capacity of fan units with air economizers is less than the minimum specified in Table C403.3(1).
9. Systems that include a heat recovery system in accordance with Section C403.4.5.

❖ Section C403.3 covers economizer requirements for all cooling systems, not by building or by complex/simple system designation. Cooling systems shall have either an air- or water-side economizer unless covered by exceptions. Section C403.3 covers the general triggers for where an economizer is required and exceptions from requiring an economizer. Section 403.3.1 covers integrated controls; Section 403.3.2 covers heating system impacts; Section 403.3.3 covers specific requirements for systems with air-side economizers; Section 403.3.4 covers specific requirements for systems with water-side economizers.

To take advantage of locations having favorable weather conditions that permit free cooling (without mechanical refrigeration), the code requires that mechanical systems of certain capacities be equipped with either an air-side economizer or water-side economizer.

The cost effectiveness of economizers depends on the load characteristics of the building, internal heat generation, the type and size of the HVAC system, and the local climate. For instance, certain climates do experience significant periods where the outdoor temperature is cool, but heavily laden with moisture [i.e., wet-bulb temperatures greater than 69°F (21°C)]. Still, other climates do not experience enough hours during a typical meteorological year where cool-dry conditions conductive to economizer operation prevail. That is, the number of hours between 8:00 a.m. and 4:00 p.m. where the outside dry-bulb temperature is between 55°F (13°C) and 69°F (21°C) is insignificant. Therefore, the potential for energy savings, when compared to the assessed cost, is unjustified. Because there are many cases where economizers will not be cost effective, this section includes exceptions to the economizer requirement.

Exceptions 1 and 2: Climate Zones 1A and 1B are exempt from the requirement for economizers, except for large chilled-water systems in Climate Zone 1B covered under Table C403.3(1). Small cooling units that have a cooling capacity of less than 54,000 Btu/h (15.8 kW) at design conditions are exempt. For such small systems, the added cost of economizers will be significantly higher than the annual energy savings, even in mild climates.

Exception 3 covers systems where over 25 percent of the supply airflow is to spaces that require humidification to over 35°F (1.7°C) dew point temperature. This is intended for spaces that require humidification for process loads, not comfort conditioning. In winter, economizer air can overdry the air; excess energy would be expended to rehumidify the outdoor economizer air, defeating the cooling energy savings.

Exception 4 was intended to apply to systems that have a capacity less than five times the capacity listed in Exception 2 (54,00 Btu/h). The reference in this exception to Table C403.3(1) was to the table by that number in the 2012 edition of the code. While the current referral to Table C403.3(1) would permit a significant exception, Exception 4 is intended to only exempt systems serving residential spaces up to 270,00 Btu/h.

An economizer is unlikely to have effective payback of installed cost for systems that operate less than 20 hours per week. And refrigerated cases in supermarkets expend large amounts of energy defrosting the refrigerated cases. While an economizer may provide cooling energy savings, the additional latent (moisture) loads introduced by economizer operation can result in excessive con-

TABLE C403.3(1)
MINIMUM CHILLED-WATER SYSTEM COOLING CAPACITY FOR DETERMINING ECONOMIZER COOLING REQUIREMENTS

CLIMATE ZONES (COOLING)	TOTAL CHILLED-WATER SYSTEM CAPACITY LESS CAPACITY OF COOLING UNITS WITH AIR ECONOMIZERS	
	Local Water-cooled Chilled-water Systems	Air-cooled Chilled-water Systems or District Chilled-Water Systems
1a	No economizer requirement	No economizer requirement
1b, 2a, 2b	960,000 Btu/h	1,250,000 Btu/h
3a, 3b, 3c, 4a, 4b, 4c	720,000 Btu/h	940,000 Btu/h
5a, 5b, 5c, 6a, 6b, 7, 8	1,320,000 Btu/h	1,720,000 Btu/h

For SI: 1 British thermal unit per hour = 0.2931 W.

densation in the refrigerated cases. The extra energy expended condensing and defrosting cases can exceed the economizer savings. So Exception 6 exempts systems that serve supermarkets with open refrigerated cases.

Minimum mechanical cooling system efficiencies are set in the tables in Section C403.2.3 for various equipment types. If the system efficiency is improved to the extent in Table C403.3(2) for the listed climate zones, then an economizer is not required. If a system requires 13 SEER minimum efficiency with an economizer, the economizer is exempted in Climate Zone 4 where the system is 20 percent better than code minimum. A SEER 15.6 or higher unit can be provided without an economizer.

Where a chilled-water system either does not utilize a fan for cooling (e.g., radiant system) or uses induction air (e.g., active chilled beams), the system is not required to have an economizer (water or air) if the total system capacity is less than that listed in Table C403.3(1). The table has values based both on climate zone and whether the system is air cooled or water/evaporatively cooled.

Where the cooling equipment exceeds the efficiency requirement of Table C403.2.3(1) or C403.2.3(2) by the percentage shown in Table C403.3(2), economizer controls are not required. The economizer trade-off stems from the development of the economizer trade-off tables from prior editions of ANSI/ASHRAE/IESNA 90.1 and remain unchanged. Results were based on simulations of energy cost savings for economizers in 16 different locations in the United States.

TABLE C403.3(2)
EQUIPMENT EFFICIENCY PERFORMANCE EXCEPTION FOR ECONOMIZERS

CLIMATE ZONES	COOLING EQUIPMENT PERFORMANCE IMPROVEMENT (EER OR IPLV)
2B	10% efficiency improvement
3B	15% efficiency improvement
4B	20% efficiency improvement

❖ A simple performance path for compliance appears in the economizer trade-off Table C403.3(2) and is applied under Exception 7. Economizers provide a way to improve system efficiency by using outdoor air for building cooling when appropriate. Their use is required in specific climates and where systems greater than a specified cooling capacity are used. However, although the cost for a building economizer for a given system size is essentially fixed for all regions of the country, the efficiency benefits vary significantly. Therefore, the cost effectiveness of the economizer will vary across the country. The economizer trade-off allows building HVAC designers to trade off the use of a required economizer in exchange for the installation of higher-efficiency air-conditioning equipment. This table shows the cooling efficiency required for an air conditioner or heat pump when the design forgoes the installation of an air-side economizer in a climate where the code requires one. The cooling efficiencies shown in the economizer trade-off table result in a level of energy efficiency that is somewhat greater than that from using minimum-efficiency-level cooling equipment as defined by EPAct or NAECA (Public Law 100-12) in conjunction with an air-side economizer in the climate zones identified.

C403.3.1 Integrated economizer control. Economizer systems shall be integrated with the mechanical cooling system and be capable of providing partial cooling even where additional mechanical cooling is required to provide the remainder of the cooling load. Controls shall not be capable of creating a false load in the mechanical cooling systems by limiting or disabling the economizer or any other means, such as hot gas bypass, except at the lowest stage of mechanical cooling.

Units that include an air economizer shall comply with the following:

1. Unit controls shall have the mechanical cooling capacity control interlocked with the air economizer controls such that the outdoor air damper is at the 100-percent open position when mechanical cooling is on and the outdoor air damper does not begin to close to prevent coil freezing due to minimum compressor run time until the leaving air temperature is less than 45°F (7°C).
2. Direct expansion (DX) units that control 75,000 Btu/h (22 kW) or greater of rated capacity of the capacity of the mechanical cooling directly based on occupied space temperature shall have not fewer than two stages of mechanical cooling capacity
3. Other DX units, including those that control space temperature by modulating the airflow to the space, shall be in accordance with Table C403.3.1.

❖ To meet the code, economizers must be able to supply not greater than 100 percent of the design supply air as outside air. The code also requires that economizers be integrated. Integrated economizers can reduce the cooling load with the remainder of the load being met by mechanical cooling. Economizers that cannot operate simultaneously with the mechanical cooling system are called "nonintegrated" economizers. Integration can greatly extend economizer operation in mild climates, which reduces the cooling energy cost. Examples of nonintegrated water economizers are shown in Commentary Figure C403.3.1(2). Those shown in Commentary Figures C403.3.1(3) and C403.3.1(4) are integrated economizers because the economizer and mechanical cooling may operate concurrently.

A nonintegrated air economizer will be able to reduce cooling energy when outside air temperatures are between 55°F (13°C) and 60°F (15°C), depending on the required supply air temperatures. Above those temperatures, mechanical cooling is required, so the economizer is shut off. If the economizer were integrated, it could continue to operate, reducing mechanical cooling energy usage even though it cannot provide the entire cooling load until the high-limit setpoint is reached, around 65°F (18°C) to 75°F (24°C), depending on the climate. In some climates, there are hundreds or even thousands of operating hours when the outside air temperature is in that range. Factory-supplied or factory-installed economizers supplied by equipment manufacturers include integrated controls that also prevent ice formation when outside temperatures are near 50°F (10°C); however, because integrated economizer control strategies that also prevent ice formation are the most commonly available for packaged air-conditioning systems, they are assumed as standard by the code. The controls are wired so the compressor cannot operate until the economizer has been locked out by its high-limit switch, or the economizer is interlocked to shut off when the compressor comes on.

C403.3.2 Economizer heating system impact. HVAC system design and economizer controls shall be such that economizer operation does not increase building heating energy use during normal operation.

> **Exception:** Economizers on variable air volume (VAV) systems that cause *zone* level heating to increase due to a reduction in supply air temperature.

❖ This section is an obvious energy-saving measure. When the outdoor conditions are colder than necessary for building loads, air- or water-side economizing may provide air or water to the building that is colder than necessary. This section requires that the economizer have system controls and capacity to prevent overcooling that triggers activation of heating systems for zones that would need heating.

Variable air volume systems are exempted when *zone*-level heating is activated due to a reduction in supply-air temperature. Heating at the system-wide supply-air-handling system is not exempted. The economizer shall not trigger heating at the air handling unit; the outside air intake must be at the minimum ventilation setting (see Section C403.4.4).

C403.3.3 Air economizers. Air economizers shall comply with Sections C403.3.3.1 through C403.3.3.5.

❖ Air economizers provide free cooling when the outdoor air temperature is cooler than the return air temperature. Through a combination of return, exhaust, outdoor air, and/or mixing dampers, the system exhausts the warmer return air and uses the cooler outdoor air to satisfy some or all of the system cooling needs. An example of air economizer operation is in Commentary Figure 403.3.3(1). This section determines the requirements for economizers using air for the free cooling.

C403.3.3.1 Design capacity. Air economizer systems shall be capable of modulating *outdoor air* and return air dampers to provide up to 100 percent of the design supply air quantity as *outdoor air* for cooling.

❖ This section requires air economizer systems to provide up to 100 percent of the design supply air quantity as outdoor air for cooling. This allows the HVAC system design to be optimized; during cool weather, when the economizer is in operation, the system will be at peak load.

TABLE C403.3.1
DX COOLING STAGE REQUIREMENTS FOR MODULATING AIRFLOW UNITS

RATING CAPACITY	MINIMUM NUMBER OF MECHANICAL COOLING STAGES	MINIMUM COMPRESSOR DISPLACEMENTa
≥ 65,000 Btu/h and < 240,000 Btu/h	3 stages	≤ 35% of full load
≥ 240,000 Btu/h	4 stages	≤ 25% full load

For SI: 1 British thermal unit per hour = 0.2931 W.

a. For *mechanical cooling* stage control that does not use variable compressor displacement, the percent displacement shall be equivalent to the mechanical cooling capacity reduction evaluated at the full load rating conditions for the compressor.

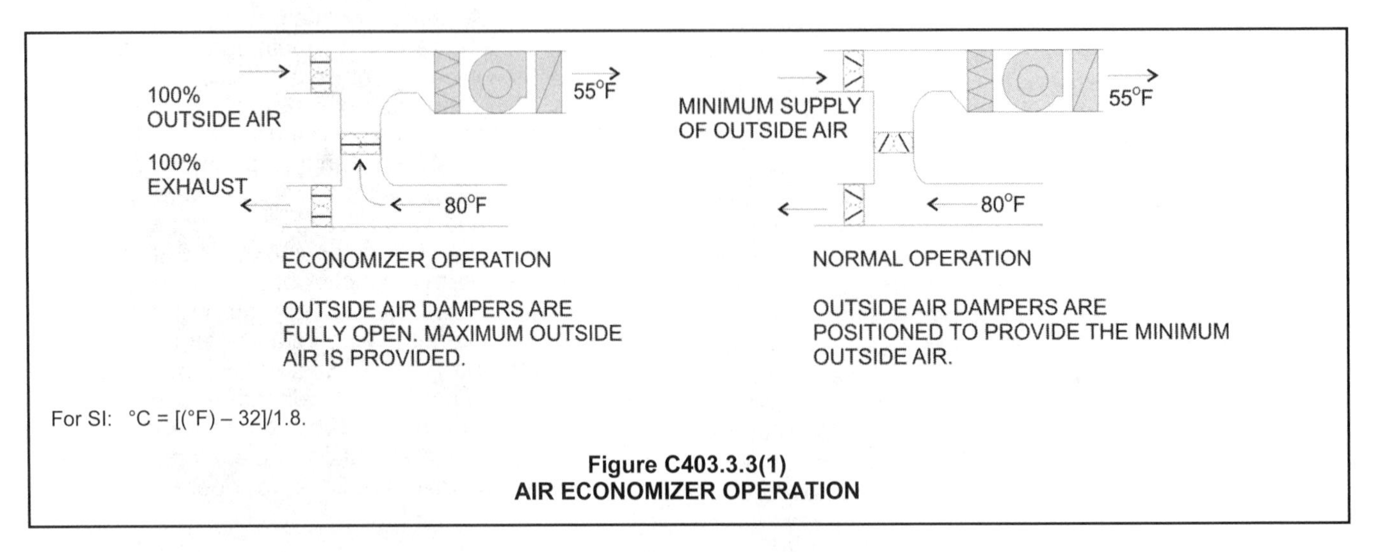

Figure C403.3.3(1)
AIR ECONOMIZER OPERATION

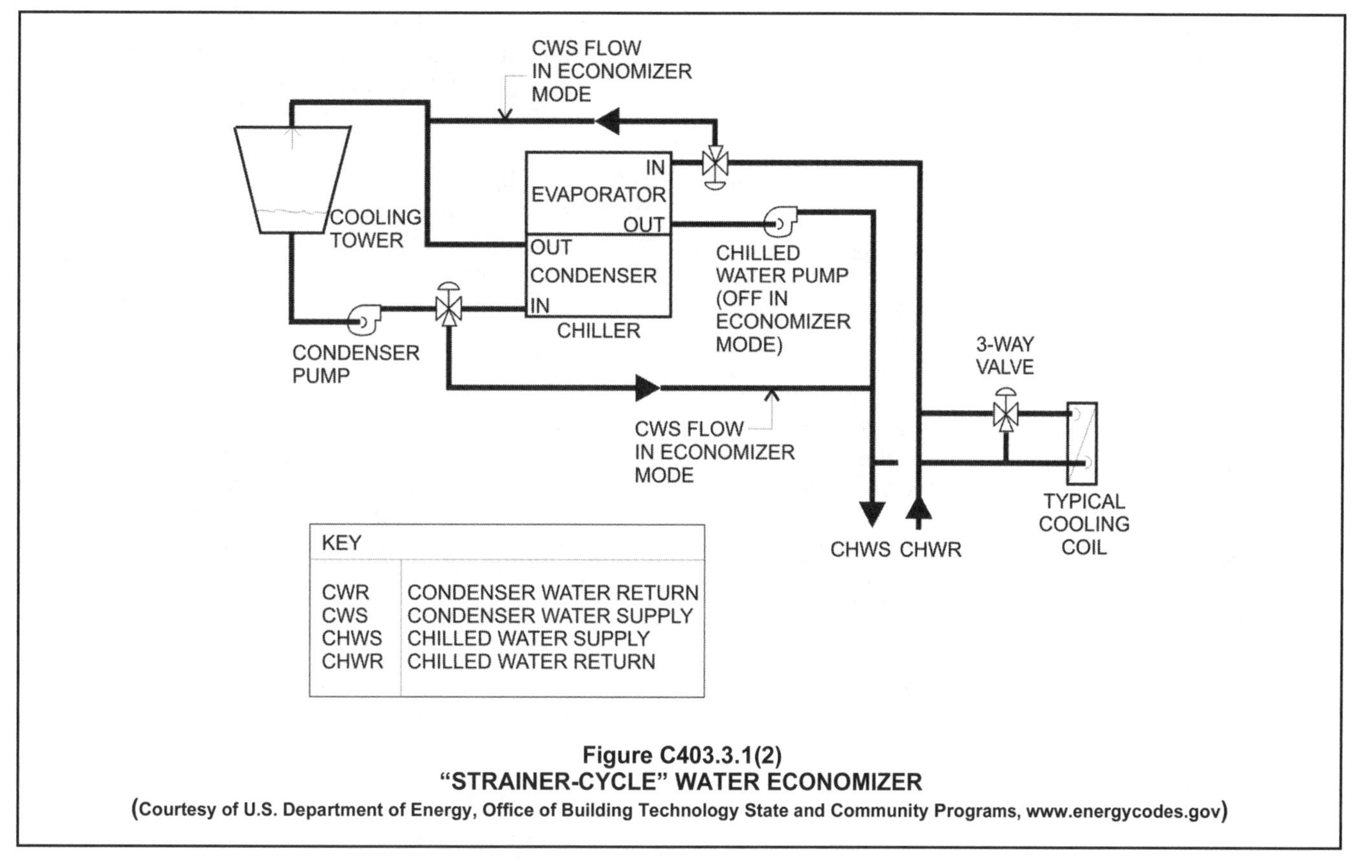

Figure C403.3.1(2)
"STRAINER-CYCLE" WATER ECONOMIZER
(Courtesy of U.S. Department of Energy, Office of Building Technology State and Community Programs, www.energycodes.gov)

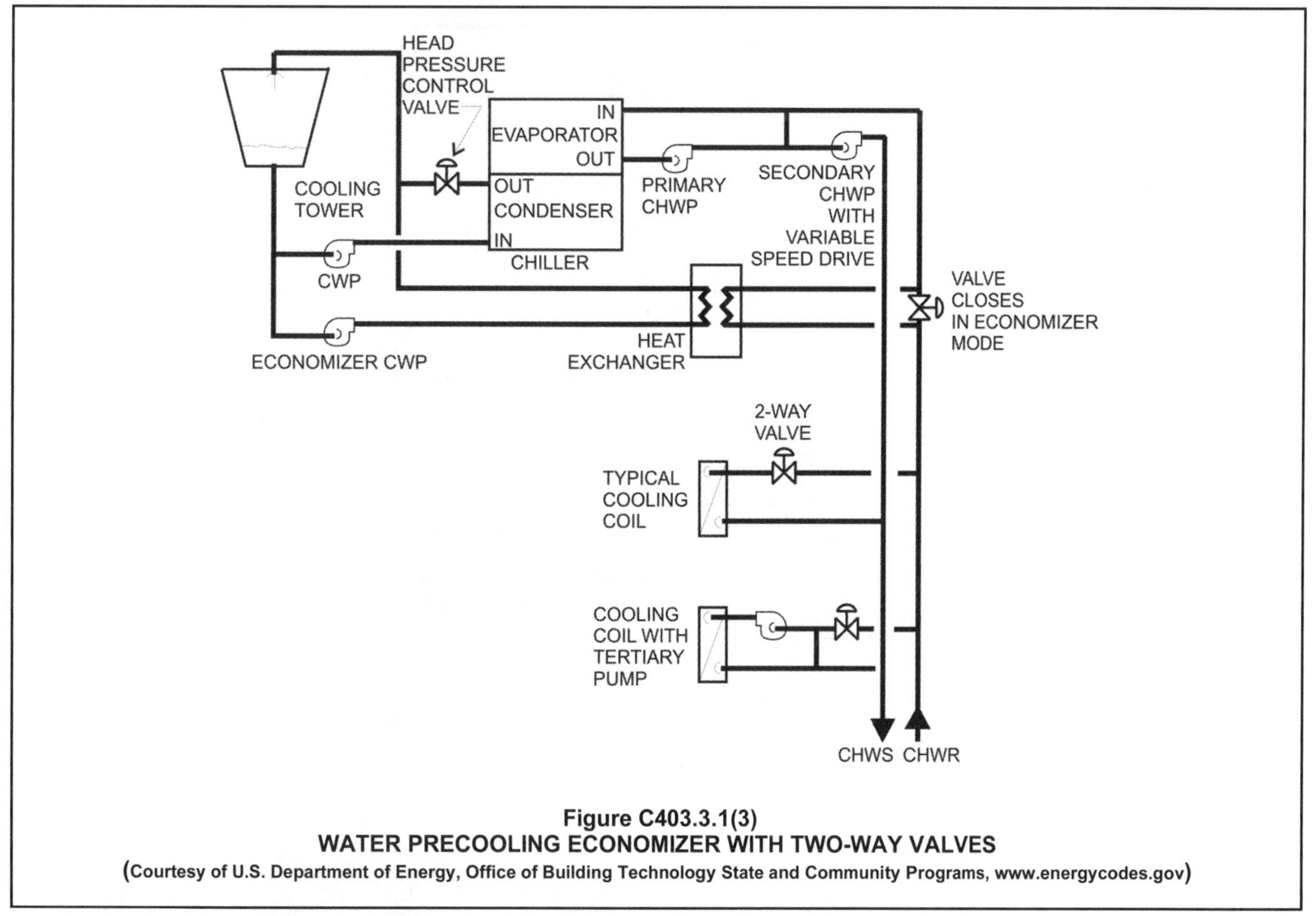

Figure C403.3.1(3)
WATER PRECOOLING ECONOMIZER WITH TWO-WAY VALVES
(Courtesy of U.S. Department of Energy, Office of Building Technology State and Community Programs, www.energycodes.gov)

C403.3.3.2 Control signal. Economizer dampers shall be capable of being sequenced with the mechanical cooling equipment and shall not be controlled by only mixed-air temperature.

> **Exception:** The use of mixed-air temperature limit control shall be permitted for systems controlled from space temperature (such as single-*zone* systems).

❖ This section requires economizer dampers to be sequenced with the mechanical cooling equipment. They are not to be controlled only by mixed air temperature, unless the system controlled is from a single-zone system. It is common for outside air dampers to be specified as low leakage dampers (those with blade and jamb seals). For an economizer, leakage through the return damper can be more important because any leakage of return air when the system is in the 100-percent outside air mode [when outside air temperatures are between about 55°F (12.7°C) until the high-limit setpoint is reached, which can be a majority of the economizer operating hours] will increase supply-air temperatures and force the mechanical cooling system to operate at colder outside air temperatures and increase the cooling load once the mechanical cooling is on. Both dampers should be the low-leakage type.

C403.3.3.3 High-limit shutoff. Air economizers shall be capable of automatically reducing *outdoor air* intake to the design minimum *outdoor air* quantity when *outdoor air* intake will no longer reduce cooling energy usage. High-limit shutoff control types for specific climates shall be chosen from Table C403.3.3.3. High-limit shutoff control settings for these control types shall be those specified in Table C403.3.3.3.

❖ This section requires the air economizers to be capable of automatically reducing outdoor air intake to the design minimum outdoor air quantity when outdoor air intake will no longer reduce cooling energy usage. Otherwise this would increase cooling energy by increasing the latent load. Common enthalpy controllers include a high-limit switch, which enables the economizer to operate when outside air enthalpy is

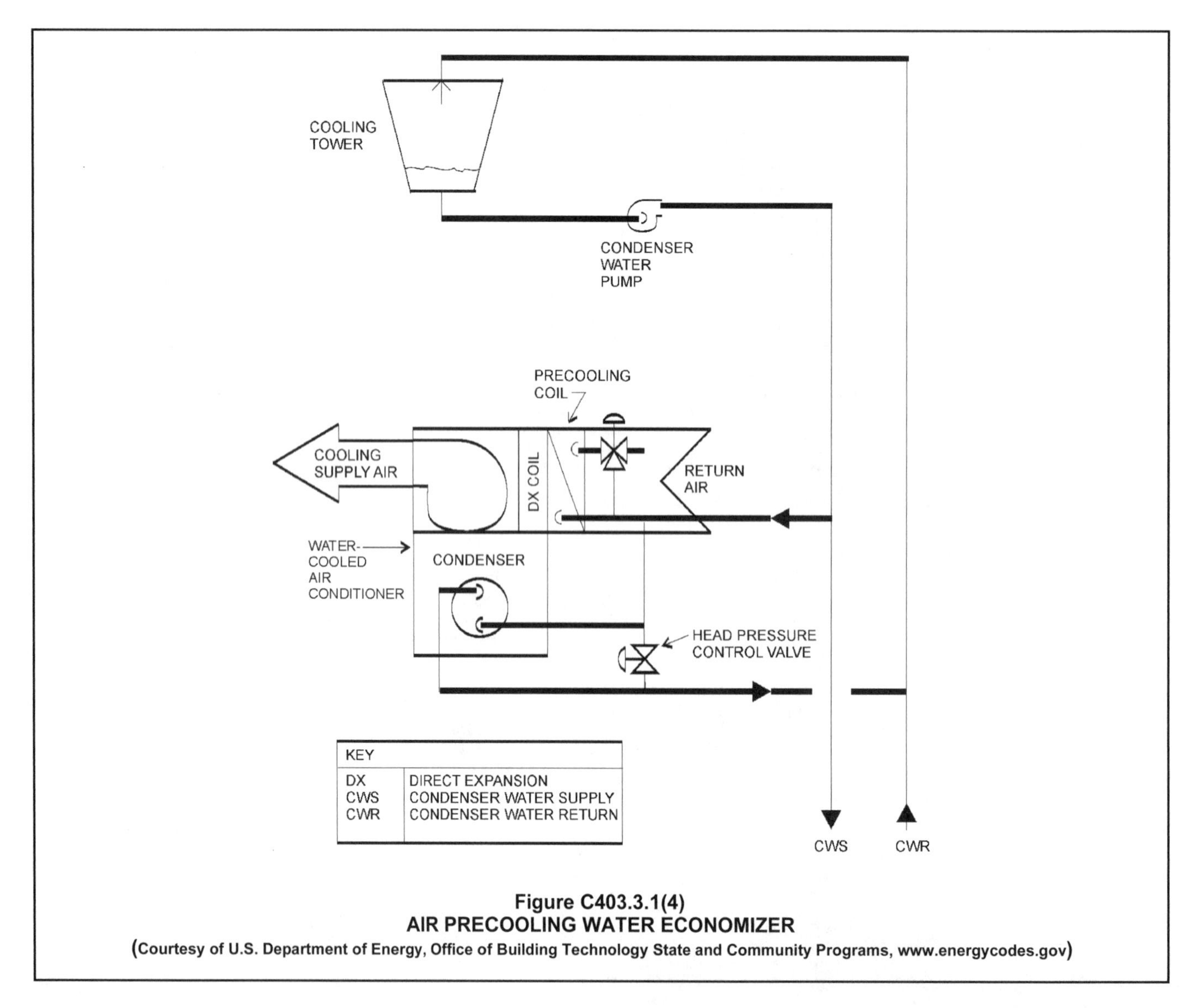

Figure C403.3.1(4)
AIR PRECOOLING WATER ECONOMIZER
(Courtesy of U.S. Department of Energy, Office of Building Technology State and Community Programs, www.energycodes.gov)

below a fixed setpoint, and differential enthalpy controls that enable the economizer to operate whenever outside air enthalpy is less than return air enthalpy. A high-limit shutoff that measures enthalpy and dry bulb should be used. In dry climates, the dry-bulb controllers will be the least expensive and most reliable. In humid climates, the added cost and complication of the enthalpy controller will be justified by the resulting energy savings.

C403.3.3.4 Relief of excess outdoor air. Systems shall be capable of relieving excess *outdoor air* during air economizer operation to prevent overpressurizing the building. The relief air outlet shall be located to avoid recirculation into the building.

❖ This section requires systems to be capable of relieving excess outdoor air during an air economizer operation to prevent overpressurizing the building. When a building is overpressurized, the doors do not open. In the economizer mode, the relief fans can save energy only by operating to maintain a slightly positive building pressure; if exfiltration paths have low-pressure drops, the fans will operate at lower volumes than return fans, which are generally controlled by airflow differentials. To maximize efficiency, relief fans should be axial or propeller fans, which are the most efficient for the low-pressure drops typical of return air paths, such as those using the ceiling attic as return air plenums.

C403.3.3.5 Economizer dampers. Return, exhaust/relief and outdoor air dampers used in economizers shall comply with Section C403.2.4.3.

❖ Section C403.2.4.3 sets maximum leakage rates for outdoor air intake and exhaust dampers, as well as requirements for automatic operation. There are limited exceptions for the use of gravity and motorized dampers in some climate zones.

C403.3.4 Water-side economizers. Water-side economizers shall comply with Sections C403.3.4.1 and C403.3.4.2.

❖ Just as air-side economizers use cool air directly to reduce cooling load, a water economizer (also called a "water-side" economizer) uses outside air to cool water that then cools supply air through a coil, thus reducing the mechanical cooling load. Water economizers are essentially indirect or direct evaporative coolers. In large commercial projects they are usually indirect coolers, using condenser water and cooling towers to chill water during cool, dry weather. Water is circulated through a tower where it is evaporatively cooled, then circulated through coils to cool supply air indirectly. This system has an advantage over air economizers in climates where wintertime humidification is required; the large amounts of dry air introduced with air economizers often increase humidification loads. However, in most climates, water economizers are generally are not as energy efficient as air economizers because they require the operation of cooling tower fans and condenser pumps. Air economizers also have the advantage of introducing large amounts of outdoor air for a majority of the building operating hours, which can improve indoor air quality.

There are three common types of water economizers:

"Strainer-cycle" or Chiller-bypass water economizer: This type of economizer, shown in Commentary Figure C403.3.1(2), has control valves that can divert condenser water from the cooling tower and run it directly into the normal chilled-water piping loop, bypassing the chiller. This bypass configuration will occur as long as the tower can cool the condenser water sufficiently to handle the cooling load, usually around 45°F (7°C) to 50°F (10°C). The

TABLE C403.3.3.3
HIGH-LIMIT SHUTOFF CONTROL SETTING FOR AIR ECONOMIZERS[b]

DEVICE TYPE	CLIMATE ZONE	REQUIRED HIGH LIMIT (ECONOMIZER OFF WHEN):	
		Equation	Description
Fixed dry bulb	1B, 2B, 3B, 3C, 4B, 4C, 5B, 5C, 6B, 7, 8	$T_{OA} > 75°F$	Outdoor air temperature exceeds 75°F
	5A, 6A	$T_{OA} > 70°F$	Outdoor air temperature exceeds 70°F
	1A, 2A, 3A, 4A	$T_{OA} > 65°F$	Outdoor air temperature exceeds 65°F
Differential dry bulb	1B, 2B, 3B, 3C, 4B, 4C, 5A, 5B, 5C, 6A, 6B, 7, 8	$T_{OA} > T_{RA}$	Outdoor air temperature exceeds return air temperature
Fixed enthalpy with fixed dry-bulb temperatures	All	$h_{OA} > 28$ Btu/lba or $T_{OA} > 75°F$	Outdoor air enthalpy exceeds 28 Btu/lb of dry air[a] or Outdoor air temperature exceeds 75°F
Differential enthalpy with fixed dry-bulb temperature	All	$h_{OA} > h_{RA}$ or $T_{OA} > 75°F$	Outdoor air enthalpy exceeds return air enthalpy or Outdoor air temperature exceeds 75°F

For SI: 1 foot = 305 mm, °C = (°F - 32)/1.8, 1 Btu/lb = 2.33 kJ/kg.

a. At altitudes substantially different than sea level, the fixed enthalpy limit shall be set to the enthalpy value at 75°F and 50-percent relative humidity. As an example, at approximately 6,000 feet elevation, the fixed enthalpy limit is approximately 30.7 Btu/lb.

b. Devices with selectable setpoints shall be capable of being set to within 2°F and 2 Btu/lb of the setpoint listed.

term "strainer-cycle," as this type of economizer is commonly called, started as a trade name of a type of inline water filter intended to clean the dirty open-circuit tower water before it flowed into the clean (normally closed circuit) chilled-water circuit. Because chilled-water control valves and coils easily can become clogged, it is essential to install good water-treatment systems with this type of economizer. To resolve the clogging problem, a heat exchanger can be used to isolate the tower and chilled water circuits, at both considerable first-cost expense as well as reduced energy savings due to higher pump heads and a nonzero heat exchanger approach. Note that the chiller-bypass water economizer is "nonintegrated," meaning the chiller provides 100 percent of the expected system cooling load and cannot operate when the condenser water is in the bypass arrangement; thus the economizer must provide all of the cooling load or it can provide none of it. Because of this characteristic, this economizer design may not meet the integration requirement of this section.

Water precooling water economizer: This type of economizer, shown in Commentary Figure C403.3.1(3), uses cold tower water when it is available to precool chilled water return (through a heat exchanger) before it enters the chiller. One advantage of this type of economizer over the strainer cycle is that it is integrated, meaning it can provide "free" cooling even when the chillers are operating by reducing chilled-water return temperatures.

It also isolates the open-circuit tower system from the chilled-water system with the heat exchanger, reducing fouling problems caused by the poor water quality of the open circuit. But the heat exchanger reduces the cooling energy savings because the water leaving the heat exchanger cannot be as cold as the tower water, and it increases pump energy during economizer operation because of the pressure drop of the heat exchanger.

Air precooling water economizer: This type of water economizer requires an additional cooling coil upstream of the normal, mechanical cooling coil, as shown in Commentary Figure C403.3.1(4). Condenser water is run through the tower and cooled as much as possible [typically to a minimum of about 50°F (10°C)]. The water then is run through the economizer coil, precooling the air before it enters the mechanical system cooling coil. This is an integrated economizer because mechanical cooling can be provided simultaneously with the economizer cooling.

Water from the cooling tower first passes through the economizer coil, precooling or fully cooling supply air, then goes on to remove condenser heat from the mechanical cooling system, with water flow modulated or bypassed to maintain head pressures, a control required because of the low water temperatures. The three-way control valve shown in Commentary Figure C403.3.1(4) operates so that, if the tower water is warmer than the return air, the water bypasses the coil to avoid warming the air and increasing the cooling load. This is similar to the high-limit control used with air economizers. Because the economizer and mechanical cooling can operate concurrently with this type of economizer, it is integrated and must meet the integration requirements of Section C403.3.1. This scheme is very popular when water-cooled air conditioners are used because the condenser water must be piped to the units anyway, so the only expense of the water economizer is the added coil and controls.

Although either air or water economizers will meet the requirements of this section, air economizers will outperform water economizers except in very dry climates (such as Nevada or Arizona) or very cold climates where humidification must be added to offset the dryness caused by the introduction of cold, dry air. Water economizers require cooling tower fans and tower pumps to operate during economizer operation, plus added fan and pump pressure drops for economizers like that shown in Commentary Figure C403.3.1(4) that will increase fan and pump energy even when the economizer is not operating. Air economizers do not have such parasitic loads, except where supplemental humidification is required in cold climates. Air economizers also can improve indoor air quality because of the many hours when outside air intake exceeds minimum levels. Thus, air economizers are preferred over water economizers for most applications if the architecture of the building and the HVAC system type are suitable.

Exceptions 2 and 3 provide economizer tradeoffs similar to that found in Section C403.3.1, Exception 1 (see commentary, Section C403.3.1). These tradeoffs are provided in terms of percentages above the minimum EER or, in the case of chiller systems, the IPLV cooling requirements. The reference to Table C403.3.1 and the stated efficiency improvement simplifies the application to a broad category of cooling equipment. This provides a simple calculation to get the allowed EER or IPLV levels for the tradeoffs, and is not allowed for the chiller equipment or condenser systems.

C403.3.4.1 Design capacity. Water economizer systems shall be capable of cooling supply air by indirect evaporation and providing up to 100 percent of the expected system cooling load at *outdoor air* temperatures of not greater than 50°F (10°C) dry bulb/45°F (7°C) wet bulb.

Exceptions:

1. Systems primarily serving computer rooms in which 100 percent of the expected system cooling load at 40°F (4°C) dry bulb/35°F (1.7°C) wet bulb is met with evaporative water economizers.
2. Systems primarily serving computer rooms with dry cooler water economizers which satisfy 100 percent of the expected system cooling load at 35°F (1.7°C) dry bulb.

3. Systems where dehumidification requirements cannot be met using outdoor air temperatures of 50°F (10°C) dry bulb/45°F (7°C) wet bulb and where 100 percent of the expected system cooling load at 45°F (7°C) dry bulb/40°F (4°C) wet bulb is met with evaporative water economizers.

❖ This section is for water economizer systems. This design is specified because, unlike air economizers, which use cold outside air directly for cooling, the performance of water economizers depends greatly on the selection of components, such as cooling towers and heat exchangers. The system sizing is based on the building design load at an outdoor temperature of 50°F dry bulb temperature and 47°F wet bulb temperature.

Exceptions 1 and 2: Computer rooms and data centers require precision temperature control and may require greater energy use when using air-side economizers. Less restrictive design temperatures to meet 100 percent of the design cooling load are allowed for systems serving primarily computer rooms. Separate design conditions are set for evaporative systems and dry-cooler type systems.

Some buildings/spaces generate excessive quantities of interior moisture (latent load). Some high-occupancy spaces or spaces with similar vapor generation will not be satisfied by water-side economizing at 50°F dry bulb/47°F wet bulb. The cooling coil temperature will not be cold enough to condense moisture in the supply air. When interior space design conditions cannot be adequately dehumidified, Exception 3 allows the economizer design condition to be reduced to 45°F dry bulb/40°F wet bulb.

C403.3.4.2 Maximum pressure drop. Precooling coils and water-to-water heat exchangers used as part of a water economizer system shall either have a water-side pressure drop of less than 15 feet (45 kPa) of water or a secondary loop shall be created so that the coil or heat exchanger pressure drop is not seen by the circulating pumps when the system is in the normal cooling (noneconomizer) mode.

❖ This section is for precooling coils and water-to-water heat exchangers used as part of a water economizer system. If the water-side pressure drop is less than 15 feet (4572 mm) of water or a secondary loop so that the coil or heat exchanger pressure drop is not seen by the circulating pumps, there will be energy wasted as the circulating pumps continue to operate as if the system is in the normal cooling mode.

C403.4 Hydronic and multiple-zone HVAC systems controls and equipment. (Prescriptive). Hydronic and multiple-zone HVAC system controls and equipment shall comply with this section.

❖ The equipment, systems and controls found in the various subsections of Section C403.4 are not required, per se, but are commonly found in larger buildings with multiple zones for space conditioning and the need for more-sophisticated controls. Where a designer includes these systems in the building, the requirements of these sections must be met.

C403.4.1 Fan control. Controls shall be provided for fans in accordance with Sections C403.4.1.1 through C403.4.1.3.

❖ Fan control must address the control of airflow, the placement of static pressure sensors and the setpoints for direct digital control.

C403.4.1.1 Fan airflow control. Each cooling system listed in Table C403.4.1.1 shall be designed to vary the indoor fan airflow as a function of load and shall comply with the following requirements:

1. Direct expansion (DX) and chilled water cooling units that control the capacity of the mechanical cooling directly based on space temperature shall have not fewer than two stages of fan control. Low or minimum speed shall not be greater than 66 percent of full speed. At low or minimum speed, the fan system shall draw not more than 40 percent of the fan power at full fan speed. Low or minimum speed shall be used during periods of low cooling load and ventilation-only operation.
2. Other units including DX cooling units and chilled water units that control the space temperature by modulating the airflow to the space shall have modulating fan control. Minimum speed shall be not greater than 50 percent of full speed. At minimum speed the fan system shall draw not more than 30 percent of the power at full fan speed. Low or minimum speed shall be used during periods of low cooling load and ventilation- only operation.
3. Units that include an airside economizer in accordance with Section C403.3 shall have not fewer than two speeds of fan control during economizer operation

Exceptions:

1. Modulating fan control is not required for chilled water and evaporative cooling units with fan motors of less than 1 hp (0.746 kW) where the units are not used to provide *ventilation* air and the indoor fan cycles with the load.
2. Where the volume of outdoor air required to comply with the *ventilation* requirements of the *International Mechanical Code* at low speed exceeds the air that would be delivered at the speed defined in Section C403.4.1, the minimum speed shall be selected to provide the required *ventilation air*.

❖ Fans generally are the largest energy-using components of HVAC systems. Most HVAC systems spend most of their operating hours at part loads, where little or no cooling is required. Few systems require full cooling capacity at all operating hours. For cooling systems, multiple stages of cooling at part-load conditions allows lower fan speed and higher efficiency for DX cooling. Significant energy savings can be realized. This section sets limits for multiple-stage cooling that align with requirements found in ASHRAE 90.1.

Under Item 1, when a call for cooling by the space thermostat cycles (increases or decreases) the stages of DX cooling or control of chilled water supply, the fan shall have a minimum of two speeds. The minimum fan speed shall be no more than 66 percent of the peak fan speed. Under these systems, the fan reacts to the amount of cooling generated.

Under Item 2, when a call for cooling by the space thermostat modulates (increases/decreases) the supply fan speed, the fan minimum speed at the lowest call for cooling shall be no more than 50 percent of the maximum fan speed at peak cooling. Under these systems, the cooling capacity modulates to keep track of the fan speed.

Exception 1 is for fan speed modulation for units not used for providing ventilation air. The fan must cycle on-off with a call for cooling or heating. With no call for heating/cooling, the fan is off.

Exception 2 is relative to spaces requiring high outdoor air ventilation rates, such as an assembly occupancy. The fan minimum speed shall not be required to operate at a speed lower than required to deliver code minimum ventilation air. Section C403.4.1.1 may be exceeded to meet ventilation.

Note that after January 1, 2016, the limit for DX cooling systems requiring fan control decreases from 75,000 Btu/h to 65,000 Btu/h.

TABLE C403.4.1.1
EFFECTIVE DATES FOR FAN CONTROL

COOLING SYSTEM TYPE	FAN MOTOR SIZE	MECHANICAL COOLING CAPACITY
DX cooling	Any	≥ 75,000 Btu/h (before 1/1/2016)
		≥ 65,000 Btu/h (after 1/1/2016)
Chilled water and evaporative cooling	≥ 5 hp	Any
	≥ $^1/_4$ hp	Any

For SI: 1 British thermal unit per hour = 0.2931 W; 1 hp = 0.746 kW.

C403.4.1.2 Static pressure sensor location. Static pressure sensors used to control VAV fans shall be located such that the controller set point is not greater than 1.2 inches w.c. (299 Pa). Where this results in one or more sensors being located downstream of major duct splits, not less than one sensor shall be located on each major branch to ensure that static pressure can be maintained in each branch.

❖ This section describes where the static pressure sensors should be located. Their location was determined by studies conducted by the New Buildings Institute (NBI), the American Institute of Architects (AIA) and the U.S. Department of Energy (DOE) to ensure the static pressure sensor can be maintained and functional for a long time. Accurate control of VAV fan systems increases energy savings. The sensor location for control of VAV fan speed is required to be where the duct system static pressure is no greater than 1.2 inches w.c. (299 Pa). This locates the sensor where it will see greater changes in duct pressure and will result in fan speed reduction over greater ranges of operation. It is also more flexible than prior code cycles, which may have set the controller position where it would not sense flow variations in the duct.

This section describes where the static pressure sensors should be located. Their location was determined by NBI, AIA and U.S. DOE studies to ensure the static pressure sensor can be maintained and functional for a long time.

C403.4.1.3 Set points for direct digital control. For systems with direct digital control of individual zones reporting to the central control panel, the static pressure set point shall be reset based on the *zone* requiring the most pressure. In such case, the set point is reset lower until one *zone* damper is nearly wide open. The direct digital controls shall be capable of monitoring *zone* damper positions or shall have an alternative method of indicating the need for static pressure that is capable of all of the following:

1. Automatically detecting any *zone* that excessively drives the reset logic.
2. Generating an alarm to the system operational location.
3. Allowing an operator to readily remove one or more *zones* from the reset algorithm.

❖ This section notes that the static pressure setpoints must be reset. This is potentially a major energy saver in buildings with VAV systems. Most VAV systems already have DDC systems and adding this capability to a DDC system requires minimal cost for the additional programming. This requirement only applies for systems with DDC.

The control system shall be capable of identifying problematic VAV zones. An example may be a zone that is calling for full cooling (100 percent open) for too many hours of operation relative to the remaining VAV zones. This activity may cause the static pressure to be reset higher than it should be set; this results in excess fan energy to be expended based upon an "outlier" zone. The control sequence and required alarms alert the system operator, who can address the issue.

C403.4.2 Hydronic systems controls. The heating of fluids that have been previously mechanically cooled and the cooling of fluids that have been previously mechanically heated shall be limited in accordance with Sections C403.4.2.1 through C403.4.2.3. Hydronic heating systems comprised of multiple-packaged boilers and designed to deliver conditioned water or steam into a common distribution system shall include automatic controls capable of sequencing operation of the boilers. Hydronic heating systems comprised of a single boiler and greater than 500,000 Btu/h (146.5 kW) input design capacity shall include either a multistaged or modulating burner.

❖ Most simultaneous heating and cooling in contemporary mechanical systems occurs on the air side. However, some wasted energy also can result from poor design practices in hydronic systems. This requirement, in effect, prohibits the so-called three-pipe systems that use a common return where the heated water and cooled water are mixed. It is evident that both heating and cooling energy usage is needlessly wasted to bring this tempered mixture back to a useful state (thermally).

The provisions addressing either multiple packaged boilers or a single boiler greater than 500,000 Btu/h (146 550 W) input capacity are simply intended to ensure that the boilers are only operated in an efficient manner regardless of the heating-load demand.

C403.4.2.1 Three-pipe system. Hydronic systems that use a common return system for both hot water and chilled water are prohibited.

❖ This section explicitly prohibits the use of a three-pipe system. As stated previously, three-pipe systems use a common return where the heated and cooled water are mixed. It is evident that both heating and cooling energy are needlessly wasted to either reheat or recool this tempered mixture back to a useful state (thermally).

C403.4.2.2 Two-pipe changeover system. Systems that use a common distribution system to supply both heated and chilled water shall be designed to allow a dead band between changeover from one mode to the other of not less than 15°F (8.3°C) outside air temperatures; be designed to and provided with controls that will allow operation in one mode for not less than 4 hours before changing over to the other mode; and be provided with controls that allow heating and cooling supply temperatures at the changeover point to be not more than 30°F (16.7°C) apart.

❖ A common distribution system is used in two-pipe changeover systems to supply heated or chilled water to fan coils or air handlers. No energy is wasted by this design so long as the system operates in one mode (heating or cooling) for a long period. Energy is wasted, however, during the changeover period (heating to cooling and cooling to heating) where the mass of water in the system is thermally reversed and brought to temperature.

The control measures described are established to operate the changeover cycle more efficiently such that the system is not rapidly switching over and fighting itself. Two-pipe systems can be efficient when the proper controls, such as those listed, are used.

C403.4.2.3 Hydronic (water loop) heat pump systems. Hydronic heat pump systems shall comply with Sections C403.4.2.3.1 through C403.4.2.3.2.

❖ Hydronic heat pumps typically are connected to a common condenser water loop. Also, connected to the loop are boilers and chillers to either add heat or cool the water, depending on the demand. The intent of this section is to reduce the energy usage

C403.4.2.3.1 Temperature dead band. Hydronic heat pumps connected to a common heat pump water loop with central devices for heat rejection and heat addition shall have controls that are capable of providing a heat pump water supply temperature dead band of not less than 20°F (11°C) between initiation of heat rejection and heat addition by the central devices.

Exception: Where a system loop temperature optimization controller is installed and can determine the most efficient operating temperature based on realtime conditions of demand and capacity, dead bands of less than 20°F (11°C) shall be permitted.

❖ This section requires hydronic heat pumps that are connected to a common heat pump water loop to have controls capable of providing a temperature dead band of not less than 20°F (11.1°C) between heat rejection and heat addition by the central devices. The dead band may be reduced by system loop temperature-optimization controllers that determine the most efficient operating temperature based on real-time demand and capacity conditions. The purpose of this section is to include a closed-circuit cooling tower for shutting down the pumps when a heat exchanger is to be used to control heat loss.

C403.4.2.3.2 Heat rejection. Heat rejection equipment shall comply with Sections C403.4.2.3.2.1 and C403.4.2.3.2.2.

Exception: Where it can be demonstrated that a heat pump system will be required to reject heat throughout the year.

❖ The purpose of this section is to provide additional energy requirements in Sections C403.4.3.3.2.1 and C403.4.3.3.2.2 for heat rejection equipment.

C403.4.2.3.2.1 Climate Zones 3 and 4. For *Climate Zones* 3 and 4:

1. Where a closed-circuit cooling tower is used directly in the heat pump loop, either an automatic valve shall be installed to bypass all but a minimal flow of water around the tower, or lower leakage positive closure dampers shall be provided.
2. Where an open-circuit tower is used directly in the heat pump loop, an automatic valve shall be installed to bypass all heat pump water flow around the tower.
3. Where an open- or closed-circuit cooling tower is used in conjunction with a separate heat exchanger to isolate the cooling tower from the heat pump loop, then heat loss shall be controlled by shutting down the circulation pump on the cooling tower loop.

❖ This section addresses both closed- and open-circuit cooling towers used in hydronic (water loop) heat pump systems. When used directly in the heat pump loop or with a heat exchanger in Climate Zones 3 and 4, requirements for controlling heat loss are provided. This ensures the elimination of flow in the cooling tower loop when heat rejection is not required. By just requiring the shut experience, a temperature differential of 4°F to 7°F (2.2°C to 3.9°C) has been observed through a heat exchanger on a heat pump loop due to the fluid flow from convection with the tower pumps shut down. The boiler must compensate for the additional heat rejection, which is not energy efficient and could be considerable depending on the system size. The requirement that the valve, when used with the heat exchanger, bypass the tower is no different than the requirement for the valve, when the tower is installed within the heat pump loop. This will ensure an energy-efficient system.

C403.4.2.3.2.2 Climate Zones 5 through 8. For *Climate Zones* 5 through 8, where an open- or closed-circuit cooling tower is used, a separate heat exchanger shall be provided to isolate the cooling tower from the heat pump loop, and heat loss shall be controlled by shutting down the circulation pump on the cooling tower loop and providing an automatic valve to stop the flow of fluid.

❖ This section describes both closed- and open-circuit cooling towers used in hydronic (water loop) heat pump systems. When used directly in the heat pump loop or with a heat exchanger in Climate Zones 3 and 4, requirements for controlling heat loss are provided. If a closed-circuit cooling tower is proposed in Climate Zones 5 to 8, a heat exchanger is required along with the requirements to control heat loss. This ensures the elimination of flow in the cooling tower loop when heat rejection is not required. By just requiring the shut experience, a temperature differential of 4°F to 7°F (2.2°C to 3.9°C) has been observed through a heat exchanger on a heat pump loop due to the fluid flow from convection with the tower pumps shut down. The boiler must compensate for the additional heat rejection, which is not energy efficient and could be considerable depending on the system size. The requirement that the valve, when used with the heat exchanger, bypass the tower is no different than the requirement for the valve when the tower is installed within the heat pump loop. This will ensure an energy-efficient system.

C403.4.2.3.3 Two-position valve. Each hydronic heat pump on the hydronic system having a total pump system power exceeding 10 hp (7.5 kW) shall have a two-position valve.

❖ The variable flow requirements of Section C403.4.2.4, Item 2, are to be consulted where individual heat pumps served by the common condenser loop exceed 10 hp (7.5 kW). This section requires each hydronic heat pump on the hydronic system having a total pump system power exceeding 10 hp (7.5 kW) to have a two-position valve that must be interlocked to shut off flow to the heat pump when the compressor is off. This two-position valve is required to isolate the system components to facilitate their repair, maintenance or replacement.

C403.4.2.4 Part-load controls. Hydronic systems greater than or equal to 500,000 Btu/h (146.5 kW) in design output capacity supplying heated or chilled water to comfort conditioning systems shall include controls that have the capability to do all of the following:

1. Automatically reset the supply-water temperatures in response to varying building heating and cooling demand using coil valve position, zone-return water temperature, building-return water temperature or outside air temperature. The temperature shall be capable of being reset by not less than 25 percent of the design supply-to-return water temperature difference.
2. Automatically vary fluid flow for hydronic systems with a combined motor capacity of 10 hp (7.5 kW) or larger with three or more control valves or other devices by reducing the system design flow rate by not less than 50 percent by designed valves that modulate or step open and close, or pumps that modulate or turn on and off as a function of load.
3. Automatically vary pump flow on chilled-water systems and heat rejection loops serving water- cooled unitary air conditioners with a combined motor capacity of 10 hp (7.5 kW) or larger by reducing pump design flow by not less than 50 percent, utilizing adjustable speed drives on pumps, or multiple-staged pumps where not less than one-half of the total pump horsepower is capable of being automatically turned off. Pump flow shall be controlled to maintain one control valve nearly wide open or to satisfy the minimum differential pressure.

Exceptions:

1. Supply-water temperature reset for chilled-water systems supplied by off-site district chilled water or chilled water from ice storage systems.
2. Minimum flow rates other than 50 percent as required by the equipment manufacturer for proper operation of equipment where using flow bypass or end-of-line 3-way valves.
3. Variable pump flow on dedicated equipment circulation pumps where configured in primary/secondary design to provide the minimum flow requirements of the equipment manufacturer for proper operation of equipment.

❖ The code requires that hydronic systems with a design output capacity greater than or equal to 500,000 Btu/h (146.5 kW) be designed for variable flow. All three requirements outlined must be met. The requirements don't exclude one another but are additive. As written all three of the following control requirements must be addressed:

- Supply Water Temperature Reset.
- Variable Flow Control.
- Variable or Stepped Pumping.

Variable flow control (Item 2) in hydronic systems is needed in order to implement variable or stepped pumping (Item 3). Therefore, Item 2 is described prior to the requirements in Item 3. Item 2 applies to all other hydronic systems since two-way valve control is less expensive than three-way valve control. This requirement also aligns with Section C403.4.2.3.3, which requires two-way valve control on heat pump hydronic systems.

Cooling systems with pump capacity 10 hp or greater should have variable flow using variable speed drives or stepped pumping. Allowing cooling pumps to vary flow and ride the pump curve should not be allowed on larger pumping systems. Heating-only hydronic systems of any size are excluded from this requirement since pump inefficiencies are recaptured as a heat source in the hydronic heating system. A cost-effective analysis, as shown in the table below, indicates cooling systems with a pump capacity of 10 hp to be cost effective. The analysis assumes an average pump run time of 2,000 hours. This analysis only accounts for pump motor energy savings and doesn't account for the reduced heat rejected from the cooling pump into the chilled water system.

C403.4.2.5 Boiler turndown. *Boiler systems* with design input of greater than 1,000,000 Btu/h (293 kW) shall comply with the turndown ratio specified in Table C403.4.2.5.

The system turndown requirement shall be met through the use of multiple single input boilers, one or more *modulating boilers* or a combination of single input and *modulating boilers*.

❖ A boiler system is any combination of one or more boilers and their piping and controls that work together to supply heating water or steam for delivery to remote heating devices. Where a boiler is only able to operate at a single rate (single-input boiler) it is either off or on. If the needs of the building are less than what that single-input boiler is providing, energy is being wasted. Therefore, where the boiler has an input rating of 1,000,000 Btu/h or greater, it must be provided with a turndown mechanism to allow operation at a reduced level. Even with the turndown capacity, the demand for heat may be less than the turndown level, in which case the boiler will cycle between the turndown capacity and being off (not firing). The code specifies boiler system options that can be used to comply with this section. The list is not intended to be exclusive.

TABLE C403.4.2.5
BOILER TURNDOWN

BOILER SYSTEM DESIGN INPUT (Btu/h)	MINIMUM TURNDOWN RATIO
≥ 1,000,000 and less than or equal to 5,000,000	3 to 1
> 5,000,000 and less than or equal to 10,000,000	4 to 1
> 10,000,000	5 to 1

For SI: 1 British thermal unit per hour = 0.2931 W.

C403.4.2.6 Pump isolation. Chilled water plants including more than one chiller shall have the capability to reduce flow automatically through the chiller plant when a chiller is shut down. Chillers piped in series for the purpose of increased temperature differential shall be considered as one chiller.

Boiler plants including more than one boiler shall have the capability to reduce flow automatically through the boiler plant when a boiler is shut down.

❖ These requirements apply to chilled water plants (more than one chiller) and hot water plants (more than one boiler). In the case of a chiller plant, the purpose is to provide for a reduction in the cooling flow rate, thereby minimizing pumping energy when full plant-chilled water flow is not necessary (i.e., a chiller piped in series is shut down). This section essentially requires flow through chillers (or boilers) piped in parallel to be shut off when the chillers (or boilers) are inactive.

C403.4.3 Heat rejection equipment. Each fan powered by a motor of 7.5 hp (5.6 kW) or larger shall have the capability to operate that fan at two-thirds of full speed or less, and shall have controls that automatically change the fan speed to control the leaving fluid temperature or condensing temperature/pressure of the heat rejection device.

Exception: Factory-installed heat rejection devices within HVAC equipment tested and rated in accordance with Tables C403.2.3(6) and C403.2.3(7).

❖ Since fans are one of the major energy-using components of an HVAC system, this section provides one or more methods for improving the efficiency where the savings benefit can be justified. Heat rejection equipment covered by this section would include cooling systems, such as air-cooled condensers, cooling towers (open and closed circuit) and evaporative condensers. In each circumstance, the fans powered by motors of 7.5 hp (5.6 kW) or larger require controls to automatically change fan speed in response to part-load conditions. The exception excludes the "factory-installed" devices that are indicated because the fan energy is included in the efficiency ratings for that equipment.

C403.4.3.1 General. Heat rejection equipment such as air-cooled condensers, dry coolers, open-circuit cooling towers, closed-circuit cooling towers and evaporative condensers used for comfort cooling applications shall comply with this section.

Exception: Heat rejection devices where energy usage is included in the equipment efficiency ratings listed in Tables C403.2.3(6) and C403.2.3(7).

❖ See the commentary to Section C403.4.3.

C403.4.3.2 Fan speed control. The fan speed shall be controlled as provided in Sections C403.4.3.2.1 and C403.4.3.2.2.

❖ Fan speed must be controlled in accordance with the requirements of Sections C403.4.3.2.1 and C403.4.3.2.2.

C403.4.3.2.1 Fan motors not less than 7.5 hp. Each fan powered by a motor of 7.5 hp (5.6 kW) or larger shall have the capability to operate that fan at two-thirds of full speed or less, and shall have controls that automatically change the fan speed to control the leaving fluid temperature or condensing temperature/pressure of the heat rejection device.

Exception: The following fan motors over 7.5 hp (5.6 kW) are exempt:

1. Condenser fans serving multiple refrigerant circuits.

2. Condenser fans serving flooded condensers.
3. Installations located in *Climate Zones* 1 and 2.

❖ Multispeed fans provide another method by which energy is saved. Systems are designed for heating on the coldest day and cooling on the warmest day, therefore lower settings in the system will allow savings on days where extremes are not occurring.

C403.4.3.2.2 Multiple-cell heat rejection equipment. Multiple-cell heat rejection equipment with variable speed fan drives shall be controlled in both of the following manners:

1. To operate the maximum number of fans allowed that comply with the manufacturer's requirements for all system components.
2. So all fans can operate at the same fan speed required for the instantaneous cooling duty, as opposed to staged (on/off) operation.

Minimum fan speed shall be the minimum allowable speed of the fan drive system in accordance with the manufacturer's recommendations.

❖ Optimal airflow must be maintained across the whole heat rejection surface area regardless of load. Heat rejection equipment that is composed of multiple cells typically has at least one fan per cell. The cells must be linked to operate as a unit so that increases and decreases of airflow are consistent. The minimum controlled fan speed can not be set lower than the minimum speed recommended by the heat rejection equipment manufacturer for the fan drive system. Finally, staging the fans will not satisfy this requirement as this does not maintain uniform airflow across the heat rejection surfaces.

C403.4.3.3 Limitation on centrifugal fan open-circuit cooling towers. Centrifugal fan open-circuit cooling towers with a combined rated capacity of 1,100 gpm (4164 L/m) or greater at 95°F (35°C) condenser water return, 85°F (29°C) condenser water supply, and 75°F (24°C) outdoor air wet-bulb temperature shall meet the energy efficiency requirement for axial fan open-circuit cooling towers listed in Table C403.2.3(8).

Exception: Centrifugal open-circuit cooling towers that are designed with inlet or discharge ducts or require external sound attenuation.

❖ Centrifugal fans with the rated capacities set in this section must be at least as efficient as specified in Table C403.2.3(8). For the same types of installations, centrifugal fan cooling towers will use about twice the energy as axial (or propeller) fan towers. In general, designers should use axial fan cooling towers when possible.

C403.4.3.4 Tower flow turndown. Open-circuit cooling towers used on water-cooled chiller systems that are configured with multiple- or variable-speed condenser water pumps shall be designed so that all open- circuit cooling tower cells can be run in parallel with the larger of the flow that is produced by the smallest pump at its minimum expected flow rate or at 50 percent of the design flow for the cell.

❖ This requirement applies only to open-circuit cooling towers and does not apply to any closed-loop systems. The provision requires multiple pump installations to be linked so they run in parallel. The flow from the smallest pump dictates the flow from the connected system. Where the flow rate is below that recommended by a manufacturer, the design may require use of towers with gravity-flow water distribution systems, weir dams or other mechanisms to allow the pumps to balance. There is no exception for any category of this equipment.

C403.4.4 Requirements for complex mechanical systems serving multiple zones. Sections C403.4.4.1 through C403.4.6.4 shall apply to complex mechanical systems serving multiple zones. Supply air systems serving multiple zones shall be variable air volume (VAV) systems that, during periods of occupancy, are designed and capable of being controlled to reduce primary air supply to each *zone* to one of the following before reheating, recooling or mixing takes place:

1. Thirty percent of the maximum supply air to each *zone*.
2. Three hundred cfm (142 L/s) or less where the maximum flow rate is less than 10 percent of the total fan system supply airflow rate.
3. The minimum ventilation requirements of Chapter 4 of the *International Mechanical Code.*
4. Any higher rate that can be demonstrated to reduce overall system annual energy use by offsetting reheat/recool energy losses through a reduction in *outdoor air* intake for the system, as *approved* by the *code official*.
5. The airflow rate required to comply with applicable codes or accreditation standards, such as pressure relationships or minimum air change rates.

Exception: The following individual *zones* or entire air distribution systems are exempted from the requirement for VAV control:

1. Zones or supply air systems where not less than 75 percent of the energy for reheating or for providing warm air in mixing systems is provided from a site-recovered or site-solar energy source.
2. *Zones* where special humidity levels are required to satisfy process needs.
3. *Zones* with a peak supply air quantity of 300 cfm (142 L/s) or less and where the flow rate is less than 10 percent of the total fan system supply airflow rate.
4. *Zones* where the volume of air to be reheated, recooled or mixed is not greater than the volume of outside air required to provide the minimum ventilation requirements of Chapter 4 of the *International Mechanical Code*.
5. *Zones* or supply air systems with thermostatic and humidistatic controls capable of operating in

sequence the supply of heating and cooling energy to the *zones* and which are capable of preventing reheating, recooling, mixing or simultaneous supply of air that has been previously cooled, either mechanically or through the use of economizer systems, and air that has been previously mechanically heated.

❖ When air-conditioning system designs were developed in the late 1950s and early l960s, energy costs were a minor concern. The systems were designed primarily for precise temperature control with little regard for energy costs. Zone temperature control was achieved by reheating cold supply air (constant volume reheat system), recooling warm supply air (such as perimeter induction systems), or mixing hot and cold air (constant volume dual-duct and multizone system). Although these systems provided fine temperature control, they did so at great expense of energy.

To curb the use of these wasteful energy design practices, this section requires systems serving multiple zones to be VAV systems equipped with zone thermostatic and humidistat controls designed and specifically capable of sequencing the supply of heating and cooling energy to each zone to a minimum before reheating, recooling or mixing takes place. The requisite controls must thereby minimize:

- Reheating.
- Recooling.
- Mixing or simultaneous supply of air that has been previously mechanically heated and air that has been previously cooled, either mechanically or by economizer systems.

The five items listed in this section establish what level of air supply must be reached before it is permissible for the system to either reheat, recool or mix heated and cooled air. The first three are straightforward specifications, easily determined for each building for each zone. Item 4 allows the designer to demonstrate a greater rate of supplying air where a system reduces overall energy use. This alternative threshold design must be approved by the code official. Item 5 allows special spaces to be addressed where there are other standards at play. Examples of such spaces include areas of hospitals (operating rooms, patient rooms and laboratories) where pressures need to be maintained to prevent contaminants from entering or leaving the areas. The potential from release of contaminants, or in other cases dangerous chemicals, must be allowed to be balanced with the energy saving needs for the HVAC system.

There are five exceptions that address where either individual zones or the entire air distribution system is exempt from the VAV requirement or controls affecting the simultaneous heating and cooling requirements.

Exception 1 covers zones or systems where not less than 75 percent of the energy for reheating or producing warm air supply in mixing systems is supplied by heat recovered from some process or equipment in the building (such as chiller condenser heat) or from a solar-heating system installed on site.

Exception 2 acknowledges zones where specific humidity levels must be maintained for purposes other than personal comfort, such as some areas of museums or archives where sensitive materials are displayed or stored, or some computer rooms where equipment must be maintained in precise humidity ranges to function properly. Note that much of the computer equipment manufactured today, such as common personal computers and most minicomputers, does not require this precise humidity control, so this exception would not apply.

For typical multiple-zone systems, Exception 3 can be applied to, at best, a limited number of zones served by the system. The exception applies to zones with a peak supply air quantity of 300 cfm (142 L/s) or less, and also where the flow rate is less than 10 percent of the total fan system supply airflow rate.

This criterion was included specifically to address reheat systems and heating problems that can occur in small zones or subzones that have low peak supply air volumes, but a larger zone requires additional heat, such as spaces with large glass areas facing north, areas shaded by overhangs and fins, or entry vestibules. These spaces often can require higher heating volumes than allowed by Exception 4 using reasonable supply air temperatures. In most cases, 300 cfm (142 L/s) will be sufficient for heating. If not, a fan-powered VAV box could be used to improve the circulation rate.

The exception allows reheat to be used for small subzones of a larger zone. For instance, an air conditioner might serve an office space in which more heat is needed at the front entry area to offset infiltration when exterior doors are opened. If the supply to the entry is less than 300 cfm (142 L/s), a reheat coil may be installed to meet this special heating requirement.

Exception 4 is most common for multiple-zone systems. In this case, the VAV control strategy also could reduce the airflow to each zone that is being reheated, recooled or mixed to a minimum rate not exceeding the volume meeting the minimum ventilation requirements of Chapter 4 of the IMC. A conservative interpretation of this IMC chapter might lead to very high minimum flow rates to guarantee each zone receives minimum outside air, regardless of the percentage of outside air brought in at the system level. Instead, to reduce energy costs and first costs, the designer could take advantage of the direct transfer of air from adjacent overventilated zones.

Simultaneous heating and cooling is allowed by Exception 5 if it is minimized by the use of VAV controls.

Although there is little empirical evidence and no code-related requirement, many designers feel that a minimum supply air circulation rate of 0.4 cfm/ft^3

[0.002032 $m^3/(s \times m^2)$] of the zone-conditioned floor area must be maintained for comfort. In fact, ASHRAE 55 (a standard for thermal environmental conditions for human occupancy) states there is no minimum air velocity required for comfort. Nevertheless, with little or no air movement, some designers feel occupants will complain of stuffiness. Moreover, typical air outlet performance drops off at low flows, reducing its ability to effectively mix supply air with room air. This is a particular problem with reheat boxes because the supply air temperature at the minimum volume will be warm and the buoyancy of the supply air will further decrease mixing, unless the outlet velocities are maintained.

As in Exception 5, single-zone (simple) systems, with controls interlocked or otherwise capable of sequencing typical heating and cooling, will inherently meet these requirements. But most common multiple-zone (complex) systems require the use of simultaneous heating and cooling for zone temperature control.

C403.4.4.1 Single-duct VAV systems, terminal devices. Single-duct VAV systems shall use terminal devices capable of reducing the supply of primary supply air before reheating or recooling takes place.

❖ VAV systems control temperatures by varying the quantity (cfm) of supply air rather than varying supply air temperature. The VAV terminal device, therefore, varies that quantity of air supplied to the space at the zone level. Although this basic principle is typical of the VAV terminal device, VAV terminal devices are available in a variety of configurations. There is, however, one prevailing theme found in Sections C403.4.4.1 through C403.4.4.3: all VAV terminal devices equipped with a reheating, recooling or mixing function must permit the reduction of primary airflow (cfm) to a minimum as the first step in control; heating, cooling or mixing of air previously heated or cooled mechanically is the second step. Some typical VAV terminal unit configurations include:

Reheat. In a reheat terminal, heating (electric, hot water) is integrated at the unit. A predetermined throttling ratio establishes the lowest air quantity needed as required ventilation air to meet air movement and maximum humidity constraints before reheat is engaged.

Induction. VAV induction terminals reduce the primary (cooled) air supply with load while simultaneously inducing room or ceiling air to maintain relatively constant air movement. The ceiling air enhanced by these induction boxes generally is tempered by heat recovered from lights.

Fan powered. Fan-powered terminal units generally are selected to maintain higher circulation rates (cfm) at low loads; these VAV boxes often accommodate minimum (near zero) flow conditions while still maintaining constant airflow to the space. The energy savings is realized in that, as the primary (core) air closes to a minimum position (as necessary for minimum ventilation), the unit recirculates more plenum air. Both parallel (intermittent fan operation) and series (constant fan induction) arrangements are offered primarily for perimeter and interior/perimeter zone applications, respectively.

C403.4.4.2 Dual-duct and mixing VAV systems, terminal devices. Systems that have one warm air duct and one cool air duct shall use terminal devices that are capable of reducing the flow from one duct to a minimum before mixing of air from the other duct takes place.

❖ Dual-duct systems condition all air at a central air-handling unit via a hot deck and cold deck, and then distribute that air to the conditioned space(s) using two main duct trunk lines, one carrying cold air and the other carrying warm air. The dual-duct and mixing VAV systems blend these two air systems in various combinations at the terminal units. Accordingly, dual-duct and mixing VAV terminal devices must be capable of reducing flow from not less than one of the two ducts to a minimum, before reheating or recooling takes place (see commentary, Section C403.4.4.1).

C403.4.4.3 Single-fan dual-duct and mixing VAV systems, economizers. Individual dual-duct or mixing heating and cooling systems with a single fan and with total capacities greater than 90,000 Btu/h [(26.4 kW) 7.5 tons] shall not be equipped with air economizers.

❖ In dual-duct, single-fan systems a single primary air fan is sized for the total coincident peak of the hot and cold decks (see commentary, Section C403.4.4.2). Generally, the cold deck is maintained at a constant temperature, but during periods of high humidity, the hot deck temperature often is adjusted higher to increase airflow through the cold deck for dehumidification. This function, occurring at the primary air-handler level, can be energy wasteful if controls are not used to prevent the introduction of large amounts of outdoor air, such as may be admitted by an economizer, when the outdoor air is more humid than the return air. Therefore, the code structures an acceptable economizer trade-off around this particular scenario.

C403.4.4.4 Fractional hp fan motors. Motors for fans that are not less than $^1/_{12}$ hp (0.082 kW) and less than 1 hp (0.746 kW) shall be electronically commutated motors or shall have a minimum motor efficiency of 70 percent, rated in accordance with DOE 10 CFR 431. These motors shall also have the means to adjust motor speed for either balancing or remote control. The use of belt-driven fans to sheave adjustments for airflow balancing instead of a varying motor speed shall be permitted.

Exceptions: The following motors are not required to comply with this section:

1. Motors in the airstream within fan coils and terminal units that only provide heating to the space served.

2. Motors in space-conditioning equipment that comply with Section 403.2.3 or C403.2.12.

3. Motors that comply with Section C405.8.

❖ The efficiency of motors used in fan applications that are $^{1}/_{12}$ hp (62.1 W) or greater but less than 1 hp (0.75 kW) is addressed in this section. Motors in this range of horsepower must either be electronically commutated motors (ECMs) or be at least 70-percent efficient where rated in accordance with DOE 10 CFR 431.

C403.4.4.5 Supply-air temperature reset controls. Multiple-*zone* HVAC systems shall include controls that automatically reset the supply-air temperature in response to representative building loads, or to outdoor air temperature. The controls shall be capable of resetting the supply air temperature not less than 25 percent of the difference between the design supply-air temperature and the design room air temperature.

Exceptions:

1. Systems that prevent reheating, recooling or mixing of heated and cooled supply air.

2. Seventy-five percent of the energy for reheating is from site-recovered or site-solar energy sources.

3. *Zones* with peak supply air quantities of 300 cfm (142 L/s) or less.

❖ The purpose of this section is to require supply-air temperature reset controls on multiple-zone systems. Multiple-zone HVAC systems typically deliver supply air at a constant temperature necessary to provide cooling for a zone with the worst-case peak-cooling load. Any zone requiring less cooling than this must reduce the amount of cool air to that zone (if a VAV system) and eventually reheat that cool air if necessary to prevent zone overcooling. This is known as simultaneous heating and cooling. Building control systems can easily incorporate logic and controls that continually poll the various zones and raise the supply air temperature as high as possible while still satisfying the cooling demand. This reduces the necessity for simultaneous heating and cooling and saves energy. Systems that prevent reheating, recooling or mixing of heated and cooled supply air are exempt, as they cannot benefit by the addition of supply-air temperature reset controls. Seventy-five percent of the energy from site-recovered or site solar energy sources for reheating would be exempt, as the controls installed should be capable of resetting the supply-air temperature to not less than 25 percent of the difference between the design supply-air temperature and the design room-air temperature, and that would be marginal. For zones with peak supply air quantities of 300 cfm (142 L/s) or less, it would not be cost effective to install supply-air temperature reset controls, as the supply air is a small quantity compared to the building load.

C403.4.4.6 Multiple-zone VAV system ventilation optimization control. Multiple-zone VAV systems with direct digital control of individual zone boxes reporting to a central control panel shall have automatic controls configured to reduce outdoor air intake flow below design rates in response to changes in system *ventilation* efficiency (E_v) as defined by the *International Mechanical Code*.

Exceptions:

1. VAV systems with zonal transfer fans that recirculate air from other zones without directly mixing it with outdoor air, dual-duct dual-fan VAV systems, and VAV systems with fan-powered terminal units.

2. Systems having exhaust air energy recovery complying with Section C403.2.7.

3. Systems where total design exhaust airflow is more than 70 percent of total design outdoor air intake flow requirements.

❖ Multiple-zone VAV systems with direct digital control (DDC) to the zone level are required to provide an automated means to reduce outdoor air intake flows below design rates in response to changes in system ventilation efficiency as defined by the IMC. Essentially, this requires each VAV system to communicate with a central system to determine what is the optimal intake of air needed at a given time and adjust the intake mechanism accordingly. Each VAV box controller senses the current primary airflow (V_{pz}) and calculates the current zone outdoor-air fraction (Z_d). The building automation system totals the primary airflows and required outdoor airflows (V_{oz}) from all VAV boxes and determines the highest outdoor-air fraction reported.

C403.4.5 Heat recovery for service water heating. Condenser heat recovery shall be installed for heating or reheating of service hot water provided that the facility operates 24 hours a day, the total installed heat capacity of water-cooled systems exceeds 6,000,000 Btu/hr (1 758 kW) of heat rejection, and the design service water heating load exceeds 1,000,000 Btu/h (293 kW).

The required heat recovery system shall have the capacity to provide the smaller of the following:

1. Sixty percent of the peak heat rejection load at design conditions.

2. The preheating required to raise the peak service hot water draw to 85°F (29°C).

Exceptions:

1. Facilities that employ condenser heat recovery for space heating or reheat purposes with a heat recovery design exceeding 30 percent of the peak water-cooled condenser load at design conditions.

2. Facilities that provide 60 percent of their service water heating from site solar or site recovered energy or from other sources.

❖ The code requires heat recovery from the condenser side of water-cooled systems for preheating service hot water in large 24-hour facilities. Typical applications are homes; dormitories; mixed-use residential projects; commercial kitchens; and institutions, such as prisons and hospitals. Note that a facility must comply with the requirement if all the following are true:

- The facility operates 24 hours a day.
- The total installed heat rejection capacity of the water-cooled system exceeds 6,000,000 Btu/h (1759 kW). This equates to roughly 400 tons (362 874 kg) of electrical chiller capacity or 250 tons to 300 tons (226 796 kg to 272 155 kg) of gas- or thermally fired chiller capacity.
- The design service water-heating load exceeds 1,000,000 Btu/h (293 kW). An example of this would be a 1,000-bed nursing home [at 1.5 gallons per hour per bed (0.001577 L/s)].

Exception 1 would accept systems that use the recovered heat for space-heating purposes instead of the normally required water heating. Exception 2 simply excludes locations where not less than 60 percent of the energy for service water heating comes from either a solar-heating system that is installed on site or is recovered from some process or equipment within the building.

C403.4.6 Hot gas bypass limitation. Cooling systems shall not use hot gas bypass or other evaporator pressure control systems unless the system is designed with multiple steps of unloading or continuous capacity modulation. The capacity of the hot gas bypass shall be limited as indicated in Table C403.4.6, as limited by Section C403.3.1.

❖ This section was added to the code to restrict hot gas bypass and evaporator control systems, which will, in turn, save energy. Hot gas bypass is a control strategy used in commercial cooling and refrigeration equipment that allows cooling compressors to remain online at low load and in colder weather by raising the condenser pressure. Many new commercial cooling units have hot gas bypass systems, and have the capacity to modulate or unload capacity, which saves energy.

TABLE C403.4.6
MAXIMUM HOT GAS BYPASS CAPACITY

RATED CAPACITY	MAXIMUM HOT GAS BYPASS CAPACITY (% of total capacity)
≤ 240,000 Btu/h	50
> 240,000 Btu/h	25

For SI: 1 British thermal unit per hour = 0.2931 W.

C403.5 Refrigeration systems. Refrigerated display cases, *walk-in coolers* or *walk-in freezers* that are served by remote compressors and remote condensers not located in a *condensing unit*, shall comply with Sections C403.5.1 and C403.5.2.

Exception: Systems where the working fluid in the refrigeration cycle goes through both subcritical and supercritical states (transcritical) or that use ammonia refrigerant are exempt.

❖ Where the condenser serving one of the refrigeration systems addressed in this section is a separate piece of equipment, it is essential for energy savings to select the condenser that is optimal for both the refrigeration equipment being cooled and the installation location's climate. Sections C403.5.1 and C403.5.2 provide the tools by which the project designer can select the appropriate equipment.

C403.5.1 Condensers serving refrigeration systems. Fan-powered condensers shall comply with the following:

1. The design *saturated condensing temperatures* for air-cooled condensers shall not exceed the design dry-bulb temperature plus 10°F (5.6°C) for low-*temperature refrigeration systems*, and the design dry- bulb temperature plus 15°F (8°C) for *medium temperature refrigeration systems* where the *saturated condensing temperature* for blend refrigerants shall be determined using the average of liquid and vapor temperatures as converted from the condenser drain pressure.
2. Condenser fan motors that are less than 1 hp (0.75 kW) shall use electronically commutated motors, permanent split-capacitor-type motors or 3-phase motors.
3. Condenser fans for air-cooled condensers, evaporatively cooled condensers, air- or water-cooled fluid coolers or cooling towers shall reduce fan motor demand to not more than 30 percent of design wattage at 50 percent of design air volume, and incorporate one of the following continuous variable speed fan control approaches:
   - 3.1. Refrigeration system condenser control for air-cooled condensers shall use variable setpoint control logic to reset the condensing temperature setpoint in response to ambient dry-bulb temperature.
   - 3.2. Refrigeration system condenser control for evaporatively cooled condensers shall use variable setpoint control logic to reset the condensing temperature setpoint in response to ambient wet-bulb temperature.
4. Multiple fan condensers shall be controlled in unison.
5. The minimum condensing temperature setpoint shall be not greater than 70°F (21°C).

❖ See the commentary to Section C403.5.

C403.5.2 Compressor systems. Refrigeration compressor systems shall comply with the following:

1. Compressors and multiple-compressor system suction groups shall include control systems that use floating suction pressure control logic to reset the target suction pressure temperature based on the temperature requirements of the attached refrigeration display cases or walk-ins.

   Exception: Controls are not required for the following:

   1. Single-compressor systems that do not have variable capacity capability.
   2. Suction groups that have a design saturated suction temperature of 30°F (-1.1°C) or

higher, suction groups that comprise the high stage of a two-stage or cascade system, or suction groups that primarily serve chillers for secondary cooling fluids.

2. Liquid subcooling shall be provided for all low-temperature compressor systems with a design cooling capacity equal to or greater than 100,000 Btu/hr (29.3 kW) with a design-saturated suction temperature of -10°F (-23°C) or lower. The sub-cooled liquid temperature shall be controlled at a maximum temperature setpoint of 50°F (10°C) at the exit of the subcooler using either compressor economizer (interstage) ports or a separate compressor suction group operating at a saturated suction temperature of 18°F (-7.8°C) or higher.
    2.1. Insulation for liquid lines with a fluid operating temperature less than 60°F (15.6°C) shall comply with Table C403.2.10.
3. Compressors that incorporate internal or external crankcase heaters shall provide a means to cycle the heaters off during compressor operation.

❖ See the commentary to Section C403.

SECTION C404 SERVICE WATER HEATING (MANDATORY)

C404.1 General. This section covers the minimum efficiency of, and controls for, service water-heating equipment and insulation of service hot water piping.

❖ This section restates the general purpose of the requirements for service water-heating equipment (see Commentary Figure C404.1). "Service water heating" is defined as the supply of hot water for purposes other than comfort heating. Therefore, this section applies to equipment and piping used to produce and distribute hot water for the following:

- Restrooms.
- Showers.
- Laundries.
- Kitchens.
- Pools and spas.
- Defrosting of sidewalks and driveways.
- Car washes, beauty salons and other commercial enterprises.

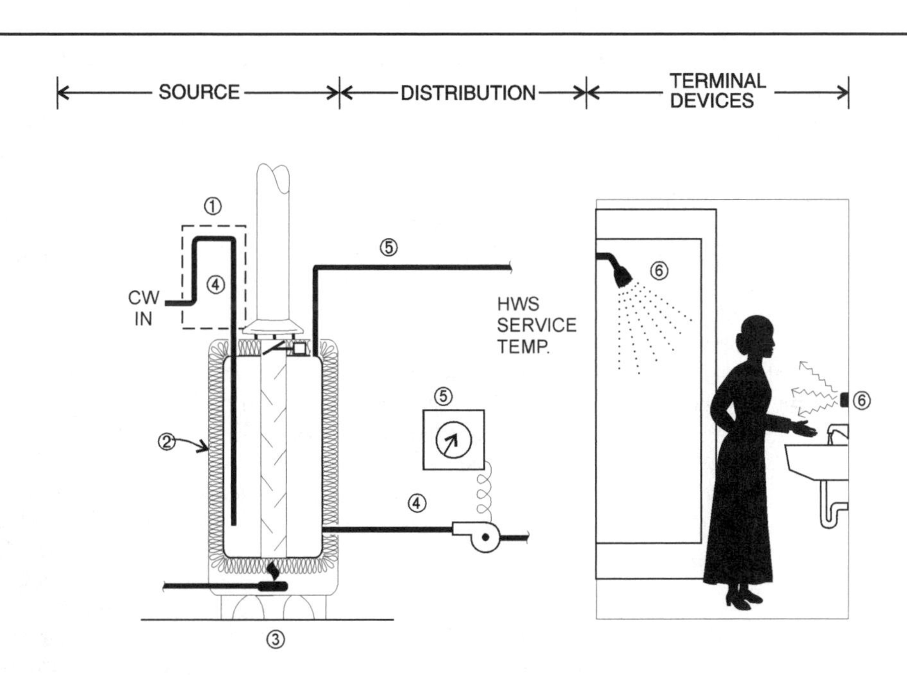

① HEAT TRAPS TO REDUCE STANDBY LOSSES

② INSULATED TANKS TO REDUCE STANDBY LOSSES

③ HIGH EFFICIENCY SOURCES

④ PIPE INSULATION TO REDUCE DISTRIBUTION AND STANDBY LOSSES

⑤ CIRCULATION LOOP TEMPERATURE CONTROLS TO REDUCE DISTRIBUTION LOSSES

⑥ FLOW LIMITING DEVICES SUCH AS LOW FLOW SHOWER HEADS AND OCCUPANT SENSORS TO REDUCE WASTE ARE ENCOURAGED BUT NOT REQUIRED BY THE IECC (FOR FURTHER DISCUSSION OF WATER CONSERVATION MEASURES, SEE COMMENTARY TO SECTION 604.4 OF THE INTERNATIONAL PLUMBING CODE)

Figure C404.1
ELEMENTS COVERED BY SECTION C404
(Courtesy of U.S. Department of Energy, Office of Building Technology State and Community Programs, www.energycodes.gov)

- Dwelling units and lowest rooms in high-rise residential buildings and hotels, respectively.

The requirements of this section are for service water-heating equipment and systems only; however, the principles presented could apply to energy-efficient process water-heating systems and equipment, as well. Space-conditioning boilers and distribution systems for human comfort are covered under the requirements of Section C403. The service water-heating portion of combined space-conditioning and water-heating systems must meet the requirements of this chapter as well.

Compliance with this section is based on the application of some basic, cost-effective design practices:

- Minimize standby losses with heat traps, thermal insulation and temperature controls.
- Reduce distribution losses with thermal insulation and system temperature controls, or eliminate through point-of-use heaters.
- Reduce hot water waste with flow-limiting or metering terminal devices.
- Increase overall system performance by utilizing high-efficiency sources.
- Increase efficiency of supply piping by limiting flow rate, and limiting pipe length or pipe volume.

Although compliance with the code ensures a minimum level of water-heating system performance, the designer is encouraged to view the code as a minimum standard of care and investigate designs that exceed these requirements. Application of heat recovery, solar energy or high-efficiency equipment can create a system that is more efficient than the code requires and one that exhibits an excellent return on investment. Examples of measures that exceed code requirements include:

- Specifying low-flow shower heads that exceed the conservation requirements specified in Chapter 6 of the *International Plumbing Code*® (IPC®).
- Extending the public restroom low-flow requirements found in the IPC to all restrooms.
- Recovering heat from gray water.
- Specifying low-water usage or low-temperature appliances, including residential and commercial clothes washers and dishwashers.

For instance, Energy Star®-labeled dishwashers save energy by using both improved technology for the primary wash cycle and by using less hot water to clean. Federal law also requires all new dishwashers to be equipped with a no-heat drying option to save even more energy. A typical household does nearly 400 loads of laundry per year, using about 40 to 47 gallons (151 to 178 L) of water per full load with a conventional washer. In contrast, commercial laundries equipped with a full-sized Energy Star-qualified clothes washer use 10 to 25 gallons (68 to 95 L) per load, and many newer residential washers use only 11 to 32 gallons (42 to 121 L) per load.

C404.2 Service water-heating equipment performance efficiency. Water-heating equipment and hot water storage tanks shall meet the requirements of Table C404.2. The efficiency shall be verified through data furnished by the manufacturer of the equipment or through certification under an *approved* certification program. Water-heating equipment also intended to be used to provide space heating shall meet the applicable provisions of Table C404.2.

❖ Table C404.2 presents the minimum required efficiencies for water-heating equipment used for hot water supply and space heating. Equipment must meet the requirement for a minimum heater efficiency [energy factor (EF)], thermal efficiency (E_t) or a maximum standby loss (SL). Smaller equipment, which falls under NAECA, must satisfy the federal energy factor requirements before it can be sold. This minimum energy factor is a combined measure of thermal efficiency and standby loss. NAECA-covered water-heating equipment includes:

- Electric heaters of all types with input ratings less than 12 kW.
- Fuel-fired storage heaters at or below 75,000 Btu/h (21 983 W) input for gas, or 105,000 Btu/h (30 776 W) input for oil.
- Fuel-fired instantaneous heaters at or below 200,000 Btu/h (58 620 W) input for gas, or 210,000 Btu/h (61 551 W) for oil.
- All fuel-fired pool and spa heaters.

The efficiency requirements for equipment not covered by NAECA are summarized throughout the remaining rows of Table C404.2. Equipment in the table is classified by type (instantaneous gas, storage electric, etc.) capacity (input rating) and subcategory or rating condition. Data from nationally recognized testing and certification programs using standard DOE/ANSI test procedures often are preferred or used to demonstrate compliance with these requirements.

Pool heaters must meet the efficiency requirements for service water-heating in Table C404.2. Electric pool heaters and pumps must be installed with time switches so that they can be shut down during periods of limited use or inactivity (see Section C404.9.2).

TABLE C404.2. See page C4-87.

❖ This table provides the water heater efficiency requirements in Chapter 4 [CE]. The minimum equipment performance requirements in this table will meet the requirements found in ANSI/ASHRAE/IESNA 90.1 with a few limited exceptions. The efficiency equation for electric-resistance water heaters in the code is intended to meet the existing federal minimum manufacturing standard, and, therefore, differs from the ASHRAE level.

TABLE C404.2
MINIMUM PERFORMANCE OF WATER-HEATING EQUIPMENT

EQUIPMENT TYPE	SIZE CATEGORY (input)	SUBCATEGORY OR RATING CONDITION	PERFORMANCE REQUIRED[a, b]	TEST PROCEDURE
Water heaters, electric	≤ 12 kW[d]	Resistance	0.97 - 0.00 132V, EF	DOE 10 CFR Part 430
	> 12 kW	Resistance	(0.3 + 27/V_m), %/h	ANSI Z21.10.3
	≤ 24 amps and ≤ 250 volts	Heat pump	0.93 - 0.00 132V, EF	DOE 10 CFR Part 430
Storage water heaters, gas	≤ 75,000 Btu/h	≥ 20 gal	0.67 - 0.0019V, EF	DOE 10 CFR Part 430
	> 75,000 Btu/h and ≤ 155,000 Btu/h	< 4,000 Btu/h/gal	80% E_t (Q/800 + 110$\sqrt{V}$)SL, Btu/h	ANSI Z21.10.3
	> 155,000 Btu/h	< 4,000 Btu/h/gal	80% E_t (Q/800 + 110$\sqrt{V}$)SL, Btu/h	
Instantaneous water heaters, gas	> 50,000 Btu/h and < 200,000 Btu/h[c]	≥ 4,000 (Btu/h)/gal and < 2 gal	0.62 - 0.00 19V, EF	DOE 10 CFR Part 430
	≥ 200,000 Btu/h	≥ 4,000 Btu/h/gal and < 10 gal	80% E_t	ANSI Z21.10.3
	≥ 200,000 Btu/h	≥ 4,000 Btu/h/gal and ≥ 10 gal	80% E_t (Q/800 + 110$\sqrt{V}$)SL, Btu/h	
Storage water heaters, oil	≤ 105,000 Btu/h	≥ 20 gal	0.59 - 0.0019V, EF	DOE 10 CFR Part 430
	≥ 105,000 Btu/h	< 4,000 Btu/h/gal	80% E_t (Q/800 + 110$\sqrt{V}$)SL, Btu/h	ANSI Z21.10.3
Instantaneous water heaters, oil	≤ 210,000 Btu/h	≥ 4,000 Btu/h/gal and < 2 gal	0.59 - 0.0019V, EF	DOE 10 CFR Part 430
	> 210,000 Btu/h	≥ 4,000 Btu/h/gal and < 10 gal	80% E_t	ANSI Z21.10.3
	> 210,000 Btu/h	≥ 4,000 Btu/h/gal and ≥ 10 gal	78% E_t (Q/800 + 110$\sqrt{V}$)SL, Btu/h	
Hot water supply boilers, gas and oil	≥ 300,000 Btu/h and < 12,500,000 Btu/h	≥ 4,000 Btu/h/gal and < 10 gal	80% E_t	ANSI Z21.10.3
Hot water supply boilers, gas	≥ 300,000 Btu/h and < 12,500,000 Btu/h	≥ 4,000 Btu/h/gal and ≥ 10 gal	80% E_t (Q/800 + 110$\sqrt{V}$)SL, Btu/h	
Hot water supply boilers, oil	> 300,000 Btu/h and < 12,500,000 Btu/h	> 4,000 Btu/h/gal and > 10 gal	78% E_t (Q/800 + 110$\sqrt{V}$)SL, Btu/h	
Pool heaters, gas and oil	All	—	82% E_t	ASHRAE 146
Heat pump pool heaters	All	—	4.0 COP	AHRI 1160
Unfired storage tanks	All	—	Minimum insulation requirement R-12.5 (h · ft^2 · °F)/Btu	(none)

For SI: °C = [(°F) - 32]/1.8, 1 British thermal unit per hour = 0.2931 W, 1 gallon = 3.785 L, 1 British thermal unit per hour per gallon = 0.078 W/L.

a. Energy factor (EF) and thermal efficiency (E_t) are minimum requirements. In the EF equation, V is the rated volume in gallons.

b. Standby loss (SL) is the maximum Btu/h based on a nominal 70°F temperature difference between stored water and ambient requirements. In the SL equation, Q is the nameplate input rate in Btu/h. In the equations for electric water heaters, V is the rated volume in gallons and V_m is the measured volume in gallons. In the SL equation for oil and gas water heaters and boilers, V is the rated volume in gallons.

c. Instantaneous water heaters with input rates below 200,000 Btu/h shall comply with these requirements where the water heater is designed to heat water to temperatures 180°F or higher.

d. Electric water heaters with an input rating of 12 kW (40,950 Btu/hr) or less that are designed to heat water to temperatures of 180°F or greater shall comply with the requirements for electric water heaters that have an input rating greater than 12 kW (40,950 Btu/h).

C404.2.1 High input-rated service water-heating systems. Gas-fired water-heating equipment installed in new buildings shall be in compliance with this section. Where a singular piece of water-heating equipment serves the entire building and the input rating of the equipment is 1,000,000 Btu/h (293 kW) or greater, such equipment shall have a thermal efficiency, E_t, of not less than 90 percent. Where multiple pieces of water-heating equipment serve the building and the combined input rating of the water-heating equipment is 1,000,000 Btu/h (293 kW) or greater, the combined input-capacity-weighted-average thermal efficiency, E_t, shall be not less than 90 percent.

Exceptions:

1. Where 25 percent of the annual *service water-heating* requirement is provided by site-solar or site-recovered energy, the minimum thermal efficiency requirements of this section shall not apply.
2. The input rating of water heaters installed in individual dwelling units shall not be required to be included in the total input rating of *service water-heating* equipment for a building.
3. The input rating of water heaters with an input rating of not greater than 100,000 Btu/h (29.3 kW) shall not be required to be included in the total input rating of *service water-heating* equipment for a building.

❖ Where the total input rating of water-heating equipment installed in a new building is 1,000,000 Btu/h or greater, the thermal efficiency of the equipment (E_t) must be not less than 90 percent. Where multiple pieces of equipment are used, this can be a combined input capacity weighted average. The requirement only applies to gas-fired water-heating equipment. In determining the total input rating of the service water-heating equipment, the code exempts water heaters installed for individual dwelling units and water heaters with an individual rating less than 100,000 Btu/h. Where 25 percent of the annual service water-heating requirements of the building are supplied by site-solar or site-recovered energy systems, the minimum thermal efficiency for the remaining gas-fired equipment is waived.

C404.3 Heat traps. Water-heating equipment not supplied with integral heat traps and serving noncirculating systems shall be provided with heat traps on the supply and discharge piping associated with the equipment.

❖ A "heat trap" is a means to counteract the natural convection of heated water in a vertical pipe run. The means is either a device specifically designed for the purpose or an arrangement of tubing that forms a 360-degree (6.3 rad) loop of piping that forms the point of connection to the water heater (inlet or outlet), which includes a length of piping directed downward before the connection to the vertical piping of the supply water or hot water distribution system.

Storage hot water tanks in noncirculating systems are required to have either integral heat traps or external heat traps [which may be a separate device, a 360-degree (6.3 rad) loop of plumbers' flex or simply an inverted pipe trap] on both the inlet and outlet piping located as close as practical to the water heater [see Commentary Figures C404.4(1) through C404.4(3)]. A heat trap is an inexpensive, one-way

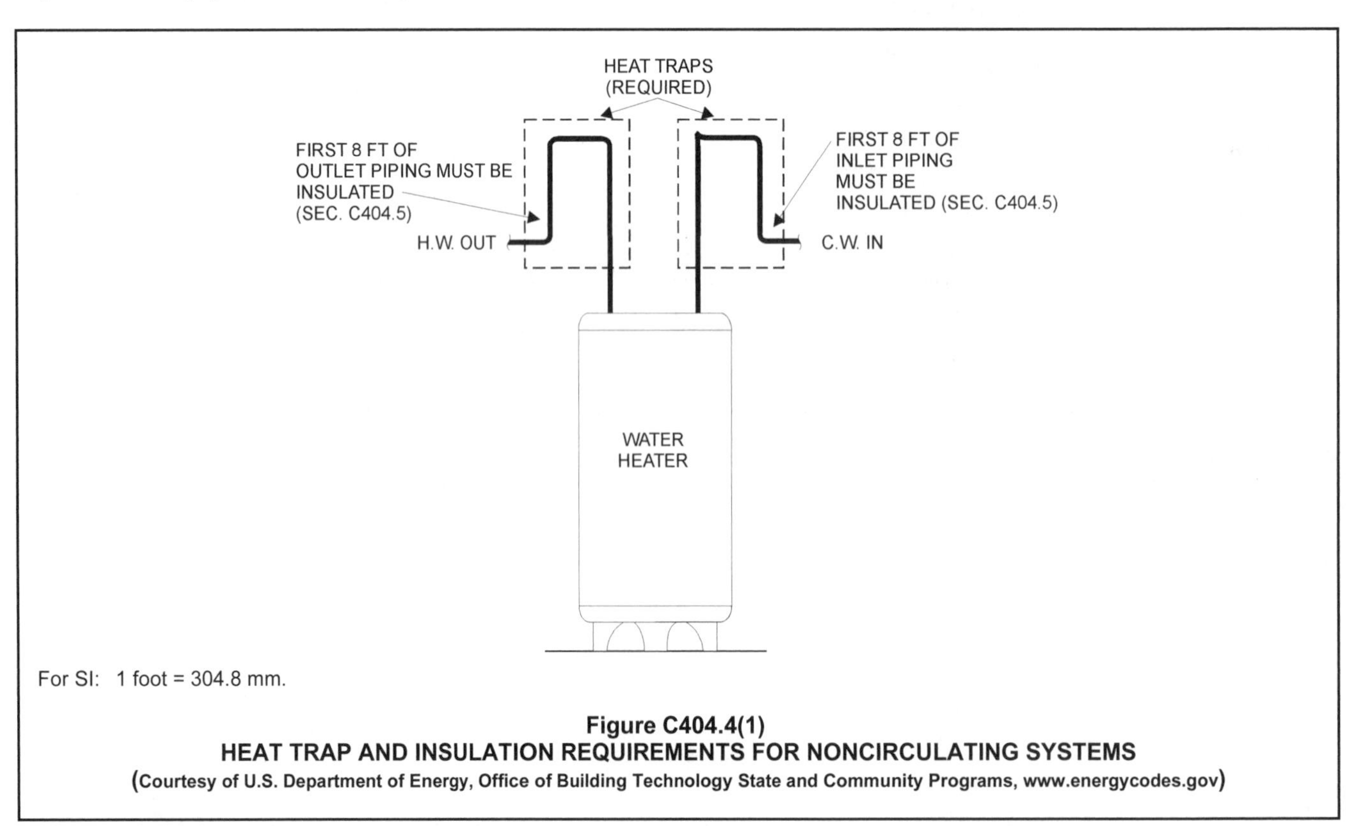

For SI: 1 foot = 304.8 mm.

Figure C404.4(1)
HEAT TRAP AND INSULATION REQUIREMENTS FOR NONCIRCULATING SYSTEMS
(Courtesy of U.S. Department of Energy, Office of Building Technology State and Community Programs, www.energycodes.gov)

check valve fitting [see Commentary Figure C404.4(1)] or an arrangement of piping [see Commentary Figures C404.4(2) and C404.4(3)] to keep buoyant hot water from circulating through a piping distribution system by natural convection. Integral heat traps are sometimes designed into the water heater by the manufacturer.

C404.4 Insulation of piping. Piping from a water heater to the termination of the heated water fixture supply pipe shall be insulated in accordance with Table C403.2.10. On both the inlet and outlet piping of a storage water heater or heated water storage tank, the piping to a heat trap or the first 8 feet (2438 mm) of piping, whichever is less, shall be insulated. Piping that is heat traced shall be insulated in accordance with Table C403.2.10 or the heat trace manufacturer's instructions. Tubular pipe insulation shall be installed in accordance with the insulation manufacturer's instructions. Pipe insulation shall be continuous except where the piping passes through a framing member. The minimum insulation thickness requirements of this section shall not supersede any greater insulation thickness requirements necessary for the protection of piping from freezing temperatures or the protection of personnel against external surface temperatures on the insulation.

> **Exception:** Tubular pipe insulation shall not be required on the following:
>
> 1. The tubing from the connection at the termination of the fixture supply piping to a plumbing fixture or plumbing appliance.

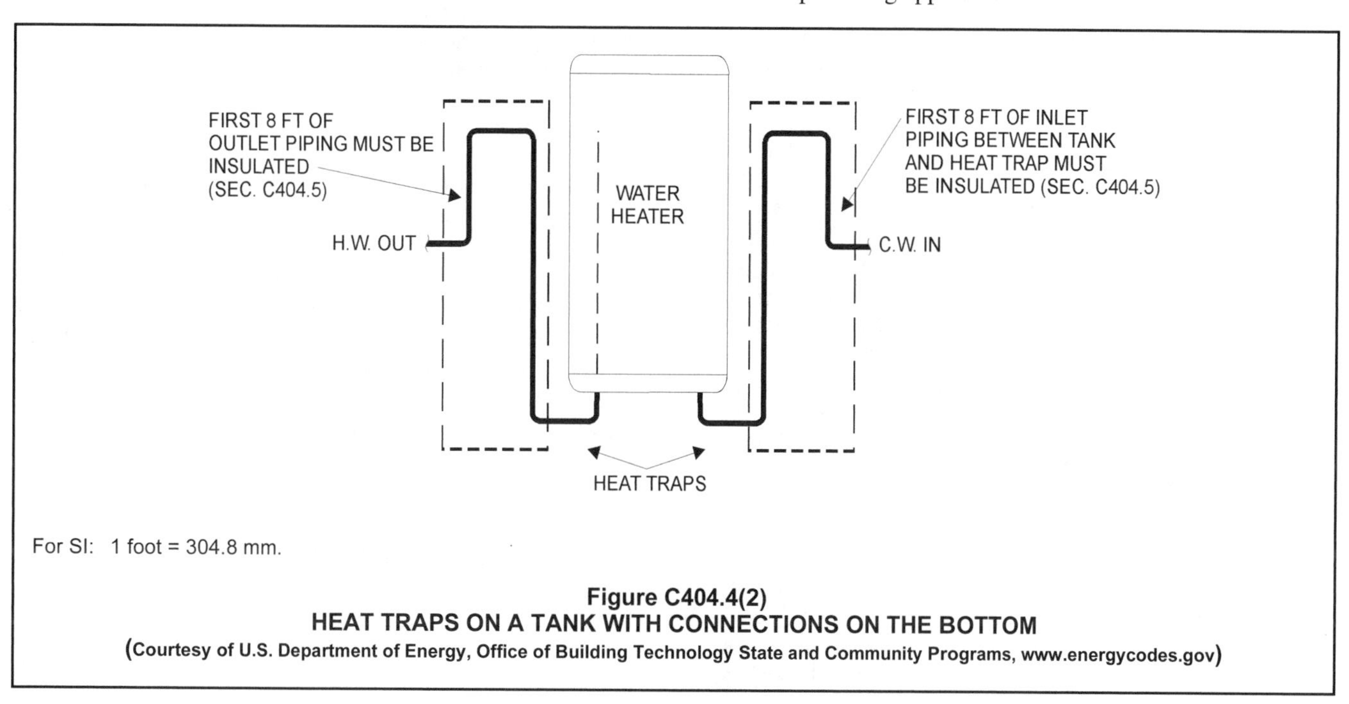

Figure C404.4(2)
HEAT TRAPS ON A TANK WITH CONNECTIONS ON THE BOTTOM
(Courtesy of U.S. Department of Energy, Office of Building Technology State and Community Programs, www.energycodes.gov)

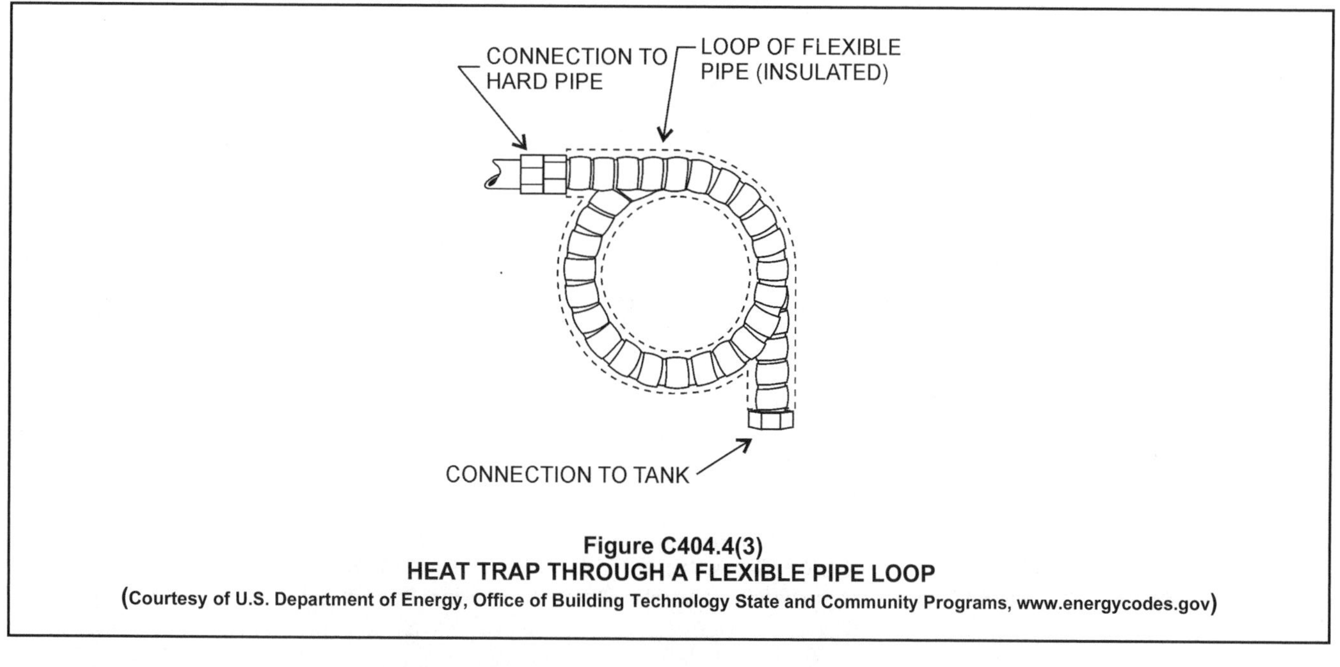

Figure C404.4(3)
HEAT TRAP THROUGH A FLEXIBLE PIPE LOOP
(Courtesy of U.S. Department of Energy, Office of Building Technology State and Community Programs, www.energycodes.gov)

2. Valves, pumps, strainers and threaded unions in piping that is 1 inch (25 mm) or less in nominal diameter.
3. Piping from user-controlled shower and bath mixing valves to the water outlets.
4. Cold-water piping of a demand recirculation water system.
5. Tubing from a hot drinking-water heating unit to the water outlet.
6. Piping at locations where a vertical support of the piping is installed.
7. Piping surrounded by building insulation with a thermal resistance (*R*-value) of not less than R-3.

❖ Distribution losses impact building energy use both in the energy required to make up for the lost heat and in the additional load that can be placed on the space-cooling system if the heat is released to air-conditioned space. These losses can be limited by insulating the hot water storage vessel and pipes. The required amount of insulation is given in Table C403.2.10, which specifies different thicknesses of insulation depending on the size of the pipe or tube, the operating temperature range of the fluid, in this case water, and the conductivity of the insulation. For typical hot water supply, the operating temperature range in the table to consult is 140°F to 200°F.

The list of items where pipe insulation is not required is almost common sense but still, these items need to be stated to avoid confusion and possible misinterpretations by the code officials. Insulating valves is time consuming and, if the right type of valve is not used, nearly impossible (think ball valve without a raised handle). A few uninsulated valves in the system are not going to lose a lot of heat. Pumps are also difficult to insulate and in some cases insulation might cause overheating of the pump motor. Threaded unions usually only occur in smaller diameter piping systems and are time consuming to insulate. Such unions represent a small amount of heat loss compared to the entire system. Piping or tubing from a small tankless water heater serving one sink is too small to easily insulate. The heat loss is negligible [see Commentary Figure C404.4(4)].

C404.5 Efficient heated water supply piping. Heated water supply piping shall be in accordance with Section C404.5.1 or C404.5.2. The flow rate through $^1/_4$-inch (6.4 mm) piping shall be not greater than 0.5 gpm (1.9 L/m). The flow rate through $^5/_{16}$-inch (7.9 mm) piping shall be not greater than 1 gpm (3.8 L/m). The flow rate through $^3/_8$-inch (9.5 mm) piping shall be not greater than 1.5 gpm (5.7 L/m).

❖ The provisions of this section are simply to limit the volume of water in a hot water supply pipe and therefore reduce the amount of time it takes to get hot water to the faucet. The volume is limited by the method in either Section C404.5.1, which simply limits the length of pipe between the source and the outlet, or Section C404.5.2, which requires a calculation of the volume of water in the specific piping system. The flow rate limitations reflect federal limitations on flow rate based on water conservation measures. Higher flow rates will obviously reduce the time-to-tap performance; however, the maximum volumes in Sections C404.5.1 and C404.5.2 are based on the maximum flow rates listed.

C404.5.1 Maximum allowable pipe length method. The maximum allowable piping length from the nearest source of heated water to the termination of the fixture supply pipe shall be in accordance with the following. Where the piping contains more than one size of pipe, the largest size of pipe within the piping shall be used for determining the maximum allowable length of the piping in Table C404.5.1.

1. For a public lavatory faucet, use the "Public lavatory faucets" column in Table C404.5.1.

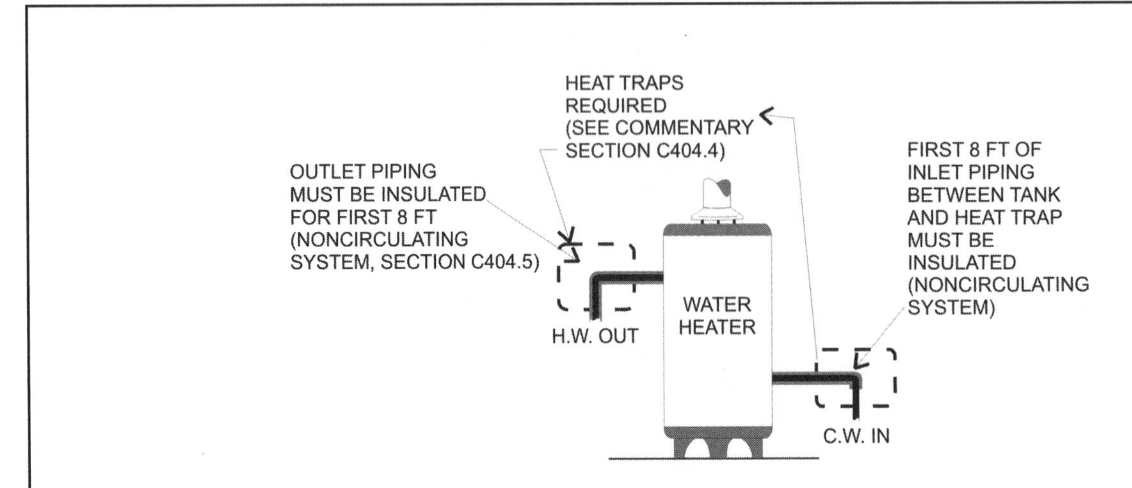

For SI: 1 foot = 304.8 mm.

Figure C404.4(4)
HEAT TRAPS ON A TANK WITH CONNECTIONS ON SIDES
(Courtesy of U.S. Department of Energy, Office of Building Technology State and Community Programs, www.energycodes.gov)

2. For all other plumbing fixtures and plumbing appliances, use the "Other fixtures and appliances" column in Table C404.5.1.

❖ This section provides one of the two methods called out in Section 404.5 to limit the volume of water in the portion of the hot water supply pipe from the source of the hot water to the outlet, or faucet. Quite simply, this limits the volume by limiting the length of pipe, based on the size of the pipe. Therefore, a smaller pipe diameter has less volume per foot and is allowed to have more length. Although not stated in this section, the source of hot water supply can be a water heater, a circulating water system or a heat trace maintenance system, as stated in Section C404.5.2. The maximum allowable lengths of pipe from a source to a public lavatory system are restrictive. In most designs, this will certainly require use of a circulating water system or a heat trace system.

C404.5.2 Maximum allowable pipe volume method. The water volume in the piping shall be calculated in accordance with Section C404.5.2.1. Water heaters, circulating water systems and heat trace temperature maintenance systems shall be considered sources of heated water.

The volume from the nearest source of heated water to the termination of the fixture supply pipe shall be as follows:

1. For a public lavatory faucet: not more than 2 ounces (0.06 L).
2. For other plumbing fixtures or plumbing appliances; not more than 0.5 gallon (1.89 L).

❖ This section provides one of the two methods called out in Section 404.5 to limit the volume of water in the portion of the hot water supply pipe from the source of the hot water to the outlet, or faucet. For calculation of a system where there is merely a run of pipe involved, the method in Section C404.5.1 is easier, and, for larger pipe diameters, somewhat less restrictive. For instance, if a $^1/_2$-inch supply pipe were to be used, Section C404.5.1 would allow 2 feet of pipe from the hot water source to a public faucet. Using this method, only $1^1/_3$ feet of pipe would be allowed. Where the piping system contains a number of valves, fittings, meters or manifolds, this method is needed (see Section 404.5.2.1). This section contains a provision that also applies to Section 404.5.1: the source of hot water can be the water heater, a circulating water system or a heat trace system. Given the very small volume of water allowed for the supply from the source to a public lavatory faucet, this section would, in most cases, require the use of a circulating water system or a heat trace system.

C404.5.2.1 Water volume determination. The volume shall be the sum of the internal volumes of pipe, fittings, valves, meters and manifolds between the nearest source of heated water and the termination of the fixture supply pipe. The volume in the piping shall be determined from the "Volume" column in Table C404.5.1. The volume contained within fixture shutoff valves, within flexible water supply connectors to a fixture fitting and within a fixture fitting shall not be included in the water volume determination. Where heated water is supplied by a recirculating system or heat-traced piping, the volume shall include the portion of the fitting on the branch pipe that supplies water to the fixture.

❖ This section provides information regarding what needs to be included in calculating the volume of water when using Section 404.5.2 as the design option for limiting the volume of water for the piping and system connecting the hot water source to the faucet or other outlet.

C404.6 Heated-water circulating and temperature maintenance systems. Heated-water circulation systems shall be in accordance with Section C404.6.1. Heat trace temperature maintenance systems shall be in accordance with Section C404.6.2. Controls for hot water storage shall be in accor-

TABLE C404.5.1
PIPING VOLUME AND MAXIMUM PIPING LENGTHS

NOMINAL PIPE SIZE (inches)	VOLUME (liquid ounces per foot length)	MAXIMUM PIPING LENGTH (feet)	
		Public lavatory faucets	Other fixtures and appliances
$^1/_4$	0.33	6	50
$^5/_{16}$	0.5	4	50
$^3/_8$	0.75	3	50
$^1/_2$	1.5	2	43
$^5/_8$	2	1	32
$^3/_4$	3	0.5	21
$^7/_8$	4	0.5	16
1	5	0.5	13
$1^1/_4$	8	0.5	8
$1^1/_2$	11	0.5	6
2 or larger	18	0.5	4

For SI: 1 inch = 25.4 mm, 1 foot = 304.8 mm, 1 liquid ounce = 0.030 L, 1 gallon = 128 ounces.

dance with Section C404.6.3. Automatic controls, temperature sensors and pumps shall be *accessible*. Manual controls shall be *readily accessible*.

❖ When the distribution piping is heated to maintain usage temperatures, such as in circulating hot water systems, the system pump or heat-trace cable must have conveniently located manual or automatic switches, or other controls, that can be set to optimize system operation or turn off the system during periods of reduced demand.

Most automatic time clocks allow the operator or building engineer to enter different operating schedules for each of the seven days of the week. Thus, pipes are not heated during hours of inactivity. More sophisticated devices, such as a combination time and temperature and demand controls, also comply with the requirements of this section.

C404.6.1 Circulation systems. Heated-water circulation systems shall be provided with a circulation pump. The system return pipe shall be a dedicated return pipe or a cold water supply pipe. Gravity and thermo-syphon circulation systems shall be prohibited. Controls for circulating hot water system pumps shall start the pump based on the identification of a demand for hot water within the occupancy. The controls shall automatically turn off the pump when the water in the circulation loop is at the desired temperature and when there is no demand for hot water.

❖ Circulation systems for heated water must be demand activated. Demand-activated systems provide significantly more energy efficiency than any other type of circulation system. Given that the piping is required to be insulated, the water in the recirculation loop will stay hot for a long time, thus making continuous circulation of the water unnecessary.

C404.6.2 Heat trace systems. Electric heat trace systems shall comply with IEEE 515.1. Controls for such systems shall be able to automatically adjust the energy input to the heat tracing to maintain the desired water temperature in the piping in accordance with the times when heated water is used in the occupancy. Heat trace shall be arranged to be turned off automatically when there is no hot water demand.

❖ Automatic controls are necessary to adjust water temperature based on use of hot water and to turn off the heat trace system when there is no demand. These automatic controls are intended to make the heat trace system as energy efficient as possible when it is used.

C404.6.3 Controls for hot water storage. The controls on pumps that circulate water between a water heater and a heated-water storage tank shall limit operation of the pump from heating cycle startup to not greater than 5 minutes after the end of the cycle.

❖ This provision is the same as in the IPC. This section addresses water-heating systems that utilize a separate water-heating unit to generate hot water for storage in tanks. These systems have a circulating pump to move hot water into the storage tank (and cooler water back to the heating unit). This section requires that the circulation pump not continue to operate after the hot water storage tank is at the desired temperature. The run time of up to 5 minutes after the end of the heating cycle is to allow for the heating unit's heat exchanger to cool down and equalize temperature with the stored hot water.

C404.7 Demand recirculation controls. A water distribution system having one or more recirculation pumps that pump water from a heated-water supply pipe back to the heated-water source through a cold-water supply pipe shall be a *demand recirculation water system*. Pumps shall have controls that comply with both of the following:

1. The control shall start the pump upon receiving a signal from the action of a user of a fixture or appliance, sensing the presence of a user of a fixture or sensing the flow of hot or tempered water to a fixture fitting or appliance.
2. The control shall limit the temperature of the water entering the cold-water piping to 104°F (40°C).

❖ This section applies only to heated water (hot or tempered water) recirculation systems that pump water back to the heated water source through a cold-water supply pipe. Specialized equipment for recirculating hot water to remote fixture systems and back to the heated water source through a cold-water supply pipe has been used for several decades. They were most often used as a retrofit for buildings where routing of dedicated return piping was too costly or impossible. See the commentary to IPC Section 604.2 for additional understanding of "no dedicated return line" recirculation systems.

Extensive research by many organizations, including the Oak Ridge National Laboratory and the California Energy Commission, has determined that having hot water "ready and waiting" at fixtures is a significant energy waste. On the other hand, where a hot (or tempered) water circulation system is not provided, there is a significant waste of energy of heating water that never arrives at the intended point of use except where a large volume of cooler water is flushed down the drain in order to have the heated water arrive at the point of use. Certainly, such water waste is of concern (because of rising water and sewer charges), but the waste of "embedded energy" (utility treatment and pumping energy) in that water is of equal or greater concern. The focus of the research was to determine how to best install and operate hot-water (or tempered-water) recirculation systems for the lowest possible cost.

C404.8 Drain water heat recovery units. Drain water heat recovery units shall comply with CSA B55.2. Potable water-side pressure loss shall be less than 10 psi (69 kPa) at maximum design flow. For Group R occupancies, the efficiency of drain water heat recovery unit efficiency shall be in accordance with CSA B55.1.

❖ Drain water heat recovery is often a cost-effective way to add to energy efficiency by recapturing hot water energy that is literally "going down the drain." While it is not required for hot-water heating systems in this code, it can be used in the performance design of a system, as provided by Section C407 and Table C407.5.1(1). Drain water heat recovery (DWHR) works particularly well where heated water flows down the drain at the same time as water that needs to be heated flows in. This "coincident flow" occurs in occupancies with showering and lavatory use. Performance of a DWHR unit is characterized by both efficiency and pressure loss. It is important to ensure that DWHR devices do not impose large pressure losses in the piping in order to minimize the impact on water flow in the building. Given the available DWHR efficiencies, savings are typically 10 to 35 percent of the energy used for heating water. Two standards, CSA 55.1 and CSA 55.2, provide criteria for these units. CSA B55.2 is for fabrication and material quality of DWHR units. CSA B55.1 is for testing and labeling of DWHR units efficiency and pressure loss at 2.5 gpm.

C404.9 Energy consumption of pools and permanent spas. (Mandatory). The energy consumption of pools and permanent spas shall be controlled by the requirements in Sections C404.9.1 through C404.9.3.

❖ Because of the heating and filtering operations involved in pools and inground, permanently installed spas, they provide a good opportunity to save energy by limiting heat loss or pump operation. This section provides the scoping requirements for the pool heaters, time switches and pool covers that make operation of the pool and permanently installed spas more energy efficient. These requirements also can be found in ANSI/ASHRAE/IESNA 90.1, and can help reduce the energy used by pools and permanently installed spas.

C404.9.1 Heaters. The electric power to all heaters shall be controlled by a readily accessible on-off switch that is an integral part of the heater, mounted on the exterior of the heater, or external to and within 3 feet (914 mm) of the heater. Operation of such switch shall not change the setting of the heater thermostat. Such switches shall be in addition to a circuit breaker for the power to the heater. Gas-fired heaters shall not be equipped with continuously burning ignition pilots.

❖ All heaters must have an accessible on-off switch that shuts off the heater without adjusting the thermostat setting because the thermostat may be broken and the wrong temperature could cause injury to personnel. Heaters fired by natural gas or liquefied petroleum gas (LP-gas) must not have continuously burning pilot lights because pilot lights are an energy-wasting technology. The accessible on-off switch allows the heater to be completely turned off where there are periods when the heat is not needed or when the pool and spa may not be used.

C404.9.2 Time switches. Time switches or other control methods that can automatically turn off and on heaters and pump motors according to a preset schedule shall be installed for heaters and pump motors. Heaters and pump motors that have built-in time switches shall be in compliance with this section.

Exceptions:

1. Where public health standards require 24-hour pump operation.
2. Pumps that operate solar- and waste-heat-recovery pool heating systems.

❖ Time switches or other control methods provide an easy system for pool operations and energy savings. The application of Exception 1 is dependent on the requirements of the local health department in the jurisdiction. Because the pools are often in public areas, the health department may require continuous filtering or circulation. Exception 2 grants a credit for using other systems that help the pool and the inground, permanently installed spa operate more efficiently. Therefore, when solar- and waste-heat recovery systems are used to heat the pool and spa, the exception eliminates the time-switch requirement.

C404.9.3 Covers. Outdoor heated pools and outdoor permanent spas shall be provided with a vapor-retardant cover or other *approved* vapor-retardant means.

Exception: Where more than 70 percent of the energy for heating, computed over an operating season, is from site-recovered energy such as from a heat pump or solar energy source, covers or other vapor-retardant means shall not be required.

❖ This section states outdoor heated pools and permanently installed spas are required to have a vapor-retardant cover or other means to retard evaporation from the pool or spa. Covers are not required to provide any minimum level of insulation value. The cover or other system simply will help hold some of the heat in, much like placing a lid on a pot, by not letting it dissipate into the air. An exception is provided to waive the cover or other vapor-retardant means where more than 70 percent of the heating energy for the operating season is derived from site-recovered energy, such as a heat pump or solar-energy source.

C404.10 Energy consumption of portable spas (Mandatory). The energy consumption of electric-powered portable spas shall be controlled by the requirements of APSP 14.

❖ APSP 14, *American National Standard for Portable Electric Spa Energy Efficiency,* is a standard promulgated by the Association of Pool and Spa Professionals. The standard addresses water heating equipment, pump motors, and covers.

C404.11 Service water-heating system commissioning and completion requirements. *Service water-heating systems,* swimming pool water-heating systems, spa water-heating systems and the controls for those systems shall be commissioned and completed in accordance with Section C408.2.

❖ These provisions are intended to ensure that the building has been "tuned" prior to occupancy to make sure it is properly operating and capable of continuing to operate properly. Many hot water systems have recirculation or heat trace systems that need to be checked to verify that time or other controls are in place to avoid excessive unoccupied piping heat loss. This extends the value and validity of the code provisions because there is little value in requiring something to be provided in a building if it is not properly installed and ready to perform its intended function.

SECTION C405
ELECTRICAL POWER AND LIGHTING SYSTEMS

C405.1 General (Mandatory). This section covers lighting system controls, the maximum lighting power for interior and exterior applications and electrical energy consumption.

Exception: Dwelling units within commercial buildings shall not be required to comply with Sections C405.2 through C405.5, provided that they comply with Section R404.1.

❖ Section C405 contains requirements for electric power and lighting systems. Most of this section addresses electric lighting systems; however, it is worth noting that this section also contains significant new requirements that apply to electrical distribution systems, motors, and vertical and horizontal transportation systems. Section C405 is organized as follows:

- Section C405.2 addresses minimum requirements for both interior and exterior lighting controls.
- Section C405.3 describes minimum efficiency requirements for exit signs.
- Sections C405.4 and C405.5 contain requirements for interior and exterior lighting power.
- Section C405.6 requires individual submetering of most dwelling units located in commercial buildings.
- Section C405.7 describes minimum efficiency standards for transformers.
- Section C405.8 describes minimum efficiency standards for motors.
- Section C405.9 includes a number of requirements related to vertical and horizontal transportation systems.

If the energy-cost-budget method is used (see commentary, Section C407), the interior lighting power requirements of Sections C404.4 and C404.5 do not apply, but all other requirements of Section C405 must still be satisfied as applicable (see commentary, Section C401.2).

Electric lighting accounts for approximately 20 percent of all commercial building electricity consumption, accounting for about 7 percent of total electricity consumption in the United States. Electric lighting also adds heat to building spaces that must be removed by the cooling system. As a rule of thumb, an air conditioner will use an additional 15 to 20 kW (51 193 to 68 257 Btu/h) to remove the heat produced by each 100 kW (341 287.4 Btu/h) of installed lighting power during the cooling season.

The code utilizes two strategies to encourage energy-efficient lighting systems: first, limit the total overall wattage of lighting that can be installed, and second, reduce the utilization of that lighting with automatic control systems. Both strategies are equally important, as a 30-percent reduction in lighting power saves exactly as much energy as a 30-percent reduction in operating hours.

Significant improvement in lighting source efficacy (lumens per watt) is resulting in much lower installed lighting power (watts) to achieve the same light levels. Light-emitting diodes (LEDs) are responsible for much of this improvement, and LED lamps that will use 80 percent less energy than the incandescent lamps they are intended to replace are widely available. LED sources have also achieved significant market penetration in replacing compact fluorescent, and are on the cusp of replacing most linear fluorescent and high intensity discharge (metal halide and high pressure sodium) sources as well.

Automatic controls required by this code fall into three general categories. Occupant sensors are intended to automatically detect room occupancy, and turn lights off when rooms are unoccupied. Time-switch systems are not capable of detecting whether occupants are present, but will automatically turn lights off at times when occupants are not expected to be present. Occupant sensors save more energy than time-switch systems, and when properly configured should be less of an inconvenience for building occupants. Daylight responsive controls automatically turn off or dim lights near windows and skylights when sufficient daylight is present so that the electric lights are not needed.

A primary goal in most building designs is to minimize initial construction costs. Consequently, without codes or incentives, inefficient and ineffective lighting systems are often installed to lower first costs. This approach is short sighted as the long-term operational costs of bad lighting systems are greater than the initial construction cost. To cite just one example, a 100W incandescent light fixture operated 12 hours a day, 5 days a week in a commercial building will use $600 of electricity over 20 years (at 10 cents per kWh), and cost $600 to maintain (at $10 per re-lamping). Even more important are the indirect costs of bad lighting: occupant complaints, reduced sales, increased absenteeism, slower performance, poor work accuracy, and decreased safety.

The code encourages the use of energy-efficient lighting equipment in design practice by assigning lighting power allowances for interior and exterior lighting systems. The project complies with the code when the total installed power (watts) of new lighting is less than the interior lighting power allowance, and the total installed power (watts) of exterior lighting is less than the exterior lighting power allowance. This approach promotes design flexibility while ensuring a minimum level of efficiency.

As previously noted, lighting accounts for about 20 percent of all commercial building electricity consumption in the United States. Lighting also is expensive, particularly in areas where utilities assess peak demand charges for afternoon electricity use. For example, the cost of lighting a 10,000-square-foot (929 m^2) office building with connected lighting power of 10,000 watts (1 W/ft^2) ranges from $1,000 to more than $4,000 per year (you would rarely get much above 1 W/sf in an office building complying with the this code, and operating hours would be reduced to 8-10 per day by automatic controls), not including air-conditioning and environmental costs. When cost analysis is used to assess equipment options in new buildings, lighting measures usually exceed all other types of building improvements, showing rapid (often less than 1 year) payback and impressive life-cycle cost savings.

The cost of lighting may be loosely defined as:

Lighting Cost = Lighting Power (kW) × Time of Use (hour) × Average Cost of Electricity ($/kWh).

In many locations the utility will also apply a demand charge ($ per kW).

In addition to saving energy, compliance with the lighting requirements in the code encourages reduced lighting costs in commercial buildings.

The lighting system requirements in this code apply to almost all new lighting installations regardless of whether the project is new construction or an alteration to an existing building. The only exception is where less than 50 percent of the luminaires in a space are being replaced (see Section C503.1, Exception 7), or where the entire energy code is not applicable (see Chapter 1 [CE]). For existing buildings, where luminaires are being replaced or added, Sections C405.4 and C405.5 must be complied with, and where lighting controls are being altered in existing buildings, Section C405.2 must be complied with. Alteration of lighting control systems is not a trigger for applicability of Sections C405.4 and C405.5, just as luminaire replacement does not trigger applicability of Section C405.2.

For the most part, applying Section C405 is relatively straightforward. However, there are a number of specific cases pertaining to lighting systems where additional information may be helpful in interpreting code requirements.

- **Exterior and interior lighting.** The code contains separate requirements for exterior and interior lighting systems. Exterior and interior lighting must comply separately with their respective requirements. Trade-offs between the two are not allowed.
- **Interior lighting power.** The code includes two different approaches to determine compliance with the interior lighting power requirement: the Building Area Method, and the Space-By-Space Method. One method must be chosen for each project and used to show compliance for the entire project; a project cannot use different methods for different components. However, different projects in the same building that are filed separately can use different compliance paths.
- **Dwelling Units.** There are two different ways of determining compliance for lighting in dwelling units in commercial buildings. Dwelling units that comply with the provisions of Section R404.1 are exempt from the requirements of Sections C405.2 through C405.5. These dwelling units would not be counted in the area used to determine the allowed interior and exterior lighting power for the rest of the building, and the lighting in these units would not be included in the calculation of total installed lighting power for the rest of the building. Dwelling units that do not comply with Section R404.1 must be included in the lighting power calculations required by Sections C405.4 and C405.5. Since there is no space-type category for dwelling units in Table C405.4.2(2), these projects must use the Building Area Method.
- **Lighting in multiple-building facilities or campuses.** Each building in campus-like facilities must comply separately with the interior lighting power requirements, even if they are covered under a single building permit. The exterior lighting power requirement, however, applies for the entire site.
- **Lighting in speculative buildings.** Speculative buildings often are built before the tenants are known. The initial building permit application usually includes just the shell and core with lighting installed only for common areas of the building, such as corridors, toilets, stairwells and lobbies. Lighting for tenant spaces is installed as part of the tenant improvements and often is customized for each tenant. Lighting systems included in each permit application must comply independently with all provisions of the code that is in effect at the time the permit is issued, and cannot rely on tradeoffs with systems that are not included in the same permit. For example, if lighting in core toilet rooms is

included in the base building permit, those toilet rooms must be provided with code-compliant lighting controls at the time they are built; they cannot rely on future tenants to install lighting control systems that will "bring them up to code."

- **Lighting in shell buildings.** Shell buildings are built before the building's use is known. The space could become light manufacturing, office, warehouse or any other use depending on the requirements of the tenant. In shell buildings, the lighting system rarely is installed before the space is leased. Leasing a building to a tenant effectively defines its use and allows for a determination of the interior lighting power allowance. When a developer wants to install a lighting system before the use is known, the interior lighting power allowance must be set based on discussions between the code official and the designer. In these situations it may be prudent to use a minimal lighting power density or that of the lowest anticipated possible uses.
- **Garages and parking areas.** A covered parking garage is treated as interior space and is included as part of the interior lighting power [for parking garages, see Table C405.4.2(1) in the row "Parking garage," or Table C405.4.2(2) in the row "Parking area, interior")]. Interior lighting power also would include covered areas of open parking garages. Uncovered areas (including rooftop parking) and outdoor parking areas, however, are treated as exterior lighting and would use Table C405.5.2(2).

Example of Application to First Tenant Spaces

The core and shell of a high-rise office was completed prior to the effective date of the code. The construction included the building envelope, the base HVAC system and lighting improvements for the common areas only. Lighting improvements for each tenant space will be made on a tenant-by-tenant basis when each space is leased.

Tenant spaces on two floors of the building remain empty and unimproved until they are leased a year later, after the code takes effect. At this time, the tenant files a permit application for the construction of a lighting system along with other tenant improvements. Does the code apply to the design of the lighting system?

Yes. The first tenant improvements in a building are considered new construction, and the lighting requirements of the code apply. In Section C405.2, the phrase "all areas in the building covered in this permit" would support this requirement.

Glossary of Lighting Terms used in Section C405.

A glossary of terms is provided for the benefit of the user of this commentary.

Floor area or building area. (As used to determine interior lighting power for Section C405.) These terms describe the area of the building measured to inside wall surfaces. This area is used to calculate the interior lighting power allowed for the building.

Lamp. The term "lamp" is used to describe a device that creates light. The most common term for this is the "light bulb." Common lamp types include:

- Light Emitting Diode (LED).
- Incandescent (including halogen, krypton, and xenon).
- Fluorescent (including compact fluorescent).
- Metal halide.
- High-pressure sodium.
- Low-pressure sodium.
- Specialty sources, such as neon, carbon, arc, xenon strobe, etc.

Lighting power density (LPD). LPD is the maximum watts per square foot of lighting for a general building type or a specific building space, function or activity as found in Tables C405.4.2(1) and C405.4.2(2).

Luminaire. The luminaire is the entire light fixture, including the lamp, the ballast (if any) and the housing, designed to distribute the light and protect the lamps. A luminaire will typically bear a label from a testing agency such as UL indicating that it meets electrical safety requirements.

C405.2 Lighting controls (Mandatory). Lighting systems shall be provided with controls as specified in Sections C405.2.1, C405.2.2, C405.2.3, C405.2.4 and C405.2.5.

Exceptions: Lighting controls are not required for the following:

1. Areas designated as security or emergency areas that are required to be continuously lighted.
2. Interior exit stairways, interior exit ramps and exit passageways.
3. Emergency egress lighting that is normally off.

❖ Lighting systems, must have controls that comply with this section. Acceptable control devices include manual switches and dimmers, occupant sensor controls, timeswitch controls, daylight responsive controls, and preset dimming systems. Circuit breakers, fuses and other devices not intended or suited for lighting control do not satisfy this requirement. The exceptions are worth considering individually.

Exception 1 refers to "areas" of the building that require continuous lighting for security or emergencies, not to specific light fixtures. This exception is not terribly specific, but should certainly be considered to be applicable where a building code, zoning requirement, or other life safety ordinance such as a health code, banking code, etc., specifically states that lighting cannot be turned off for safety reasons. It would also be applicable where a lighting control system malfunction could reasonably be considered to pose a risk of immediate harm to building occupants if the lighting were suddenly shut off: a hospital operating room during surgery, for example, or certain areas of prisons. Finally, there are certain spaces such as 911 call centers, hospital emergency rooms and certain transportation facilities that are continuously occupied and where lights should never be turned off.

Exception 2 uses language that has a very specific meaning in most building codes. In jurisdictions that have adopted the IBC, Chapter 10 defines different elements of the means of egress.

Exception 3 refers to emergency lighting equipment that is normally off, but incorporates an integral transfer switch that automatically turns the light "on" (usually from an integral battery pack) when the power feed to the fixture is lost.

Lighting control requirements are established to reduce energy use by automatically turning lights off when they are not needed, and by allowing building occupants to turn off lights when occupants are present but do not require that the lights be turned on.

Although there are many unique approaches to lighting controls, including task tuning, lumen maintenance, demand response load shedding, and others, the code primarily relies on occupant sensors, time-switch controls, and daylight responsive controls as the most proven and robust strategies for achieving energy savings.

C405.2.1 Occupant sensor controls. Occupant *sensor controls* shall be installed to control lights in the following space types:

1. Classrooms/lecture/training rooms.
2. Conference/meeting/multipurpose rooms.
3. Copy/print rooms.
4. Lounges.
5. Employee lunch and break rooms.
6. Private offices.
7. Restrooms.
8. Storage rooms.
9. Janitorial closets.
10. Locker rooms.
11. Other spaces 300 square feet (28 m^2) or less that are enclosed by floor-to-ceiling height partitions.
12. Warehouses.

❖ Occupant sensors (more commonly referred to as "occupancy" or "motion" sensors) are automatic control devices that detect occupancy and adjust lights in response to the presence or absence of people in the space. The vast majority of these devices work by sensing motion, but some utilize "microphonic" technology that listens for occupant noises rather than motion. The most common technologies are:

- Passive infrared (PIR) sensors, which perceive and respond to the heat patterns of motion. The body-heat patterns of humans in the infrared range can be detected fairly easily relative to other forms of heat. This same technology is used in most residential and commercial security systems. PIR sensors often use a dual-element sensing device so that when one element senses infrared energy before the other, the sensor assumes occupancy and is turned to the occupied state. The chief advantage of PIR sensors is that they are relatively inexpensive and reliable. They very rarely false trigger by responding to nonoccupant motion in a space. The major limitation to PIR sensors is that they are strictly line-of-sight devices. They cannot "see" around corners or over partitions. The sensitivity of these devices also drops significantly with distance so that they may be good at detecting "large" motion (such as walking) at greater distances, and "small" motion (such as typing) at close distances. Because of this, they are best in smaller spaces without partitions, or in large spaces where major motion is common (like warehouses and gymnasiums). Two important advantages of PIR technology are that it can be used outdoors and in cold spaces like walk-in freezers, and the "coverage pattern" can be limited by covering part of the lens, so that it does not detect motion in adjacent areas outside the intended coverage pattern.
- Ultrasonic sensors (US), which radiate ultrasonic waves into a space, then read the frequency of the reflected waves. Motion causes a slight shift in frequency, which the detector interprets as occupancy. US use a quartz crystal oscillator to generate an inaudible signal. The ultrasonic signal is broadcast into a space, bounces off surfaces and returns to the sensor's receiver. Changes in the signal's return time indicate occupancy, US are typically more expensive than PIR but offer greater coverage and have the ability to "see" behind partitions, which is both an advantage and a disadvantage. However, the increased sensitivity means that they are more susceptible to false triggering from such things as HVAC systems, open windows and people walking by open doors to controlled spaces. They often are used effectively in partitioned spaces, but are more prone to false triggering because of their sensitivity to air movement. Proper design and installation minimizes this potential problem. US are best used in bathrooms, office spaces with partitions, larger open areas, etc.
- Dual-technology (DT) sensors, which integrate both PIR and US technology into one package. Typically, these are designed to avoid false triggering by holding the lights off unless both detectors sense motion in the space. DT sensors typically use both a passive infrared and an ultrasonic technology. They offer the combined strengths of both technologies without

the weaknesses of either. In addition, the control options usually are configurable by the user. For example, the unit may be set so that both technologies are required to trigger "occupied" and then either technology would be required to trigger "hold on." DT sensors are especially well suited for classrooms or spaces that are difficult to cover with a single technology.

- Microwave sensors, which are similar to ultrasonic in that a signal is generated and moving objects will cause a shift in the frequency of the return signal. Microwave sensors primarily are limited to the security and alarm industries.

The code does not specify which technologies are acceptable.

Most devices can be calibrated for length-of-time delay between the last detected occupancy and extinguishing of the lights. The code specifically requires that this time delay cannot be longer than 30 minutes, and to comply with this requirement it is recommended that all installed sensors have a maximum 30-minute time delay. The shorter the time delay setting, the greater the energy savings, as the lights will spend more time "off." If occupants complain about the lights shutting off in occupied spaces with a 30-minute time delay, then the occupant sensor is not working properly and must be reconfigured, relocated, or replaced. Some devices include a "sensitivity" setting, but most now incorporate some sort of automatic self-calibration mechanism instead of a manual sensitivity setting.

Occupant sensors save energy by turning lights off automatically. Turning lights on automatically never saves energy, because there are many instances where a user would not choose to turn lights on. For this reason, full automatic-on is only permitted in

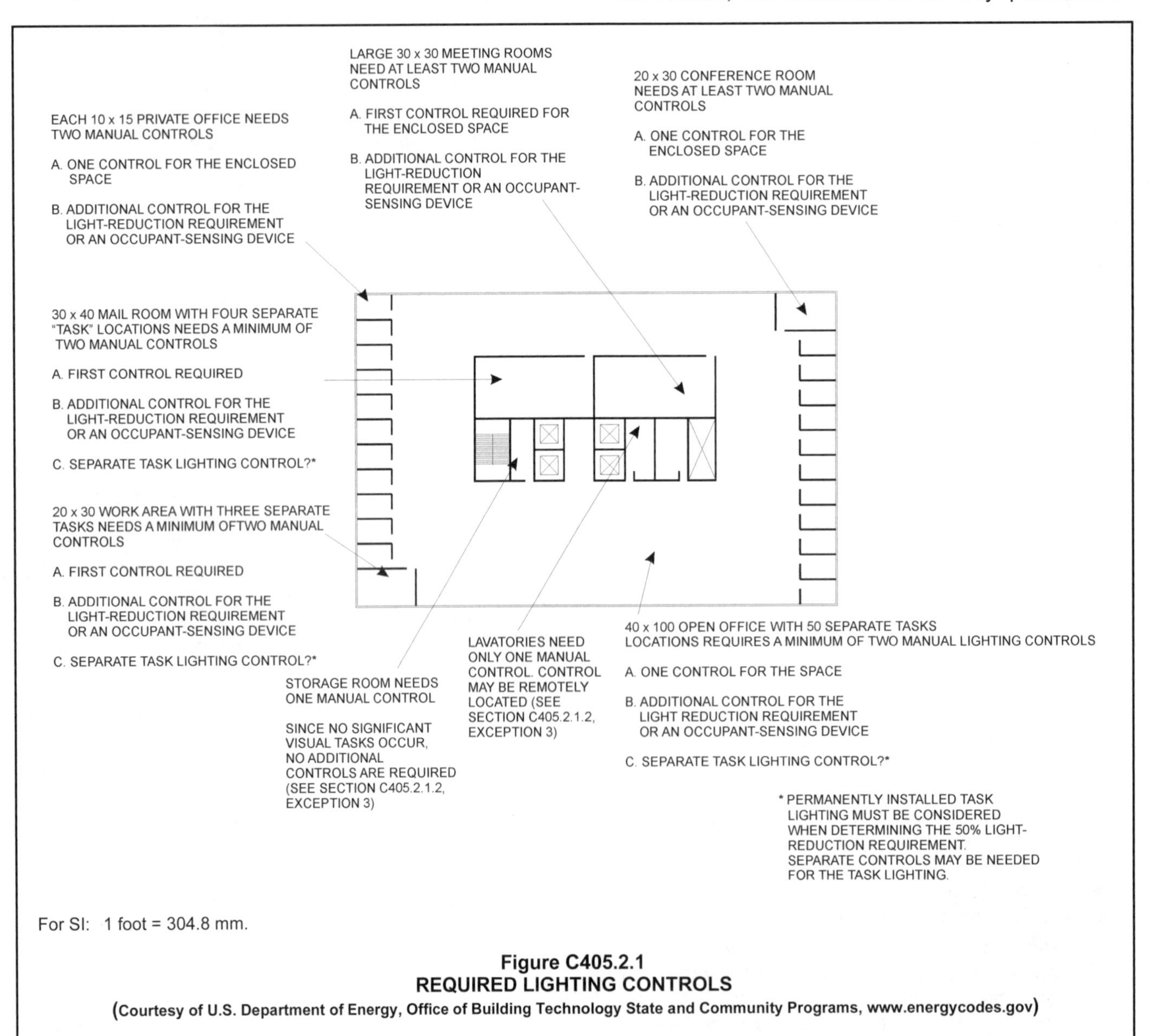

Figure C405.2.1
REQUIRED LIGHTING CONTROLS
(Courtesy of U.S. Department of Energy, Office of Building Technology State and Community Programs, www.energycodes.gov)

"public corridors, stairways, restrooms, primary building entrance areas and lobbies, and areas where manual-on operation would endanger the safety or security of the room or building occupants" (see the exception to Section C405.2.1.1). Remember that even in these spaces, an accessible manual switch must still be provided to allow occupants to turn lights off if they choose to.

In all other spaces, occupant sensors must be configured to turn on automatically to no more than 50-percent power, and an accessible switch must be provided to bring lights to full output. The simplest way to achieve this is with a "combination" occupant sensor/wall switch that controls all lights in "manual-on, automatic-off mode." This is also sometimes referred to as a "vacancy sensor." However, in some larger rooms the wall switch may need to be located separately from the occupant sensing device, or several occupant sensing devices may need to be connected together to provide coverage for the entire room. Occupant sensor systems typically consist of a motion detector, a control unit and a relay. Usually, two or more of the components are integrated into one package. Many systems also require a power supply in the form of a transformer that steps down the building voltage to 24 volts. The detector collects information and sends it to the controller, where it is processed. Output from the controller activates the relay, which in turn switches the light circuit.

It is difficult to generalize about the amount of lighting savings that is attributable to the use of occupant sensors. Applications differ widely, and actual savings depend on occupancy patterns, lighting schedules, employee habits and many other factors. Sensors are most effective in building spaces where occupancy is sporadic or unpredictable and in spaces, such as storage areas, where the lights are likely to be left on inadvertently. Typical savings range from 10 percent in large open offices to 60 percent in some warehouse applications.

In fact, a pilot study was done at a federal site in Boulder, Colorado, involving occupant sensors and desktop dimmers. Offices with just occupant sensors averaged a 50-percent reduction in lighting energy used during normal operating hours. Offices that also had manual dimmers averaged a 75-percent reduction.

Despite the unpredictability of the savings obtained with occupant-sensor lighting control, it is a control strategy that consistently pays off in terms of its cost reduction. In many cases, occupant sensors pay for themselves in less than a year (see Commentary Figure C405.2.1).

C405.2.1.1 Occupant sensor control function. Occupant sensor controls in spaces other than warehouses specified in Section C405.2.1 shall comply with the following:

1. Automatically turn off lights within 30 minutes of all occupants leaving the space.
2. Be manual on or controlled to automatically turn the lighting on to not more than 50 percent power.

   Exception: Full automatic-on controls shall be permitted to control lighting in public corridors, stairways, restrooms, primary building entrance areas and lobbies, and areas where manual-on operation would endanger the safety or security of the room or building occupants.
3. Shall incorporate a *manual control* to allow occupants to turn lights off.

❖ See the commentary to Section C405.2.1.

C405.2.1.2 Occupant sensor control function in warehouses. In warehouses, the lighting in aisleways and open areas shall be controlled with occupant sensors that automatically reduce lighting power by not less than 50 percent when the areas are unoccupied. The occupant sensors shall control lighting in each aisleway independently and shall not control lighting beyond the aisleway being controlled by the sensor.

❖ See the commentary to Section C405.2.1.

C405.2.2 Time-switch controls. Each area of the building that is not provided with *occupant sensor controls* complying with Section C405.2.1.1 shall be provided with *time switch controls* complying with Section C405.2.2.1.

Exception: Where a *manual control* provides light reduction in accordance with Section C405.2.2.2, automatic controls shall not be required for the following:

1. *Sleeping units.*
2. Spaces where patient care is directly provided.
3. Spaces where an automatic shutoff would endanger occupant safety or security.
4. Lighting intended for continuous operation.
5. Shop and laboratory classrooms.

❖ Time-switch controls use a time-scheduling device to switch lighting systems on or off according to predetermined schedules. This is the oldest form of automatic lighting control, and in most cases the least efficient. For this reason, time-switch control systems are only required in spaces where occupant sensors are not provided. In many smaller projects, the simplest, most efficient, and lowest-cost approach to compliance with this code will simply be to provide manual-on, automatic-off occupant sensors in all spaces. However, time-switch controls can make sense in spaces with tightly controlled occupancy schedules (like circulation areas in schools), or where occupant sensor controls would be prohibitively expensive (like "big box" retailers, airports, or shopping mall arcades). Time scheduling of interior lighting systems is, for the most part, based on occupancy schedules. In some cases, time-switch controls are used to energize additional lighting control systems, such as daylighting responsive controls, which are held off during unoccupied periods. It is worth remembering that time-switch controls save energy when they turn lights off automatically, not when they turn lights on automatically. The code requires shutoff, but does not require turn-on. In fact, it is best in most cases to let space occupants turn lights on when they arrive in the morning, and allowing the time-switch system to be used to provide only automatic-off in the evening.

Time-switch controls include the following components:

- A central processor, usually capable of controlling several output channels, each of which may be assigned to one or more lighting circuits.
- Relays, series wired to lighting control zones and controlled by the central processor.
- Override switches that are required to accommodate individuals who use the space during scheduled off hours. Individuals who use the space during scheduled off hours can activate manual switches or telephone overrides to regain temporary control of the lights in a given space.

In most cases, Class II (low-voltage) wiring links all the components in the system, and the system uses a flashing or "quick flicker" warning system to let individuals know that the lights are going off. This allows occupants to either vacate the space or activate an override to keep the lights on.

The crucial component in any time-scheduling system is the programmable central processor, which is essentially a multiple-circuit controller. The central processor can be programmed by building maintenance personnel to schedule on and off loads on each of its output channels. If desired, several different on-off sequences may be programmed on each channel.

A central processor typically consists of the following components:

- A programmable microprocessor with an electronic clock that is capable of separately scheduling weekday, weekend and holiday operations. Astronomical time-keeping ability means the processor is able to make seasonal and daylight savings adjustments. Typically, the processor has a built-in battery backup so that the programmed schedule remains in memory during power outages. The processor usually is able to "sweep" at regular intervals during its off hours. The processor remembers when overrides have been used to keep lights on in any particular area and then will repeat the operation to turn off the lights.
- Switch inputs allow occupants to override the shutoff function of the processor. Usually the switches and wiring to the controller are low voltage. Inputs also may be wired to photocells or occupancy sensors for additional flexibility.
- Output channels are required for each lighting control zone. Sophisticated designs sometimes include two or more outputs for each control zone to allow for stepped control of the zone. In some systems, output channels can be designed to provide a variable signal, allowing for dimming applications. These are a considerable expense, however, because each dimming channel requires a digital-to-analog signal converter.

Generally, time scheduling is most effective when occupancy patterns are relatively regular or when lighting operating hours are easy to predict.

C405.2.2.1 Time-switch control function. Each space provided with *time-switch controls* shall also be provided with a *manual control* for light reduction in accordance with Section C405.2.2.2. Time-switch *controls* shall include an override switching device that complies with the following:

1. Have a minimum 7-day clock.
2. Be capable of being set for seven different day types per week.
3. Incorporate an automatic holiday "shutoff" feature, which turns off all controlled lighting loads for at least 24 hours and then resumes normally scheduled operations.
4. Have program backup capabilities, which prevent the loss of program and time settings for at least 10 hours, if power is interrupted.
5. Include an override switch that complies with the following:
   - 5.1. The override switch shall be a manual control.
   - 5.2. The override switch, when initiated, shall permit the controlled lighting to remain on for not more than 2 hours.
   - 5.3. Any individual override switch shall control the lighting for an area not larger than 5,000 square feet (465 m^2).

Exceptions:

1. Within malls, arcades, auditoriums, single-tenant retail spaces, industrial facilities and arenas:
   - 1.1. The time limit shall be permitted to be greater than 2 hours, provided that the override switch is a captive key device.
   - 1.2. The area controlled by the override switch is permitted to be greater than 5,000 square feet (465 m^2), but shall not be greater than 20,000 square feet (1860 m^2).
2. Where provided with *manual control*, the following areas are not required to have light reduction control:
   - 2.1. Spaces that have only one luminaire with a rated power of less than 100 watts.

2.2. Spaces that use less than 0.6 watts per square foot (6.5 W/m^2).

2.3. Corridors, equipment rooms, public lobbies, electrical or mechanical rooms.

❖ See the commentary to Section 405.2.2.

C405.2.2.2 Light-reduction controls. Spaces required to have light-reduction controls shall have a *manual control* that allows the occupant to reduce the connected lighting load in a reasonably uniform illumination pattern by at least 50 percent. Lighting reduction shall be achieved by one of the following or another *approved* method:

1. Controlling all lamps or luminaires.
2. Dual switching of alternate rows of luminaires, alternate luminaires or alternate lamps.
3. Switching the middle lamp luminaires independently of the outer lamps.
4. Switching each luminaire or each lamp.

Exception: Light reduction controls are not required in *daylight zones* with *daylight responsive controls* complying with Section C405.2.3.

❖ Light-reduction controls, also known as "two-level switching" or "bi-level switching", save energy by allowing occupants to reduce light levels. Light reduction controls are required in most spaces where lights are controlled by a time-switch system, but are never required in spaces with occupant sensors, or for lights connected to daylight responsive controls. The exceptions to the light-reduction control requirement are found in Section C405.2.2.1 (Exceptions 2.1, 2.2, and 2.3).

For the most part, the exceptions are self-evident in that they recognize the rather modest opportunity to save lighting energy in spaces used as a whole and other unique, infrequently used spaces. Section C405.4.2 and the table that goes with it establish that parking garages or other spaces with a low power density would be exempt from the lighting-reduction requirement. These types of areas would have little energy conservation benefit from the lighting-reduction controls because of the low amount of illumination in use.

The more uniform the reduced light levels are, the more likely it is that these controls will actually be used to reduce light levels and save energy. Therefore, Method 1, "Controlling all lamps or luminaires," is the best choice, and merits a little further consideration, as there are two ways of accomplishing this:

1. Make all lights dimmable, and connect them to a dimmer. Note that dimming is not universally available for all light sources, but is available for most LED and fluorescent sources, at different cost points and performance levels. The lowest cost option will often be 0- to 10-volt dimming, and this will often dim the light source to 10 to 20 percent of full light output. More expensive dimming options will usually reduce light levels to 1 percent or lower of full light output. Some LED products are now available that will dim to 0.1 percent of full light output, which effectively matches the dimming range of incandescent, but at a higher cost. For the purposes of this code, dimming to 50 percent of maximum power is required.
2. Utilize bi-level switching ballasts or drivers. Most linear fluorescent luminaires can be provided with ballasts that switch to 50 percent of maximum power. Most LED luminaires that are intended to compete with linear fluorescent are also available with drivers that offer the same performance. These products are often less expensive than dimming, and work using a simpler protocol, where each luminaire receives one neutral wire and two switched hot legs. When both legs are hot, the luminaire operates at 100 percent power. When one leg is hot, the luminaire operates at 50 percent power. And when both leg are not hot, the luminaire is off (see Commentary Figure C405.2.2.2).

C405.2.2.3 Manual controls. *Manual controls* for lights shall comply with the following:

1. Shall be readily accessible to occupants.
2. Shall be located where the controlled lights are visible, or shall identify the area served by the lights and indicate their status.

❖ This section provides requirements for location of manual controls for lighting. It does not create any requirements for the location of occupant sensing devices, the central microprocessor in a time-switch system, or any other automatic lighting control devices; rather, it creates a requirement for the location of any manual controls that must be accessed by occupants for proper operation of these systems. For example, consider a multipurpose room with four ceiling-mounted occupant sensors and two manual switches, all wired to a switchpack. The ceiling-mounted occupant sensors and the switchpack can be located wherever is most appropriate, but the location of the two manual switches is specified by this section of the code.

Proper location of manual controls is critical to their success in saving energy. If controls are all banked together in one central location, unlabeled, and where the lights being controlled cannot be seen, then frustrated users will simply turn all of the lights on, even when they only need some of the lights.

The code addresses this with a simple set of requirements. When using this section, it is important to remember that the term "readily accessible" is defined in the code and has a very specific meaning.

C405.2.3 Daylight-responsive controls. *Daylight-responsive controls* complying with Section C405.2.3.1 shall be provided to control the electric lights within *daylight zones* in the following spaces:

1. Spaces with a total of more than 150 watts of *general lighting* within sidelight *daylight zones* complying with Section C405.2.3.2. *General lighting* does not include lighting that is required to have specific application control in accordance with Section C405.2.4.
2. Spaces with a total of more than 150 watts of *general lighting* within toplight *daylight zones* complying with Section C405.2.3.3.

Exceptions: Daylight responsive controls are not required for the following:

1. Spaces in health care facilities where patient care is directly provided.
2. Dwelling units and sleeping units.
3. Lighting that is required to have specific application control in accordance with Section C405.2.4.
4. Sidelight daylight zones on the first floor above grade in Group A-2 and Group M occupancies.

❖ The code includes an expanded requirement that most lights installed in areas with regular access to daylight be provided with automatic controls to dim or switch those lights off when sufficient daylight is present. The actual potential for energy savings from daylight responsive controls will vary dramatically based on the building. Some buildings receive ample daylight in most interior spaces, and other buildings have very limited access to daylight. However, on average, buildings that incorporate daylighting strategies are estimated to see a 28-percent lighting energy savings.

This section has been crafted to require daylight responsive controls based on where they will achieve the most meaningful energy savings. Hence, the code has the following requirements to ensure sufficient energy savings where automatic controls are required:

1. Daylight responsive control are required only for lights located in daylight zones, and the description of these daylight zones only includes areas that receive ample daylight.
2. Daylight responsive controls are required only where there is at least 150 watts of lighting to be controlled in the daylight zone.

When interpreting this section it is important to remember that "general lighting" is a defined term in the code with a specific meaning, and does not include decorative or accent lighting. The exceptions to these requirements are worth considering in more detail:

1. Light has a direct biological impact on human beings, through regulation of melatonin production and other pathways. However, the biological effects of light only become apparent at light levels that are much higher than those required for visual acuity so healthcare facilities are excluded.

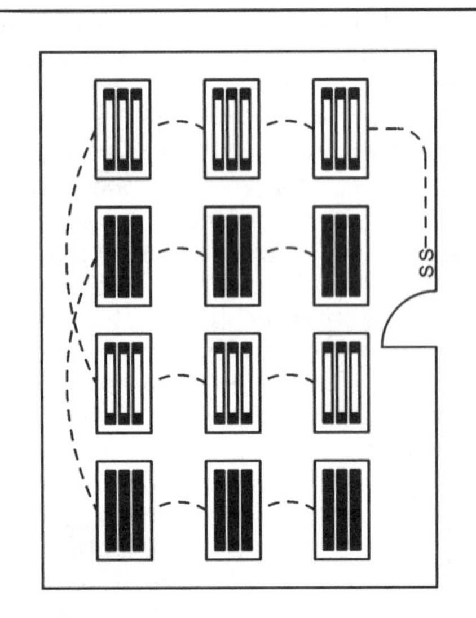

ALTERNATE ROW SWITCHING

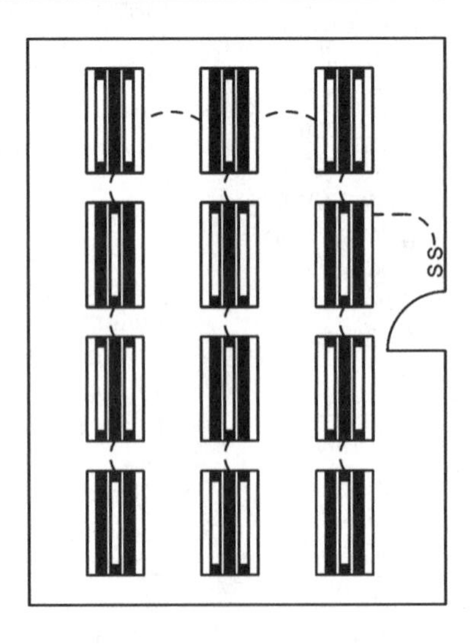

ALTERNATE LAMP SWITCHING

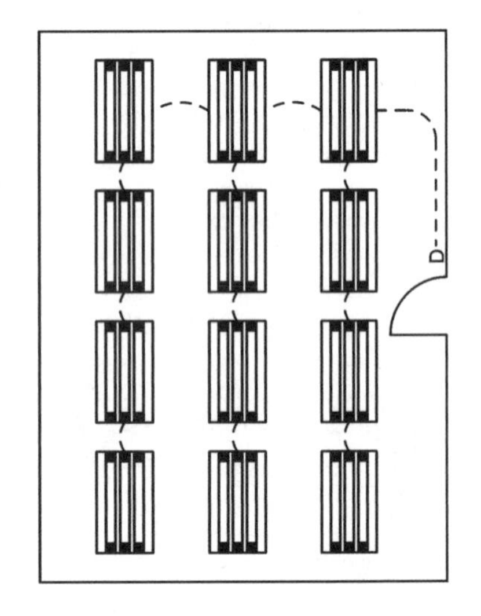

ALTERNATE DIMMING CONTROL

SS = SEPARATE SWITCHING
D = DIMMING CONTROL

Figure C405.2.2.2
LIGHT-REDUCTION CONTROLS
(Courtesy of U.S. Department of Energy, Office of Building Technology State and Community Programs, www.energycodes.gov)

2. Dwelling units and sleeping units conventionally are provided with manual lighting controls, and residents in these settings are likely to use those controls to turn lights off when they are not needed.

3. Certain specific lighting applications are already provided with separate controls. Connecting these lights to daylight responsive controls in addition to the special application controls would be difficult, and in some cases (e.g., plant survival lighting) would be contrary to the purpose and function of the special application lighting.

4. Per the 2015 IBC, Occupancy Group A-2 includes restaurants, and Occupancy Group M includes retail stores. Both of these building types often use lighting in windows to signify that they are "open for business," and switching these lights off during the day would defeat their primary purpose.

C405.2.3.1 Daylight-responsive control function. Where required, *daylight-responsive controls* shall be provided within each space for control of lights in that space and shall comply with all of the following:

1. Lights in toplight *daylight zones* in accordance with Section C405.2.3.3 shall be controlled independently of lights in sidelight *daylight zones* in accordance with Section C405.2.3.2.
2. *Daylight responsive controls* within each space shall be configured so that they can be calibrated from within that space by authorized personnel.
3. Calibration mechanisms shall be *readily accessible*.
4. Where located in offices, classrooms, laboratories and library reading rooms, *daylight responsive controls* shall dim lights continuously from full light output to 15 percent of full light output or lower.
5. *Daylight responsive controls* shall be capable of a complete shutoff of all controlled lights.
6. Lights in sidelight *daylight zones* in accordance with Section C405.2.3.2 facing different cardinal orientations [i.e., within 45 degrees (0.79 rad) of due north, east, south, west] shall be controlled independently of each other.

   Exception: Up to 150 watts of lighting in each space is permitted to be controlled together with lighting in a daylight zone facing a different cardinal orientation.

❖ This section describes how daylight responsive controls required by the code must function. Other daylight responsive controls that are installed in the project, but not required by code (i.e., "above code controls") do not need to comply with this section.

The goal of this section is to establish minimum standards for the required daylight controls, and avoid a lack of functionality that would result in the installed control system not working properly. This is not a design guide or a specification. For example, there is no mention in the code of important issues such as sensor location, setpoints, dead bands or response times, all of which are important considerations when designing and specifying daylight responsive controls. Further, there are many instances where it might make sense to exceed these requirements to achieve a system that functions better for the intended use (e.g., dimming to 5 percent instead of 15 percent). These requirements are worth considering in more detail:

1. The intensity and direction of daylight coming through skylights (i.e., "toplighting") is different than the intensity and direction of daylight coming through window (i.e., "sidelighting"). It would not be at all surprising if spaces adjacent to an east-facing window received the most daylight at 9:00 AM, while spaces underneath a rooftop skylight received the most daylight at 1:00 PM. Because of this lack of synchronicity, the lights in these two daylight zones must be separately controlled.

2. Calibration of daylight responsive controls always requires that light level readings be taken in the space with the controlled lights at the time that the calibration is done. Systems without this capability are highly unlikely to be properly commissioned.

3. The term "readily accessible" is defined in the code and has a very specific meaning. However, the readily accessible controls can be in the form of a hand-held remote control that is pointed at the photosensor on the ceiling to make adjustments. Other systems can be programmed from a hand-held smart phone or wifi-enabled laptop. These are examples of calibration mechanisms that would be permitted.

4. Occupants are more likely to accept daylight responsive controls if they are not conscious of lights abruptly being turned on and off. This is not a problem in transient spaces like corridors, lobbies, warehouses, or parking garages, but it is a problem in any space with fixed task locations. Therefore, continuous dimming is required in "offices, classrooms, laboratories, and library reading rooms." Dimming systems are more costly but they always provide superior performance to switching systems.

5. Research has shown that the energy savings of daylight responsive controls that dim lights is dramatically improved if those systems also switch the lights completely off when sufficient daylight is present. In other words, the electric lights would dim to 15 percent of full light output (or whatever their lowest setting is), and would switch off completely when sufficient daylight is present. A lower bound on the dimming range is

less likely to lead occupants to object to the lights being switched completely off.

6. This item is best explained by an example. In a room with windows facing east and west, the east side of the room will be brightest in the morning, and the west side will be brightest in the afternoon. In order for the daylight responsive controls to function properly, the lights in the east side of the room must be controlled separately from those in the west side of the room.

   An exception allows up to 150 watts of lighting in each space to be controlled together with lighting in a daylight zone facing a different cardinal orientation. The intention of this exception is to allow for lights in a corner room with less than 300 watts of lighting in daylight zones to be controlled by one (sensor) system. While the code is not specific that this applies to a corner, if this exception were applied to the space in the example cited above, with east- and west-facing fenestration, the results would not be satisfactory.

C405.2.3.2 Sidelight daylight zone. The sidelight *daylight zone* is the floor area adjacent to vertical *fenestration* which complies with all of the following:

1. Where the fenestration is located in a wall, the daylight zone shall extend laterally to the nearest full-height wall, or up to 1.0 times the height from the floor to the top of the fenestration, and longitudinally from the edge of the fenestration to the nearest full-height wall, or up to 2 feet (610 mm), whichever is less, as indicated in Figure C405.2.3.2(1).
2. Where the *fenestration* is located in a rooftop monitor, the *daylight zone* shall extend laterally to the nearest obstruction that is taller than 0.7 times the ceiling height, or up to 1.0 times the height from the floor to the bottom of the *fenestration*, whichever is less, and longitudinally from the edge of the *fenestration* to the nearest obstruction that is taller than 0.7 times the ceiling height, or up to 0.25 times the height from the floor to the bottom of the *fenestration*, whichever is less, as indicated in Figures C405.2.3.2(2) and C405.2.3.2(3).
3. The area of the *fenestration* is not less than 24 square feet (2.23 m^2).
4. The distance from the *fenestration* to any building or geological formation which would block access to daylight is greater than the height from the bottom of the *fenestration* to the top of the building or geologic formation.
5. Where located in existing buildings, the *visible transmittance* of the *fenestration* is not less than 0.20.

❖ There are several reasons why a user of the code might use this section:

1. Section C103.2 requires that the construction documents submitted for filing show the "location of daylight zones on floor plans." This is a very broad requirement that is intended to be applicable to all projects, assist in the layout of lighting controls to comply with the code, and make enforcement of the code easier for plan reviewers.
2. Section C402.4.1.1 allows certain prescriptive path envelope requirements to be relaxed if a specified minimum percentage of the floor area is in a daylight zone. In this case, designers would refer to this section to understand the criteria that must be satisfied to establish a daylight zone.
3. Section C405.2.3 requires that daylight responsive controls be provided for some electric lights installed in daylight zones. This exercise becomes quite simple once the daylight zones have been overlaid on the floor plans, as required by Section C103.2.

(a)

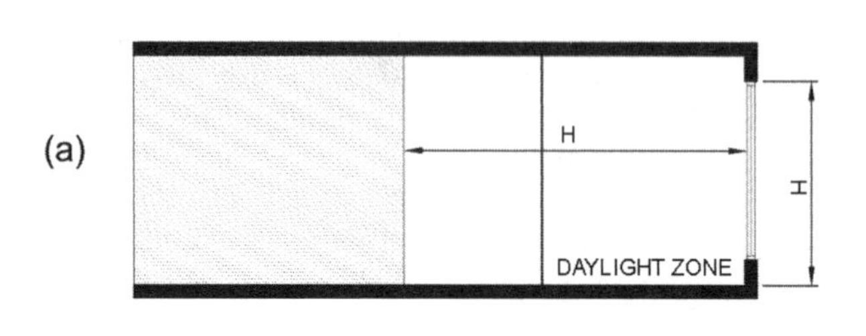

(a) Section view
(b) Plan view of daylight zone under a rooftop monitor

(b)

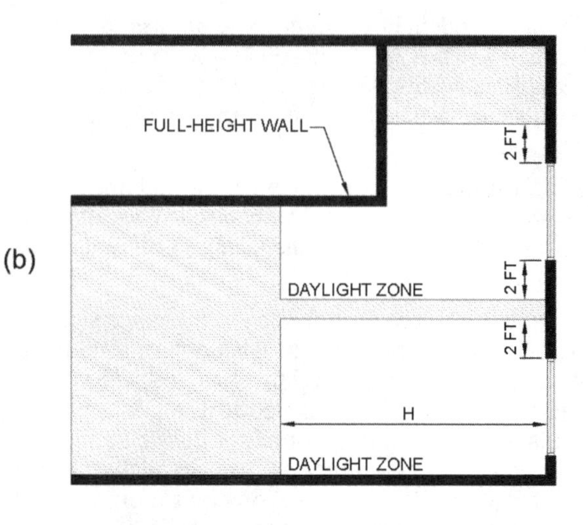

FIGURE C405.2.3.2(1)
DAYLIGHT ZONE ADJACENT TO FENESTRATION IN A WALL

The concept of "daylight zones" has existed in architecture since people first started designing buildings. A geometric relationship exists between the building floor area and the fenestration assembly. This relationship describes the potential for daylight to enter the building, and is most easily understood by looking at Figures C405.2.3.2(1), C405.2.3.2(2), C405.2.3.2(3) and C405.2.3.3. This concept is generalized for the code and is not intended as a design guide for good daylighting. The generalizations fail to consider:

1. The latitude of the building.
2. The orientation of the fenestration.
3. The climate at the building site.
4. How interior window treatments (e.g., blinds, shades, drapes, etc.) modify daylight penetration into a building.
5. The impact of daylight shading devices such as light shelves and screens.
6. The control of glare, which is critical to any successful daylighting installation.
7. The type of interior space, which often determines the quantity of light that is needed, and the tolerance for glare.

Nevertheless, this concept is quite robust and, once it has been used a few times, relatively easy to use and understand. It is worth remembering that "fenestration" consists of more than just windows: movable storefront systems, glass doors and curtain walls are other examples of fenestration systems that can transmit daylight and establish a daylight zone.

The daylight zone criteria described in the code include several important provisions intended to ensure that all daylight zones complying with the code actually have sufficient access to daylight.

1. The code establishes a minimum area for the fenestration of 24 square feet (2.23 m^2). All things being equal, there is a direct relationship between the size of a window and the amount

(a)

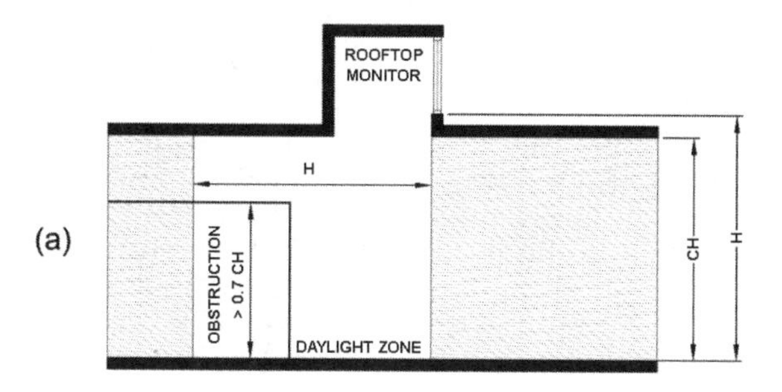

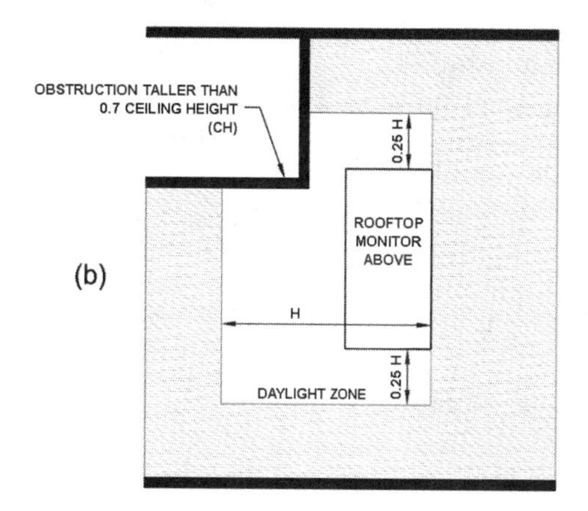

(b)

(a) Section view
(b) Plan view of daylight zone under a rooftop monitor

FIGURE C405.2.3.2(2)
DAYLIGHT ZONE UNDER A ROOFTOP MONITOR

(a)

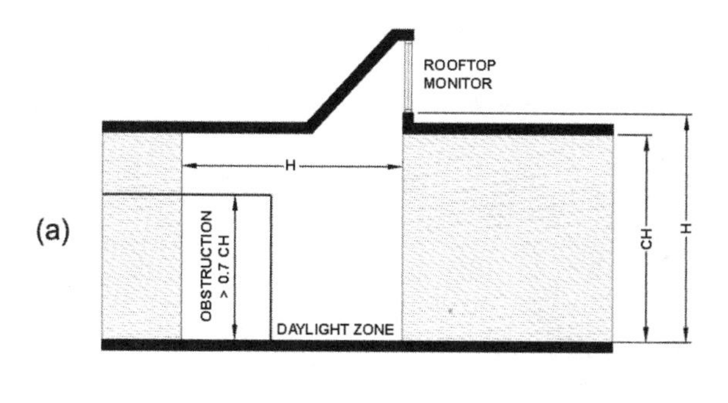

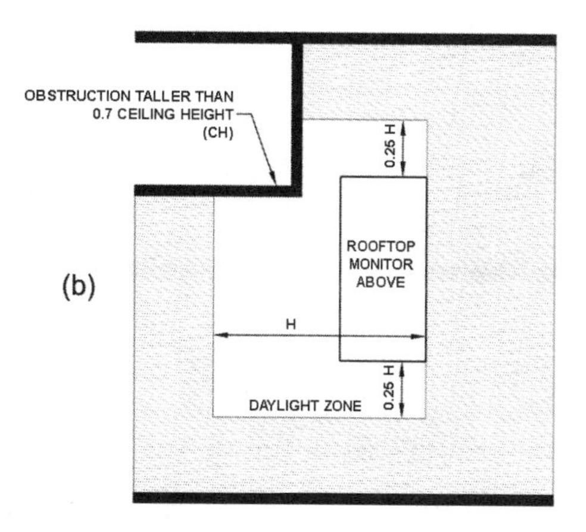

(b)

(a) Section view
(b) Plan view of daylight zone under a rooftop monitor

FIGURE C405.2.3.2(3)
DAYLIGHT ZONE UNDER A SLOPED ROOFTOP MONITOR

of daylight that will enter the building, and the window must be large enough to allow sufficient daylight to enter.

2. Per Item 4, "The distance from the fenestration to any building or geological formation that would block access to daylight is greater than the height from the bottom of the fenestration to the top of the building or geologic formation." In urban areas there can be significant overshadowing from adjacent buildings and structures that prevents windows from receiving enough daylight to effectively illuminate the building interior. Often these windows are provided for other reasons, such as view or ventilation, but do not provide abundant daylight.

3. Per Item 5, "Where located in existing buildings, the visible transmittance of the fenestration is not less than 0.20." Many older buildings have been constructed with extremely low-transmittance fenestration, which does not deliver sufficient daylight.

Once these criteria have been considered, projecting the daylight zones on the building plans is very straightforward.

C405.2.3.3 Toplight daylight zone. The toplight *daylight zone* is the floor area underneath a roof fenestration assembly which complies with all of the following:

1. The *daylight zone* shall extend laterally and longitudinally beyond the edge of the roof *fenestration* assembly to the nearest obstruction that is taller than 0.7 times the ceiling height, or up to 0.7 times the ceiling height, whichever is less, as indicated in Figure C405.2.3.3.
2. No building or geological formation blocks direct sunlight from hitting the roof *fenestration* assembly at the peak solar angle on the summer solstice.
3. Where located in existing buildings, the product of the *visible transmittance* of the roof *fenestration* assembly and the area of the rough opening of the roof *fenestration* assembly divided by the area of the *daylight zone* is not less than 0.008.

❖ The fenestration criteria for "toplight daylight zones" differ from those for "sidelight daylight zones." In addition to the general geometric criteria, the code requires:

1. "No building or geological formation blocks direct sunlight from hitting the roof fenestration assembly at the peak solar angle on the summer solstice." Outside of the tropics, the "peak solar angle on the summer solstice" would be the highest solar angle at any time throughout the year. In the continental United States this usually occurs on average at 1:00 PM on June 21 (daylight savings time shifts clock time one hour ahead of the sun in most jurisdictions). The exact time is dependent on several factors, including the location of the building within its time zone. This determination can be made by 3D modeling software used by many architects, or it can be done by hand, after consulting a solar angle calculator (these are easily found on the internet).

2. Per Item 3, "Where located in existing buildings, the product of the visible transmittance of the roof fenestration assembly and the area of the rough opening of the roof fenestration assembly divided by the area of the daylight zone is not less than 0.008." This calculation is straightforward and derives a value known as "effective aperture" that combines the area of the skylight with the transmittance of the skylight glazing to determine the potential of the skylight to deliver light to the building interior.

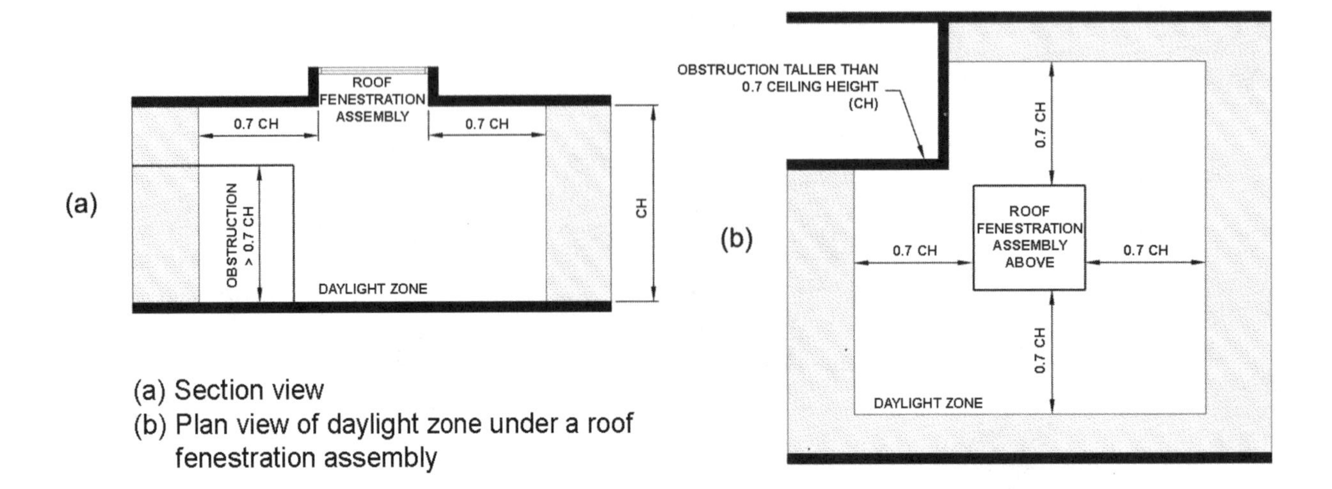

FIGURE C405.2.3.3
DAYLIGHT ZONE UNDER A ROOF FENESTRATION ASSEMBLY

C405.2.4 Specific application controls. Specific application controls shall be provided for the following:

1. Display and accent light shall be controlled by a dedicated control that is independent of the controls for other lighting within the room or space.
2. Lighting in cases used for display case purposes shall be controlled by a dedicated control that is independent of the controls for other lighting within the room or space.
3. Hotel and motel sleeping units and guest suites shall have a master control device that is capable of automatically switching off all installed luminaires and switched receptacles within 20 minutes after all occupants leave the room.

Exception: Lighting and switched receptacles controlled by captive key systems.

4. Supplemental task lighting, including permanently installed under-shelf or under-cabinet lighting, shall have a control device integral to the luminaires or be controlled by a wall-mounted control device provided that the control device is readily accessible.
5. Lighting for nonvisual applications, such as plant growth and food warming, shall be controlled by a dedicated control that is independent of the controls for other lighting within the room or space.
6. Lighting equipment that is for sale or for demonstrations in lighting education shall be controlled by a dedicated control that is independent of the controls for other lighting within the room or space.

❖ The "specific application controls" requirements listed here are almost always applicable. The only exceptions would be found in Section C405.2. However, we should keep in mind that occupant sensor and time-switch control requirements are generally applicable to lights receiving specific application control, while daylight responsive control would not be required for these lights. A more detailed discussion follows:

1. Display and accent lighting are not clearly defined by the code. However, these would certainly include lighting installed in retail sales areas to highlight merchandise as described in Item 1 of Section C405.4.2.2.1, and "display lighting for exhibits" as described in Exception 2.2 to Section C405.4.1. These lights must be connected to occupant sensors or time-switch control systems as applicable and must be zoned separately from other lights in the space for separate control.
2. This includes any lighting installed in a display case, regardless of how the case is configured, what it displays, or the type of space in which it is located. These lights must be connected to occupant sensors or time-switch control systems as applicable and must be zoned separately from other lights in the space for separate control.
3. This section provides a complete description of all lighting control requirements for sleeping units. Occupant sensor controls, time-switch controls and daylight responsive controls are not specifically required here. However, this section does allow that an occupant sensor or similar system could be provided to comply with the code, if a captive key control is not used. The term "captive key control" is not defined in the code, but is understood to mean a control device where the room key must be deposited in order for lighting circuits in the room to be energized. The idea is that the room key must be taken when the occupant exits the room (to be used to regain access), and thus the lights will be de-energized when the occupant leaves the room.
4. The scope of the code does not include portable lights that are plugged in after occupancy. Rather, it includes only lights that are permanently installed using mechanical or electrical connections. The provisions of this section therefore apply to permanently installed "under cabinet" or "under shelf" lights that provide higher light levels in a limited portion of the space, but not general lighting or portable lamps. These lights must be connected to occupant sensors or time-switch control systems as applicable and must be zoned separately from other lights in the space for separate control.
5. These lights must be connected to occupant sensors or time-switch control systems as applicable and must be zoned separately from other lights in the space for separate control.
6. These lights must be connected to occupant sensors or time-switch control systems as applicable and must be zoned separately from other lights in the space for separate control.

C405.2.5 Exterior lighting controls. Lighting for exterior applications other than emergency lighting that is intended to be automatically off during building operation, lighting specifically required to meet health and life safety requirements or decorative gas lighting systems shall:

1. Be provided with a control that automatically turns off the lighting as a function of available daylight.
2. Where lighting the building façade or landscape, the lighting shall have controls that automatically shut off the lighting as a function of dawn/dusk and a set opening and closing time.
3. Where not covered in Item 2, the lighting shall have controls configured to automatically reduce the connected lighting power by not less than 30 percent from not later than midnight to 6 a.m., from one hour after business closing to one hour before business opening or during any period when activity has not been detected for a time of longer than 15 minutes.

All time switches shall be able to retain programming and the time setting during loss of power for a period of at least 10 hours.

Exception: Lighting for covered vehicle entrances or exits from buildings or parking structures where required for safety, security or eye adaptation.

❖ Exterior lighting must be automatically switched by photocells, time switches or a combination of the two. The type of controls required will depend on whether the lights are intended to be on during the full dusk-to-dawn period or turned off during some of the night-time hours. Where timers are used, they must have some means of seasonal daylight adjustment. Timers must be equipped with power back-up provisions to allow accurate timekeeping and prevent the loss of lighting programming through a minimum 10-hour power loss.

Traditionally, exterior lights have been controlled by electro-mechanical clocks with mechanical trippers that toggle circuit switches. These devices typically are equipped with a manual override. Many of these traditional devices have neither seasonal correction nor 10-hour backup. They, therefore, do not meet the requirements of this section. The following devices will meet the requirements:

1. Directional photocells.
2. Electrically driven mechanical clocks with trippers, astronomical dial and 10-hour spring-wound storage.
3. Electronic programmable time switches with astronomic correction and 10-hour battery backup.
4. Either of the times above with a directional photocell (in places of astronomical correction).

Of these devices, a photocell/time clock combination is the most effective because its integral photocell automatically and continuously compensates for changes in the seasons. In addition, the redundancy of the time clock allows for continued operation if the photocell fails.

C405.3 Exit signs (Mandatory). Internally illuminated exit signs shall not be more than 5 watts per side.

❖ The amount of power presecribed is adequate to properly illuminate the exit signs and conserve energy from older signs.

Note that the annual energy savings for light-emitting diode (LED) exit signs is about $32.50 per year compared to incandescent signs and $5.25 per year compared to fluorescent signs. In addition, the average incremental cost of maintenance for incandescent signs compared to LEDs was about $30. The average incremental maintainance cost for the fluorescent signs was about $15. Another advantage of these newer technologies, such as LEDs, is that the signs tend to have a much higher eventual light output than the technologies that were being used earlier than the mid-1990s. They can often provide nearly 80 percent of the original output after about 10 years compared with 50 percent after the same amount of time for the older systems.

C405.4 Interior lighting power requirements (Prescriptive). A building complies with this section where its total connected lighting power calculated under Section C405.4.1 is not greater than the interior lighting power calculated under Section C405.4.2.

❖ This section identifies the lighting equipment that must be included in the calculations for interior lighting power and describes lighting equipment that is exempt. It describes the method used to calculate the maximum total interior lighting power allowed. It also explains how to calculate the interior connected lighting power.

Interior lighting power complies with the requirements of the code when the total connected interior lighting power of all nonexempt interior lighting is less than or equal to the maximum interior lighting power allowed. Determining compliance with the interior lighting power requirements is a two-step calculation procedure (see commentary, Section C405.4.2).

Example Applying the Lighting Requirements to a Private Office

A 13-foot, 4-inch by 18-foot (4064 mm by 5486 mm) private office (see Commentary Figure C405.4) has two recessed, three-lamp parabolic luminaires for general and task lighting, as well as two recessed, low-voltage down lights that accent artwork on the wall. Each parabolic fixture has a two-lamp rapid-start electronically ballasted F40T5 twin-tube (40 W), low-wattage, reduced output fluorescent lamp. The parabolic fixtures are controlled by a manual wall-box dimmer. Does this space comply with the lighting requirements?

Yes. This space complies with all applicable requirements for lighting power, controls and ballasts.

Using Section C405.4.2, the most appropriate LPD for this space is 1.0 W/ft^2, based on the "building-area type" listing for "office" in Table C405.4.2(2). Based on the room area of 240 square feet (22 m^2), a total of 240 watts is permitted (1.0 W/ft^2 × 240 ft^2 = 240 W). Therefore, this is the maximum interior lighting power allowed for this space. The two low-voltage down lights use 50W MR 16 lamps, and the input wattage of the parabolic fixtures is 60 watts per lamp. The total connected lighting power, therefore, is only 220 watts. The 20 watts not used may be used elsewhere in the building, if needed.

This room is required to have multiple controls. Because the main lights are connected to the occupant sensor, they do not require an additional control (see Section C405.2.2). The down lights in this space must have an additional control and can use the dimmer to satisfy the light reduction control requirement of Section C405.2.2.2.

Because two two-lamp ballasts are used to control

the parabolic luminaires, the requirement for tandem wiring does not apply.

C405.4.1 Total connected interior lighting power. The total connected interior lighting power shall be determined in accordance with Equation 4-9.

$TCLP = [SL + LV + LTPB + \text{Other}]$ **(Equation 4-9)**

where:

TCLP = Total connected lighting power (watts).

SL = Labeled wattage of luminaires for screw-in lamps.

LV = Wattage of the transformer supplying low-voltage lighting.

LTPB = Wattage of line-voltage lighting tracks and plug-in busways as the specified wattage of the luminaires, but at least 30 W/lin. ft. (100 W/lin m), or the wattage limit of the system's circuit breaker, or the wattage limit of other permanent current-limiting devices on the system.

Other = The wattage of all other luminaires and lighting sources not covered previously and associated with interior lighting verified by data supplied by the manufacturer or other *approved* sources.

Exceptions:

1. The connected power associated with the following lighting equipment is not included in calculating total connected lighting power.
   1.1. Professional sports arena playing field lighting.
   1.2. Lighting in sleeping units, provided that the lighting complies with Section R404.1.
   1.3. Emergency lighting automatically off during normal building operation.
   1.4. Lighting in spaces specifically designed for use by occupants with special lighting needs, including those with visual impairment and other medical and age-related issues.
   1.5. Lighting in interior spaces that have been specifically designated as a registered interior historic landmark.
   1.6. Casino gaming areas.
   1.7. Mirror lighting in dressing rooms.
2. Lighting equipment used for the following shall be exempt provided that it is in addition to general lighting and is controlled by an independent control device:
   2.1. Task lighting for medical and dental purposes.
   2.2. Display lighting for exhibits in galleries, museums and monuments.
3. Lighting for theatrical purposes, including performance, stage, film production and video production.
4. Lighting for photographic processes.
5. Lighting integral to equipment or instrumentation and installed by the manufacturer.
6. Task lighting for plant growth or maintenance.
7. Advertising signage or directional signage.
8. In restaurant buildings and areas, lighting for food warming or integral to food preparation equipment.
9. Lighting equipment that is for sale.
10. Lighting demonstration equipment in lighting education facilities.

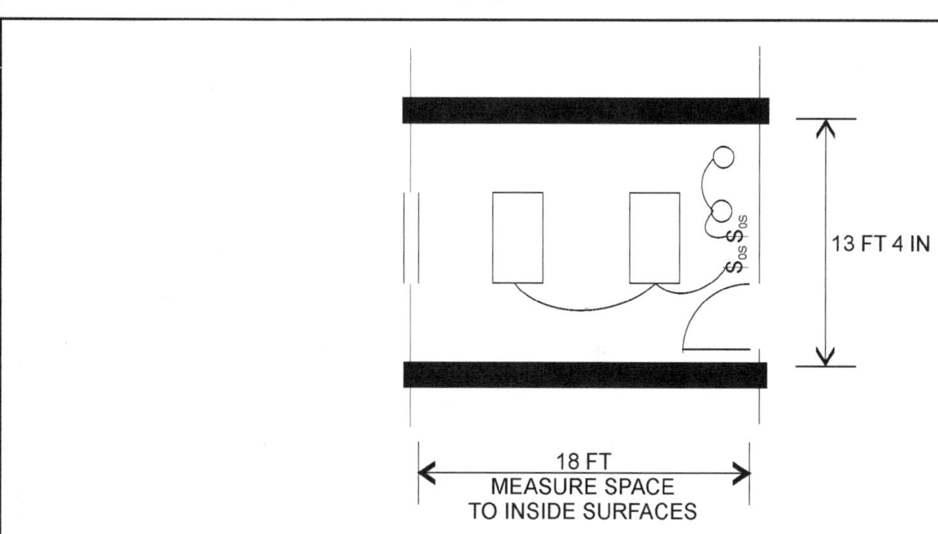

For SI: 1 inch = 25.4, 1 foot = 304.8 mm.

Figure C405.4
APPLYING THE LIGHTING REQUIREMENTS TO A PRIVATE OFFICE
(Courtesy of U.S. Department of Energy, Office of Building Technology State and Community Programs, www.energycodes.gov)

11. Lighting *approved* because of safety or emergency considerations, inclusive of exit lights.
12. Lighting integral to both open and glass-enclosed refrigerator and freezer cases.
13. Lighting in retail display windows, provided the display area is enclosed by ceiling-height partitions.
14. Furniture-mounted supplemental task lighting that is controlled by automatic shutoff.
15. Exit signs.

❖ Most interior lighting, including both permanent and portable luminaires, must be included in the calculation of the total connected interior lighting power. However, certain specialized lighting is exempt as referenced above. Exempt lighting can be ignored when determining the installed lighting power for comparison against the lighting power allowance. The building area method is the easiest way of determining the lighting power allowance. This method must be used for an entire building or a separate occupancy in the building. It may not be used, for instance, for large offices where the lighting is installed in phases. The allowance is appropriate for the entire building, but not for subareas. The LPD is multiplied by the gross lighted area of the building to determine the interior lighting power allowance.

Exempt lighting must be independently controlled. Interior lighting required for safety, security or special commercial needs is exempt from this section as listed in Exceptions 1 through 15. The code is relatively clear about what must be included in the calculations for interior lighting power. Nevertheless, the following examples describe common situations that can be confusing:

Covered or enclosed parking garages. Although the exterior lighting power requirement (see commentary, Section C405.5) applies to rooftop decks of parking garages and surface parking areas outdoors, covered or enclosed parking garages, including "open" parking garages, are included in the requirements for interior lighting.

Portable equipment. Although the designer cannot prevent users from plugging in portable lighting of their own choosing, the designer must account for portable lighting intended to serve a specific space function, such as furniture-mounted task lights and lighting equipment specifically designed, directed and installed to highlight merchandise. Even if the designer of the project is not responsible for specifying portable lighting, in some instances, the code permits an accommodation for the expected use of this equipment.

Example of Exempt Interior Lighting—Laboratory Test Lights

A laboratory is studying the effects of lighting on the treatment of a new chemical process. Ordinary fluorescent luminaires are arranged over the test areas and connected to timers. In addition, general lighting is used throughout the laboratory. What lighting is exempt?

The lighting arranged for the test is exempt. However, the lighting should be installed in a manner consistent with the permeance of the experiments. If only temporary, the lighting should not be recessed or otherwise installed in a relatively permanent fashion.

The total connected interior lighting power is the sum of the input wattages of all nonexempt interior luminaires in a building. The building complies with the requirement for interior power if the total connected interior lighting power is less than or equal to the maximum interior lighting power allowed (see Section C405.4.2). If the total connected interior lighting power

TABLE C405.4.2(1)
INTERIOR LIGHTING POWER ALLOWANCES: BUILDING AREA METHOD

BUILDING AREA TYPE	LPD (w/ft^2)
Automotive facility	0.80
Convention center	1.01
Courthouse	1.01
Dining: bar lounge/leisure	1.01
Dining: cafeteria/fast food	0.9
Dining: family	0.95
Dormitory	0.57
Exercise center	0.84
Fire station	0.67
Gymnasium	0.94
Health care clinic	0.90
Hospital	1.05
Hotel/Motel	0.87
Library	1.19
Manufacturing facility	1.17
Motion picture theater	0.76
Multifamily	0.51
Museum	1.02
Office	0.82
Parking garage	0.21
Penitentiary	0.81
Performing arts theater	1.39
Police station	0.87
Post office	0.87
Religious building	1.0
Retail	1.26
School/university	0.87
Sports arena	0.91
Town hall	0.89
Transportation	0.70
Warehouse	0.66
Workshop	1.19

exceeds the maximum interior lighting power allowed, the total connected interior lighting power may be adjusted by eliminating some luminaires, using more efficient lighting systems or demonstrating compliance in accordance with Section C407.

TABLE C405.4.2(2)
INTERIOR LIGHTING POWER ALLOWANCES: SPACE-BY-SPACE METHOD

COMMON SPACE TYPES[a]	LPD (watts/sq.ft)
Atrium	
Less than 40 feet in height	0.03 per foot in total height
Greater than 40 feet in height	0.40 + 0.02 per foot in total height
Audience seating area	
In an auditorium	0.63
In a convention center	0.82
In a gymnasium	0.65
In a motion picture theater	1.14
In a penitentiary	0.28
In a performing arts theater	2.43
In a religious building	1.53
In a sports arena	0.43
Otherwise	0.43
Banking activity area	1.01
Breakroom (See Lounge/Breakroom)	
Classroom/lecture hall/training room	
In a penitentiary	1.34
Otherwise	1.24
Conference/meeting/multipurpose room	1.23
Copy/print room	0.72
Corridor	
In a facility for the visually impaired (and not used primarily by the staff)b	0.92
In a hospital	0.79
In a manufacturing facility	0.41
Otherwise	0.66
Courtroom	1.72
Computer room	1.71
Dining area	
In a penitentiary	0.96
In a facility for the visually impaired (and not used primarily by the staff)[b]	1.9
In bar/lounge or leisure dining	1.07
In cafeteria or fast food dining	0.65
In family dining	0.89
Otherwise	0.65
Electrical/mechanical room	0.95
Emergency vehicle garage	0.56

(continued)

TABLE C405.4.2(2)—continued
INTERIOR LIGHTING POWER ALLOWANCES: SPACE-BY-SPACE METHOD

COMMON SPACE TYPES[a]	LPD (watts/sq.ft)
Food preparation area	1.21
Guest room	0.47
Laboratory	
In or as a classroom	1.43
Otherwise	1.81
Laundry/washing area	0.6
Loading dock, interior	0.47
Lobby	
In a facility for the visually impaired (and not used primarily by the staff)[b]	1.8
For an elevator	0.64
In a hotel	1.06
In a motion picture theater	0.59
In a performing arts theater	2.0
Otherwise	0.9
Locker room	0.75
Lounge/breakroom	
In a healthcare facility	0.92
Otherwise	0.73
Office	
Enclosed	1.11
Open plan	0.98
Parking area, interior	0.19
Pharmacy area	1.68
Restroom	
In a facility for the visually impaired (an not used primarily by the staff[b])	1.21
Otherwise	0.98
Sales area	1.59
Seating area, general	0.54
Stairway (See space containing stairway)	
Stairwell	0.69
Storage room	0.63
Vehicular maintenance area	0.67
Workshop	1.59
BUILDING TYPE SPECIFIC SPACE TYPES[a]	**LPD (watts/sq.ft)**
Facility for the visually impaired[b]	
In a chapel (and not used primarily by the staff)	2.21
In a recreation room (and not used primarily by the staff)	2.41
Automotive (See Vehicular Maintenance Area above)	
Convention Center—exhibit space	1.45
Dormitory—living quarters	0.38
Fire Station—sleeping quarters	0.22
Gymnasium/fitness center	
In an exercise area	0.72
In a playing area	1.2

(continued)

TABLE C405.4.2(2)—continued
INTERIOR LIGHTING POWER ALLOWANCES: SPACE-BY-SPACE METHOD

BUILDING TYPE SPECIFIC SPACE TYPESa	LPD (watts/sq.ft)
Healthcare facility	
In an exam/treatment room	1.66
In an imaging room	1.51
In a medical supply room	0.74
In a nursery	0.88
In a nurse's station	0.71
In an operating room	2.48
In a patient room	0.62
In a physical therapy room	0.91
In a recovery room	1.15
Library	
In a reading area	1.06
In the stacks	1.71
Manufacturing facility	
In a detailed manufacturing area	1.29
In an equipment room	0.74
In an extra high bay area (greater than 50′ floor-to-ceiling height)	1.05
In a high bay area (25-50′ floor-to-ceiling height)	1.23
In a low bay area (less than 25' floor-to-ceiling height)	1.19
Museum	
In a general exhibition area	1.05
In a restoration room	1.02
Performing arts theater—dressing room	0.61
Post Office—Sorting Area	0.94
Religious buildings	
In a fellowship hall	0.64
In a worship/pulpit/choir area	1.53
Retail facilities	
In a dressing/fitting room	0.71
In a mall concourse	1.1
Sports arena—playing area	
For a Class I facility	3.68
For a Class II facility	2.4
For a Class III facility	1.8
For a Class IV facility	1.2
Transportation facility	
In a baggage/carousel area	0.53
In an airport concourse	0.36
At a terminal ticket counter	0.8
Warehouse—storage area	
For medium to bulky, palletized items	0.58
For smaller, hand-carried items	0.95

a. In cases where both a common space type and a building area specific space type are listed, the building area specific space type shall apply

b. A 'Facility for the Visually Impaired' is a facility that is licensed or will be licensed by local or state authorities for senior long-term care, adult daycare, senior support or people with special visual needs.

C405.4.2 Interior lighting power. The total interior lighting power allowance (watts) is determined according to Table C405.4.2(1) using the Building Area Method, or Table C405.4.2(2) using the Space-by-Space Method, for all areas of the building covered in this permit.

❖ The interior lighting power allowed is the yardstick by which the total connected interior lighting power is measured to determine whether a building complies with the requirements for interior lighting power. Two methods for determining the interior lighting power allowance are provided—the Building Area Method and the Space-by-Space Method.

C405.4.2.1 Building Area Method. For the Building Area Method, the interior lighting power allowance is the floor area for each building area type listed in Table C405.4.2(1) times the value from Table C405.4.2(1) for that area. For the purposes of this method, an "area" shall be defined as all contiguous spaces that accommodate or are associated with a single building area type, as listed in Table C405.4.2(1). Where this method is used to calculate the total interior lighting power for an entire building, each building area type shall be treated as a separate area.

❖ LPD allowances in W/ft^2 are listed for 33 common space functions or activities in Table C405.4.2(1). For a given activity type, for example "Convention Center," the Interior Lighting Power (ILP) allowed is equal to the LPD multiplied by the "area" of the specific space or "building area type." The result is a lighting power budget for the respective room or group of rooms. Thus:

$$ILP = A \times LPD$$

The total ILP allowed for the building is the sum of the ILPs for each building area type or group of related building spaces.

Applying the LPD values to different specific spaces to determine the lighting power budget is relatively straightforward. Nevertheless, some of the terms used can be confusing to the occasional user. The following explanations address some of the more troublesome terms and issues.

Floor area or areas. The floor area of each specific building area type, tenant area or building space is measured to the surface of the perimeter walls or partitions, and does not include interior or exterior wall thickness. Note that area of the space is not the same thing as gross conditioned floor area previously discussed.

Total ILP Allowance Multiuse Facility. The total interior ILP allowance for a multiuse facility as shown in Commentary Figure C405.4.2 can be determined by the building-area method using Table C405.4.2(1). Commentary Figure C405.4.2(1) shows these calculations.

C405.4.2.2 Space-by-Space Method. For the Space-by-Space Method, the interior lighting power allowance is determined by multiplying the floor area of each space times the value for the space type in Table C405.4.2(2) that most closely represents the proposed use of the space, and then summing the lighting power allowances for all spaces. Trade-offs among spaces are permitted.

❖ The space-by-space method utilizes the specific space types in buildings as shown in Table C405.4.2(2).

Example of Total ILP Allowance Multiuse Facility
The total interior ILP allowance for a multiuse facility as shown in Commentary Figure C405.4.2 can be determined by the space-by-space method using Table C405.4.2(2). Commentary Figure C405.4.2(2) shows these calculations.

This table lists the ILP allowances using the space-by-space method. Note that under "Warehouse," the fine material storage typically is referring to spaces that function like a "pick" warehouse operation where small parts commonly in bins are picked to fill orders

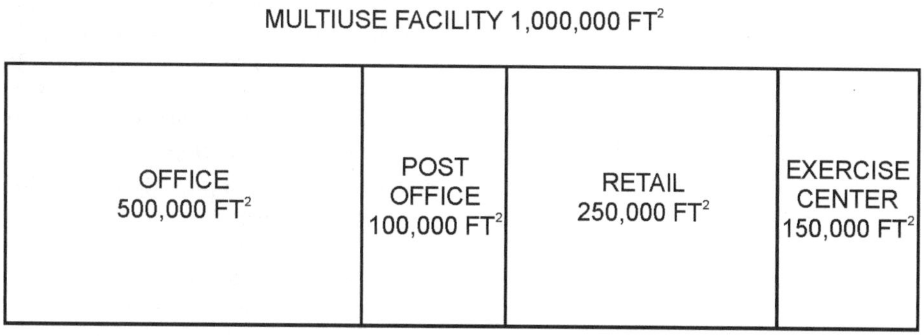

For SI: 1 square foot = 0.0929 m².

Figure C405.4.2
MULTIUSE FACILITY

Figure C405.4.2(1)
TOTAL LIGHTING POWER ALLOWANCE CALCULATION USING THE BUILDING AREA METHOD

USAGE CATEGORY	GROSS CONDITIONED (LIGHTED) FLOOR AREA (SQUARE FEET)		LIGHTING POWER DENSITY (WATTS PER SQUARE FEET)		LIGHTING POWER ALLOWANCE (WATTS)
Office	500,000	X	0.82	=	410,000
Post office	100,000	X	0.87	=	87,000
Department store—sales and dressing areas	221,000	X	1.26	=	278,460
Department store: ware-house	29,000	X	0.66	=	19,140
Exercise center	150,000	X	0.84	=	126,000
Total for building	1,000,000	X	—	—	920,600

For SI: 1 square foot = 0.0929 m², 1 watt per square foot = 11 W/m².

Figure C405.4.2(2)
TOTAL LIGHTING POWER ALLOWANCE CALCULATION USING THE SPACE-BY-SPACE METHOD

USAGE CATEGORY	GROSS CONDITIONED (LIGHTED) FLOOR AREA (SQUARE FEET)		LIGHTING POWER DENSITY (WATTS PER SQUARE FEET)		LIGHTING POWER ALLOWANCE (WATTS)
Office—enclosed	100,000	X	1.11	=	111,000
Office—open plan	400,000	X	0.98	=	392,000
Post office	100,000	X	0.94	=	94,000
Retail—sales area	220,000	X	1.59	=	349,800
Retail—warehouse	29,000	X	0.58	=	16,820
Retail—dressing area	1,000	X	0.71	=	710
Exercise center—fitness	150,000	X	0.72	=	108,000
Total for building	1,000,000	—	—	—	922,330

For SI: 1 square foot = 0.0929 m², 1 watt per square foot = 11 W/m².

and, therefore, more lighting is needed to discern small items and specific labels. The medium/bulky material is considered the more typical warehouse/storage where boxes or larger items are stored on the floor or racks.

C405.4.2.2.1 Additional interior lighting power. Where using the Space-by-Space Method, an increase in the interior lighting power allowance is permitted for specific lighting functions. Additional power shall be permitted only where the specified lighting is installed and automatically controlled separately from the general lighting, to be turned off during nonbusiness hours. This additional power shall be used only for the specified luminaires and shall not be used for any other purpose. An increase in the interior lighting power allowance is permitted in the following cases:

1. For lighting equipment to be installed in sales areas specifically to highlight merchandise, the additional lighting power shall be determined in accordance with Equation 4-10.

   Additional interior lighting power allowance = 500 watts + (Retail Area 1 • 0.6 W/ft^2) + (Retail Area 2 • 0.6 W/ft2) + (Retail Area 3 • 1.4 W/ft2) + (Retail Area 4 • 2.5 W/ft^2)

   (Equation 4-10)

   where:

   Retail Area 1 = The floor area for all products not listed in Retail Area 2, 3 or 4.

   Retail Area 2 = The floor area used for the sale of vehicles, sporting goods and small electronics.

   Retail Area 3 = The floor area used for the sale of furniture, clothing, cosmetics and artwork.

   Retail Area 4 = The floor area used for the sale of jewelry, crystal and china.

   Exception: Other merchandise categories are permitted to be included in Retail Areas 2 through 4, provided that justification documenting the need for additional lighting power based on visual inspection, contrast, or other critical display is *approved* by the code official.

2. For spaces in which lighting is specified to be installed in addition to the general lighting for the purpose of decorative appearance or for highlighting art or exhibits, provided that the additional lighting power shall be not more than 1.0 w/ft^2 (10.7 w/m^2) of such spaces.

❖ This section provides allowances for increasing in the total lighting power budget for sales areas or for exhibits such as art or museum exhibits. This is only allowed for these spaces when the space-by-space method in Section 405.4.2 is employed in calculation of allowable lighting.

C405.5 Exterior lighting (Mandatory). Where the power for exterior lighting is supplied through the energy service to the building, all exterior lighting shall comply with Section C405.5.1.

Exception: Where *approved* because of historical, safety, signage or emergency considerations.

❖ The code separates requirements for exterior and interior lighting systems. Commentary Figure C405.5.1 illustrates the basic efficacy requirements for exterior lighting and shows the performances of typical luminaires. A building or facility complies with the exterior lighting power requirement where the total amount of lighting power is limited in accordance with Section C405.5.1. The section only applies to exterior lighting that is "supplied through the energy service to the building," which means that lighting not powered by the building electrical system, such as municipal street lights, is exempt.

C405.5.1 Exterior building lighting power. The total exterior lighting power allowance for all exterior building applications is the sum of the base site allowance plus the individual allowances for areas that are to be illuminated and are permitted in Table C405.5.2(2) for the applicable lighting zone. Trade-offs are allowed only among exterior lighting applications listed in Table C405.5.2(2), in the Tradable Surfaces section. The lighting zone for the building exterior is determined from Table C405.5.2(1) unless otherwise specified by the local jurisdiction.

Exception: Lighting used for the following exterior applications is exempt where equipped with a control device independent of the control of the nonexempt lighting:

1. Specialized signal, directional and marker lighting associated with transportation.
2. Advertising signage or directional signage.
3. Integral to equipment or instrumentation and is installed by its manufacturer.
4. Theatrical purposes, including performance, stage, film production and video production.
5. Athletic playing areas.
6. Temporary lighting.
7. Industrial production, material handling, transportation sites and associated storage areas.
8. Theme elements in theme/amusement parks.
9. Used to highlight features of public monuments and registered historic landmark structures or buildings.

❖ This section was added to conserve energy in outdoor lighting. All outdoor light that is misdirected to areas outside the intended area for illumination is wasted energy. When calculating exterior building lighting power, the designer may use Tables C405.5.2(1) and C405.5.2(2). The advantage of using both of these tables together is that it allows the designer to adjust the lighting levels of tradable surfaces based on the lighting zone as defined in Table C405.5.2(1). This typically will allow the designer, in some instances, to reduce the lighting wattage that is used and thus save energy.

Building-mounted exterior lighting. Lighting mounted on the building, less specific exceptions as noted, is governed. This means that all lanterns, soffit lights, floodlights, step lights, wall packs and additional decorative lighting, such as neon outlining, ornamental pendants and globes, are subject to both the lighting power allowance and the source efficacy requirement.

Grounds, roads, parking lots and other exterior lighting. All lighting on the building site, less specific exceptions as noted, is governed. This generally includes pole-mounted lighting, landscape lighting, bollards, step lights, wall packs and all other lighting for the roads, walks, parking lots, gardens, trees and other portions of the site. Note that lighting not powered by the building electrical system, such as municipal street lights, is exempt. Lighting for the playing area of outdoor athletic facilities is covered by Exception 5.

Parking areas. Open-air parking lots, rooftop parking and drives are included in the exterior lighting requirements. Covered or enclosed parking areas, such as enclosed parking garages or carports, are part of the interior lighting requirements and are not subject to the exterior lighting efficacy or LPD requirements.

The exterior spaces regulated by this section are divided into two classifications: "tradable surfaces" and "nontradable surfaces." The items included in each can clearly be seen in Table C405.5.2(2). The total exterior lighting allowed is simply the sum of all of the various items plus the code and will allow an additional 5 percent to be used for any purpose. The 5-percent additional power allowance can be used in any way for any of the various surfaces that are being illuminated, including those in the "nontradable surfaces" section.

When applying the provisions, the tradable surface power allowance can be used for any of the items listed (uncovered parking areas, building grounds, building entrances and exits, canopies and overhangs or outdoor sales). As an example, the lighting level of the parking lot could be reduced slightly to allow additional lighting on a walkway or at a building entrance as long as the total power level did not exceed what was originally calculated. The tradable section, therefore, establishes a maximum power allowance but it provides flexibility to the designer to use that power for a variety of locations. When dealing with the "nontradable surfaces," the power allowance that is determined may only be used for the specific application listed. Therefore, the "building façade" allowance could not be reduced in order to permit the automated teller machine or drive-up windows at fast food restaurants to be increased.

The purposes of the 5-percent unrestricted increase are to provide the designer with some flexibility and to limit the uncontrolled use of exterior lights that was permitted when the code only regulated the efficacy of the exterior lighting.

The exceptions to both Sections C405.5 and C405.5.1 are important to the proper application of the code when addressing exterior lighting. The code does not regulate lighting used for historical, safety, security, emergency, exterior manufacturing, or similar commercial or industrial process needs. In addition, the provisions of the last sentence of this section are important because they also eliminate the efficacy requirements of Section C405.5.1 for the lighting covered by the exceptions.

The following are specific exceptions to the exterior lighting power requirements:

Exception 2: Both self-contained and exterior illumination integral to advertising signage is exempt.

Exception 5: Lighting for outdoor athletic activity areas of all types is exempt.

Exception 7: This includes items such as lighting for outdoor manufacturing, commercial greenhouses and processing facilities. This exemption, while intuitive, is rooted primarily in the historical scoping requirements of the code, where it applied only to "those portions of factory and industrial occupancies designed primarily for human occupancy (comfort)." Thus, this section does not apply to exterior lighting systems for outdoor commercial, agricultural and industrial work areas, such as refineries.

Exception 9: Exterior lighting intended primarily for the display of public monuments, statues or other items of historical interest or importance are exempt.

TABLE C405.5.2(1)
EXTERIOR LIGHTING ZONES

LIGHTING ZONE	DESCRIPTION
1	Developed areas of national parks, state parks, forest land, and rural areas
2	Areas predominantly consisting of residential zoning, neighborhood business districts, light industrial with limited nighttime use and residential mixed-use areas
3	All other areas not classified as lighting zone 1, 2 or 4
4	High-activity commercial districts in major metropolitan areas as designated by the local land use planning authority

C405.6 Electrical energy consumption (Mandatory). Each dwelling unit located in a Group R-2 building shall have a separate electrical meter.

❖ People are more likely to conserve energy when provided with the means to track how much they are using. This requirement does not mandate that the units be billed separately for the electrical service; it simply helps people to understand how their use of the service and selection of equipment can affect energy consumption.

TABLE C405.5.2(2)
INDIVIDUAL LIGHTING POWER ALLOWANCES FOR BUILDING EXTERIORS

		LIGHTING ZONES			
		Zone 1	**Zone 2**	**Zone 3**	**Zone 4**
Base Site Allowance (Base allowance is usable in tradable or nontradable surfaces.)		500 W	600 W	750 W	1300 W
Tradable Surfaces (Lighting power densities for uncovered parking areas, building grounds, building entrances and exits, canopies and overhangs and outdoor sales areas are tradable.)	**Uncovered Parking Areas**				
	Parking areas and drives	0.04 W/ft^2	0.06 W/ft^2	0.10 W/ft^2	0.13 W/ft^2
	Building Grounds				
	Walkways less than 10 feet wide	0.7 W/linear foot	0.7 W/linear foot	0.8 W/linear foot	1.0 W/linear foot
	Walkways 10 feet wide or greater, plaza areas special feature areas	0.14 W/ft^2	0.14 W/ft^2	0.16 W/ft^2	0.2 W/ft^2
	Stairways	0.75 W/ft^2	1.0 W/ft^2	1.0 W/ft^2	1.0 W/ft^2
	Pedestrian tunnels	0.15 W/ft^2	0.15 W/ft^2	0.2 W/ft^2	0.3 W/ft^2
	Building Entrances and Exits				
	Main entries	20 W/linear foot of door width	20 W/linear foot of door width	30 W/linear foot of door width	30 W/linear foot of door width
	Other doors	20 W/linear foot of door width	20 W/linear foot of door width	20 W/linear foot of door width	20 W/linear foot of door width
	Entry canopies	0.25 W/ft^2	0.25 W/ft^2	0.4 W/ft^2	0.4 W/ft^2
	Sales Canopies				
	Free-standing and attached	0.6 W/ft^2	0.6 W/ft^2	0.8 W/ft^2	1.0 W/ft^2
	Outdoor Sales				
	Open areas (including vehicle sales lots)	0.25 W/ft^2	0.25 W/ft^2	0.5 W/ft^2	0.7 W/ft^2
	Street frontage for vehicle sales lots in addition to "open area" allowance	No allowance	10 W/linear foot	10 W/linear foot	30 W/linear foot
Nontradable Surfaces (Lighting power density calculations for the following applications can be used only for the specific application and cannot be traded between surfaces or with other exterior lighting. The following allowances are in addition to any allowance otherwise permitted in the "Tradable Surfaces" section of this table.)	Building facades	No allowance	0.075 W/ft2 of gross above-grade wall area	0.113 W/ft^2 of gross above-grade wall area	0.15 W/ft^2 of gross above-grade wall area
	Automated teller machines (ATM) and night depositories	270 W per location plus 90 W per additional ATM per location	270 W per location plus 90 W per additional ATM per location	270 W per location plus 90 W per additional ATM per location	270 W per location plus 90 W per additional ATM per location
	Entrances and gatehouse inspection stations at guarded facilities	0.75 W/ft^2 of covered and uncovered area	0.75 W/ft^2 of covered and uncovered area	0.75 W/ft^2 of covered and uncovered area	0.75 W/ft^2 of covered and uncovered area
	Loading areas for law enforcement, fire, ambulance and other emergency service vehicles	0.5 W/ft^2 of covered and uncovered area	0.5 W/ft^2 of covered and uncovered area	0.5 W/ft^2 of covered and uncovered area	0.5 W/ft^2 of covered and uncovered area
	Drive-up windows/doors	400 W per drive-through	400 W per drive-through	400 W per drive-through	400 W per drive-through
	Parking near 24-hour retail entrances	800 W per main entry	800 W per main entry	800 W per main entry	800 W per main entry

For SI: 1 foot = 304.8 mm, 1 watt per square foot = W/0.0929 m^2.
W = watts.

C405.7 Electrical transformers (Mandatory). Electric transformers shall meet the minimum efficiency requirements of Table C405.7 as tested and rated in accordance with the test procedure listed in DOE 10 CFR 431. The efficiency shall be verified through certification under an approved certification program or, where a certification program does not exist, the equipment efficiency ratings shall be supported by data furnished by the transformer manufacturer.

Exceptions: The following transformers are exempt:

1. Transformers that meet the *Energy Policy Act of 2005* exclusions based on the DOE 10 CFR 431 definition of special purpose applications.
2. Transformers that meet the *Energy Policy Act of 2005* exclusions that are not to be used in general purpose applications based on information provided in DOE 10 CFR 431.
3. Transformers that meet the *Energy Policy Act of 2005* exclusions with multiple voltage taps where the highest tap is at least 20 percent more than the lowest tap.
4. Drive transformers.
5. Rectifier transformers.
6. Auto-transformers.
7. Uninterruptible power system transformers.
8. Impendance transformers.
9. Regulating transformers.
10. Sealed and nonventilating transformers.
11. Machine tool transformers.
12. Welding transformers.
13. Grounding transformers.
14. Testing transformers.

❖ New to the 2015 code is regulation of energy efficiency for electric low-voltage dry-type transformers. This is an important component of electrical energy consumption. The provisions are consistent with ASHRAE Standard 90.1-2013.

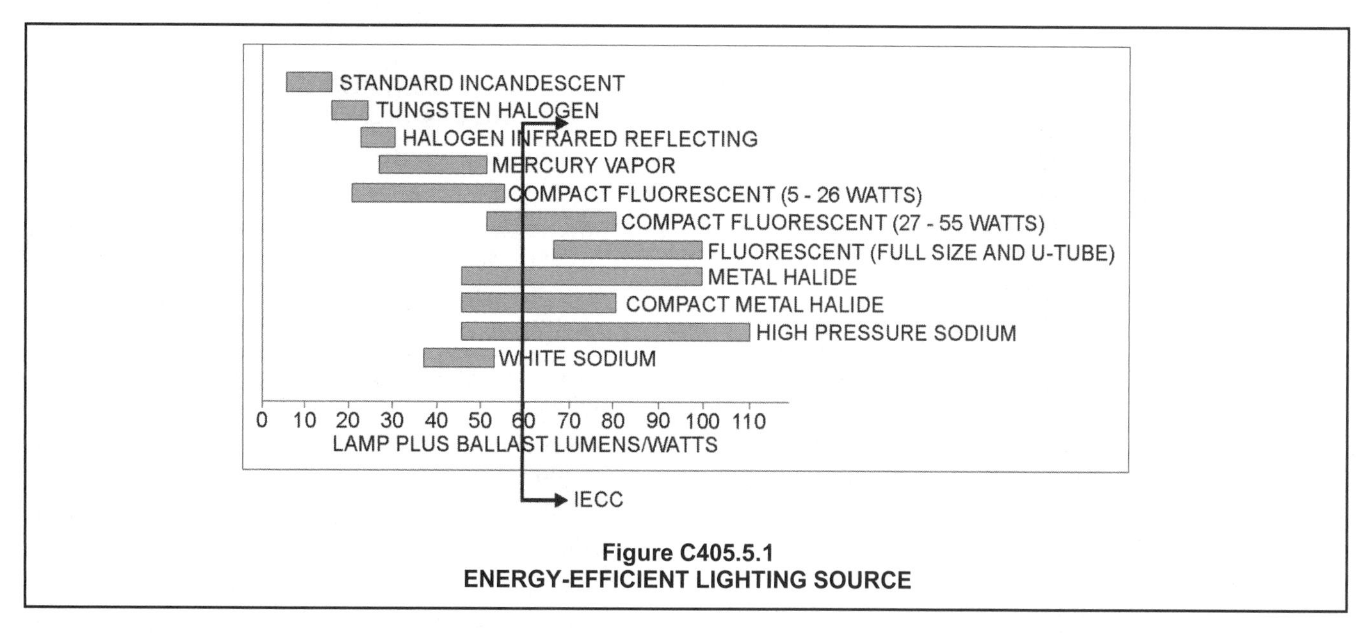

Figure C405.5.1
ENERGY-EFFICIENT LIGHTING SOURCE

TABLE C405.7
MINIMUM NOMINAL EFFICIENCY LEVELS FOR 10 CFR 431 LOW-VOLTAGE DRY-TYPE DISTRIBUTION TRANSFORMERS

SINGLE-PHASE TRANSFORMERS		THREE-PHASE TRANSFORMERS	
kVA[a]	Efficiency (%)[b]	kVA[a]	Efficiency (%)[b]
15	97.7	15	97.0
25	98.0	30	97.5
37.5	98.2	45	97.7
50	98.3	75	98.0
75	98.5	112.5	98.2
100	98.6	150	98.3
167	98.7	225	98.5
250	98.8	300	98.6
333	98.9	500	98.7
		750	98.8
		1000	98.9

a. kiloVolt-Amp rating.
b. Nominal efficiencies shall be established in accordance with the DOE 10 CFR 431 test procedure for low-voltage dry-type transformers.

C405.8 Electrical motors (Mandatory). Electric motors shall meet the minimum efficiency requirements of Tables C405.8(1) through C405.8(4) when tested and rated in accordance with the DOE 10 CFR 431. The efficiency shall be verified through certification under an approved certification program or, where a certification program does not exist, the equipment efficiency ratings shall be supported by data furnished by the motor manufacturer.

❖ New to the 2015 code is regulation of energy efficiency for electrical motors. This is an important component of electrical energy consumption. The provisions are consistent with ASHRAE Standard 90.1-2013.

C405.9 Vertical and horizontal transportation systems and equipment. Vertical and horizontal transportation systems and equipment shall comply with this section.

❖ This section provides energy reduction measures for the operation of elevators and escalators. The provisions are new to the 2015 code. They are consistent with provisions of ASHRAE Standard 90.1.

C405.9.1 Elevator cabs. For the luminaires in each elevator cab, not including signals and displays, the sum of the lumens divided by the sum of the watts shall be not less than 35 lumens per watt. Ventilation fans in elevators that do not have their own air-conditioning system shall not consume more than 0.33 watts/cfm at the maximum rated speed of the fan. Controls shall be provided that will de-energize ventilation fans and lighting systems when the elevator is stopped, unoccupied and with its doors closed for over 15 minutes.

❖ This section provides requirements for energy reduction on the two components in an elevator where the opportunity exists to make reductions: lighting of the cab and operation of ventilating equipment.

C405.9.2 Escalators and moving walks. Escalators and moving walks shall comply with ASME A17.1/CSA B44 and shall have automatic controls configured to reduce speed to the minimum permitted speed in accordance with ASME A17.1/CSA B44 or applicable local code when not conveying passengers.

❖ This section is intended to provide for reduced energy use and longer equipment life due to reduced wear and tear during the hours on standby mode or light loading conditions of escalators. These escalator controls have been standard in Canada, Europe and most of Asia for many years. The 2010 ANSI/ASME A17.1 safety standard for elevators and escalators now allows use of escalators and moving walks with "sleep mode" for reducing speed during unoccupied periods and provides for their safe operation. Sensors detect approaching passengers and bring the escalator or walk to full speed before the passenger steps on.

The energy consumed by a typical pair of escalators is approximately 24,000–36,000 kWh per year, and the predicted energy savings ranges from 25 to 60 percent. The higher figure applies to escalators that have bursts of usage at wide intervals, as occurs with performing arts or transportation facilities. The lower figure would apply where usage is scattered throughout the day, as in shopping malls or office buildings.

C405.9.2.1 Regenerative drive. An escalator designed either for one-way down operation only or for reversible operation shall have a variable frequency regenerative drive that supplies electrical energy to the building electrical system when the escalator is loaded with passengers whose combined weight exceeds 750 pounds (340 kg).

❖ By using regenerative technology, the escalator utilizes surplus energy, which is the energy needed to accelerate the escalator when a load comes aboard and decelerate the escalator when the load transfers off, and transfers it back to the building's electrical system for use in other areas or to further power the lift at later intervals. This represents a significant saving both financially and environmentally. The 750-pound threshold for activation of the regenerative drive is derived from the 5-passenger threshold mentioned in typical manufacturers' literature (5 passengers x 150 pounds = 750 pounds).

SECTION C406
ADDITIONAL EFFICIENCY PACKAGE OPTIONS

C406.1 Requirements. Buildings shall comply with at least one of the following:

1. More efficient HVAC performance in accordance with Section C406.2.
2. Reduced lighting power density system in accordance with Section C406.3.
3. Enhanced lighting controls in accordance with Section C406.4.
4. On-site supply of renewable energy in accordance with Section C406.5.
5. Provision of a dedicated outdoor air system for certain HVAC equipment in accordance with Section C406.6.
6. High-efficiency service water heating in accordance with Section C406.7.

❖ Section C406 requires that additional energy savings measures be incorporated in the design and construction of each commercial building. Section C406 is in addition to the provisions found in Sections C402, C403, C404 and C405. The section provides six methods for compliance. Only one of the six needs to be included in the design and construction of a building. The purpose of this section is to provide flexibility to achieve the energy savings needed to meet the overall energy savings goal.

Section 406 only comes into play in a design where the designer has chosen in Section C401.2 to comply with Item 2. Where a designer pursues compliance via meeting ANSI/ASHRAE/IESNA 90.1 (Item 1 of Section C401.2) or is using the total building performance approach (Item 3 of Section C401.2), then Section C406 does not need to be addressed.

TABLE C405.8(1)
MINIMUM NOMINAL FULL-LOAD EFFICIENCY FOR 60 HZ NEMA GENERAL PURPOSE ELECTRIC MOTORS (SUBTYPE I) RATED 600 VOLTS OR LESS (Random Wound)[a]

MOTOR HORSEPOWER	NUMBER OF POLES	OPEN DRIP-PROOF MOTORS			TOTALLY ENCLOSED FAN-COOLED MOTORS		
		2	4	6	2	4	6
	Synchronous Speed (RPM)	3600	1800	1200	3600	1800	1200
1		77.0	85.5	82.5	77.0	85.5	82.5
1.5		84.0	86.5	86.5	84.0	86.5	87.5
2		85.5	86.5	87.5	85.5	86.5	88.5
3		85.5	89.5	88.5	86.5	89.5	89.5
5		86.5	89.5	89.5	88.5	89.5	89.5
7.5		88.5	91.0	90.2	89.5	91.7	91.0
10		89.5	91.7	91.7	90.2	91.7	91.0
15		90.2	93.0	91.7	91.0	92.4	91.7
20		91.0	93.0	92.4	91.0	93.0	91.7
25		91.7	93.6	93.0	91.7	93.6	93.0
30		91.7	94.1	93.6	91.7	93.6	93.0
40		92.4	94.1	94.1	92.4	94.1	94.1
50		93.0	94.5	94.1	93.0	94.5	94.1
60		93.6	95.0	94.5	93.6	95.0	94.5
75		93.6	95.0	94.5	93.6	95.4	94.5
100		93.6	95.4	95.0	94.1	95.4	95.0
125		94.1	95.4	95.0	95.0	95.4	95.0
150		94.1	95.8	95.4	95.0	95.8	95.8
200		95.0	95.8	95.4	95.4	96.2	95.8
250		95.0	95.8	95.4	95.8	96.2	95.8
300		95.4	95.8	95.4	95.8	96.2	95.8
350		95.4	95.8	95.4	95.8	96.2	95.8
400		95.8	95.8	95.8	95.8	96.2	95.8
450		95.8	96.2	96.2	95.8	96.2	95.8
500		95.8	96.2	96.2	95.8	96.2	95.8

a. Nominal efficiencies shall be established in accordance with DOE 10 CFR 431.

TABLE C405.8(2)
MINIMUM NOMINAL FULL-LOAD EFFICIENCY OF GENERAL PURPOSE ELECTRIC MOTORS (SUBTYPE II) AND ALL DESIGN B MOTORS GREATER THAN 200 HORSEPOWER[a]

MOTOR HORSEPOWER	NUMBER OF POLES	OPEN DRIP-PROOF MOTORS				TOTALLY ENCLOSED FAN-COOLED MOTORS			
		2	4	6	8	2	4	6	8
	Synchronous Speed (RPM)	3600	1800	1200	900	3600	1800	1200	900
1		NR	82.5	80.0	74.0	75.5	82.5	80.0	74.0
1.5		82.5	84.0	84.0	75.5	82.5	84.0	85.5	77.0
2		84.0	84.0	85.5	85.5	84.0	84.0	86.5	82.5
3		84.0	86.5	86.5	86.5	85.5	87.5	87.5	84.0
5		85.5	87.5	87.5	87.5	87.5	87.5	87.5	84.0
7.5		87.5	88.5	88.5	88.5	88.5	89.5	89.5	85.5
10		88.5	89.5	90.2	89.5	89.5	89.5	89.5	88.5
15		89.5	91.0	90.2	89.5	90.2	91.0	90.2	88.5
20		90.2	91.0	91.0	90.2	90.2	91.0	90.2	89.5
25		91.0	91.7	91.7	90.2	91.0	92.4	91.7	89.5
30		91.0	92.4	92.4	91.0	91.0	92.4	91.7	91.0
40		91.7	93.0	93.0	91.0	91.7	93.0	93.0	91.0
50		92.4	93.0	93.0	91.7	92.4	93.0	93.0	91.7
60		93.0	93.6	93.6	92.4	93.0	93.6	93.6	91.7
75		93.0	94.1	93.6	93.6	93.0	94.1	93.6	93.0
100		93.0	94.1	94.1	93.6	93.6	94.5	94.1	93.0
125		93.6	94.5	94.1	93.6	94.5	94.5	94.1	93.6
150		93.6	95.0	94.5	93.6	94.5	95.0	95.0	93.6
200		94.5	95.0	94.5	93.6	95.0	95.0	95.0	94.1
250		94.5	95.4	95.4	94.5	95.4	95.0	95.0	94.5
300		95.0	95.4	95.4	NR	95.4	95.4	95.0	NR
350		95.0	95.4	95.4	NR	95.4	95.4	95.0	NR
400		95.4	95.4	NR	NR	95.4	95.4	NR	NR
450		95.8	95.8	NR	NR	95.4	95.4	NR	NR
500		95.8	95.8	NR	NR	95.4	95.8	NR	NR

NR = No requirement.

a. Nominal efficiencies shall be established in accordance with DOE 10 CFR 431.

**TABLE C405.8(3)
MINIMUM AVERAGE FULL-LOAD EFFICIENCY FOR POLYPHASE SMALL ELECTRIC MOTORS[a]**

MOTOR HORSEPOWER	OPEN MOTORS			
	Number of Poles	2	4	6
	Synchronous Speed (RPM)	3600	1800	1200
0.25		65.6	69.5	67.5
0.33		69.5	73.4	71.4
0.50		73.4	78.2	75.3
0.75		76.8	81.1	81.7
1		77.0	83.5	82.5
1.5		84.0	86.5	83.8
2		85.5	86.5	N/A
3		85.5	86.9	N/A

a. Average full load efficiencies shall be established in accordance with 10 CFR 431.

**TABLE C405.8(4)
MINIMUM AVERAGE FULL LOAD EFFICIENCY FOR
CAPACITOR-START CAPACITOR-RUN AND CAPACITOR-START INDUCTION-RUN SMALL ELECTRIC MOTORS[a]**

MOTOR HORSEPOWER	OPEN MOTORS			
	Number of Poles	2	4	6
	Synchronous Speed (RPM)	3600	1800	1200
0.25		66.6	68.5	62.2
0.33		70.5	72.4	66.6
0.50		72.4	76.2	76.2
0.75		76.2	81.8	80.2
1		80.4	82.6	81.1
1.5		81.5	83.8	N/A
2		82.9	84.5	N/A
3		84.1	N/A	N/A

a. Average full load efficiencies shall be established in accordance with 10 CFR 431.

C406.1.1 Tenant spaces. Tenant spaces shall comply with Section C406.2, C406.3, C406.4, C406.6 or C406.7. Alternatively, tenant spaces shall comply with Section C406.5 where the entire building is in compliance.

❖ Separately permitted tenant spaces are allowed to choose any of the paths that are applicable to the space being improved, but are able to choose the on-site renewable energy path only if the entire building meets the minimum renewable energy capacity requirements described in Section 406.5.

C406.2 More efficient HVAC equipment performance. Equipment shall exceed the minimum efficiency requirements listed in Tables C403.2.3(1) through C403.2.3(7) by 10 percent, in addition to the requirements of Section C403. Where multiple performance requirements are provided, the equipment shall exceed all requirements by 10 percent. *Variable refrigerant flow systems* shall exceed the energy efficiency provisions of ANSI/ASHRAE/IES 90.1 by 10 percent. Equipment not listed in Tables C403.2.3(1) through C403.2.3(7) shall be limited to 10 percent of the total building system capacity.

❖ The HVAC equipment requirements in this compliance path option call for a 10-percent increase in efficiency over the base requirements provided in Section C403. The HVAC equipment efficiency levels contained in this compliance path directly reference, and then replace, equipment efficiencies contained in Section C403, or in ASHRAE 90.1 for the case of VRF systems, with a value that is 10 percent more efficient. This path cannot be used where the use of heating and cooling equipment not listed in this section exceeds 10 percent of the total building capacity. This would allow some systems, such as electric resistance heat, to be used in a limited capacity for the proposed project and still allow the code user to use this option.

C406.3 Reduced lighting power density. The total interior lighting power (watts) of the building shall be determined by using 90 percent of the lighting power values specified in Table C405.4.2(1) times the floor area for the building types, or by using 90 percent of the interior lighting power allowance calculated by the Space-by-Space Method in Section C405.4.2.

❖ The LPD requirements in this compliance path option call for a 10-percent increase in efficiency over the base requirements found in Section C405. This will ensure that the LPD levels contained in this section provide the necessary energy savings over the LPDs contained in Section C405. This section allows the use of space-by-space LPDs or whole building LPDs to provide more flexibility in this compliance option. The values contained in this section are similar to those included as part of ASHRAE Standard 189.1.

C406.4 Enhanced digital lighting controls. Interior lighting in the building shall have the following enhanced lighting controls that shall be located, scheduled and operated in accordance with Section C405.2.2.

1. Luminaires shall be capable of continuous dimming.
2. Luminaires shall be capable of being addressed individually. Where individual addressability is not available for the luminaire class type, a controlled group of not more than four luminaries shall be allowed.
3. Not more than eight luminaires shall be controlled together in a *daylight zone*.
4. Fixtures shall be controlled through a digital control system that includes the following function:
   4.1. Control reconfiguration based on digital addressability.
   4.2. Load shedding.
   4.3. Individual user control of overhead general illumination in open offices.
   4.4. Occupancy sensors shall be capable of being reconfigured through the digital control system.
5. Construction documents shall include submittal of a Sequence of Operations, including a specification outlining each of the functions in Item 4 of this section.
6. Functional testing of lighting controls shall comply with Section C408.

❖ Many new buildings are incorporating central control systems for lighting that include direct digital control over lighting sensors and luminaires. These lighting systems usually offer many degrees of controllability and opportunities for saving lighting energy through reducing power or shutting off fixtures. Such systems will likely satisfy the requirements of this compliance option. The enhanced lighting controls package requires a digital control system to allow continuous dimming and a significant level of controllability on individual luminaires. If individual addressability is not possible, up to four or eight luminaires can be grouped together for control purposes, depending on the type of control.

C406.5 On-site renewable energy. Total minimum ratings of on-site renewable energy systems shall comply with one of the following:

1. Provide not less than 0.50 watts per square foot (5.4 W/m^2) of conditioned floor area.
2. Provide not less than 3 percent of the energy used within the building for building mechanical and service water heating equipment and lighting regulated in Chapter 4.

❖ The on-site renewable energy option provides three straightforward compliance approaches: electricity generation, thermal collection and a calculation method for any type or combination of energy production. A path to include the purchase of renewable power or credits was carefully considered, but not included based on concerns regarding the verification and permanence of the transaction after the certificate of occupancy has been issued. If sufficient renewable energy is not available to the site, the project now has five alternative compliance options to choose from.

C406.6 Dedicated outdoor air system. Buildings covered by Section C403.4 shall be equipped with an independent ventilation system designed to provide not less than the minimum 100-percent outdoor air to each individual occupied space, as specified by the *International Mechanical Code*. The ventilation system shall be capable of total energy recovery. The HVAC system shall include supply-air temperature controls that automatically reset the supply-air temperature in response to representative building loads, or to outdoor air temperatures. The controls shall reset the supply-air temperature at least 25 percent of the difference between the design supply-air temperature and the design room-air temperature.

❖ The Dedicated Outdoor Air System (DOAS) package is based on technical specifications from the Pacific Northwest National Laboratory's *50% Advanced Energy Design Guides* series of technical documents. The DOAS compliance path option requires that an adequate quantity of outside air is delivered separately to spaces in the buildings while employing energy recovery. A DOAS design for ventilation requirements reduces the need for excess outdoor air or supply air to run the mechanical system, and it uses less energy for terminal reheating.

C406.7 Reduced energy use in service water heating. Buildings shall be of the following types to use this compliance method:

1. Group R-1: Boarding houses, hotels or motels.
2. Group I-2: Hospitals, psychiatric hospitals and nursing homes.
3. Group A-2: Restaurants and banquet halls or buildings containing food preparation areas.
4. Group F: Laundries.
5. Group R-2: Buildings with residential occupancies.
6. Group A-3: Health clubs and spas.
7. Buildings showing a service hot water load of 10 percent or more of total building energy loads, as shown with an energy analysis as described in Section C407.

❖ The service water heating compliance option is modified from similar language in the *International Green Construction Code*® (IgCC®) and the 2012 *North Carolina Energy Conservation Code*. The section identifies which building types have internal processes and hot water uses that in combination may provide enough recoverable thermal energy to qualify for this compliance path.

C406.7.1 Load fraction. The building service water-heating system shall have one or more of the following that are sized to provide not less than 60 percent of hot water requirements, or sized to provide 100 percent of hot water requirements if the building shall otherwise comply with Section C403.4.7:

1. Waste heat recovery from service hot water, heat-recovery chillers, building equipment, process equipment, or a combined heat and power system.
2. Solar water-heating systems.

❖ The buildings identified in Section C406.7.1 were found to be building types having internal processes and hot water uses that in combination may provide enough recoverable thermal energy to qualify for this compliance path. Because Section C403.4.5 already requires some heat recovery in certain circumstances, this section requires heat recovery sizing for 100 percent of the hot water used in the building instead of the 60-percent minimum required for buildings not subject to Section 403.4.5. Solar thermal water heating systems may also be used independently for this section, or in combination with a heat recovery system. This package is independent of the compliance path offered in Section C406.5. Only one package is required for compliance with Section 406, but either Section 406.5 or Section 406.7, if applicable, can be used to calculate the minimum capacity required for a solar thermal system.

SECTION C407
TOTAL BUILDING PERFORMANCE

C407.1 Scope. This section establishes criteria for compliance using total building performance. The following systems and loads shall be included in determining the total building performance: heating systems, cooling systems, service water heating, fan systems, lighting power, receptacle loads and process loads.

❖ This section establishes criteria for compliance using total building performance. This section only comes into play when the designer chooses to comply with Item 3 of Section C401.2. If either of the other two compliance options are selected in Section C401.2, then this section would not apply. When compliance with Section C407 is chosen, Section C401.2 requires that the building's energy costs be 85 percent or less of the costs of the standard reference design building. Section C407 provides the methodology to make the determination of compliance with the 85-percent requirement.

The following systems and loads must be included to determine the total building performance: heating systems, cooling systems, service water heating, fan systems, lighting power, receptacle loads and process loads. This section is based on the mandatory and prescriptive provisions of this chapter. Provisions of ANSI/ASHRAE/IESNA 90.1 are not applicable.

Section C407 offers maximum design flexibility by using the annual energy cost as the performance metric when evaluating the total performance of a building in its entirety. Using this approach, a building design is evaluated based on the cost of various types of energy used, rather than the units (Btu, kW) of energy used. The general procedure is to show that the annual energy cost for the proposed building is less than the annual energy cost of a standard design building that meets the minimum prescriptive requirements of the code. The designer has to estimate the annual energy costs for two buildings: the one the designer wants to build and the standard design. The two are compared on the basis of annual energy cost. Designs that have lower demand changes or use energy when rates are lower are favored. This allows for the calculation of energy use of different fuel sources at different times.

One of the advantages of the total-building-performance method is that it gives credit for innovative energy-efficient designs not accounted for in the prescriptive paths in this chapter, such as passive solar heating and daylighting. Another advantage in using the total-building-performance approach is that it allows the designer to make trade-offs among the various energy-using systems of the building because the prescriptive methods in Sections C402, C403, C404 and C405 set minimum performance levels for building components or systems, while the total-building-performance method takes into account how various building components interact. For instance, a particular designer may want to deliberately incorporate lower efficiency levels (when compared to prescriptive requirements) to achieve specific design objectives; for example, aesthetics in selecting a construction assembly, larger glass areas for special views or extra lighting power for visual effect.

The total-building-performance method allows the designer to not only trade off within each section (e.g., wall insulation for ceiling insulation) but also permits trading between the various sections (e.g., reduced insulation for increased mechanical system efficiency). The code does, however, contain certain requirements that may not be reduced or eliminated. These include requirements for air leakage, moisture control, mechanical systems, service water heating, lighting controls, tandem wiring, exit signs, exterior lighting and separate metering. These sections are marked as "mandatory" in the code in order to distinguish that these sections must be complied with regardless of whether the performance or prescriptive method of compliance is used.

When using this section and other performance options, the terms "standard design" and "proposed design" are used extensively. These terms can create confusion for some users, but really are very easy to understand. The "proposed design" is exactly that. It is the building as it is proposed to be built, including building size and room configurations, window and wall assemblies, mechanical equipment, orientation, lighting systems, etc. It is essentially the building as it is shown on the construction documents (plans and specifications). The "standard design" is the same building configuration and orientation as the proposed design but instead of matching the plans for the building as it actually will be built, the standard design is shown only to meet the minimum requirements of the code. For example, the standard design building in Climate Zone 4 would require that the walls and roof be insulated to a level of R-13 (+ continuous insulation) and R-30, respectively, based on the requirements in Table C402.1.3. On the other hand, the proposed design may show that the intent is to install R-19 insulation in the walls and R-38 in the roof. Therefore, if everything else in the two buildings was identical, it would be easy to see in this example that the proposed design does exceed what is required by the code in the standard design.

Total-building-performance criteria

Section C407 references the total-building-performance criteria, which offer more flexibility than the prescriptive criteria of this chapter, especially with regard to making trade-offs between building envelope components.

Section C407 allows a designer to make trade-offs between heating and cooling requirements for the building envelope. All components of the building envelope may be considered, including fenestration, floors, ceilings, below-grade walls, opaque walls, and opaque doors. For instance, SHGC can be traded for better wall *R*-values or vice versa. Often, the analysis is performed by a simulation tool considering the overall performance of the building envelope and taking account of the following factors:

1. Orientation.
2. Fenestration area.
3. Fenestration SHGC.
4. PF (size of overhangs).
5. Fenestration *U*-factor.
6. Visible light transmittance, if automatic daylighting controls are installed.
7. Daylighting control fraction, if automatic daylighting controls are installed.
8. *U*-factors of building envelope components.
9. Heat capacity; for example, credit for mass walls.
10. Insulation position, for mass walls only.

Some more-sophisticated building-energy simulation tools can even be used to make trade-offs against the lighting or mechanical requirements. All lighting must comply with the requirements of Section C405, regardless of the values entered under any performance-based analysis. The switching, wiring and exterior efficacy in the total-building-performance method is the only way to make envelope trade-offs against lighting or HVAC improvements.

To summarize, design by total-building performance may be used either when the proposed building cannot meet the prescriptive criteria of this chapter or when more flexibility is desired for an innovative design.

Another important advantage of designing by total-building performance is that the evaluation is based not only on compliance but also on the margin by which a project complies. This margin is used by some utilities as a basis for energy-efficiency rebates.

The primary disadvantage of designing by total-building performance is its complexity. The designer must be familiar with building-energy simulation programs, some of which are difficult to master. The commentary and discussion that follow provide an explanation and some guidelines for modeling a building and for determining compliance, including instructions for using these building-energy simulation programs.

C407.2 Mandatory requirements. Compliance with this section requires that the criteria of Sections C402.5, C403.2, C404 and C405 be met.

❖ Although design by total building performance permits trade-offs in energy use between different building systems, the proposed design still must comply with the mandatory requirements. Sections C402.5, C403.2, C404 and C405 must be met for several reasons:

- Many requirements cannot be accurately modeled.
- Many requirements are inherently cost effective.
- Some requirements are calculation methodologies that establish a fair basis for comparing component performance (such as *U*-factor calculations).
- Some requirements are not intended for trade-offs (such as exterior lighting).

C407.3 Performance-based compliance. Compliance based on total building performance requires that a proposed building (*proposed design*) be shown to have an annual energy cost that is less than or equal to the annual energy cost of the *standard reference design*. Energy prices shall be taken from a source *approved* by the *code official*, such as the Department of Energy, Energy Information Administration's *State Energy Price and Expenditure Report*. *Code officials* shall be permitted to require time-of-use pricing in energy cost calculations. Nondepletable energy collected off site shall be treated and priced the same as purchased energy. Energy from nondepletable energy sources collected on site shall be omitted from the annual energy cost of the *proposed design*.

> **Exception:** Jurisdictions that require site energy (1 kWh = 3413 Btu) rather than energy cost as the metric of comparison.

❖ The intent of this section is there are a variety of programs and systems that can be used for reviewing the energy performance of a proposed building design with an annual energy cost that is less than or equal to the annual energy cost of the standard reference design. Energy prices should be taken from an approved source by the code official. The code official can determine what programs are acceptable. One program that can be used is the COMcheck™ program (which is based on ANSI/ASHRAE/IESNA 90.1). COMcheck™ can demonstrate compliance with commercial building energy codes. Nondepletable energy collected off site shall be treated and priced the same as purchased energy. Energy from nondepletable energy sources collected on site shall be omitted from the annual energy cost of the proposed design.

C407.4 Documentation. Documentation verifying that the methods and accuracy of compliance software tools conform to the provisions of this section shall be provided to the *code official*.

❖ The construction documents must contain the data necessary to confirm that the results of the design have been incorporated into the construction documents and also to allow assessment of compliance. This is of particular importance when coupled with performing a comprehensive plan review, given the lengthy and sophisticated submittals required by the total-building-performance method (see commentary, Section C104).

Sections C407.4.1 through C407.6.2 contain specific requirements for construction documents in support of a total-building-performance analysis. These requirements are intended to reflect the minimum scope of information needed to determine code compliance. The documentation that is required to support the analysis must provide the following information:

- Annual energy use and cost.
- List of building features.
- Output files showing energy use totals.
- Energy use by source and end use.
- Total hours that the space-conditioning loads were not met.
- Software error messages or warnings.
- Written explanations of any error messages or warnings.

C407.4.1 Compliance report. Permit submittals shall include a report documenting that the proposed design has annual energy costs less than or equal to the annual energy costs of the *standard reference design*. The compliance documentation shall include the following information:

1. Address of the building.
2. An inspection checklist documenting the building component characteristics of the *proposed design* as specified in Table C407.5.1(1). The inspection checklist shall show the estimated annual energy cost for both the *standard reference design* and the *proposed design*.

3. Name of individual completing the compliance report.
4. Name and version of the compliance software tool.

❖ Compliance software tools, for example the COMcheck program, can generate a report documenting that the proposed design has annual energy costs less than or equal to the annual energy costs of the standard reference design. State laws in certain locations require that the design of building construction and the development of a comparative building-energy analysis be done by a registered design professional in accordance with the licensing laws of the state where the work will take place.

C407.4.2 Additional documentation. The *code official* shall be permitted to require the following documents:

1. Documentation of the building component characteristics of the *standard reference design.*
2. Thermal zoning diagrams consisting of floor plans showing the thermal zoning scheme for *standard reference design* and *proposed design.*
3. Input and output reports from the energy analysis simulation program containing the complete input and output files, as applicable. The output file shall include energy use totals and energy use by energy source and end-use served, total hours that space conditioning loads are not met and any errors or warning messages generated by the simulation tool as applicable.
4. An explanation of any error or warning messages appearing in the simulation tool output.
5. A certification signed by the builder providing the building component characteristics of the *proposed design* as given in Table C407.5.1(1).

❖ Additional documentation may be requested by the code official to ensure the building component characteristics of the proposed design are complete and correct for that location. Various simulation tools often use jargon or code language that may not be readily understood. Error messages or warning messages often occur during the course of normal user interface with nearly all building-energy simulation tools. Nevertheless, it is up to the applicant to confirm to the code official that such errors or warnings have no substantial or material effect on the validity of the simulation results.

C407.5 Calculation procedure. Except as specified by this section, the *standard reference design* and *proposed design* shall be configured and analyzed using identical methods and techniques.

❖ Calculation procedures used to comply with this section must be software tools capable of calculating the annual energy consumption of all building elements between the standard reference design and the proposed design using identical methods and techniques. The calculation procedure will not allow the user to modify the building component characteristics that would help reduce the energy costs of the standard design.

C407.5.1 Building specifications. The *standard reference design* and *proposed design* shall be configured and analyzed as specified by Table C407.5.1(1). Table C407.5.1(1) shall include by reference all notes contained in Table C402.1.4.

❖ This section provides a performance approach for designers in meeting energy code requirements. The standard reference design and proposed design must be configured and analyzed using the specifications for the standard reference and proposed designs as listed in Table C407.5.1(1) and the building envelope requirements for opaque assemblies in Table C402.1.3. The column labeled "Standard Reference Design" in Table C407.5.1(1) provides the details used as the basis for determining the annual energy cost representing the maximum energy cost that the proposed design could use. This column simply states how the standard design building is to be configured and treated in the simulations comparing it to the proposed design. Further, this column ensures that comparisons between the two designs truly are based on a plan that meets the intended performance of the code. The "Proposed Design" column simply represents the building for which compliance is being evaluated. As indicated, most of the items in the proposed design will be "as proposed." There are a few entries in this column where the proposed design uses the same values as the standard design. These items are restricted so that a fair comparison is provided and the design parameters may not simply be changed in an attempt to show improved efficiency.

C407.5.2 Thermal blocks. The *standard reference design* and *proposed design* shall be analyzed using identical thermal blocks as specified in Section C407.5.2.1, C407.5.2.2 or C407.5.2.3.

❖ The standard reference design and the proposed design are analyzed using the same thermal blocks as specified in Section C407.5.2.1, C407.5.2.2 or C407.5.2.3, as follows. This section requires that the heating and cooling equipment in the standard reference design meet the minimum efficiency requirements that are currently required at the time of the evaluation. A thermal block is a simulation program term. It is similar to an HVAC zone, except that often in simulation practice a number of zones, which are similar in loads and are served by similar systems, are combined into a single thermal block for modeling purposes.

TABLE C407.5.1(1)
SPECIFICATIONS FOR THE STANDARD REFERENCE AND PROPOSED DESIGNS

BUILDING COMPONENT CHARACTERISTICS	STANDARD REFERENCE DESIGN	PROPOSED DESIGN
Space use classification	Same as proposed	The space use classification shall be chosen in accordance with Table C405.5.2 for all areas of the building covered by this permit. Where the space use classification for a building is not known, the building shall be categorized as an office building.
Roofs	Type: Insulation entirely above deck	As proposed
	Gross area: same as proposed	As proposed
	U-factor: as specified in Table C402.1.4	As proposed
	Solar absorptance: 0.75	As proposed
	Emittance: 0.90	As proposed
Walls, above-grade	Type: Mass wall where proposed wall is mass; otherwise steel-framed wall	As proposed
	Gross area: same as proposed	As proposed
	U-factor: as specified in Table C402.1.4	As proposed
	Solar absorptance: 0.75	As proposed
	Emittance: 0.90	As proposed
Walls, below-grade	Type: Mass wall	As proposed
	Gross area: same as proposed	As proposed
	U-Factor: as specified in Table C402.1.4 with insulation layer on interior side of walls	As proposed
Floors, above-grade	Type: joist/framed floor	As proposed
	Gross area: same as proposed	As proposed
	U-factor: as specified in Table C402.1.4	As proposed
Floors, slab-on-grade	Type: Unheated	As proposed
	F-factor: as specified in Table C402.1.4	As proposed
Opaque doors	Type: Swinging	As proposed
	Area: Same as proposed	As proposed
	U-factor: as specified in Table C402.1.4	As proposed
Vertical fenestration other than opaque doors	Area 1. The proposed glazing area; where the proposed glazing area is less than 40 percent of above-grade wall area. 2. 40 percent of above-grade wall area; where the proposed glazing area is 40 percent or more of the above-grade wall area.	As proposed
	U-factor: as specified in Table C402.4	As proposed
	SHGC: as specified in Table C402.4 except that for climates with no requirement (NR) SHGC = 0.40 shall be used	As proposed
	External shading and PF: None	As proposed
Skylights	Area 1. The proposed skylight area; where the proposed skylight area is less than 3 percent of gross area of roof assembly. 2. 3 percent of gross area of roof assembly; where the proposed skylight area is 3 percent or more of gross area of roof assembly	As proposed
	U-factor: as specified in Table C402.4	As proposed
	SHGC: as specified in Table C402.4 except that for climates with no requirement (NR) SHGC = 0.40 shall be used.	As proposed
Lighting, interior	The interior lighting power shall be determined in accordance with Section C405.4.2. Where the occupancy of the building is not known, the lighting power density shall be 1.0 Watt per square foot (10.7 W/m^2) based on the categorization of buildings with unknown space classification as offices.	As proposed
Lighting, exterior	The lighting power shall be determined in accordance with Table C405.5.2(2). Areas and dimensions of tradable and nontradable surfaces shall be the same as proposed.	As proposed

(continued)

TABLE C407.5.1(1)—continued
SPECIFICATIONS FOR THE STANDARD REFERENCE AND PROPOSED DESIGNS

BUILDING COMPONENT CHARACTERISTICS	STANDARD REFERENCE DESIGN	PROPOSED DESIGN
Internal gains	Same as proposed	Receptacle, motor and process loads shall be modeled and estimated based on the space use classification. All end-use load components within and associated with the building shall be modeled to include, but not be limited to, the following: exhaust fans, parking garage ventilation fans, exterior building lighting, swimming pool heaters and pumps, elevators, escalators, refrigeration equipment and cooking equipment.
Schedules	Same as proposed	Operating schedules shall include hourly profiles for daily operation and shall account for variations between weekdays, weekends, holidays and any seasonal operation. Schedules shall model the time-dependent variations in occupancy, illumination, receptacle loads, thermostat settings, mechanical ventilation, HVAC equipment availability, service hot water usage and any process loads. The schedules shall be typical of the proposed building type as determined by the designer and approved by the jurisdiction.
Mechanical ventilation	Same as proposed	As proposed, in accordance with Section C403.2.6.
Heating systems	Fuel type: same as proposed design	As proposed
	Equipment type[a]: as specified in Tables C407.5.1(2) and C407.5.1(3)	As proposed
	Efficiency: as specified in Tables C403.2.3(4) and C403.2.3(5)	As proposed
	Capacity[b]: sized proportionally to the capacities in the proposed design based on sizing runs, and shall be established such that no smaller number of unmet heating load hours and no larger heating capacity safety factors are provided than in the proposed design.	As proposed
Cooling systems	Fuel type: same as proposed design	As proposed
	Equipment type[c]: as specified in Tables C407.5.1(2) and C407.5.1(3)	As proposed
	Efficiency: as specified in Tables C403.2.3(1), C403.2.3(2) and C403.2.3(3)	As proposed
	Capacity[b]: sized proportionally to the capacities in the proposed design based on sizing runs, and shall be established such that no smaller number of unmet cooling load hours and no larger cooling capacity safety factors are provided than in the proposed design.	As proposed
	Economizer[d]: same as proposed, in accordance with Section C403.3.	As proposed
Service water heating[e]	Fuel type: same as proposed	As proposed
	Efficiency: as specified in Table C404.2	For Group R, as proposed multiplied by SWHF. For other than Group R, as proposed multiplied by efficiency as provided by the manufacturer of the DWHR unit.
	Capacity: same as proposed	As proposed
	Where no service water hot water system exists or is specified in the proposed design, no service hot water heating shall be modeled.	

SWHF = Service water heat recovery factor, DWHR = Drain water heat recovery.

a. Where no heating system exists or has been specified, the heating system shall be modeled as fossil fuel. The system characteristics shall be identical in both the standard reference design and proposed design.

b. The ratio between the capacities used in the annual simulations and the capacities determined by sizing runs shall be the same for both the standard reference design and proposed design.

c. Where no cooling system exists or no cooling system has been specified, the cooling system shall be modeled as an air-cooled single-zone system, one unit per thermal zone. The system characteristics shall be identical in both the standard reference design and proposed design.

d. If an economizer is required in accordance with Table C403.3 and where no economizer exists or is specified in the proposed design, then a supply-air economizer shall be provided in the standard reference design in accordance with Section C403.3.

e. The SWHF shall be applied as follows:

1. Where potable water from the DWHR unit supplies not less than one shower and not greater than two showers, of which the drain water from the same showers flows through the DWHR unit then SWHF = [1 – (DWHR unit efficiency · 0.36)].
2. Where potable water from the DWHR unit supplies not less than three showers and not greater than four showers, of which the drain water from the same showers flows through the DWHR unit then SWHF = [1 – (DWHR unit efficiency · 0.33)].
3. Where potable water from the DWHR unit supplies not less than five showers and not greater than six showers, of which the drain water from the same showers flows through the DWHR unit, then SWHF = [1 – (DWHR unit efficiency · 0.26)].
4. Where Items 1 through 3 are not met, SWHF = 1.0.

TABLE C407.5.1(2)
HVAC SYSTEMS MAP

CONDENSER COOLING SOURCE[a]	HEATING SYSTEM CLASSIFICATION[b]	STANDARD REFERENCE DESIGN HVC SYSTEM TYPE[c]		
		Single-zone Residential System	Single-zone Nonresidential System	All Other
Water/ground	Electric resistance	System 5	System 5	System 1
	Heat pump	System 6	System 6	System 6
	Fossil fuel	System 7	System 7	System 2
Air/none	Electric resistance	System 8	System 9	System 3
	Heat pump	System 8	System 9	System 3
	Fossil fuel	System 10	System 11	System 4

a. Select "water/ground" where the proposed design system condenser is water or evaporatively cooled; select "air/none" where the condenser is air cooled. Closed-circuit dry coolers shall be considered air cooled. Systems utilizing district cooling shall be treated as if the condenser water type were "water." Where no mechanical cooling is specified or the mechanical cooling system in the proposed design does not require heat rejection, the system shall be treated as if the condenser water type were "Air." For proposed designs with ground-source or groundwater-source heat pumps, the standard reference design HVAC system shall be water-source heat pump (System 6).

b. Select the path that corresponds to the proposed design heat source: electric resistance, heat pump (including air source and water source), or fuel fired. Systems utilizing district heating (steam or hot water) and systems with no heating capability shall be treated as if the heating system type were "fossil fuel." For systems with mixed fuel heating sources, the system or systems that use the secondary heating source type (the one with the smallest total installed output capacity for the spaces served by the system) shall be modeled identically in the standard reference design and the primary heating source type shall be used to determine *standard* reference design HVAC system type.

c. Select the standard reference design HVAC system category: The system under "single-zone residential system" shall be selected where the HVAC system in the proposed design is a single-zone system and serves a residential space. The system under "single-zone nonresidential system" shall be selected where the HVAC system in the proposed design is a single-zone system and serves other than residential spaces. The system under "all other" shall be selected for all other cases.

**TABLE C407.5.1(3)
SPECIFICATIONS FOR THE STANDARD REFERENCE DESIGN HVAC SYSTEM DESCRIPTIONS**

SYSTEM NO.	SYSTEM TYPE	FAN CONTROL	COOLING TYPE	HEATING TYPE
1	Variable air volume with parallel fan-powered boxes[a]	VAV[d]	Chilled water[e]	Electric resistance
2	Variable air volume with reheat[b]	VAV[d]	Chilled water[e]	Hot water fossil fuel boiler[f]
3	Packaged variable air volume with parallel fan-powered boxes[a]	VAV[d]	Direct expansion[c]	Electric resistance
4	Packaged variable air volume with reheat[b]	VAV[d]	Direct expansion[c]	Hot water fossil fuel boiler[f]
5	Two-pipe fan coil	Constant volume[i]	Chilled water[e]	Electric resistance
6	Water-source heat pump	Constant volume[i]	Direct expansion[c]	Electric heat pump and boiler[g]
7	Four-pipe fan coil	Constant volume[i]	Chilled water[e]	Hot water fossil fuel boiler[f]
8	Packaged terminal heat pump	Constant volume[i]	Direct expansion[c]	Electric heat pump[h]
9	Packaged rooftop heat pump	Constant volume[i]	Direct expansion[c]	Electric heat pump[h]
10	Packaged terminal air conditioner	Constant volume[i]	Direct expansion	Hot water fossil fuel boiler[f]
11	Packaged rooftop air conditioner	Constant volume[i]	Direct expansion	Fossil fuel furnace

For SI: 1 foot = 304.8 mm, 1 cfm/ft^2 = 0.4719 L/s, 1 Btu/h = 0.293/W, °C = [(°F) -32/1.8].

a. **VAV with parallel boxes:** Fans in parallel VAV fan-powered boxes shall be sized for 50 percent of the peak design flow rate and shall be modeled with 0.35 W/cfm fan power. Minimum volume setpoints for fan-powered boxes shall be equal to the minimum rate for the space required for ventilation consistent with Section C403.4.4, Exception 4. Supply air temperature setpoint shall be constant at the design condition.

b. **VAV with reheat:** Minimum volume setpoints for VAV reheat boxes shall be 0.4 cfm/ft^2 of floor area. Supply air temperature shall be reset based on zone demand from the design temperature difference to a 10°F temperature difference under minimum load conditions. Design airflow rates shall be sized for the reset supply air temperature, i.e., a 10°F temperature difference.

c. **Direct expansion:** The fuel type for the cooling system shall match that of the cooling system in the proposed design.

d. VAV: Where the proposed design system has a supply, return or relief fan motor 25 hp or larger, the corresponding fan in the VAV system of the standard reference design shall be modeled assuming a variable-speed drive. For smaller fans, a forward-curved centrifugal fan with inlet vanes shall be modeled. Where the proposed design's system has a direct digital control system at the zone level, static pressure setpoint reset based on zone requirements in accordance with Section C403.4.1 shall be modeled.

e. **Chilled water:** For systems using purchased chilled water, the chillers are not explicitly modeled and chilled water costs shall be based as determined in Sections C407.3 and C407.5.2. Otherwise, the standard reference design's chiller plant shall be modeled with chillers having the number as indicated in Table C407.5.1(4) as a function of standard reference building chiller plant load and type as indicated in Table C407.5.1(5) as a function of individual chiller load. Where chiller fuel source is mixed, the system in the standard reference design shall have chillers with the same fuel types and with capacities having the same proportional capacity as the proposed design's chillers for each fuel type. Chilled water supply temperature shall be modeled at 44°F design supply temperature and 56°F return temperature. Piping losses shall not be modeled in either building model. Chilled water supply water temperature shall be reset in accordance with Section C403.4.3.3. Pump system power for each pumping system shall be the same as the proposed design; where the proposed design has no chilled water pumps, the standard reference design pump power shall be 22 W/gpm (equal to a pump operating against a 75-foot head, 65-percent combined impeller and motor efficiency). The chilled water system shall be modeled as primary-only variable flow with flow maintained at the design rate through each chiller using a bypass. Chilled water pumps shall be modeled as riding the pump curve or with variable-speed drives when required in Section C403.4.3.3. The heat rejection device shall be an axial fan cooling tower with two-speed fans where required in Section C403.4.3. Condenser water design supply temperature shall be 85°F or 10°F approach to design wet-bulb temperature, whichever is lower, with a design temperature rise of 10°F. The tower shall be controlled to maintain a 70°F leaving water temperature where weather permits, floating up to leaving water temperature at design conditions. Pump system power for each pumping system shall be the same as the proposed design; where the proposed design has no condenser water pumps, the standard reference design pump power shall be 19 W/gpm (equal to a pump operating against a 60-foot head, 60-percent combined impeller and motor efficiency). Each chiller shall be modeled with separate condenser water and chilled water pumps interlocked to operate with the associated chiller.

f. **Fossil fuel boiler:** For systems using purchased hot water or steam, the boilers are not explicitly modeled and hot water or steam costs shall be based on actual utility rates. Otherwise, the boiler plant shall use the same fuel as the proposed design and shall be natural draft. The standard reference design boiler plant shall be modeled with a single boiler where the standard reference design plant load is 600,000 Btu/h and less and with two equally sized boilers for plant capacities exceeding 600,000 Btu/h. Boilers shall be staged as required by the load. Hot water supply temperature shall be modeled at 180°F design supply temperature and 130°F return temperature. Piping losses shall not be modeled in either building model. Hot water supply water temperature shall be reset in accordance with Section C403.4.3.3. Pump system power for each pumping system shall be the same as the proposed design; where the proposed design has no hot water pumps, the standard reference design pump power shall be 19 W/gpm (equal to a pump operating against a 60-foot head, 60-percent combined impeller and motor efficiency). The hot water system shall be modeled as primary only with continuous variable flow. Hot water pumps shall be modeled as riding the pump curve or with variable speed drives when required by Section C403.4.3.3.

g. **Electric heat pump and boiler:** Water-source heat pumps shall be connected to a common heat pump water loop controlled to maintain temperatures between 60°F and 90°F. Heat rejection from the loop shall be provided by an axial fan closed-circuit evaporative fluid cooler with two-speed fans where required in Section C403.4.1. Heat addition to the loop shall be provided by a boiler that uses the same fuel as the proposed design and shall be natural draft. Where no boilers exist in the proposed design, the standard reference building boilers shall be fossil fuel. The standard reference design boiler plant shall be modeled with a single boiler where the standard reference design plant load is 600,000 Btu/h or less and with two equally sized boilers for plant capacities exceeding 600,000 Btu/h. Boilers shall be staged as required by the load. Piping losses shall not be modeled in either building model. Pump system power shall be the same as the proposed design; where the proposed design has no pumps, the standard reference design pump power shall be 22 W/gpm, which is equal to a pump operating against a 75-foot head, with a 65-percent combined impeller and motor efficiency. Loop flow shall be variable with flow shutoff at each heat pump when its compressor cycles off as required by Section C403.4.3.3. Loop pumps shall be modeled as riding the pump curve or with variable speed drives when required by Section C403.4.3.3.

h. **Electric heat pump:** Electric air-source heat pumps shall be modeled with electric auxiliary heat. The system shall be controlled with a multistage space thermostat and an outdoor air thermostat wired to energize auxiliary heat only on the last thermostat stage and when outdoor air temperature is less than 40°F.

i. **Constant volume:** Fans shall be controlled in the same manner as in the proposed design; i.e., fan operation whenever the space is occupied or fan operation cycled on calls for heating and cooling. Where the fan is modeled as cycling and the fan energy is included in the energy efficiency rating of the equipment, fan energy shall not be modeled explicitly.

TABLE C407.5.1(4)
NUMBER OF CHILLERS

TOTAL CHILLER PLANT CAPACITY	NUMBER OF CHILLERS
≤ 300 tons	1
> 300 tons, < 600 tons	2, sized equally
≥ 600 tons	2 minimum, with chillers added so that no chiller is larger than 800 tons, all sized equally

For SI: 1 ton = 3517 W.

TABLE C407.5.1(5)
WATER CHILLER TYPES

INDIVIDUAL CHILLER PLANT CAPACITY	ELECTRIC CHILLER TYPE	FOSSIL FUEL CHILLER TYPE
≤ 100 tons	Reciprocating	Single-effect absorption, direct fired
> 100 tons, < 300 tons	Screw	Double-effect absorption, direct fired
≥ 300 tons	Centrifugal	Double-effect absorption, direct fired

For SI: 1 ton = 3517 W.

C407.5.2.1 HVAC zones designed. Where HVAC *zones* are defined on HVAC design drawings, each HVAC *zone* shall be modeled as a separate thermal block.

Exception: Different HVAC *zones* shall be allowed to be combined to create a single thermal block or identical thermal blocks to which multipliers are applied provided:

1. The space use classification is the same throughout the thermal block.
2. All HVAC *zones* in the thermal block that are adjacent to glazed exterior walls face the same orientation or their orientations are within 45 degrees (0.79 rad) of each other.
3. All of the *zones* are served by the same HVAC system or by the same kind of HVAC system.

❖ Each HVAC zone must be designed according to Section C403 and must be modeled as a separate thermal block. An HVAC zone is physically determined by the design of the HVAC system. It includes some number of thermodynamically similar spaces, whose loads can be satisfied through the use of a single thermostat or other type of temperature control. The duct outlets or other terminal units controlled by a single thermostat serve the zone.

For example, the interior HVAC zones of a multistory building may be physically separate zones on each floor, but they often may be combined into a single thermal block in the simulation model of the building because they have similar loads and are served by similar systems. However, a cafeteria or computer room in an office building would need to be modeled separately, as would lower-floor retail uses.

Zones may be combined, or multipliers may be used, if all the following conditions are met:

1. All of the space use classifications must be the same throughout the thermal block. This ensures that they have the same load and schedule characteristics.
2. For exterior HVAC zones with glazing, the glazing for all zones included in the thermal block must have the same orientation, or their orientations must be within 45 degrees (0.79 rad) of each other. This ensures that they have the same solar heat gain characteristics. This is not to say that the zones may not have two or more glazing orientations—a corner office could easily have two—but that the zones must have similar orientations. It would be acceptable, for example, to group all of the northeast corner offices on the intermediate floors of an office tower into a single thermal block.

All of the HVAC zones in the thermal block must be served either by the same HVAC system or by the same kind of HVAC system. This is so the simulation program can accurately model the performance of the system(s) serving the block.

C407.5.2.2 HVAC zones not designed. Where HVAC *zones* have not yet been designed, thermal blocks shall be defined based on similar internal load densities, occupancy, lighting, thermal and temperature schedules, and in combination with the following guidelines:

1. Separate thermal blocks shall be assumed for interior and perimeter spaces. Interior spaces shall be those located more than 15 feet (4572 mm) from an exterior wall. Perimeter spaces shall be those located closer than 15 feet (4572 mm) from an *exterior wall.*
2. Separate thermal blocks shall be assumed for spaces adjacent to glazed exterior walls: a separate *zone* shall be provided for each orientation, except orientations that differ by not more than 45 degrees (0.79 rad) shall be permitted to be considered to be the same orientation. Each *zone* shall include floor area that is 15 feet (4572 mm) or less from a glazed perimeter wall, except that floor area within 15 feet (4572 mm) of glazed perimeter walls having more than one orientation shall be divided proportionately between *zones.*

3. Separate thermal blocks shall be assumed for spaces having floors that are in contact with the ground or exposed to ambient conditions from *zones* that do not share these features.
4. Separate thermal blocks shall be assumed for spaces having exterior ceiling or roof assemblies from *zones* that do not share these features.

❖ When HVAC zones are not designed at the time of the permit application for which the total-building-performance method is being used, then the configuration of the thermal blocks must be assumed. This is quite common in commercial buildings where the future tenants will determine the zoning of spaces in the building. A default heating system consisting of thermal blocks must be assumed and modeled on similar internal load densities, occupancy, lighting, and thermal and temperature schedules; and in combination with the previously listed guidelines. This default heating system should be a simple, fossil-fuel heating system with sufficient capacity to meet the design heating loads for the building as follows:

- When the HVAC zones are not yet designed, assign separate thermal blocks to the interior spaces located more than 15 feet (4572 mm) from an exterior wall and to the perimeter spaces within 15 feet (4572 mm) of the exterior.
- Glazed exterior walls should be assigned to different perimeter thermal blocks for each major orientation. Orientations within 45 degrees (0.79 rad) of each other may be combined. Spaces with two or more glazed orientations, such as corner offices, should be divided proportionately between zones having different orientations.
- Spaces exposed to ambient conditions, such as the top floor or an overhanging floor, and spaces in contact with the ground, such as the ground floor, must be zoned separately from zones that are not exposed to ambient conditions, such as intermediate floors in a multistory building.

C407.5.2.3 Multifamily residential buildings. Residential spaces shall be modeled using one thermal block per space except that those facing the same orientations are permitted to be combined into one thermal block. Corner units and units with roof or floor loads shall only be combined with units sharing these features.

❖ In multifamily residential buildings, the residential spaces must be treated as separate thermal blocks, except that some combinations are allowed. Units all facing the same orientation, and having similar conditions at the top, bottom and sides, may be combined. Similar corner units may be combined, and units with similar roof or floor loads may be combined.

C407.6 Calculation software tools. Calculation procedures used to comply with this section shall be software tools capable of calculating the annual energy consumption of all building elements that differ between the *standard reference design* and the *proposed design* and shall include the following capabilities.

1. Building operation for a full calendar year (8,760 hours).
2. Climate data for a full calendar year (8,760 hours) and shall reflect approved coincident hourly data for temperature, solar radiation, humidity and wind speed for the building location.
3. Ten or more thermal zones.
4. Thermal mass effects.
5. Hourly variations in occupancy, illumination, receptacle loads, thermostat settings, mechanical ventilation, HVAC equipment availability, service hot water usage and any process loads.
6. Part-load performance curves for mechanical equipment.
7. Capacity and efficiency correction curves for mechanical heating and cooling equipment.
8. Printed code official inspection checklist listing each of the proposed design component characteristics from Table C407.5.1(1) determined by the analysis to provide compliance, along with their respective performance ratings including, but not limited to, *R*-value, *U*-factor, SHGC, HSPF, AFUE, SEER, EF.

❖ Software tools, such as DOE-2, BLAST, COMcheck or Energy/Plus, capable of calculating the annual energy consumption of all building elements can be used to differentiate between the standard reference design and the proposed design. The calculation software tool must have the ability to model all of the following:

- Building operation and climate data for 8,760 hours per year.
- Ten or more thermal zones.
- Thermal mass effects.
- Part-load performance curves for mechanical heating and cooling equipment.
- Capacity and efficiency correction curves for mechanical heating and cooling equipment.
- Printed code official inspection checklist with each of the proposed design component's characteristics.

C407.6.1 Specific approval. Performance analysis tools complying with the applicable subsections of Section C407 and tested according to ASHRAE Standard 140 shall be permitted to be *approved.* Tools are permitted to be *approved* based on meeting a specified threshold for a jurisdiction. The *code official* shall be permitted to approve tools for a specified application or limited scope.

❖ Performance analysis tools complying with Section C407 and tested according to ASHRAE 140 must be permitted to be approved if they meet the specific requirements of the jurisdiction. The code official must be permitted to approve tools for a specific application or limited scope for total building performance.

C407.6.2 Input values. Where calculations require input values not specified by Sections C402, C403, C404 and C405, those input values shall be taken from an *approved* source.

❖ Input values must be taken from an approved source. This section was intended to prevent the unscrupulous gaming of sizing discrepancies between the proposed design and the standard design. The requirement has the effect of preventing the systems proposed for the standard design from being conveniently oversized to make the proposed building systems result in a more favorable net energy cost credit.

C407.6.3 Exceptional calculation methods. Where the simulation program does not model a design, material or device of the *proposed design*, an exceptional calculation method shall be used where approved by the *code official*. Where there are multiple designs, materials or devices that the simulation program does not model, each shall be calculated separately and exceptional savings determined for each. The total exceptional savings shall not constitute more than half of the difference between the baseline building performance and the proposed building performance. Applications for approval of an exceptional method shall include all of the following:

1. Step-by-step documentation of the exceptional calculation method performed, detailed enough to reproduce the results.
2. Copies of all spreadsheets used to perform the calculations.
3. A sensitivity analysis of energy consumption where each of the input parameters is varied from half to double the value assumed.
4. The calculations shall be performed on a time step basis consistent with the simulation program used.
5. The performance rating calculated with and without the exceptional calculation method.

❖ This section provides both guidance and limitation on the use of exceptional calculation methods. While the analysis programs continue to grow in sophistication, not every design concept can be modeled. Therefore, it is sometimes necessary to go outside established simulation programs to use other calculation methods. While this provides flexibility of review and guidance to the code official, it limits the extent to which such calculations are used in any one design. ANSI/ASHRAE/IESNA 90.1 also addresses exceptional calculation methods that can be used as an example of this approach.

SECTION C408
SYSTEM COMMISSIONING

C408.1 General. This section covers the commissioning of the building mechanical systems in Section C403 and electrical power and lighting systems in Section C405.

❖ Commissioning as provided in this section ensures that the mechanical systems and electrical systems (lighting controls) perform as designed for the intended purpose. Commissioning is a systematic process of verification and documentation that ensures the selected building systems have been designed, installed and function properly and can be maintained in accordance with the contract documents in order to satisfy the building owner's design intent and operational requirements. This section of the code details the commissioning plan that must be developed by a registered design professional or an approved agency, and checks that the completed work complies with the design documents and also details what documentation needs to be provided to the code official. The code requirements for commissioning are: a preliminary commissioning report; drawings and manuals; a system balancing report; a final commissioning report; and verification of HVAC, lighting and electrical systems. Functional performance testing of equipment, controls, economizers and lighting control systems is necessary to ensure such systems function as designed and operate in the intended manner. The code requires the owner to receive documentation of the mechanical and lighting systems commissioning and completion requirements. The code official can obtain a copy of this documentation on request. Manuals must be provided for the overall building system so the building operator can understand the operation, maintain it as designed, and troubleshoot any future problems to keep it operating as intended.

C408.2 Mechanical systems and service water-heating systems commissioning and completion requirements. Prior to the final mechanical and plumbing inspections, the *registered design professional* or *approved agency* shall provide evidence of mechanical systems *commissioning* and completion in accordance with the provisions of this section.

Construction document notes shall clearly indicate provisions for *commissioning* and completion requirements in accordance with this section and are permitted to refer to specifications for further requirements. Copies of all documentation shall be given to the owner or owner's authorized agent and made available to the *code official* upon request in accordance with Sections C408.2.4 and C408.2.5.

Exceptions: The following systems are exempt:

1. Mechanical systems and service water heater systems in buildings where the total mechanical equip-

ment capacity is less than 480,000 Btu/h (140.7 kW) cooling capacity and 600,000 Btu/h (175.8 kW) combined service water-heating and space-heating capacity.

2. Systems included in Section C403.3 that serve individual *dwelling units* and *sleeping units*.

❖ See the commentary to Section C408.1.

C408.2.1 Commissioning plan. A *commissioning plan* shall be developed by a *registered design professional* or *approved agency* and shall include the following items:

1. A narrative description of the activities that will be accomplished during each phase of *commissioning,* including the personnel intended to accomplish each of the activities.
2. A listing of the specific equipment, appliances or systems to be tested and a description of the tests to be performed.
3. Functions to be tested including, but not limited to, calibrations and economizer controls.
4. Conditions under which the test will be performed. Testing shall affirm winter and summer design conditions and full outside air conditions.
5. Measurable criteria for performance.

❖ See the commentary to Section C408.1.

C408.2.2 Systems adjusting and balancing. HVAC systems shall be balanced in accordance with generally accepted engineering standards. Air and water flow rates shall be measured and adjusted to deliver final flow rates within the tolerances provided in the product specifications. Test and balance activities shall include air system and hydronic system balancing.

❖ See the commentary to Section C408.1.

C408.2.2.1 Air systems balancing. Each supply air outlet and *zone* terminal device shall be equipped with means for air balancing in accordance with the requirements of Chapter 6 of the *International Mechanical Code*. Discharge dampers used for air-system balancing are prohibited on constant-volume fans and variable- volume fans with motors 10 hp (18.6 kW) and larger. Air systems shall be balanced in a manner to first minimize throttling losses then, for fans with system power of greater than 1 hp (0.746 kW), fan speed shall be adjusted to meet design flow conditions.

Exception: Fans with fan motors of 1 hp (0.74 kW) or less are not required to be provided with a means for air balancing.

❖ See the commentary to Section C408.1.

C408.2.2.2 Hydronic systems balancing. Individual hydronic heating and cooling coils shall be equipped with means for balancing and measuring flow. Hydronic systems shall be proportionately balanced in a manner to first minimize throttling losses, then the pump impeller shall be trimmed or pump speed shall be adjusted to meet design flow conditions. Each hydronic system shall have either the capability to measure pressure across the pump, or test ports at each side of each pump.

Exceptions: The following equipment is not required to be equipped with a means for balancing or measuring flow:

1. Pumps with pump motors of 5 hp (3.7 kW) or less.
2. Where throttling results in no greater than 5 percent of the nameplate horsepower draw above that required if the impeller were trimmed.

❖ See the commentary to Section C408.1.

C408.2.3 Functional performance testing. Functional performance testing specified in Sections C408.2.3.1 through C408.2.3.3 shall be conducted.

❖ See the commentary to Section C408.1.

C408.2.3.1 Equipment. Equipment functional performance testing shall demonstrate the installation and operation of components, systems, and system-to-system interfacing relationships in accordance with approved plans and specifications such that operation, function, and maintenance serviceability for each of the commissioned systems is confirmed. Testing shall include all modes and *sequence of operation*, including under full-load, part-load and the following emergency conditions:

1. All modes as described in the *sequence* of *operation.*
2. Redundant or *automatic* back-up mode.
3. Performance of alarms.
4. Mode of operation upon a loss of power and restoration of power.

Exception: Unitary or packaged HVAC equipment listed in Tables C403.2.3(1) through C403.2.3(3) that do not require supply air economizers.

❖ See the commentary to Section C408.1.

C408.2.3.2 Controls. HVAC and service water-heating control systems shall be tested to document that control devices, components, equipment and systems are calibrated and adjusted and operate in accordance with approved plans and specifications. Sequences of operation shall be functionally tested to document they operate in accordance with *approved* plans and specifications.

❖ See the commentary to Section C408.1.

C408.2.3.3 Economizers. Air economizers shall undergo a functional test to determine that they operate in accordance with manufacturer's specifications.

❖ See the commentary to Section C408.1.

C408.2.4 Preliminary commissioning report. A preliminary report of *commissioning* test procedures and results shall be completed and certified by the *registered design professional* or *approved agency* and provided to the building owner or owner's authorized agent. The report shall be organized with mechanical and service hot water findings in separate sections to allow independent review. The report shall be identified as "Preliminary Commissioning Report" and shall identify:

1. Itemization of deficiencies found during testing required by this section that have not been corrected at the time of report preparation.
2. Deferred tests that cannot be performed at the time of report preparation because of climatic conditions.
3. Climatic conditions required for performance of the deferred tests.

❖ See the commentary to Section C408.1.

C408.2.4.1 Acceptance of report. Buildings, or portions thereof, shall not be considered acceptable for a final inspection pursuant to Section C104.3 until the *code official* has received a letter of transmittal from the building owner acknowledging that the building owner or owner's authorized agent has received the Preliminary Commissioning Report.

❖ See the commentary to Section C408.1.

C408.2.4.2 Copy of report. The *code official* shall be permitted to require that a copy of the Preliminary Commissioning Report be made available for review by the *code official.*

❖ See the commentary to Section C408.1.

C408.2.5 Documentation requirements. The *construction documents* shall specify that the documents described in this section be provided to the building owner or owner's authorized agent within 90 days of the date of receipt of the *certificate of occupancy.*

❖ See the commentary to Section C408.1.

C408.2.5.1 Drawings. *Construction documents* shall include the location and performance data on each piece of equipment.

❖ See the commentary to Section C408.1.

C408.2.5.2 Manuals. An operating and maintenance manual shall be provided and include all of the following:

1. Submittal data stating equipment size and selected options for each piece of equipment requiring maintenance.
2. Manufacturer's operation manuals and maintenance manuals for each piece of equipment requiring maintenance, except equipment not furnished as part of the project. Required routine maintenance actions shall be clearly identified.
3. Name and address of at least one service agency.
4. HVAC and service hot water controls system maintenance and calibration information, including wiring diagrams, schematics and control sequence descriptions. Desired or field-determined set points shall be permanently recorded on control drawings at control devices or, for digital control systems, in system programming instructions.
5. Submittal data indicating all selected options for each piece of lighting equipment and lighting controls.
6. Operation and maintenance manuals for each piece of lighting equipment. Required routine maintenance actions, cleaning and recommended relamping shall be clearly identified.
7. A schedule for inspecting and recalibrating all lighting controls.
8. A narrative of how each system is intended to operate, including recommended set points.

❖ See the commentary to Section C408.1.

C408.2.5.3 System balancing report. A written report describing the activities and measurements completed in accordance with Section C408.2.2.

❖ See the commentary to Section C408.1.

C408.2.5.4 Final commissioning report. A report of test procedures and results identified as "Final Commissioning Report" shall be delivered to the building owner or owner's authorized agent. The report shall be organized with mechanical system and service hot water system findings in separate sections to allow independent review. The report shall include the following:

1. Results of functional performance tests.
2. Disposition of deficiencies found during testing, including details of corrective measures used or proposed.
3. Functional performance test procedures used during the commissioning process including measurable criteria for test acceptance, provided herein for repeatability.

Exception: Deferred tests that cannot be performed at the time of report preparation due to climatic conditions.

❖ See the commentary to Section C408.1.

C408.3 Lighting system functional testing. Controls for automatic lighting systems shall comply with this section.

❖ See the commentary to Section C408.1.

C408.3.1 Functional testing. Prior to passing final inspection, the *registered design professional* shall provide evidence that the lighting control systems have been tested to ensure that control hardware and software are calibrated, adjusted, programmed and in proper working condition in accordance with the *construction documents* and manufacturer's instructions. Functional testing shall be in accordance with Sections C408.3.1.1 and C408.3.1.2 for the applicable control type.

❖ See the commentary to Section C408.1.

C408.3.1.1 Occupant sensor controls. Where *occupant sensor controls* are provided, the following procedures shall be performed:

1. Certify that the *occupant sensor* has been located and aimed in accordance with manufacturer recommendations.
2. For projects with seven or fewer *occupant sensors,* each sensor shall be tested.
3. For projects with more than seven *occupant sensors*, testing shall be done for each unique combination of

sensor type and space geometry. Where multiples of each unique combination of sensor type and space geometry are provided, not less than 10 percent, but in no case less than one, of each combination shall be tested unless the *code official* or design professional requires a higher percentage to be tested. Where 30 percent or more of the tested controls fail, all remaining identical combinations shall be tested.

For *occupant sensor controls* to be tested, verify the following:

3.1. Where *occupant sensor controls* include status indicators, verify correct operation.

3.2. The controlled lights turn off or down to the permitted level within the required time.

3.3. For auto-on *occupant sensor controls*, the lights turn on to the permitted level when an occupant enters the space.

3.4. For manual-on *occupant sensor controls*, the lights turn on only when manually activated.

3.5. The lights are not incorrectly turned on by movement in adjacent areas or by HVAC operation.

❖ See the commentary to Section C408.1.

C408.3.1.2 Time-switch controls. Where *time-switch controls* are provided, the following procedures shall be performed:

1. Confirm that the *time-switch control* is programmed with accurate weekday, weekend and holiday schedules.
2. Provide documentation to the owner of *time- switch controls* programming including weekday, weekend, holiday schedules, and set-up and preference program settings.
3. Verify the correct time and date in the time switch.
4. Verify that any battery back-up is installed and energized.
5. Verify that the override time limit is set to not more than 2 hours.
6. Simulate occupied condition. Verify and document the following:

   6.1. All lights can be turned on and off by their respective area control switch.

   6.2. The switch only operates lighting in the enclosed space in which the switch is located.

7. Simulate unoccupied condition. Verify and document the following:

   7.1. Nonexempt lighting turns off.

   7.2. Manual override switch allows only the lights in the enclosed space where the override switch is located to turn on or remain on until the next scheduled shutoff occurs.

8. Additional testing as specified by the *registered design professional.*

❖ See the commentary to Section C408.1.

C408.3.1.3 Daylight responsive controls. Where *daylight responsive controls* are provided, the following shall be verified:

1. Control devices have been properly located, field calibrated and set for accurate setpoints and threshold light levels.
2. Daylight controlled lighting loads adjust to light level set points in response to available daylight.
3. The locations of calibration adjustment equipment are readily accessible only to authorized personnel.

❖ See the commentary to Section C408.1.

C408.3.2 Documentation requirements. The *construction documents* shall specify that documents certifying that the installed lighting controls meet documented performance criteria of Section C405 are to be provided to the building owner within 90 days from the date of receipt of the *certificate of occupancy.*

❖ See the commentary to Section C408.1.

Bibliography

The following resource materials were used in the preparation of the commentary for this chapter of the code.

AABC-02, *National Standards for Total System Balance*. Washington, DC: Associated Air Balance Council, 2002.

Advanced Lighting Guidelines. Sacramento, CA: California Energy Commission, 1993.

ASHRAE-07 *Handbook—HVAC Applications*. Atlanta, GA: American Society of Heating, Refrigerating and Air-Conditioning Engineers, Inc., 2007

ASHRAE-08 *Handbook—HVAC Systems and Equipment*. Atlanta, GA: American Society of Heating, Refrigerating and Air-Conditioning Engineers, Inc., 2008.

ASHRAE-09, *Handbook of Fundamentals*. Atlanta, GA: American Society of Heating, Refrigerating and Air-Conditioning Engineers, Inc., 2009.

ASHRAE/IES 93, *Energy Code for Commercial and High Rise Residential Buildings—Codification of ASHRAE/IES Standard 90.1-89, Energy-efficient Design of New Buildings Except Low-rise Residential Buildings*. Atlanta, GA: American Society of Heating, Refrigerating and Air-Conditioning Engineers, Inc., 1993.

ASHRAE 55-10, *Thermal Environmental Conditions for Human Occupancy*. Atlanta, GA: American Society of Heating, Refrigeration and Air-Conditioning Engineers, Inc., 2010.

ASHRAE 62.1-10, *Ventilation for Acceptable Indoor Air Quality*. Atlanta, GA: American Society of Heating, Refrigerating and Air-Conditioning Engineers, Inc., 2010.

ASHRAE 189.1-11, *Standard for the Design of High-performance, Green Buildings, Except Low-rise Residential Buildings*. Atlanta, GA: American Society of Heating, Refrigerating and Air-Conditioning Engineers, Inc., 2011.

ASTM C 518-10, *Standard Test Method for Steady-state Thermal Transmission Properties by Means of the Heat Flow Meter Apparatus*. West Conshohocken, PA: ASTM International, 2010.

Baylon, D. and J. Heller. *Super Good Cents Heat Loss Reference* Volumes I through IV. Portland, OR: Ecotope for the Bonneville Power Administration, 1998.

Certified Products Directory, 5th Edition. Silver Spring, MD: National Fenestration Rating Council, 1995.

Clean Air Act 1992.

Collier, R.K., Jr. *Desiccant Dehumidification and Cooling Systems: Assessment and Analysis*. Richland, WA: Pacific Northwest National Laboratory, 1997; PNNL-11694.

COMcheck™ Manual, Version 3.4. Richland, WA: Building Energy Codes Program, Pacific Northwest National Laboratory, U.S. Department of Energy, 2007.

COMcheck™ Prescriptive Package. Richland, WA: Building Energy Codes Program, Pacific Northwest National Laboratory, U.S. Department of Energy, 2007.

COMcheck™ Software User's Guide. Richland, WA: Building Energy Codes Program, Pacific Northwest National Laboratory, U.S. Department of Energy, 2005.

Commercial Buildings Characteristics 1992: Energy Consumption Survey, DOE/EIA-0246 (92). Washington, DC: Energy Information Administration, U.S. Department of Energy, 1994.

County and City Data Book—1988—Places. Washington DC: Census Bureau, U.S. Department of Commerce, 1988.

Crawley, D.B., P.K. Riesen and R.S. Briggs. *User's Guide for EN VS TD Program Version 2.0 and LTGSTD Program Version 2.0*, PNL-6839 (includes software). Richland, WA: Pacific Northwest National Laboratory, U.S. Department of Energy, 1989.

DOE 2.1E Simulation Tool, Computerized Program. Washington, DC: U.S. Department of Energy.

Elegant Lighting for Elegant Dining. Seattle, WA: Design Lighting Laboratory, Bonneville Power Administration, 1991.

Energy Efficiency Standards for Residential and Non-residential Buildings. Sacramento, CA: California Energy Commission, 1995.

Energy Policy Act of 1992 (EPAct). Public Law 102-486, 106 Stat 2776, 16 USC, Part 1531 et seq., as amended. Washington, DC: U.S. Department of Energy, 1992.

Federal User's Manual—Performance Standards for New Commercial and Multi-family High-rise Residential Buildings. Washington, DC: Office of Codes and Standards, U.S. Department of Energy, 1994.

Hogan, J.F. "Code Compliance Considerations in the Development of the Building Envelope Requirements for ANSI/ASHRAE/IESNA 90.1-89R." *Thermal VI Thermal Performance of the Exterior Envelopes of Buildings VI*. Clearwater Beach, FL, Office of Building Technologies, U.S. Department of Energy, Oak Ridge National Laboratory; American Society of Heating, Refrigerating and Air-Conditioning Engineers, Inc.; and Building Environment and Thermal Envelope Council, 1996.

ICC A117.1-09, *Accessible and Usable Buildings and Facilities*. Washington, DC: International Code Council, 2009.

IRC-15, *International Residential Code*. Washington, DC: International Code Council, 2014.

Jones, J.W. "Special Project 41: Development of Recommendations to Upgrade ASHRAE 90A-80, *Energy Conservation in New Building Design*." *ASHRAE Journal*, October 1983.

Model Energy Code, 1993 Edition. Falls Church, VA: Council of American Building Officials, 1993.

NAECA-87, National Appliance Energy Conservation Act of 1987. Public Law 100-12, 42 USC, Part 6291 et seq., as amended. Washington, DC: U.S. Department of Energy, 1987.

NAIMA AH 116-09, *Fibrous Glass Duct Construction Standards, Fifth Edition*. Alexandria, VA: North American Insulation Manufacturers Association, 2009.

NEBB-05, *Procedural Standards*. Gaithersburg, MD: National Environmental Balancing Bureau, 2005.

NFPA 70-14, *National Electrical Code*. Quincy, MA: National Fire Prevention Association, 2014.

Owenby, J.R., D.S. Ezell and R.R. Heim, Jr. *Annual Degree Days to Selected Bases Derived from the 1961 to 1990 Normals*. Climatography of the United States No. 81—Supplement No. 2. Asheville, NC: U.S. Department of Commerce; National Oceanic and Atmospheric Administration; and National Climatic Data Center, 1992.

"Prescriptive Requirements for Lighting—Calculation of Allowed Lighting Power Density—Area Category Method." *California Code of Regulations*. Subchapter 5, Section 146(b)2. Sacramento, CA: 1995.

SMACNA-05, *HVAC Duct Construction Standards: Metal and Flexible*. Chantilly, VA: Sheet Metal and Air Conditioning Contractors National Association, Inc., 2005.

Stucky, D.S., R.X. Shankle, R.X. Schultz, E.E. Richman and J. Dirks. "Pacific Northwest Laboratory's Lighting Technology Screening Matrix: Let There Be Energy-Efficient Light." *Energy Engineering*. Richland, WA: Pacific Northwest National Laboratory, 1994.

Technical Support Document for COMcheck-EZ Version 1.0. Letter Report prepared for the U.S. Department of Energy under contact NE-AC06-76RLO 1830. Richland, WA: Pacific Northwest National Laboratory, 1997.

The Basics of Efficient Lighting. Conyers, GA: Lithonia Lighting and U.S. Department of Energy, 2000.

UL 1784-01, *Standard for Air Leakage Test of Door Assemblies*. Northbrook, IL: Underwriter's Laboratories, Inc., 2001.

UL 181B-05, *Standard for Closure Systems for use with Flexible Air Ducts and Air Connectors*. Northbrook, IL: Underwriter's Laboratories, Inc., 2005.

Unitary and Applied Certification Directories (Online report). Arlington, VA: Air-Conditioning Heating and Refrigeration Institute, 1996. Available URL: http// www.ahridirectory.org/ahridirectury.

Chapter 5[CE]: Existing Buildings

General Comments

New for the 2015 IECC Commercial Provisions is a separate chapter dealing with alterations, repairs, additions and change of occupancy of existing buildings. These provisions for the most part, originated in Chapter 1 of the 2012 Commercial Provisions; however, this 2015 edition now contains options to deal with situations where compliance with the code for additions is difficult. Very simply, the provisions now allow an energy-neutral method for demonstrating compliance by basically saying that the building with the addition uses no more energy than the existing building. This will allow projects to take advantage of energy-efficient alterations on the existing building to offset difficult to comply with features on the addition. Requirements for alterations are provided, giving specific details and exceptions for the building envelope, heating and cooling systems, hot water systems, and lighting. Finally, the code allows that repairs be done without consideration of compliance with this code.

Purpose

The purpose of this chapter is to provide requirements for the unique circumstances involved with existing building additions, alterations, and repairs.

SECTION C501 GENERAL

C501.1 Scope. The provisions of this chapter shall control the *alteration*, *repair*, *addition* and change of occupancy of existing buildings and structures.

❖ The scope of this chapter encompasses specific circumstances related to changes to existing buildings. This chapter provides a roadmap when dealing with different types of projects—alterations, repairs, additions, and changes of occupancy.

C501.2 Existing buildings. Except as specified in this chapter, this code shall not be used to require the removal, *alteration* or abandonment of, nor prevent the continued use and maintenance of, an existing building or building system lawfully in existence at the time of adoption of this code.

❖ This section addresses the fact that, in general, the code does not affect existing buildings. It will permit an addition to be made to an existing building without requiring the existing building to conform to the code. In such a case, the addition is expected to comply with the current code, but the code will not require changes for the existing portion. Therefore, the code does not apply retroactively to existing buildings. When an existing building is modified by an addition, alteration, renovation or repair, Section C502, C503, or C504 will provide the guidance and requirements for such changes.

C501.3 Maintenance. Buildings and structures, and parts thereof, shall be maintained in a safe and sanitary condition. Devices and systems that are required by this code shall be maintained in conformance to the code edition under which installed. The owner or the owner's authorized agent shall be responsible for the maintenance of buildings and structures. The requirements of this chapter shall not provide the basis for removal or abrogation of energy conservation, fire protection and safety systems and devices in existing structures.

❖ The code is designed to regulate new construction and new work and is not intended to be applied retroactively to existing buildings except where existing envelope, lighting, mechanical or service water heating systems are specifically affected by Section C502, C503 or C504. Maintenance of building systems and components is required, but the code is not intended to require upgrade or modification to meet the code when maintenance is dealt with.

C501.4 Compliance. *Alterations*, *repairs*, *additions* and changes of occupancy to, or relocation of, existing buildings and structures shall comply with the provisions for *alterations*, *repairs*, *additions* and changes of occupancy or relocation, respectively, in the *International Building Code*, *International Fire Code*, *International Fuel Gas Code*, *International Mechanical Code*, *International Plumbing Code*, *International Property Maintenance Code*, *International Private Sewage Disposal Code* and NFPA 70.

❖ The code is not intended to suggest that compliance with the provisions of other *International Codes®* (I-Codes®), such as the IBC, be ignored. The I-Codes are intended to be a coordinated set of construction codes. This section clarifies the relationship between this chapter of the code and the *International Residential Code®* (IRC®), *International Building Code®* (IBC®), *International Fire Code®* (IFC®), *International Fuel Gas Code®* (IFGC®), *International Mechanical Code®* (IMC®), *International Plumbing Code®* (IPC®), *International Property Maintenance Code®* (IPMC®),

International Private Sewage Disposal Code® (IPSDC®) and NFPA 70. When alterations and repairs are made to existing mechanical and plumbing systems, the provisions of the I-Codes and NFPA 70 for alterations and repairs must be followed. Those codes indicate the extent to which existing systems must comply with the stated requirements. Where portions of existing building systems, such as plumbing, mechanical and electrical systems, are not being altered or repaired, those systems may continue to exist without being upgraded as long as they are not hazardous or unsafe to the building occupants.

C501.5 New and replacement materials. Except as otherwise required or permitted by this code, materials permitted by the applicable code for new construction shall be used. Like materials shall be permitted for *repairs*, provided hazards to life, health or property are not created. Hazardous materials shall not be used where the code for new construction would not permit use of these materials in buildings of similar occupancy, purpose and location.

❖ There are two options for materials used in repairs to an existing building. Generally, the materials used for repairs should be those that are presently required or permitted for new construction under the I-Codes. It is also acceptable to use materials consistent with those that are already present, except where those materials pose a hazard. This allowance follows the general concept that any repair should not make a building more hazardous than it was prior to the repair. It is generally possible to repair a structure, its components and its systems with materials consistent with those materials that were used previously. However, where materials that are now deemed hazardous are involved in the repair work, they may no longer be used. For example, the code identifies asbestos and lead-based paint as two common hazardous building materials that cannot be used in the repair process. Certain materials previously considered acceptable for building construction are now a threat to the health of the occupants.

C501.6 Historic buildings. No provisions of this code relating to the construction, *repair, alteration*, restoration and movement of structures, and *change of occupancy* shall be mandatory for *historic buildings* provided a report has been submitted to the *code official* and signed by a *registered design professional*, or a representative of the State Historic Preservation Office or the historic preservation authority having jurisdiction, demonstrating that compliance with that provision would threaten, degrade or destroy the historic form, fabric or function of the building.

❖ In some aspects, this is a bit of a continuation of the "existing building" provisions, but it goes even further in that historic buildings are exempt. In earlier editions of the code, this exemption applied only to the exterior envelope of such buildings, and to the interior only in those cases where the ordinance explicitly designated elements of the interior. With the current text, historic buildings are exempt from all aspects of the code. This exemption, however, is not without conditions. The most important criterion for application of this section is that the building must be specifically classified as being of historic significance by a qualified party or agency. Usually this is done by a state or local authority after considerable scrutiny of the historical value of the building. Most, if not all, states have authorities, such as a landmark commission, as do many local jurisdictions.

Because of the unique issues involved, historic buildings are exempt from the requirements of the code. This exemption could be extended to include all parts that are "historic," including additions, alterations and repairs, that would normally be addressed by Section C502 or C503. If the addition, alteration or renovation is not "historic," then the provisions of Section C502 or C503 should be applied. Consideration of energy conservation and compliance with the code is still of value in historic buildings. In exempting historic buildings, the code is simply recognizing that energy efficiency may be difficult to accomplish while maintaining the "historic" nature of the building.

SECTION C502
ADDITIONS

C502.1 General. *Additions* to an existing building, building system or portion thereof shall conform to the provisions of this code as those provisions relate to new construction without requiring the unaltered portion of the existing building or building system to comply with this code. *Additions* shall not create an unsafe or hazardous condition or overload existing building systems. An *addition* shall be deemed to comply with this code if the *addition* alone complies or if the existing building and *addition* comply with this code as a single building. *Additions* shall comply with Section C502.2.

Additions complying with ANSI/ASHRAE/IESNA 90.1. need not comply with Sections C402, C403, C404 and C405.

❖ When an addition is added to an existing building, only the addition must comply with the requirements of the code, as outlined in Section C502.2.

Simply stated, new work must comply with the current requirements for new work. Any addition to an existing system involving new work is subject to the requirements of the code. Additions can place additional loads or different demands on an existing system, and those loads or demands could necessitate changing all or part of the existing system. Additions and alterations must not cause an existing system to be any less in compliance with the code than it was before the changes.

Additions to existing buildings must comply with the code when the addition is within the scope of the code and would not otherwise be exempted (see code text and commentary, Sections C101.4 and C101.5.2). Additions include new construction, such as a conditioned bedroom, sun space or enclosed porch added to an existing building. Additions also include existing spaces converted from unconditioned or exempt spaces to conditioned spaces. For example, a finished basement, an attic converted to a bedroom or a carport converted to a den are additions. The addition of an unconditioned garage would not be considered within the scope of the code because the code applies to heated or cooled (conditioned) spaces.

Although not specifically defined in the code, building codes typically define an "Addition" as any increase in a building's habitable floor area (which can be interpreted as any increase in the conditioned floor area). For example, an unconditioned garage converted to a bedroom is an addition. If a conditioned floor area is expanded, such as a room made larger by moving out a wall, only the newly conditioned space must meet the code. A flat window added to a room does not increase the conditioned space and thus is not an addition by this definition. If several changes are made to a building at the same time, only the changes that expand the conditioned floor area are required to meet the code. The addition (the newly conditioned floor space) complies with the code if it complies with all of the applicable requirements in Chapter 4[CE]. For example, requirements applicable to the addition of a new room would most likely include insulating the exterior walls, ceiling and floor to the levels specified in the code; sealing all joints and penetrations; installing a vapor retarder in unventilated frame walls, floors and ceilings; identifying installed insulation *R*-values and window *U*-factors; and insulating and sealing any ducts passing through unconditioned portions or within exterior envelope components (walls, ceilings or floors) of the new space. Compliance approaches for additions include:

1. The entire building (the existing building plus the addition) complies with the code. If the building inclusive of the addition complies with the code, the addition will also comply, regardless of whether the addition complies alone. For example, a sunroom that does not comply with the code is added to a house. If the entire house (with the sunroom) complies, the addition also complies.

2. Where approved by the code official, the addition, including possible concurrent renovation, does not result in any increase in the building's overall area-weighted thermal transmittance (UA), or otherwise any increase in annual demand for either fossil fuel or electrical energy supply. The change in UA or energy use can be quantified using any of the commonly used hourly, full-year simulation tools. For example, additions that add rooms while simultaneously upgrading existing HVAC systems, windows and insulation often reduce the annual energy use or UA of the existing part of the home, more than offsetting the energy use attributed to the added space in the home.

3. The addition itself can comply with the prescriptive methods found in Chapter 4[CE]. These provisions provide a simple, prescriptive specification menu for each climate zone that, if followed, will yield a building envelope meeting the requirements of Table C402.1.3. The components of the building addition must meet the insulation *R*-values, fenestration *U*-factors and solar heat gain coefficient (SHGC) requirements shown in Tables C402.1.3 and C402.1.4.

An existing energy-using system (envelope, mechanical, service water heating, electrical distribution or lighting) is generally considered to be "grandfathered" with code adoption if the criteria for this level are the regulations (or code) under which the existing building was originally constructed. It should be noted that a specific level of safety is dictated by provisions dealing with hazard abatement in existing buildings and maintenance provisions, as contained in the code, the IPMC and the IFC.

Additions complying with ANSI/ASHRAE/IESNA 90.1 need not comply with Sections C402, C403, C404 and C405.

C502.2 Prescriptive compliance. *Additions* shall comply with Sections C502.2.1 through C502.2.6.2.

❖ This section simply provides guidance as to how compliance with the code is met for specific systems or building components included in an addition. Section C502.1 states:

> "An addition shall be deemed to comply with this code if the addition alone complies or if the existing building and addition comply with this code as a single building."

The ensuing sections state how, specifically, to accomplish this for different systems and building components.

C502.2.1 Vertical fenestration. New *vertical fenestration* area that results in a total building *fenestration* area less than or equal to that specified in Section C402.4.1 shall comply with Section C402.4. *Additions* with *vertical fenestration* that result in a total building *fenestration* area greater than Section C402.4.1 or *additions* that exceed the fenestration area greater than Section C402.4.1 shall comply with Section C402.4.1.1 for the *addition* only. *Additions* that result in a total building vertical glass area exceeding that specified in Section C402.4.1.1 shall comply with Section C407.

❖ This section provides specific requirements for compliance with the code for vertical fenestration when there is an addition. Basically, these provisions are consistent with the fundamental requirement stated in Section C502.1 that an addition can comply with the code if the addition alone complies, or if the existing building and addition comply as a single building. However, this section gives some guidance regarding the need for daylight controls as called out in Section C402.3.2.1 when the area of fenestration exceeds the maximum area of 30 percent of the total gross wall area of the addition or the addition plus the existing building. Essentially, the daylight controls would be applicable to the addition alone (see the commentary, Section C402.3.2.1). Finally, if the total fenestration area of the total building exceeds 40 percent of the gross wall area of the total building, the performance design in Section C407 must be utilized for the addition.

C502.2.2 Skylight area. New *skylight* area that results in a total building *fenestration* area less than or equal to that specified in Section C402.4.1 shall comply with Section C402.4. *Additions* with *skylight* area that result in a total building *skylight* area greater than C402.4.1 or additions that exceed the *skylight* area shall comply with Section C402.4.1.2 for the *addition* only. *Additions* that result in a total building *skylight* area exceeding that specified in Section C402.4.1.2 shall comply with Section C407.

❖ This section deals with the issue of skylights on additions, similar to the requirements for vertical fenestration (see the commentary to Section C502.2.1).

C502.2.3 Building mechanical systems. New mechanical systems and equipment that are part of the *addition* and serve the building heating, cooling and ventilation needs shall comply with Section C403.

❖ Consistent with the requirements in Section C502.1 that an addition must comply with the code for new construction without having the entire building comply, this section simply points to the requirements for new construction given in Section C403 for mechanical HVAC system.

C502.2.4 Service water-heating systems. New service water-heating equipment, controls and service water heating piping shall comply with Section C404.

❖ Consistent with the requirements in Section C502.1 that an addition must comply with the code for new construction without having the entire building comply, this section simply points to the requirements for new construction given in Section C404 for service water-heating systems.

C502.2.5 Pools and inground permanently installed spas. New pools and inground permanently installed spas shall comply with Section C404.9.

❖ Consistent with the requirements in Section C502.1 that an addition must comply with the code for new construction without having the entire building comply, this section simply points to the requirements for new construction given in Section C404.9 for pools or spas that are part of an addition.

C502.2.6 Lighting power and systems. New lighting systems that are installed as part of the addition shall comply with Section C405.

❖ Consistent with the requirements in Section C502.1 that an addition must comply with the code for new construction without having the entire building comply, this section simply points to the requirements for new construction given in Section C405 for the new lighting systems installed in an addition.

C502.2.6.1 Interior lighting power. The total interior lighting power for the *addition* shall comply with Section C405.4.2 for the *addition* alone, or the existing building and the *addition* shall comply as a single building.

❖ These provisions are consistent with the fundamental requirement stated in Section 502.1 that an addition can comply with the code if the addition alone complies, or if the existing building and addition comply as a single building.

C502.2.6.2 Exterior lighting power. The total exterior lighting power for the *addition* shall comply with Section C405.5.1 for the *addition* alone, or the existing building and the *addition* shall comply as a single building.

❖ These provisions are consistent with the fundamental requirement stated in Section C502.1 that an addition can comply with the code if the addition alone complies, or if the existing building and addition comply as a single building.

SECTION C503
ALTERATIONS

C503.1 General. *Alterations* to any building or structure shall comply with the requirements of the code for new construction. *Alterations* shall be such that the existing building or structure is no less conforming to the provisions of this code than the existing building or structure was prior to the *alteration*. *Alterations* to an existing building, building system or portion thereof shall conform to the provisions of this code as those provisions relate to new construction without requiring the unaltered portions of the existing building or building system to comply with this code. *Alterations* shall not create an unsafe or hazardous condition or overload existing building systems.

Alterations complying with ANSI/ASHRAE/IESNA 90.1. need not comply with Sections C402, C403, C404 and C405.

Exception: The following *alterations* need not comply with the requirements for new construction, provided the energy use of the building is not increased:

1. Storm windows installed over existing *fenestration*.
2. Surface-applied window film installed on existing single-pane *fenestration* assemblies reducing solar heat gain, provided the code does not require the glazing or *fenestration* to be replaced.
3. Existing ceiling, wall or floor cavities exposed during construction, provided that these cavities are filled with insulation.
4. Construction where the existing roof, wall or floor cavity is not exposed.
5. *Roof recover*.
6. *Air barriers* shall not be required for *roof recover* and roof replacement where the *alterations* or renovations to the building do not include *alterations*, renovations or *repairs* to the remainder of the building envelope.

7. *Alterations* that replace less than 50 percent of the luminaires in a space, provided that such *alterations* do not increase the installed interior lighting power.

❖ As stated for additions, new work must comply with the current requirements for new work. Any alteration to an existing system involving new work is subject to the requirements of the code. Just like additions, an alteration could place additional loads or different demands on an existing system and those loads or demands could necessitate changing all or part of the existing system. Alterations must not cause an existing building or system to be any less in compliance with the code than it was before the changes. The exceptions address common types of alterations where the need for upgrading to new code requirements is not warranted.

C503.2 Change in space conditioning. Any nonconditioned or low-energy space that is altered to become conditioned space shall be required to be brought into full compliance with this code.

❖ The type of alteration where nonconditioned space becomes conditioned space is really more of an addition than an alteration. This section does not obviate the need to check the total building to ensure that this alteration does not cause the existing building to be noncomplying, as required in Section C503.1.

C503.3 Building envelope. New building envelope assemblies that are part of the *alteration* shall comply with Sections C402.1 through C402.5.

❖ See the commentary to Sections C402.1 through C402.5.

C503.3.1 Roof replacement. *Roof replacements* shall comply with Table C402.1.3 or C402.1.4 where the existing roof assembly is part of the *building thermal envelope* and contains insulation entirely above the roof deck.

❖ "Roof replacement" is defined in the code in Chapter 2[CE]. Roof replacement involves removal of the roof covering and replacing or repairing parts of the roof deck. Consistent with Exceptions 3 and 4, the insulation in the roof assembly does not need to be upgraded to meet current code insulation requirements where there is already insulation in the roof cavity and it is also the ceiling cavity or when the insulation is not exposed. However, as is the case in many commercial roofing applications, insulation that is entirely above deck must be brought into compliance with current code roof-insulation requirements.

C503.3.2 Vertical fenestration. The addition of *vertical fenestration* that results in a total building *fenestration* area less than or equal to that specified in Section C402.4.1 shall comply with Section C402.4. The addition of *vertical fenestration* that results in a total building *fenestration* area greater than Section C402.4.1 shall comply with Section C402.4.1.1 for the space adjacent to the new fenestration only. *Alterations* that result in a total building vertical glass area exceeding that specified in Section C402.4.1.1 shall comply with Section C407.

❖ These provisions are the same as the provisions for vertical fenestration in an addition. (see the commentary to Section C502.2.1).

C503.3.3 Skylight area. The addition of *skylight* area that results in a total building *skylight* area less than or equal to that specified in Section C402.4.1 shall comply with Section C402.4. The addition of *skylight* area that results in a total building skylight area greater than Section C402.4.1 shall comply with Section C402.4.1.2 for the space adjacent to the new skylights. *Alterations* that result in a total building skylight area exceeding that specified in Section C402.4.1.2 shall comply with Section C407.

❖ These provisions are the same as the provisions for skylights in an addition (see the commentary to Section C502.2.2).

C503.4 Heating and cooling systems. New heating, cooling and duct systems that are part of the *alteration* shall comply with Sections C403.

❖ Consistent with the requirements in Section C502.1 that an alteration must comply with the code for new construction without having the entire building comply, this section simply points to the requirements for new construction given in Section C403 for mechanical HVAC systems.

C503.4.1 Economizers. New cooling systems that are part of *alteration* shall comply with Section C403.3.

❖ Consistent with the requirements in Section C502.1 that an alteration must comply with the code for new construction without having the entire building comply, this section simply points to the requirements for new construction given in Section C403.3 for economizer systems.

C503.5 Service hot water systems. New service hot water systems that are part of the *alteration* shall comply with Section C404.

❖ Consistent with the requirements in Section C502.1 that an alteration must comply with the code for new construction without having the entire building comply, this section simply points to the requirements for new construction given in Section C404 for service water-heating systems.

C503.6 Lighting systems. New lighting systems that are part of the *alteration* shall comply with Section C405.

> **Exception.** *Alterations* that replace less than 10 percent of the luminaires in a space, provided that such *alterations* do not increase the installed interior lighting power.

❖ Consistent with the requirements in Section C502.1 that an alteration must comply with the code for new construction without having the entire building comply, this section simply points to the requirements for new construction given in Section C405 for the new lighting systems installed in an addition. However, the exception gives a break in situations where the portion of luminaires being replaced is relatively small (10 percent) and there is no increase in lighting power.

SECTION C504 REPAIRS

C504.1 General. Buildings and structures, and parts thereof, shall be repaired in compliance with Section C501.3 and this section. Work on nondamaged components that is necessary for the required *repair* of damaged components shall be considered part of the *repair* and shall not be subject to the requirements for *alterations* in this chapter. Routine maintenance required by Section C501.3, ordinary *repairs* exempt from *permit* and abatement of wear due to normal service conditions shall not be subject to the requirements for *repairs* in this section.

Where a building was constructed to comply with ANSI/ASHRAE/IESNA 90.1, repairs shall comply with the standard and need not comply with Sections C402, C403, C404 and C405.

❖ Repairs do not require that the item being repaired be upgraded to meet the requirements for a new code. For instance, replacement of the glass in an existing sash and frame would not require making the window or door meet the fenestration requirements of the code for new construction.

C504.2 Application. For the purposes of this code, the following shall be considered repairs:

1. Glass-only replacements in an existing sash and frame.
2. *Roof repairs*.
3. Air barriers shall not be required for *roof repair* where the repairs to the building do not include *alterations*, renovations or *repairs* to the remainder of the building envelope.
4. Replacement of existing doors that separate conditioned space from the exterior shall not require the installation of a vestibule or revolving door, provided that an existing vestibule that separates a conditioned space from the exterior shall not be removed.
5. *Repairs* where only the bulb, the ballast or both within the existing luminaires in a space are replaced, provided that the replacement does not increase the installed interior lighting power.

❖ This section is intended to list items that have historically been a source of confusion as to whether they were a repair or an alteration. The list is not intended to consist of the only items that constitute a repair. For instance, a damaged door can be replaced by a door of equivalent characteristic to the damaged door.

SECTION C505 CHANGE OF OCCUPANCY OR USE

C505.1 General. Spaces undergoing a change in occupancy that would result in an increase in demand for either fossil fuel or electrical energy shall comply with this code. Where the use in a space changes from one use in Table C405.4.2(1) or C405.4.2(2) to another use in Table C405.4.2(1) or C405.4.2(2), the installed lighting wattage shall comply with Section **C405.4**.

❖ The application of this section is straightforward. However, note that a change in occupancy or use is not necessarily a change in occupancy classification. For instance, if a grocery store was changed to a drugstore, this would constitute a change in occupancy or use, but not a change in occupancy classification, because both a grocery store and a drugstore are mercantile occupancies. This section makes a specific reference to lighting power allowance requirements as these are driven by the use of the space.

Chapter 6[CE]: Referenced Standards

General Comments

Chapter 6[CE] contains a comprehensive list of standards that are referenced in the Commercial Provisions of the code. It is organized to make locating specific document references easy.

It is important to understand that not every document related to energy conservation is qualified to be a "referenced standard." The International Code Council (ICC)® has adopted a criterion that referenced standards in the *International Codes*® (I-Codes®) and standards intended for adoption into the I-Codes must meet to qualify as a referenced standard. The policy is summarized as follows:

- Code references: The scope and application of the standard must be clearly identified in the code text.
- Standard content: The standard must be written in mandatory language and be appropriate for the subject covered. The standard cannot have the effect of requiring proprietary materials or prescribing a proprietary testing agency.
- Standard promulgation: The standard must be readily available and developed and maintained in a consensus process such as those used by ASTM or ANSI.

The ICC Code Development Procedures, CDP 28, of which the standards policy is a part, are updated periodically. A copy of the latest version can be obtained from the ICC offices.

Once a standard is incorporated into the code through the code development process, it becomes an enforceable part of the code. When the code is adopted by a jurisdiction, the standard also is part of that jurisdiction's adopted code. It is for this reason that the criteria were developed. Compliance with this policy means that documents incorporated into the code are developed through the use of a consensus process, are written in mandatory language and do not mandate the use of proprietary materials or agencies. The requirement for a standard to be developed through a consensus process means that the standard is representative of the most current body of available knowledge on the subject as determined by a broad range of interested or affected parties without dominance by any single interest group. A true consensus process has many attributes, including but not limited to:

- An open process that has formal (published) procedures allowing for the consideration of all viewpoints.
- A definitive review period that allows for the standard to be updated and/or revised.
- A process of notification to all interested parties.
- An appeals process.

Many available documents related to mechanical system design and installation and construction, though useful, are not "standards" and are not appropriate for reference in the code. Often, these documents are developed or written with the intention of being used for regulatory purposes and are unsuitable for use as a standard because of extensive use of recommendations, advisory comments and nonmandatory terms. Typical examples include installation instructions, guidelines and practices.

The objective of ICC's standards policy is to provide regulations that are clear, concise and enforceable; thus the requirement for standards to be written in mandatory language. This requirement is not intended to mean that a standard cannot contain informational or explanatory material that will aid the user of the standard in its application. When the standard's promulgating agency wants such material to be included, however, the information must appear in a nonmandatory location, such as an annex or appendix, and be clearly identified as not being part of the standard.

Overall, standards referenced by the code must be authoritative, relevant, up to date and, most important, reasonable and enforceable. Standards that comply with the ICC's standards policy fulfill these expectations.

Purpose

As a performance-based code, the *International Energy Conservation Code*® (IECC®) contains numerous references to documents that are used to regulate materials and methods of construction. The references to these documents within the code text consist of the promulgating agency's acronym and its publication designation (for example, ASTM C 90) and a further indication that the document being referenced is the one that is listed in this chapter. This chapter contains all of the information that is necessary to identify the specific referenced document. Included is the following information on a document's promulgating agency (see Commentary Figure 6[CE]):

- The promulgating agency (the agency's title).
- The promulgating agency's acronym.
- The promulgating agency's address.

For example, a reference to ASTM E 283 indicates that this document can be found in Chapter 6[CE] under the heading ASTM. The specific standards designation is E 283. For convenience, these designations are listed in alphanumeric order. This chapter identifies that ASTM E 283 is titled *Test Method for Determining the Rate of Air Leakage Through Exterior Windows, Curtain Walls and Doors Under Specified Pressure Differences Across the Specimen*, the applicable edition (i.e., its year of publication) is 2004 and it is referenced in numerous sections of the code.

This chapter will also indicate when a document has been discontinued or replaced by its promulgating agency. When a document is replaced by a different one, a note will appear to tell the user the designation and title of the new document.

The key aspect of the manner in which standards are referenced by the code is that a specific edition of a specific standard is clearly identified. In this manner, the requirements necessary for compliance can be readily determined. The basis for code compliance is, therefore, established and available on an equal basis to the building official, contractor, designer and owner.

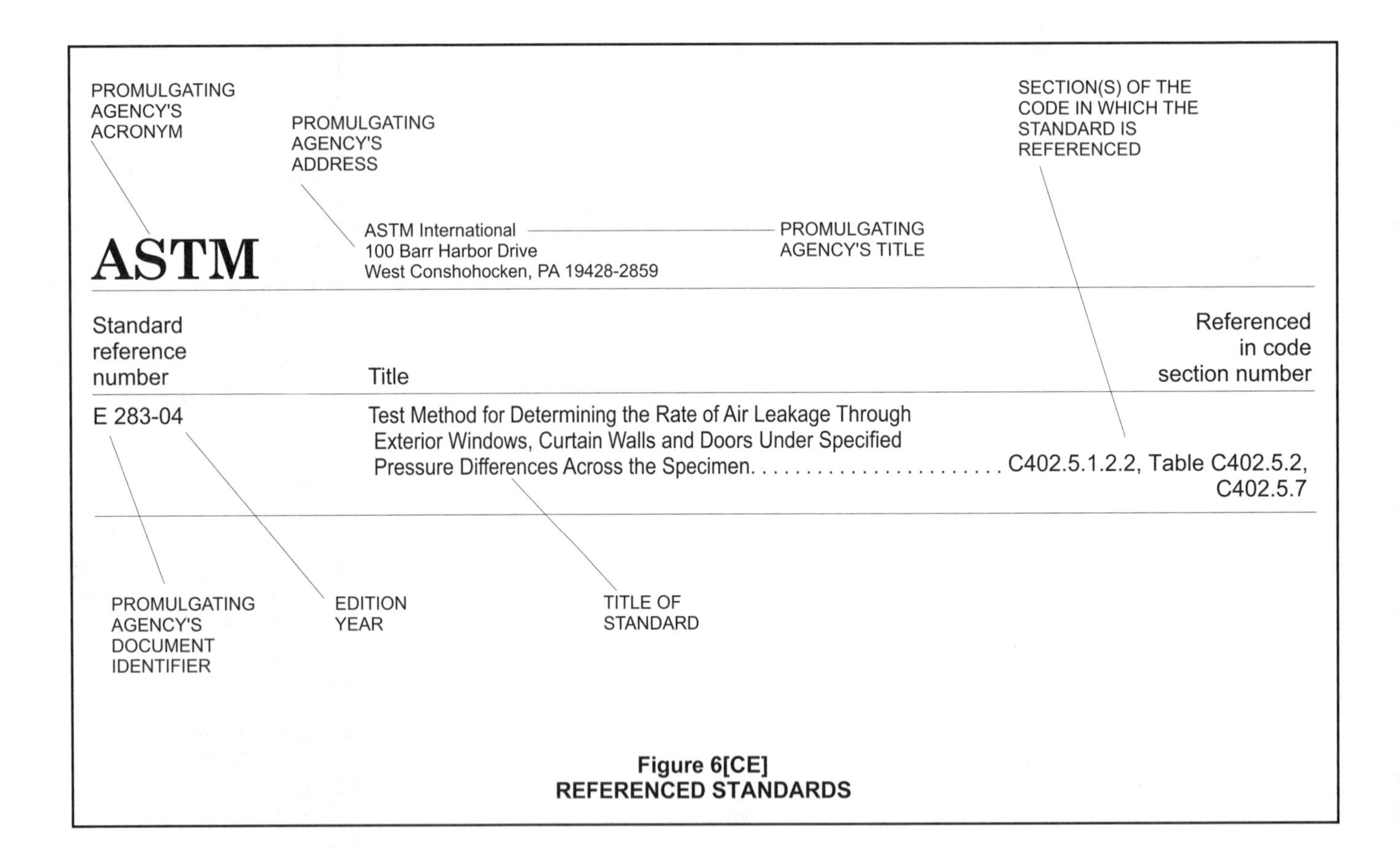

Figure 6[CE]
REFERENCED STANDARDS

This chapter lists the standards that are referenced in various sections of this document. The standards are listed herein by the promulgating agency of the standard, the standard identification, the effective date and title, and the section or sections of this document that reference the standard. The application of the referenced standards shall be as specified in Section 106.

AAMA

American Architectural Manufacturers Association
1827 Walden Office Square
Suite 550
Schaumburg, IL 60173-4268

Standard reference number	Title	Referenced in code section number
AAMA/WDMA/CSA 101/I.S.2/A C440—11	North American Fenestration Standard/ Specifications for Windows, Doors and Unit Skylights	Table C402.5.2

AHAM

Association of Home Appliance Manufacturers
1111 19th Street, NW, Suite 402
Washington, DC 20036

Standard reference number	Title	Referenced in code section number
ANSI/ AHAM RAC-1—2008	Room Air Conditioners	Table C403.2.3(3)
AHAM HRF-1—2007	Energy, Performance and Capacity of Household Refrigerators, Refrigerator-Freezers and Freezers	Table C403.2.14.1

AHRI

Air-Conditioning, Heating, and Refrigeration Institute
2111 Wilson Blvd, Suite 500
Arlington, VA 22201

Standard reference number	Title	Referenced in code section number
ISO/AHRI/ASHRAE 13256-1 (2011)	Water-to-Air and Brine-to-Air Heat Pumps—Testing and Rating for Performance	Table C403.2.3(2)
ISO/AHRI/ASHRAE 13256-2 (2011)	Water-to-Water and Brine-to-Water Heat Pumps —Testing and Rating for Performance	Table C403.2.3(2)
210/240—08 with Addenda 1 and 2	Performance Rating of Unitary Air-Conditioning and Air-Source Heat Pump Equipment	Table C403.2.3(1), Table C403.2.3(2)
310/380—04 (CSA-C744-04)	Standard for Packaged Terminal Air Conditioners and Heat Pumps	Table C403.2.3(3)
340/360—2007 with Addendum 2	Performance Rating of Commercial and Industrial Unitary Air-Conditioning and Heat Pump Equipment	Table C403.2.3(1), Table C403.2.3(2)
365(I-P)—09	Commercial and Industrial Unitary Air-Conditioning Condensing Units	Table C403.2.3(1), Table C403.2.3(6)
390—03	Performance Rating of Single Package Vertical Air-Conditioners and Heat Pumps	Table C403.2.3(3)
400—2001	Liquid to Liquid Heat Exchangers with Addendum 1 and 2	Table C403.2.3(10)
440—2008	Performance Rating of Room Fan Coils	C403.2.10
460—2005	Performance Rating of Remote Mechanical-Draft Air-Cooled Refrigerant Condensers	Table C403.2.3(8)
550/590—2011 With Addendum 1	Performance Rating of Water-Chilling and Heat Pump Water-Heating Packages Using the Vapor Compression Cycle	C403.2.3.1, Table C403.2.3(7)
560—00	Absorption Water Chilling and Water Heating Packages	Table C403.2.3(7)

AHRI—continued

1160 (I-P) —09	Performance Rating of Heat Pump Pool Heaters	Table C404.2
1200-2010	Performance Rating of Commercial Refrigerated Display Merchandisers and Storage Cabinets	C403.2.14, Table C403.2.14(1), Table C403.2.14(2)

AMCA

Air Movement and Control Association International
30 West University Drive
Arlington Heights, IL 60004-1806

Standard reference number	Title	Referenced in code section number
205—12	Energy Efficiency Classification for Fans	C403.2.12.3
220—08 (R2012)	Laboratory Methods of Testing Air Curtain Units for Aerodynamic Performance Rating	C402.5.6
500D—12	Laboratory Methods for Testing Dampers for Rating	C403.2.4.3

ANSI

American National Standards Institute
25 West 43rd Street
Fourth Floor
New York, NY 10036

Standard reference number	Title	Referenced in code section number
Z21.10.3/CSA 4.3—11	Gas Water Heaters, Volume III—Storage Water Heaters with Input Ratings Above 75,000 Btu per Hour, Circulating Tank and Instantaneous	Table C404.2
Z21.47/CSA 2.3—12	Gas-fired Central Furnaces	Table C403.2.3(4)
Z83.8/CSA 2.6—09	Gas Unit Heaters, Gas Packaged Heaters, Gas Utility Heaters and Gas-fired Duct Furnaces	Table C403.2.3(4)

APSP

The Association of Pool & Spa Professionals
2111 Eisenhower Avenue
Alexandria, VA 22314

Standard reference number	Title	Referenced in code section number
14—11	American National Standard for Portable Electric Spa Efficiency	C404.8

ASHRAE

American Society of Heating, Refrigerating and Air-Conditioning Engineers, Inc.
1791 Tullie Circle, NE
Atlanta, GA 30329-2305

Standard reference number	Title	Referenced in code section number
ASHRAE 127-2007	Method of Testing for Rating Computer	Table C403.2.3(9)
ANSI/ASHRAE/ACCA Standard 183—2007 (RA2011)	Peak Cooling and Heating Load Calculations in Buildings, Except Low-rise Residential Buildings	C403.2.1
ASHRAE—2012	ASHRAE HVAC Systems and Equipment Handbook	C403.2.1
ISO/AHRI/ASHRAE 13256-1 (2011)	Water-to-Air and Brine-to-Air Heat Pumps—Testing and Rating for Performance	Table C403.2.3(2)
ISO/AHRI/ASHRAE 13256-2 (2011)	Water-to-Water and Brine-to-Water Heat Pumps—Testing and Rating for Performance	Table C403.2.3(2)

ASHRAE—continued

90.1—2013	Energy Standard for Buildings Except Low-rise Residential Buildings	C401.2, Table C402.1.3, Table C402.1.4, C406.2, Table C407.6.1, C502.1, C503.1, C504.1
140—2011	Standard Method of Test for the Evaluation of Building Energy Analysis Computer Programs	C407.6.1
146—2011	Testing and Rating Pool Heaters	Table C404.2

ASME

American Society Mechanical Engineers
Two Park Avenue
New York, NY 10016-5990

Standard reference number	Title	Referenced in code section number
ASME A17.1/ CSA B44—2013	Safety Code for Elevators and Escalators	C405.9.2

ASTM

ASTM International
100 Barr Harbor Drive
West Conshohocken, PA 19428-2859

Standard reference number	Title	Referenced in code section number
C 90—13	Specification for Load-bearing Concrete Masonry Units	Table C401.3
C 1363—11	Standard Test Method for Thermal Performance of Building Materials and Envelope Assemblies by Means of a Hot Box Apparatus	C303.1.4.1, Table C402.1.4
C 1371—04a(2010)e1	Standard Test Method for Determination of Emittance of Materials Near Room Temperature Using Portable Emissometers	Table C402.3
C 1549—09	Standard Test Method for Determination of Solar Reflectance Near Ambient Temperature Using A Portable Solar Reflectometer	Table C402.3
D 1003—11e1	Standard Test Method for Haze and Luminous Transmittance of Transparent Plastics	C402.4.2.2
E 283—04	Test Method for Determining the Rate of Air Leakage Through Exterior Windows, Curtain Walls and Doors Under Specified Pressure Differences Across the Specimen	C402.5.1.2.2, Table C402.5.2, C402.5.7
E 408—71(2008)	Test Methods for Total Normal Emittance of Surfaces Using Inspection-meter Techniques	Table C402.3
E 779—10	Standard Test Method for Determining Air Leakage Rate by Fan Pressurization	C402.5
E 903—96	Standard Test Method Solar Absorptance, Reflectance and Transmittance of Materials Using Integrating Spheres (Withdrawn 2005)	Table C402.3
E 1677—11	Standard Specification for an Air-retarder (AR) Material or System for Low-rise Framed Building Walls	C402.5.1.2.2
E 1918—06	Standard Test Method for Measuring Solar Reflectance of Horizontal or Low-sloped Surfaces in the Field	Table C402.3
E 1980—11	Standard Practice for Calculating Solar Reflectance Index of Horizontal and Low-sloped Opaque Surfaces	Table C402.3, C402.3.2
E 2178—13	Standard Test Method for Air Permanence of Building Materials	C402.5.1.2.1
E 2357—11	Standard Test Method for Determining Air Leakage of Air Barriers Assemblies	C402.5.1.2.2

CRRC

Cool Roof Rating Council
449 15th Street, Suite 200
Oakland, CA 94612

Standard reference number	Title	Referenced in code section number
ANSI/CRRC-1—2012	CRRC-1 Standard	Table C402.3, C402.3.2, C402.3.2.1, Table C407.5.1(1)

CSA

CSA Group
8501 East Pleasant Valley
Cleveland, OH 44131-5516

Standard reference number	Title	Referenced in code section number
AAMA/WDMA/CSA 101/I.S.2/A440—11	North American Fenestration Standard/Specification for Windows, Doors and Unit Skylights	Table C402.5.2
CSA B55.1—2012	Test Method for Measuring Efficiency and Pressure Loss of Drain Water Heat Recovery Units	C404.8
CSA B55.2—2012	Drain Water Heat Recover Units	C404.8

CTI

Cooling Technology Institute
P. O. Box 73383
Houston, TX 77273-3383

Standard reference number	Title	Referenced in code section number
ATC 105 (00)	Acceptance Test Code for Water Cooling Tower	Table C403.2.3(8)
ATC 105S—11	Acceptance Test Code for Closed Circuit Cooling Towers	Table C403.2.3(8)
ATC 106—11	Acceptance Test For Mechanical Draft Evaporative Vapor Condensers	Table C403.2.3(8)
STD 201—11	Standard for Certification of Water Cooling Towers Thermal Performances	Table C403.2.3(8)

DASMA

Door and Access Systems Manufacturers Association
1300 Sumner Avenue
Cleveland, OH 44115-2851

Standard reference number	Title	Referenced in code section number
105—92 (R2004)—13	Test Method for Thermal Transmittance and Air Infiltration of Garage Doors	C303.1.3, Table C402.5.2

DOE

U.S. Department of Energy
c/o Superintendent of Documents
1000 Independence Avenue SW
Washington, DC 20585

Standard reference number	Title	Referenced in code section number
10 CFR, Part 430—1998	Energy Conservation Program for Consumer Products: Test Procedures and Certification and Enforcement Requirement for Plumbing Products; and Certification and Enforcement Requirements for Residential Appliances; Final Rule	Table C403.2.3(4), Table C403.2.3(5), Table C404.2
10 CFR, Part 430, Subpart B, Appendix N—1998	Uniform Test Method for Measuring the Energy Consumption of Furnaces and Boilers	C202
10 CFR, Part 431—2004	Energy Efficiency Program for Certain Commercial and Industrial Equipment: Test Procedures and Efficiency Standards; Final Rules	Table C403.2.3(5), C405.7, C405.8, Table C405.8
10 CFR 431 Subpart B App B	Uniform Test Method for Measuring Nominal Full Load Efficiency of Electric Motors	C403.4.4.4, Table C405.8(1), Table C405.8(2), Table C405.8(3), C405.8(4)
NAECA 87—(88)	National Appliance Energy Conservation Act 1987 [(Public Law 100-12 (with Amendments of 1988-P.L. 100-357)]	Tables C403.2.3(1), C403.2.3(2), C403.2.3(4)

ICC

International Code Council, Inc.
500 New Jersey Avenue, NW
6th Floor
Washington, DC 20001

Standard reference number	Title	Referenced in code section number
IBC—15	International Building Code®	C201.3, C303.2, C402.5.3, C501.4
IFC—15	International Fire Code®	C201.3, C501.4
IFGC—15	International Fuel Gas Code®	C201.3, C501.4
IMC—15	International Mechanical Code®	C403.2.4.3, C403.2.6, C403.2.6.1, C403.2.6.2, C403.2.7, C403.2.8, C403.2.8.1, C403.2.8.2, C403.2.9, C403.4.4, C403.4.4.6, C406.6, C501.4
IPC—15	International Plumbing Code®	C201.3, C501.4
IMPC—15	International Property Maintenance Code®	C501.4
IPSDC—15	International Private Sewage Disposal Code®	C501.4

IEEE

The Institute of Electrical and Electronic Engineers Inc.
3 Park Avenue
New York, NY 10016

Standard reference number	Title	Referenced in code section number
IEEE 515.1—2012	IEE Standard for the Testing, Design, Installation, and Maintenance of Electrical Resistance Trace Heating for Commercial Applications	C404.6.2

IES

Illuminating Engineering Society
120 Wall Street, 17th Floor
New York, NY 10005-4001

Standard reference number	Title	Referenced in code section number
ANSI/ASHRAE/IESNA 90.1—2013	Energy Standard for Buildings, Except Low-rise Residential Buildings	C401.2, Table C402.1.3, Table C402.1.4, C406.2, C502.1, C503.1, C504.1

ISO

International Organization for Standardization
1, rue de Varembe, Case postale 56, CH-1211
Geneva, Switzerland

Standard reference number	Title	Referenced in code section number
ISO/AHRI/ASHRAE 13256-1 (2011)	Water-to-Air and Brine-to-air Heat Pumps -Testing and Rating for Performance	Table C403.2.3(2)
ISO/AHRI/ASHRAE 13256-2(2011)	Water-to-Water and Brine-to-Water Heat Pumps -Testing and Rating for Performance	C403.2.3(2)

NEMA

National Electrical Manufacturers Association
1300 North 17th Street, Suite 1752
Rosslyn, VA 22209

Standard reference number	Title	Referenced in code section number
MG1—1993	Motors and Generators	C202

NFPA

National Fire Protection Association
1 Batterymarch Park
Quincy, MA 02169-7471

Standard reference number	Title	Referenced in code section number
70—14	National Electrical Code	C501.4

NFRC

National Fenestration Rating Council, Inc.
6305 Ivy Lane, Suite 140
Greenbelt, MD 20770

Standard reference number	Title	Referenced in code section number
100—2009	Procedure for Determining Fenestration Products *U*-factors—Second Edition	C303.1.3, C402.2.2
200—2009	Procedure for Determining Fenestration Product Solar Heat Gain Coefficients and Visible Transmittance at Normal Incidence—Second Edition	C303.1.3, C402.4.1.1
400—2009	Procedure for Determining Fenestration Product Air Leakage—Second Edition	Table C402.5.2

SMACNA

Sheet Metal and Air Conditioning Contractors National Association, Inc.
4021 Lafayette Center Drive
Chantilly, VA 20151-1209

Standard reference number	Title	Referenced in code section number
SMACNA—2012	HVAC Air Duct Leakage Test Manual 2nd Edition	C403.2.8.1.3

UL

UL LLC
333 Pfingsten Road
Northbrook, IL 60062-2096

Standard reference number	Title	Referenced in code section number
710—12	Exhaust Hoods for Commercial Cooking Equipment	C403.2.8
727—06	Oil-fired Central Furnaces—with Revisions through April 2010	Table C403.2.3(4)
731—95	Oil-fired Unit Heaters—with Revisions through August 2012	Table C403.2.3(4)
1784—01	Air Leakage Tests of Door Assemblies—with Revisions through July 2009	C402.5.3

US-FTC

United States-Federal Trade Commission
600 Pennsylvania Avenue NW
Washington, DC 20580

Standard reference number	Title	Referenced in code section number
CFR Title 16 (May 31, 2005)	*R*-value Rule	C303.1.4

WDMA

Window and Door Manufacturers Association
2025 M Street, NW, Suite 800
Washington, DC 20036-3309

Standard reference number	Title	Referenced in code section number
AAMA/WDMA/CSA 101/I.S.2/A440—11	North American Fenestration Standard/Specification for Windows, Doors and Unit Skylights	Table C402.5.2

INDEX

A

B

D

E

F

G

H

I

K

L

M

S

T

U

V

W

Z

IECC—RESIDENTIAL PROVISIONS

TABLE OF CONTENTS

Chapter 1 [RE]: Scope and Administration

General Comments

The 2015 edition of the *International Energy Conservation Code®*—Residential Provisions will regulate the design and construction of low-rise residential buildings for efficient energy consumption over their useful life. It includes measures for the thermal envelope, heating, ventilating and air-conditioning (HVAC) systems, and electrical systems of residential buildings up to three stories in height.

Purpose

Though not stated specifically, the code is applicable to all buildings and structures and their components and systems that use energy primarily for human comfort and needs. The code does not regulate the energy for items such as computers or coffee pots. This portion of the code addresses the design of energy-efficient building envelopes and the selection and installation of energy-efficient mechanical, service water heating, electrical distribution and illumination systems and equipment in residential buildings.

PART 1—SCOPE AND APPLICATION

SECTION R101 SCOPE AND GENERAL REQUIREMENTS

R101.1 Title. This code shall be known as the *International Energy Conservation Code* of **[NAME OF JURISDICTION]**, and shall be cited as such. It is referred to herein as "this code."

❖ This section directs the adopting jurisdiction to insert the name of the jurisdiction into the code. Because the IECC is a "model" code, it is not an enforceable document until it is adopted by a jurisdiction or agency that has enforcement powers.

R101.2 Scope. This code applies to *residential buildings* and the building sites and associated systems and equipment.

❖ The code applies to portions of the building thermal envelope that enclose conditioned space, as shown in Commentary Figure R101.2(1). Conditioned space is the area provided with heating or cooling either directly, through a positive heating/cooling supply system (such as registers located in the space), or indirectly through an opening that allows heated or cooled air to communicate directly with the space. For example, a walk-in closet connected to a master bedroom suite may not contain a positive heating supply through a register, but it would be conditioned indirectly by the free passage of heated or cooled air into the space from the bedroom.

A good example of the exception would be an unconditioned garage or attic space. In the case of a garage, if the unconditioned garage area is separated from the conditioned portions of the residence by an assembly that meets the "building envelope" criteria (meaning that the wall between them is insulated), the exterior walls of the garage would not need to be insulated to separate the garage from the exterior climate.

The building thermal envelope consists of the wall, roof/ceiling and floor assemblies that surround the conditioned space. Raised floors over a crawl space or garage, or directly exposed to the outside air are considered to be part of the floor assembly. Walls surrounding a conditioned basement (in addition to surrounding conditioned spaces above grade) are part of the building envelope. The code defines above-grade walls surrounding conditioned spaces as exterior walls. This definition includes walls between the conditioned space and unconditioned garage; roof and basement knee, dormer and gable-end walls; walls enclosing a mansard roof; and basement walls with an average below-grade area that is less than 50 percent of the total basement gross wall area. This definition would not include walls separating an unconditioned garage from the outdoors. The roof/ceiling assembly is the surface where insulation will be installed, typically on top of the gypsum board [see Commentary Figure R101.2(2)].

R101.3 Intent. This code shall regulate the design and construction of buildings for the effective use and conservation of energy over the useful life of each building. This code is intended to provide flexibility to permit the use of innovative approaches and techniques to achieve this objective. This code is not intended to abridge safety, health or environmental requirements contained in other applicable codes or ordinances.

❖ This chapter is broad in its application, yet specific to regulating the use of energy in buildings where that energy is used primarily for human comfort or heating and cooling to protect the contents. In general, the requirements of the code address the design of all building systems that affect the visual and thermal comfort of the occupants, including:

- Lighting systems and controls.
- Wall, roof and floor insulation.
- Windows and skylights.
- Cooling equipment (air conditioners, chillers and cooling towers).
- Heating equipment (boilers, furnaces and heat pumps).
- Pumps, piping and liquid circulation systems.
- Supply and return fans.
- Service hot water systems (kitchens and lavatories).

The chapter is intended to define requirements for the portions of a building and building systems that affect energy use in new construction and to promote the effective use of energy. Where a code application for a specific situation is in question, the authority having jurisdiction for the building should favor the action that will promote the effective use of energy. The code official may also consider the cost of the required action compared to the energy that will be saved over the life of that action.

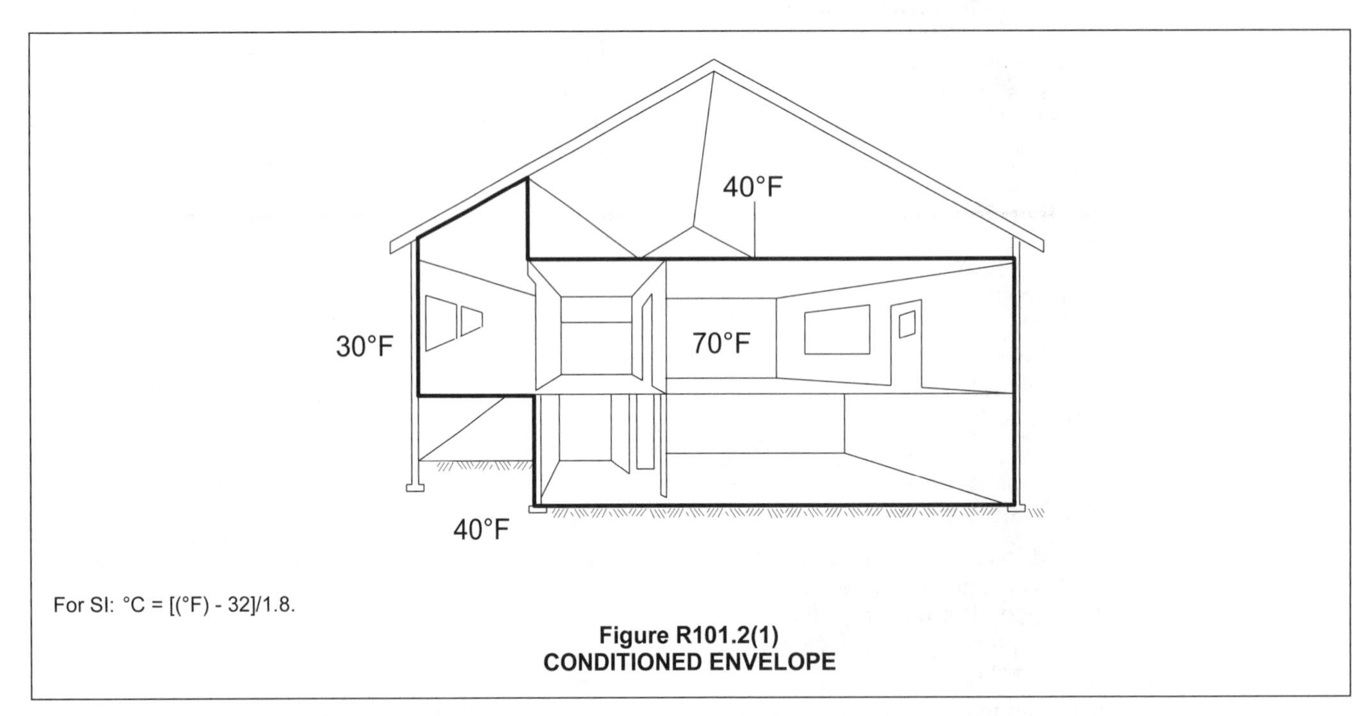

Figure R101.2(1)
CONDITIONED ENVELOPE

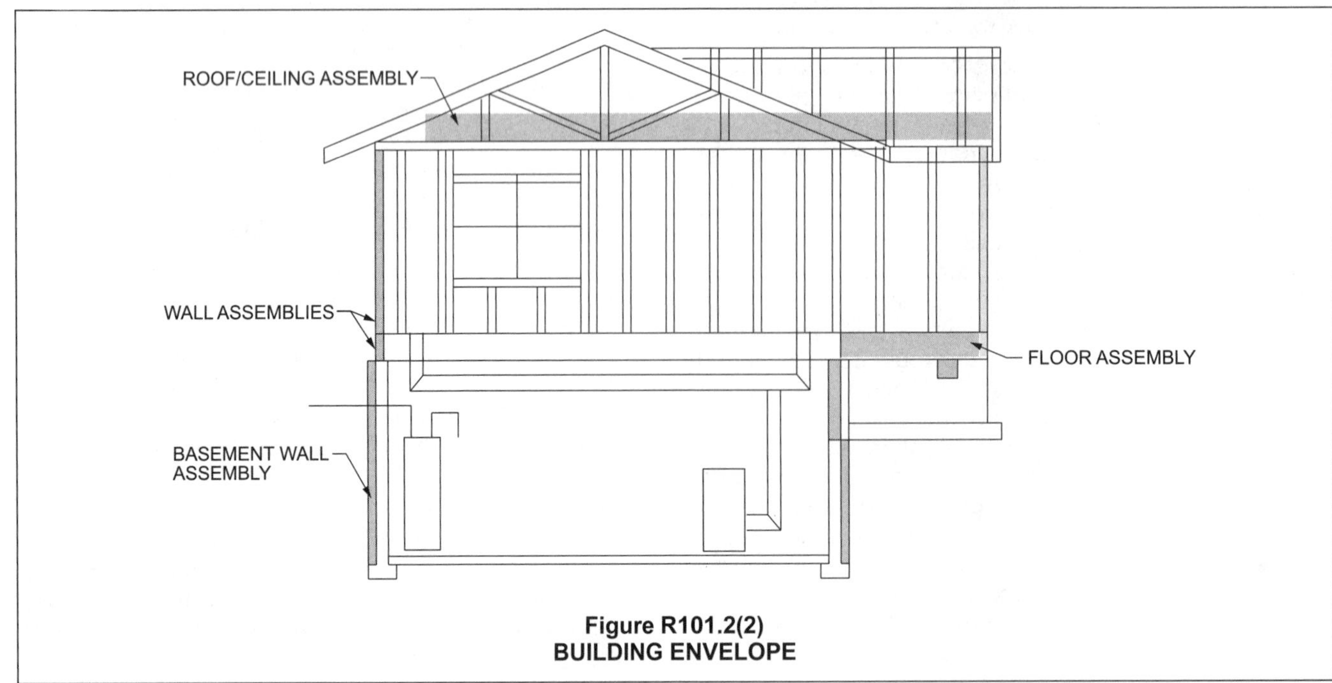

Figure R101.2(2)
BUILDING ENVELOPE

This section of the code supports flexibility in application of the code requirements. Although many of the requirements are given in a prescriptive format for ease of use, the code is not intended to stifle innovation—especially innovative techniques that conserve energy. Innovative approaches that lead to energy efficiency should be encouraged, even if the approach is not specifically listed in the code or does not meet the strict letter of the code. This principle should be applied to methods for determining compliance with the code and the building construction techniques used to meet the code.

Any design should first be evaluated to see whether it meets the code requirements directly. If an innovative approach is preferred, the applicant is responsible for demonstrating that the innovative concept promotes energy efficiency. Where the literal code requirements have not been satisfied but the applicant claims to meet the intent, the code official will likely have to exercise professional judgment to determine whether the proposed design meets the intent of the code in the interest of energy efficiency (see commentary, Section R103).

R101.4 Applicability. Where, in any specific case, different sections of this code specify different materials, methods of construction or other requirements, the most restrictive shall govern. Where there is a conflict between a general requirement and a specific requirement, the specific requirement shall govern.

❖ In general, the most restrictive requirement is to apply where there may be different requirements in the code for a specific issue. However, in cases where the code establishes a specific requirement for a certain condition, that requirement is applicable even if it is less restrictive than a general requirement elsewhere in the code.

R101.4.1 Mixed occupancy. Where a building includes both *residential* and *commercial* occupancies, each occupancy shall be separately considered and meet the applicable provisions of the IECC—Commercial Provisions or IECC—Residential Provisions.

❖ A mixed-occupancy building is one that contains both residential and commercial uses (see definitions in Chapter 2 [RE]). Where residential and commercial uses coexist in a building, each occupancy must be evaluated separately. In situations where the majority of the space is one occupancy and only a small portion of the floor is different, such evaluation is sometimes aided by the accessory area requirements found in the IBC. For example, the residential portions of that story must meet the residential requirements of the code unless 90 percent or more of the floor level is commercial, in which case the story in question (inclusive of the 10-percent residential) may, with the code official's approval, be considered commercial.

For example, consider the three-story apartment building in Commentary Figure R101.4.1, with a portion of the first story leased out to a convenience store (a commercial use). The top two stories are clearly residential because they are devoted solely to residential use and the building is not over three stories high. Though not found in the code, the first story could generally be considered all residential if 10 percent or less of the total floor area is occupied by the store. In this case, the entire first floor is subject to the residential portions of the code. When more than 10 percent of the first story is occupied by the store, the first story is considered a mixed occupancy; the portion of the first story occupied by the store is considered commercial and is subject to the applicable commercial requirements in Chapter 4[CE]. The remainder of the first story is considered residential and must meet the residential requirements found in Chapter 4 [RE].

Consider another conceivable situation in which the first story of a four-story building may be one or more retail establishments (or other commercial use). Consider that the remaining stories of this four-story building consist entirely of dwelling units and are classified as residential. This and similar situations can cause confusion over how to apply the code. Is this a commercial building because it is over three stories high, or is it a residential building because it has three stories of dwelling units?

For our current example, the definition of "Residential" makes it clear that the entire building would be considered commercial and be subject to the requirements of Chapter 4[CE]. The approach is based on the fact that the patterns of energy use generally change in buildings four stories or greater in height, and that the code, as well as its predecessor MEC versions, limit residential buildings to a maximum height of three stories above grade. Any structure over three stories is considered a commercial building for purposes of applying the code, regardless of the occupancy classification of the structure. The only exception to this distinction would be single-family or duplex detached residences four stories or greater in height, which is considered rare. See also the definitions and commentary for "Commercial building" and "Residential building" to help clarify the application of the code to mixed-occupancy buildings.

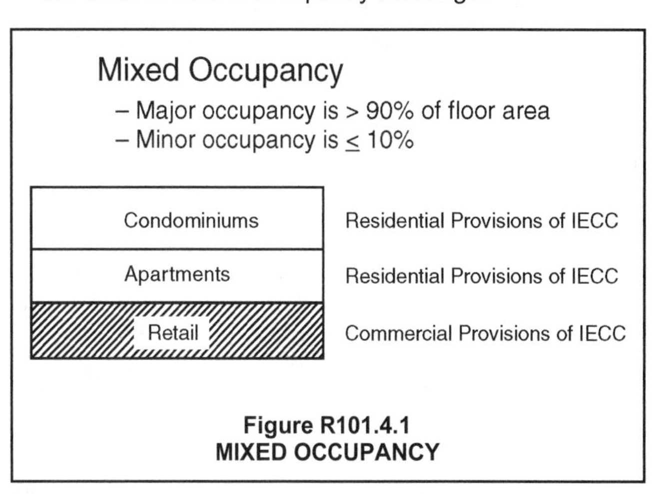

Figure R101.4.1
MIXED OCCUPANCY

R101.5 Compliance. *Residential buildings* shall meet the provisions of IECC—Residential Provisions. *Commercial buildings* shall meet the provisions of IECC—Commercial Provisions.

❖ The code contains three alternative design procedures for detached one- and two-family dwellings and residential buildings three stories or less in height. This section provides the scoping to the various sections and methods of compliance in the code.

R101.5.1 Compliance materials. The *code official* shall be permitted to approve specific computer software, worksheets, compliance manuals and other similar materials that meet the intent of this code.

❖ The code is intended to permit the use of innovative approaches and techniques, provided that they result in the effective use of energy. This section simply recognizes that there are many federal, state and local programs as well as computer software that deal with energy efficiency. Therefore, the code simply states that the code official has the authority to accept those methods of compliance, provided that they meet the intent of the code. Some of the easiest examples to illustrate this provision are the REScheck™ and COMcheck™ software put out by the U.S. Department of Energy (DOE). Another example is the ENERGY STAR program.

SECTION R102 ALTERNATIVE MATERIALS, DESIGN AND METHODS OF CONSTRUCTION AND EQUIPMENT

R102.1 General. The provisions of this code are not intended to prevent the installation of any material or to prohibit any design or method of construction not specifically prescribed by this code, provided that any such alternative has been approved. The code official shall be permitted to approve an alternative material, design or method of construction where the *code official* finds that the proposed design is satisfactory and complies with the intent of the provisions of this code, and that the material, method or work offered is, for the purpose intended, at least the equivalent of that prescribed in this code.

❖ This section reinforces Section R101.3, which states that the code is meant to be flexible, as long as the intent of the proposed alternative is to promote the effective use of energy. The code is not intended to inhibit innovative ideas or technological advances. A comprehensive regulatory document such as an energy code cannot envision and then address all future innovations in the industry. As a result, a performance code must be applicable to and provide a basis for the approval of an increasing number of newly developed, innovative materials, systems and methods for which no code text or referenced standards yet exist. The fact that a material, product or method of construction is not addressed in the code is not an indication that the material, product or method is prohibited.

The code official is expected to apply sound technical judgment in accepting materials, systems or methods that, while not anticipated by the drafters of the current code text, can be demonstrated to offer equivalent or better performance. The code regulates new and innovative construction practices while addressing the relative safety of building occupants. The code official is responsible for determining whether a requested alternative provides a level of protection of the public health, safety and welfare equal to that required by the code.

R102.1.1 Above code programs. The *code official* or other authority having jurisdiction shall be permitted to deem a national, state or local energy-efficiency program to exceed the energy efficiency required by this code. Buildings *approved* in writing by such an energy-efficiency program shall be considered in compliance with this code. The requirements identified as "mandatory" in Chapter 4 shall be met.

❖ The purpose of this section is to specifically state that the code official has the authority to review and accept compliance with another energy program that may exceed that required by the code, as long as the minimum "mandatory" requirements of the code are met. This provision is really a continuation of the provision stated in Section R101.2 and the fact that the code is intended to accept alternatives as long as the end result is an energy-efficient building that is comparable to or better than that required by the code.

This is also a good section to help reinforce the fact that the code, as a model code, is a "minimum" code. Therefore, it establishes the minimum requirement that must be met and that anything exceeding the level is permitted.

While "above code programs" are acceptable because they do exceed the "minimum" requirements of the code, it would not be proper to require compliance with such "above code programs." Besides the code being the minimum level of acceptable energy efficiency, it is also the maximum efficiency that the code official can require. A building built to the absolute minimum requirement is also the maximum that the code official can demand. It is perfectly acceptable for a designer or builder to exceed the code requirements, but it is not proper for the code official to demand such higher performance.

This section also contains language to ensure that the "mandatory" requirements of the code, such as sealing the building envelope (Section R402.4) and sealing ducts (Section R403.3.2), are complied with for all buildings. Since the code has deemed that the mandatory requirements should apply to all buildings, it is reasonable that "above code programs" not be allowed to bypass these requirements.

PART 2—ADMINISTRATION AND ENFORCEMENT

SECTION R103 CONSTRUCTION DOCUMENTS

R103.1 General. Construction documents, technical reports and other supporting data shall be submitted in one or more sets with each application for a permit. The construction documents and technical reports shall be prepared by a registered design professional where required by the statutes of the jurisdiction in which the project is to be constructed. Where special conditions exist, the *code official* is authorized to require necessary construction documents to be prepared by a registered design professional.

Exception: The *code official* is authorized to waive the requirements for construction documents or other supporting data if the *code official* determines they are not necessary to confirm compliance with this code.

❖ In most jurisdictions, the permit application must be accompanied by not less than two sets of construction documents. The code official can waive the requirements for filing construction documents when the scope of the work is minor and compliance can be verified through other means. When the quality of the materials is essential for conformity to the code, specific information must be given to establish that quality.

The code must not be cited, or the term "legal" or its equivalent used as a substitute for specific information. For example, it would be improper for the plans to simply state "windows per IECC requirements."

A detailed description of the work covered by the application must be submitted. When the work is "minor," either in scope or needed description, the code official may use judgment in determining the need for a detailed description. An example of "minor" work that may not involve a detailed description is the replacement of an existing 60-amp electrical service in a single-family residence with a 100-amp service.

The exception permits the code official to determine construction documents are not necessary when compliance can be obtained and verified without the documents.

R103.2 Information on construction documents. Construction documents shall be drawn to scale upon suitable material. Electronic media documents are permitted to be submitted where *approved* by the *code official*. Construction documents shall be of sufficient clarity to indicate the location, nature and extent of the work proposed, and show in sufficient detail pertinent data and features of the *building*, systems and equipment as herein governed. Details shall include, but are not limited to, the following as applicable:

1. Insulation materials and their *R*-values.
2. Fenestration *U*-factors and solar heat gain coefficients (SHGC).
3. Area-weighted *U*-factor and solar heat gain coefficients (SHGC) calculations.
4. Mechanical system design criteria.
5. Mechanical and service water-heating system and equipment types, sizes and efficiencies.
6. Equipment and system controls.
7. Duct sealing, duct and pipe insulation and location.
8. Air sealing details.

❖ For a comprehensive plan review, all code requirements should be incorporated in the design and construction documents. All of the project information, including specifications, scope, calculations, and detailed drawings, should be submitted to the code official so that code compliance can be verified. All parties should clearly understand what the project entails. A good plan review is essential to ensure code compliance and a successful project. A statement on the construction documents, such as "All insulation levels shall comply with the 2015 edition of the *International Energy Conservation Code®* (IECC®)," is not an acceptable substitute for showing the required information. Note also that the code official is authorized to require additional project and code-related information as necessary.

For example, insulation *R*-values and glazing and door *U*-factors must be clearly marked on the building plans, specifications or forms used to show compliance. Where two or more insulation levels exist for the same component (two insulation levels are used in ceilings), the permit applicant must record each level separately on the plans or specifications and clarify where in the building each level of insulation will be installed.

The following discussion is presented for the benefit of both the applicant and the plans examiner. This is not an all-inclusive list but rather is intended to reflect the minimum scope of information needed to determine energy code compliance:

Permit Applicant's Responsibilities. At permit application, the goal of the applicant is to provide all necessary information to show compliance with the code. If the plans examiner is able to verify compliance in a single review, the permit can be issued and construction may be started without delay.

Depending on whether the prescriptive or performance methods of compliance are used, the amount and detail of the required information may vary. For example, if using the prescriptive method of compliance, the *U*-factor and SHGC may be the only information needed to verify fenestration compliance. If the "total UA alternative" (Section R402.1.4) or the performance option (Section R405) is used, then additional information, such as the fenestration sizes and orientation, may be needed to demonstrate compliance. The envelope information that needs to be on the plans can be presented in a number of ways:

- *On the drawings.* Include elevations that indicate window, door and skylight areas and sections that show insulation position and thickness.

- *On sections and in schedules.* For instance, list *R*-values of insulation on sections and include *U*-factors, shading coefficient, visible light transmittance and air infiltration on fenestration and opaque door schedules.
- *Through notes and callouts.* Note that all exterior joints are to be caulked, gasketed, weatherstripped or otherwise sealed.
- *Through supplementary worksheets or calculations.* Provide area-weighted calculations where required, such as for projection factors and heat capacity. The permit applicant may include these calculations on the drawings, incorporate them as additional columns in the schedule or submit completed code compliance worksheets provided by the jurisdiction.

Incorrect information may be caused by a lack of understanding of the code. More likely, it indicates that the code has changed since the last project. The applicant can use a correction list to update office specifications to avoid receiving this same correction again in the future.

Plans Examiner's Responsibilities. The plans examiner must review each permit application for code compliance before a permit is issued. By letting the designer and contractor know what is expected of them early in the process, the building department can increase the likelihood that the approved drawings will comply with the code. This helps the inspector avoid the headache of correcting a contractor who is following drawings that do not meet code requirements.

The biggest challenge for the plans examiner is often determining where the necessary information is and whether the drawings are complete. The plans examiner should make sure the applicant includes a summary or checklist as part of the submittal package. When building envelope information is provided on the construction documents, it makes the job of the plans examiner easier, generally making for a more thorough review and reducing turnaround time.

A complete building envelope plan review covers all the requirements specific to the architectural building shell, but the electrical drawings may also need to be included and reviewed if the applicant seeks credit for automatic daylighting control for skylights or fenestration. A plan review should:

- Check that duct insulation thickness and conductivity (*k*-value) are on the drawings and comply with the code.
- Check that the duct insulation *R*-value is on the drawings and complies with the code.
- Check that there is a note indicating that ducts are to be constructed and sealed in accordance with the *International Mechanical Code*® (IMC®).
- Check that there is a note indicating that operating and equipment maintenance manuals will be supplied to the owner, that air and hydronic systems will be balanced and that the control system will be tested and calibrated.

R103.2.1 Building thermal envelope depiction. The *building's thermal envelope* shall be represented on the construction documents.

❖ The most important energy and performance aspects of the home are the building's thermal envelope and the alignment of the air barrier and thermal barrier systems. It is crucial that the design professional demonstrate an understanding of location of the thermal envelope and that the effort is made to draw its location so that construction personnel can successfully construct the building in accordance with the code and the specifications that have been drawn. Air sealing details help make this possible, but understanding where the details will be implemented helps ensure better implementation and enforcement.

R103.3 Examination of documents. The *code official* shall examine or cause to be examined the accompanying construction documents and shall ascertain whether the construction indicated and described is in accordance with the requirements of this code and other pertinent laws or ordinances. The *code official* is authorized to utilize a registered design professional, or other *approved* entity not affiliated with the building design or construction, in conducting the review of the plans and specifications for compliance with the code.

❖ This section describes the required action of the code official in response to a permit application. The code official can delegate review of the construction documents to subordinates. In addition, the code official can retain the services of an outside entity, such as a registered design professional, to examine the plans.

R103.3.1 Approval of construction documents. When the *code official* issues a permit where construction documents are required, the construction documents shall be endorsed in writing and stamped "Reviewed for Code Compliance." Such *approved* construction documents shall not be changed, modified or altered without authorization from the *code official*. Work shall be done in accordance with the *approved* construction documents.

One set of construction documents so reviewed shall be retained by the *code official*. The other set shall be returned to the applicant, kept at the site of work and shall be open to inspection by the *code official* or a duly authorized representative.

❖ The code official must stamp or otherwise endorse as "Reviewed for Code Compliance" the construction documents on which the permit is based. One set of approved construction documents must be kept on the construction site to serve as the basis for all subsequent inspections. To avoid confusion, the construction documents on the site must be the documents that were approved and stamped. This is because inspections are to be performed with regard to the approved documents, not the code itself. Additionally, the contractor cannot determine compliance with the approved construction documents unless they are readily available. If the approved construction documents are not available, the inspection should be postponed and work on the project halted.

R103.3.2 Previous approvals. This code shall not require changes in the construction documents, construction or designated occupancy of a structure for which a lawful permit has been heretofore issued or otherwise lawfully authorized, and the construction of which has been pursued in good faith within 180 days after the effective date of this code and has not been abandoned.

❖ If a permit is issued and construction proceeds at a normal pace and a new edition of the code is adopted by the legislative body, requiring that the building be constructed to conform to the new code is unreasonable. This section provides for the continuity of permits issued under previous codes, as long as such permits are being "actively prosecuted" subsequent to the effective date of the ordinance adopting this edition of the code.

R103.3.3 Phased approval. The *code official* shall have the authority to issue a permit for the construction of part of an energy conservation system before the construction documents for the entire system have been submitted or *approved*, provided adequate information and detailed statements have been filed complying with all pertinent requirements of this code. The holders of such permit shall proceed at their own risk without assurance that the permit for the entire energy conservation system will be granted.

❖ The code official has the authority to issue a partial permit to allow for the practice of "fast-tracking" a job. Any construction under a partial permit is "at the holder's own risk" and "without assurance that a permit for the entire structure will be granted." The code official is under no obligation to accept work or issue a complete permit in violation of the code, ordinances or statutes simply because a partial permit had been issued. Fast-tracking puts unusual administrative and technical burdens on the code official. The purpose is to proceed with construction while the design continues for other aspects of the work. Coordinating and correlating the code aspects into the project in phases requires attention to detail and project tracking so that all code issues are addressed. The coordination of these submittals is the responsibility of the registered design professional in responsible charge.

R103.4 Amended construction documents. Work shall be installed in accordance with the approved construction documents, and any changes made during construction that are not in compliance with the *approved* construction documents shall be resubmitted for approval as an amended set of construction documents.

❖ The code requires that all work be done in accordance with the approved plans and other construction documents. Where the construction will not conform to the approved construction documents, the documents must be revised and resubmitted to the code official for review and approval. Code officials should maintain a policy that all amendments be submitted for review. Otherwise, a significant change that is not approved could result in an activity that is not in compliance with the code and, therefore, cause needless delay and extra expense. The code official must retain one set of the amended and approved plans. The other set is to be kept at the construction site, ready for use by the jurisdiction's inspection staff.

R103.5 Retention of construction documents. One set of *approved* construction documents shall be retained by the *code official* for a period of not less than 180 days from date of completion of the permitted work, or as required by state or local laws.

❖ Construction documents must be retained in case a question or dispute arises after completion of the project. Unless modified because of state or local statutes, the retention period for the approved construction documents is a minimum of 180 days following the completion of the work, typically the date the certificate of occupancy is issued. Any further retention of plans by the jurisdiction as an archival record of construction activity in the community is not required by the code.

SECTION R104 INSPECTIONS

R104.1 General. Construction or work for which a permit is required shall be subject to inspection by the *code official* or his or her designated agent, and such construction or work shall remain accessible and exposed for inspection purposes until *approved*. It shall be the duty of the permit applicant to cause the work to remain accessible and exposed for inspection purposes. Neither the *code official* nor the jurisdiction shall be liable for expense entailed in the removal or replacement of any material, product, system or building component required to allow inspection to validate compliance with this code.

❖ Where a permit is required by state or local law, the building is subject to an inspection. The code official must determine whether appropriate energy-efficient features and equipment are installed in accordance with the approved construction documents and applicable code requirements.

Generally, a department's administrative rules will list required periodic inspections. Because the majority of energy-efficient construction occurs in steps or phases, periodic inspections are often necessary before portions of these systems are covered by further construction. The exact number of required inspections cannot always be specified. A reinspection may be necessary if violations are noted and corrections are required (see commentary, Section R104.4). If time permits, frequent inspections of some job sites, especially where the work is complex, can be beneficial to detect potential problems before they become too difficult to correct.

An inspector's ongoing challenge is responding to change orders during construction. In any construction project there will be field changes. The call is easy if a more efficient piece of equipment is being substituted for a less efficient one. For the opaque elements, more insulation is generally better. For fenestration, a lower *U*-factor and SHGC is generally better. Unfortunately, changing the glass almost always changes more than one characteristic, and it is not always clear whether energy efficiency is being improved. If there is any doubt concerning the impact, the inspector should confer with the plans examiner for the project.

A more difficult change order is one that reduces efficiency. For example, if the proposed substitute fenestration has a higher *U*-factor and SHGC, or if the window area is to be increased, the inspector must check with the plans examiner. The amount of information and the ease of confirming compliance will depend on whether the prescriptive or performance approach was used initially. In these cases, compliance is based on a combination of the fenestration area, *U*-factor, SHGC, projection factor, and (if a performance-based analysis has been used) opaque wall characteristics. Although there may be enough latitude to decrease the efficiency somewhat, it is not possible to make such a determination without reviewing all the elements and how compliance was initially demonstrated. Whenever there are significant changes such as described above, the inspector is expected to request that the applicant submit revised plans, so the plans examiner can verify compliance and ensure there is a correct record on file in the building department.

An even tougher case is when the contractor has already installed noncomplying equipment without checking with the inspector. For instance, ordinary double glazing may have been installed instead of double glazing with a low-emissivity coating. The inspector should be quite strict for several reasons. First, because most contracts are awarded on a cost-competitive basis, the low-bid company might win the job and then make its profit by installing noncomplying equipment. This would be unfair to the high-bid contractors.

Second, a lenient inspector's job will be more difficult in the future. If a noncomplying contractor skates by this time, that contractor will most likely have additional requests for future projects. In addition, other contractors will also begin to ask for special treatment. Self-policing, which works well if everyone is being treated fairly, will begin to decline.

Finally, there is the situation in which the approved plans do not contain all of the code requirements. If information or notes are missing from the plans, the inspector can, for instance, simply direct the contractor to make the necessary changes in the field (for example, caulk and seal joints).

The inspector's job is more difficult, however, if drawings contain information that is wrong. Perhaps the inspector in a cold climate notices the metal stud wall is not covered with insulating sheathing, as is required in that climate, and informs the contractor. The contractor responds saying that he or she is following the approved plans, and indeed he or she is. The inspector, as the representative of the code official, is clearly authorized to require the contractor to build the project to code. (If necessary, the inspector can show the contractor the building department note, which says, "Approved subject to errors and omissions." See also IBC Section 104.4.) In this case, it would be appropriate for the inspector to inform the plans examiner of the problem and ask the plans examiner to help solve the problem. The plans examiner may be able to suggest improvements in other areas that would compensate for this shortfall. It is important that the plans examiner and inspector appreciate the challenges of each other's work and the benefits of a team effort.

R104.2 Required inspections. The *code official* or his or her designated agent, upon notification, shall make the inspections set forth in Sections R104.2.1 through R104.2.5.

❖ The contractor, builder, owner or other authorized party is responsible for arranging and coordinating required inspections to prevent work from being concealed prior to inspection. For example:

- Insulation must be inspected prior to concealment. Where the insulation is concealed prior to inspection and approval, the code official has the authority to require removal of the concealing components.
- Basement wall insulation may be installed on the exterior of a below-grade basement wall. Where the insulation application is not confirmed prior to backfilling, reinspection is necessary.
- Glazing assembly *U*-factor labels are to be left on until after the building has been inspected for compliance. The applicant is responsible for giving the inspector adequate information on site to verify code-related features, such as window *U*-factor and equipment efficiencies.

After the field inspector has performed the required inspections and observed any required equipment and system tests (or has received written reports of the results of such tests), the code official must determine whether the installation or work is in compliance with all applicable sections of the code. The code official must issue a written notice of approval if the subject work or installation is in apparent compliance with the code. The notice of approval is given to the permit holder and a copy of the notice is retained on file by the code official.

R104.2.1 Footing and foundation inspection. Inspections associated with footings and foundations shall verify compliance with the code as to *R*-value, location, thickness, depth of burial and protection of insulation as required by the code and *approved* plans and specifications.

❖ Specific inspection details are called out in this section to ensure code requirements are met. In the case of footings and foundations, proper installation of the insulation is key to the performance of the building thermal envelope.

R104.2.2 Framing and rough-in inspection. Inspections at framing and rough-in shall be made before application of interior finish and shall verify compliance with the code as to types of insulation and corresponding *R*-values and their correct location and proper installation; fenestration properties (*U*-factor and SHGC) and proper installation; and air leakage controls as required by the code and approved plans and specifications.

❖ Specific inspection details are called out in this section to ensure code requirements are met. It is important to conduct inspections at the time of framing rough-in to be able to verify that the required insulation *R*-value is installed and that the installation details are followed. The code also provides detailed requirements for air barrier installation (see Section R402.4.1.1).

R104.2.3 Plumbing rough-in inspection. Inspections at plumbing rough-in shall verify compliance as required by the code and *approved* plans and specifications as to types of insulation and corresponding *R*-values and protection, and required control.

❖ Specific inspection details are called out in this section to ensure code requirements are met. In this case the integrity of the hot water supply insulation is important for meeting the energy efficiency requirements of the code.

R104.2.4 Mechanical rough-in inspection. Inspections at mechanical rough-in shall verify compliance as required by the code and *approved* plans and specifications as to installed HVAC equipment type and size, required controls, system insulation and corresponding *R*-value, system air leakage control, programmable thermostats, dampers, whole-house ventilation, and minimum fan efficiency.

Exception: Systems serving multiple dwelling units shall be inspected in accordance with Section C104.2.4.

❖ Specific inspection details are called out in this section to ensure code requirements are met. In this case the HVAC system equipment type and installation, the duct insulation, and the system controls all have requirements in the code that must be met to ensure that the building is meeting the minimum energy efficiency requirements. The exception simply points to the Commercial Provisions for a situation in which the Commercial Provisions are more appropriate than these provisions.

R104.2.5 Final inspection. The *building* shall have a final inspection and shall not be occupied until *approved*. The final inspection shall include verification of the installation of all required *building* systems, equipment and controls and their proper operation and the required number of high-efficacy lamps and fixtures.

❖ To establish compliance with all previously issued correction orders and to determine whether subsequent violations exist, a final inspection is required. The final inspection is conducted after all work is completed. Typically, the final inspection includes all items installed after the rough-in inspection and not concealed in the building construction. Subsequent reinspection is necessary if the final inspection generates a notice of violation (see commentary, Section R104.4). All violations observed during the final inspection must be noted and the permit holder must be advised of them.

R104.3 Reinspection. A *building* shall be reinspected when determined necessary by the *code official*.

❖ The provisions for reinspection could affect the entire structure or a portion thereof. For example, if no approval was given to apply interior finish that conceals ducts in an exterior wall, the code official must require removal of the interior finish to verify the ducts are insulated to code.

Reinspections generally occur when some type of violation or correction notice was issued during one of the previous inspections or where the work was not ready for the inspection. For example, if the inspector went to the project to conduct an insulation inspection and not all of the insulation was installed at that point, the inspector would need to go back to the project and "reinspect" the insulation to verify that it had been completed. After the reinspection, the inspector would issue the approval (see Section R104.2) to permit the wall or ceiling cavities to be enclosed and therefore conceal the insulation.

R104.4 Approved inspection agencies. The *code official* is authorized to accept reports of third-party inspection agencies not affiliated with the building design or construction, provided such agencies are *approved* as to qualifications and reliability relevant to the building components and systems they are inspecting.

❖ As an alternative to the code official conducting the inspection, he or she is permitted to accept inspections of and reports by approved inspection agencies. Appropriate criteria on which to base approval of inspection agencies can be found in IBC Section 1703.

R104.5 Inspection requests. It shall be the duty of the holder of the permit or their duly authorized agent to notify the *code official* when work is ready for inspection. It shall be the duty of the permit holder to provide access to and means for inspections of such work that are required by this code.

❖ It is the responsibility of the permit holder or other authorized person, such as the contractor performing the work, to arrange for the required inspections when completed work is ready and to allow for sufficient time for the code official to schedule a visit to the site to prevent work from being concealed prior to being inspected. Access to the work to be inspected must be provided, including any special means such as a ladder.

R104.6 Reinspection and testing. Where any work or installation does not pass an initial test or inspection, the necessary corrections shall be made to achieve compliance with this code. The work or installation shall then be resubmitted to the *code official* for inspection and testing.

❖ This section provides for necessary actions in the event that a tested or inspected item is not originally in compliance with the code.

R104.7 Approval. After the prescribed tests and inspections indicate that the work complies in all respects with this code, a notice of approval shall be issued by the *code official*.

❖ This section mandates that a notice of approval be issued by the code official to indicate completion of an energy conservation installation. While certificates of occupancy for construction are traditionally under the purview of one of the construction codes, the notice of approval will fill a need with regard to application and enforcement of nonbuilding codes.

R104.7.1 Revocation. The *code official* is authorized to, in writing, suspend or revoke a notice of approval issued under the provisions of this code wherever the certificate is issued in error, or on the basis of incorrect information supplied, or where it is determined that the *building* or structure, premise, or portion thereof is in violation of any ordinance or regulation or any of the provisions of this code.

❖ This section provides an important administrative tool by giving the code official the authority to revoke a certificate of completion for the reasons indicated in the text. The code official may also suspend the certificate until any code violations are corrected.

SECTION R105 VALIDITY

R105.1 General. If a portion of this code is held to be illegal or void, such a decision shall not affect the validity of the remainder of this code.

❖ This section is applicable when a court of law rules that a portion of the code (or the jurisdiction's energy code) is invalid. Only invalid sections of the code (as established by the court of jurisdiction) can be set aside. This is essential to safeguard the application of the code text to situations in which a provision of the code is declared illegal or unconstitutional. This section preserves the original legislative action that put the legal requirements (energy code) in place.

All sections of the code not judged invalid must remain in effect. Although a dispute over a particular issue (such as an appliance efficiency requirement) may have precipitated the litigation causing the requirement to be found invalid, the remainder of the code must still be considered applicable. This is sometimes called the "severability clause" and simply means that the invalid section can be removed from the code without affecting the entire document.

SECTION R106 REFERENCED STANDARDS

R106.1 Referenced codes and standards. The codes and standards referenced in this code shall be those listed in Chapter 5, and such codes and standards shall be considered as part of the requirements of this code to the prescribed extent of each such reference and as further regulated in Sections R106.1.1 and R106.1.2.

❖ The code references many standards promulgated and published by other organizations. A complete list of these standards appears in Chapter 6 [RE]. The wording of this section was carefully chosen to establish the edition of the standard that is enforceable under the code.

Although a standard is referenced, its full scope and content are not necessarily applicable. The standard is applicable only to the extent indicated in the text in which the standard is specifically referenced. A referenced standard or the portion cited in the text is an enforceable extension of the code as if the content of the standard was included in the body of the code. The use and applicability of referenced standards are limited to those portions of the standards that are specifically identified.

R106.1.1 Conflicts. Where conflicts occur between provisions of this code and referenced codes and standards, the provisions of this code shall apply.

❖ The use of referenced codes and standards to cover certain aspects of various occupancies and operations rather than write parallel or competing requirements into the code is a longstanding code development principle. Often, however, questions and potential conflicts in the use of referenced codes and standards can arise, leading to inconsistent enforcement of the code. This section of the code is intended to establish that the provisions of the code would prevail in such a conflict, regardless of the level of stringency.

R106.1.2 Provisions in referenced codes and standards. Where the extent of the reference to a referenced code or standard includes subject matter that is within the scope of this code, the provisions of this code, as applicable, shall take precedence over the provisions in the referenced code or standard.

❖ Section R106.1.2 expands on the provisions of Section R106.1.1 by making it clear that even if a referenced standard contains requirements that parallel the code [or the other referenced section(s)], the provisions of the code will always take precedence.

R106.2 Application of references. References to chapter or section numbers, or to provisions not specifically identified by number, shall be construed to refer to such chapter, section or provision of this code.

❖ This section provides some information on the use of the code and the conventions used in making references to other portions of the code. By implication then, a reference that is intended to provide information from another code would be required to state what code the reference is from.

R106.3 Other laws. The provisions of this code shall not be deemed to nullify any provisions of local, state or federal law.

❖ This provision is intended to assist the code official in dealing with situations where other laws enacted by the jurisdiction or the state or federal government may be applicable to a condition that is also governed by a requirement in the code. In such circumstances, the requirements of the code would be in addition to that other law that is still in effect, although the code official may not be responsible for its enforcement.

SECTION R107
FEES

R107.1 Fees. A permit shall not be issued until the fees prescribed in Section R107.2 have been paid, nor shall an amendment to a permit be released until the additional fee, if any, has been paid.

❖ This section requires that all fees be paid prior to permit issuance or release of an amendment to a permit. Since department operations are usually intended to be supported by fees paid by the user of department activities, it is important that these fees are received before incurring any expense.

R107.2 Schedule of permit fees. A fee for each permit shall be paid as required, in accordance with the schedule as established by the applicable governing authority.

❖ This section authorizes the establishment of a schedule of fees by the jurisdiction. The fees are usually established by law, such as in an ordinance adopting the code, a separate ordinance or legally promulgated regulation as required by state or local law, and are often based on a valuation of the work to be performed.

R107.3 Work commencing before permit issuance. Any person who commences any work before obtaining the necessary permits shall be subject to an additional fee established by the *code official* that shall be in addition to the required permit fees.

❖ The department will incur certain costs (i.e., inspection time and administrative) when investigating and citing a person who has commenced work without having obtained a permit. This section authorizes the code official to recover those costs by establishing a fee, in addition to that collected when the required permit is issued, to be imposed on the responsible party.

R107.4 Related fees. The payment of the fee for the construction, *alteration*, removal or demolition of work done in connection to or concurrently with the work or activity authorized by a permit shall not relieve the applicant or holder of the permit from the payment of other fees that are prescribed by law.

❖ This section provides the code official with a useful administrative tool establishing that all applicable fees of the jurisdiction for regulated work collateral to the work being done under the code's permit, such as sewer connections, water taps, driveways, signs, etc., must be paid.

R107.5 Refunds. The *code official* is authorized to establish a refund policy.

❖ This section authorizes the code official to establish a policy to regulate the refund of fees, which may be full or partial, typically resulting from the revocation, abandonment or discontinuance of a building project for which a permit has been issued and fees have been collected.

SECTION R108
STOP WORK ORDER

R108.1 Authority. Where the *code official* finds any work regulated by this code being performed in a manner either contrary to the provisions of this code or dangerous or unsafe, the *code official* is authorized to issue a stop work order.

❖ This section provides for the suspension of work for which a permit was issued, pending the removal or correction of a severe violation or unsafe condition identified by the code official. Stop work orders are issued when enforcement can be accomplished no other way or when a dangerous condition exists.

R108.2 Issuance. The stop work order shall be in writing and shall be given to the owner of the property involved, to the owner's authorized agent, or to the person doing the work. Upon issuance of a stop work order, the cited work shall immediately cease. The stop work order shall state the reason for the order and the conditions under which the cited work will be permitted to resume.

❖ This section makes it clear that, upon receipt of a violation notice from the code official, all construction activities identified in the notice must immediately cease, except as expressly permitted to correct the violation.

R108.3 Emergencies. Where an emergency exists, the *code official* shall not be required to give a written notice prior to stopping the work.

❖ This section gives the code official the authority to stop the work in dispute immediately when, in his or her opinion, there is an unsafe emergency condition that has been created by the work. The need for the written notice is suspended for this situation so that the work can be stopped immediately.

R108.4 Failure to comply. Any person who shall continue any work after having been served with a stop work order, except such work as that person is directed to perform to remove a violation or unsafe condition, shall be subject to a fine as set by the applicable governing authority.

❖ This section establishes consequences for disregarding a stop work order and continuing the work that is at issue, other than abatement work. The dollar amounts for the minimum and maximum fines are to be specified in the adopting ordinance.

SECTION R109 BOARD OF APPEALS

R109.1 General. In order to hear and decide appeals of orders, decisions or determinations made by the *code official* relative to the application and interpretation of this code, there shall be and is hereby created a board of appeals. The *code official* shall be an ex officio member of said board but shall not have a vote on any matter before the board. The board of appeals shall be appointed by the governing body and shall hold office at its pleasure. The board shall adopt rules of procedure for conducting its business, and shall render all decisions and findings in writing to the appellant with a duplicate copy to the *code official*.

❖ This section provides an aggrieved party that has a material interest in the decision of the code official a process to appeal such a decision before a board of appeals. This provides a forum, other than the court of jurisdiction, in which to review the code official's actions. The intent of the appeal process is not to waive or set aside a code requirement; rather it is intended to provide a means of reviewing a code official's decision on an interpretation or application of the code.

R109.2 Limitations on authority. An application for appeal shall be based on a claim that the true intent of this code or the rules legally adopted thereunder have been incorrectly interpreted, the provisions of this code do not fully apply or an equally good or better form of construction is proposed. The board shall not have authority to waive requirements of this code.

❖ This section establishes the grounds for an appeal that claims that the code official has misinterpreted or misapplied a code provision. The board is not allowed to set aside any of the technical requirements of the code; however, it is allowed to consider alternative methods of compliance with the technical requirements.

R109.3 Qualifications. The board of appeals shall consist of members who are qualified by experience and training and are not employees of the jurisdiction.

❖ This section requires that the members of the appeals board are to have experience in building construction and system matters because the decisions of the appeals board are to be based purely on the technical merits involved in an appeal.

Bibliography

The following resource materials were used in the preparation of the commentary for this chapter of the code.

ASHRAE 90A-80, *Energy Conservation in New Building Design* with Addendum 90A-2-87. Atlanta, GA: American Society of Heating, Refrigerating and Air-Conditioning Engineers, Inc., 1987.

ASHRAE 90B-75, *Energy Conservation in New Building Design*. Atlanta, GA: American Society of Heating, Refrigerating and Air-Conditioning Engineers, Inc., 1975.

ASHRAE 90C-77, *Energy Conservation in New Building Design*. Atlanta, GA: American Society of Heating, Refrigerating and Air-Conditioning Engineers, Inc., 1977.

ANSI/ASHRAE/IESNA 90.1-10, *Energy Standard for Buildings, Except Low-rise Residential Buildings*. Atlanta, GA: American Society of Heating, Refrigerating and Air-Conditioning Engineers, Inc., 2010.

ASHRAE/IESNA-93, *Energy Code for Commercial and High-rise Residential Buildings—Based on ASHRAE/IES 90.1-1989*. Atlanta, GA: American Society of Heating, Refrigerating and Air-Conditioning Engineers, Inc., 1993.

DOE-97, *Technical Support Document for COMCheck-EZTM, V1.0*. Washington, DC: U.S. Department of Energy, 1997.

DOE-09, *Methodology for Developing the REScheck™*. Software Through Version 4.2, Washington, DC: U.S. Department of Energy, August, 2009.

NFRC, *Certified Products Directory 2009*. Silver Spring, MD: National Fenestration Rating Council, Inc., 2009.

PNNL-97, *Assessment of the 1995 Model Energy Code for Adoption*. Richland, WA: Prepared for U.S. Department of Housing and Urban Development Office of Policy Development and Research by Pacific Northwest National Laboratory, 1997.

Chapter 2 [RE]: Definitions

General Comments

All terms defined in the code are listed alphabetically in Chapter 2 [RE]. The words or terms defined in this chapter are considered to be of prime importance in either specifying the subject matter of code provisions or in giving meaning to certain terms used throughout the code for administrative or enforcement purposes. The code user should be familiar with what terms are found in this chapter because the definitions are essential to the correct interpretation of the code and because the user might not be aware of the fact that a particular term found in the text is defined.

Purpose

Codes, by their nature, are technical documents. Every word, term and punctuation mark can alter a sentence's meaning and, if misused, muddy its intent. Further, the code, with its broad scope of applicability, includes terms inherent in a variety of construction disciplines. These terms can often have multiple meanings, depending on the context or discipline in which they are being used. For these reasons, maintaining a consensus on the specific meaning of terms contained in the code is essential. Chapter 2 [RE] performs this function by stating clearly what specific terms mean for the purpose of the code.

SECTION R201 GENERAL

R201.1 Scope. Unless stated otherwise, the following words and terms in this code shall have the meanings indicated in this chapter.

❖ For the purposes of the code, certain abbreviations, terms, phrases, words and their derivatives have the meanings given in Chapter 2 [RE]. The code, with its broad scope of applicability, includes terms used in a variety of construction and energy-related disciplines. These terms can often have multiple meanings, depending on their context or discipline. Therefore, Chapter 2 [RE] establishes specific meanings for these terms.

R201.2 Interchangeability. Words used in the present tense include the future; words in the masculine gender include the feminine and neuter; the singular number includes the plural and the plural includes the singular.

❖ Although the definitions contained in Chapter 2 [RE] are to be taken literally, gender, number and tense are considered to be interchangeable.

R201.3 Terms defined in other codes. Terms that are not defined in this code but are defined in the *International Building Code*, *International Fire Code*, *International Fuel Gas Code*, *International Mechanical Code*, *International Plumbing Code* or the *International Residential Code* shall have the meanings ascribed to them in those codes.

❖ When a word or term that is not defined in this chapter appears in the code, other references may be used to find its definition, such as other *International Codes*® (I-Codes®). Definitions that are applicable in other I-Codes are applicable everywhere the term is used in the code. As stated in both the "Purpose" section above and in the commentary to Section R201.1, a bit of caution is needed when looking at definitions from other codes. Because the context and discipline can vary, it is important to determine that the term does fit within the code context. As an example, the term "accessible" would have a different meaning in the *International Mechanical Code*® (IMC®) and the *International Plumbing Code*® (IPC®) versus that of the *International Building Code*® (IBC®).

R201.4 Terms not defined. Terms not defined by this chapter shall have ordinarily accepted meanings such as the context implies.

❖ Another option for defining words or terms not defined here or in other codes is their "ordinarily accepted meanings." The intent of this statement is that a dictionary definition may suffice if the definition is in context. Often, construction terms used throughout the code may not be defined in Chapter 2 [RE] or in a dictionary. In such a case, the definitions contained in the referenced standards (see Chapter 5 [RE]) and in published textbooks on the subject in question are good resources.

SECTION R202 GENERAL DEFINITIONS

ABOVE-GRADE WALL. A wall more than 50 percent above grade and enclosing *conditioned space*. This includes between-floor spandrels, peripheral edges of floors, roof and basement knee walls, dormer walls, gable end walls, walls enclosing a mansard roof and skylight shafts.

❖ This definition details which walls must be treated as above-grade walls. This will help to make the distinction between these walls and basement walls (see definition and commentary, "Basement wall"). These two wall types face a different amount of energy transfer and therefore have different insulation requirements. In order to determine the proper insulation requirements for the various walls, both definitions should be reviewed. The definition includes any wall that is a part of the building thermal envelope ("enclosing conditioned space") and meets the area requirements. For example, the wall between a dwelling and an unconditioned garage would be included in this definition (see commentary, "Building thermal envelope").

ACCESSIBLE. Admitting close approach as a result of not being guarded by locked doors, elevation or other effective means (see "Readily *accessible*").

❖ Providing access to mechanical equipment and appliances is necessary to facilitate inspection, observation, maintenance, adjustment, repair or replacement. Access to equipment means the equipment can be physically reached without having to remove a permanent portion of the structure. It is acceptable, for example, to install equipment in an interstitial space that would require removal of lay-in suspended ceiling panels to gain access. Equipment would not be considered accessible if it were necessary to remove or open any portion of a structure other than panels, doors, covers or similar obstructions intended to be removed or opened (see the definition of "Readily accessible"). Access can be described as the capability of being reached or approached for the purpose of inspection, observation, maintenance, adjustment, repair or replacement. Achieving access may first require the removal or opening of a panel, door or similar obstruction, and may require the overcoming of an obstacle such as elevation.

ADDITION. An extension or increase in the *conditioned space* floor area or height of a building or structure.

❖ The code uses this term to reflect new construction that is being added to an existing building. This definition is important when determining the applicability of the code provisions (see Section R502).

AIR BARRIER. Material(s) assembled and joined together to provide a barrier to air leakage through the building envelope. An air barrier may be a single material or a combination of materials.

❖ Building tightness against air infiltration is an important aspect of energy conservation. The term "air barrier" is defined to support the provisions of Section R402.4 regarding air leakage and building tightness. Note that an air barrier is not a single membrane but rather the system of sealants, seals, insulation and wall sheathing that prevents air infiltration.

ALTERATION. Any construction, retrofit or renovation to an existing structure other than repair or addition that requires a permit. Also, a change in a building, electrical, gas, mechanical or plumbing system that involves an extension, addition or change to the arrangement, type or purpose of the original installation that requires a permit.

❖ This definition actually includes two separate definitions: one that applies to building construction and the other to mechanical systems. An alteration is any modification or change made to an existing installation. For example, changing refrigerant types or heat transfer fluids in a system would be considered an alteration. This definition specifically excludes additions or repairs and also ties the term to situations where a permit is required. See Chapter 1 of the IBC and IMC for the information regarding when a permit is required.

APPROVED. Approval by the *code official* as a result of investigation and tests conducted by him or her, or by reason of accepted principles or tests by nationally recognized organizations.

❖ As related to the process of accepting envelope, mechanical, service water heating, lighting and electrical power installations, including materials, equipment and construction systems, this definition identifies where ultimate authority rests. Whenever this term is used, it intends that only the enforcing authority can accept a specific installation or component as complying with the code.

The definition does not force the code official to accept any third-party test. It merely permits the acceptance of the test so that the code official does not have to personally test the item. One example that demonstrates this process is an ICC Evaluation Report. An evaluation report prepared and published by the International Code Council (ICC)® is permitted to be used by a code official to aid in his or her review and approval of the material or method described in the report. Although the evaluation and report are prepared by the ICC, the code official is not obligated to accept or approve the product based on the evaluation report. The term "approved" is always tied to the code official's approval of the product or project. That the ICC published an evaluation report does not supersede the fact that the approval of the code official is still needed for the material or method described in the report.

APPROVED AGENCY. An established and recognized agency regularly engaged in conducting tests or furnishing inspection services, when such agency has been approved by the *code official*.

❖ This definition is added to the code to parallel the remaining I-Codes. It is quite often misunderstood that an authoritative agency cannot provide any services in the building construction or testing without the approval of the code official.

AUTOMATIC. Self-acting, operating by its own mechanism when actuated by some impersonal influence, as, for example, a change in current strength, pressure, temperature or mechanical configuration (see "Manual").

❖ Operation or control devices or systems operating automatically, as opposed to manually, are designed to operate safely with only periodic human intervention or supervision. A thermostat would be an example of something that is automatic. While a person would set the thermostat to the desired temperature, the thermostat would cycle the heating or cooling system on or off on its own once the temperature hits the established setpoints.

BASEMENT WALL. A wall 50 percent or more below grade and enclosing *conditioned space*.

❖ Because basement walls are in contact with the ground and ground temperatures differ from air temperatures, the amount of energy transferred through a basement wall is different than the energy transfered through a wall predominantly above grade. Therefore, the code provides different thermal requirements for basement walls and above-grade walls. An individual wall enclosing conditioned space is classified as a basement wall where the gross wall area is 50 percent or more below grade and is bounded by soil; otherwise, the wall is classified as an above-grade wall. This definition includes a below-grade interior wall separating a basement from a crawl space that meets the percentage requirement. This basement wall classification applies to the whole wall area, even if a portion of the individual wall is not below grade. Therefore, the above-grade portion of the wall is considered as part of the basement wall where the total wall is 50 percent or more below grade. Both sections of the wall (above grade and below grade) are then insulated as a "basement wall." Likewise, where an exterior wall is less than 50 percent below grade, the whole wall area is classified as an above-grade wall, including the portion underground [see Commentary Figure R202(1)]. For example, the wall of a walk-out basement that is entirely above grade is not considered a basement wall. The wall area of the walk-out wall must be considered as an above-grade wall and compared to the code requirements for above-grade walls. Thus, where the average below-grade depth of the sidewalls is 50 percent or greater, they are basement walls. If not, they are above-grade walls. The intent of this definition is to apply the provision to each wall enclosing the space. It is not intended to be applied to the aggregate of all of the walls of the basement. Therefore, this classification is done for each individual wall segment and not on an aggregate basis. The basement wall requirements apply only to the opaque basement wall area, excluding windows and doors. For purposes of meeting the code requirements, windows and doors in a basement wall are regulated as any other fenestration opening.

BUILDING. Any structure used or intended for supporting or sheltering any use or occupancy, including any mechanical systems, service water heating systems and electric power and lighting systems located on the building site and supporting the building.

❖ This definition indicates that where this term is used in the code, it means a structure intended to provide shelter or support for some activity or occupancy. Though not addressed in the code, it is important to note that the IBC does permit that a fire wall forms a demarcation in between two separate, structurally independent buildings. Therefore, the code provisions could be applied to each building separately or to the structure as a whole. This would be a designer's decision.

BUILDING SITE. A continguous area of land that is under the ownership or control of one entity.

❖ "Building site" is a key term used throughout the code. It is the area of land that is under the ownership of one entity.

BUILDING THERMAL ENVELOPE. The basement walls, exterior walls, floor, roof and any other building elements that enclose conditioned space or provide a boundary between conditioned space and *exempt* or unconditioned *space*.

❖ "Building thermal envelope" is a key term and resounding theme used throughout the code. It

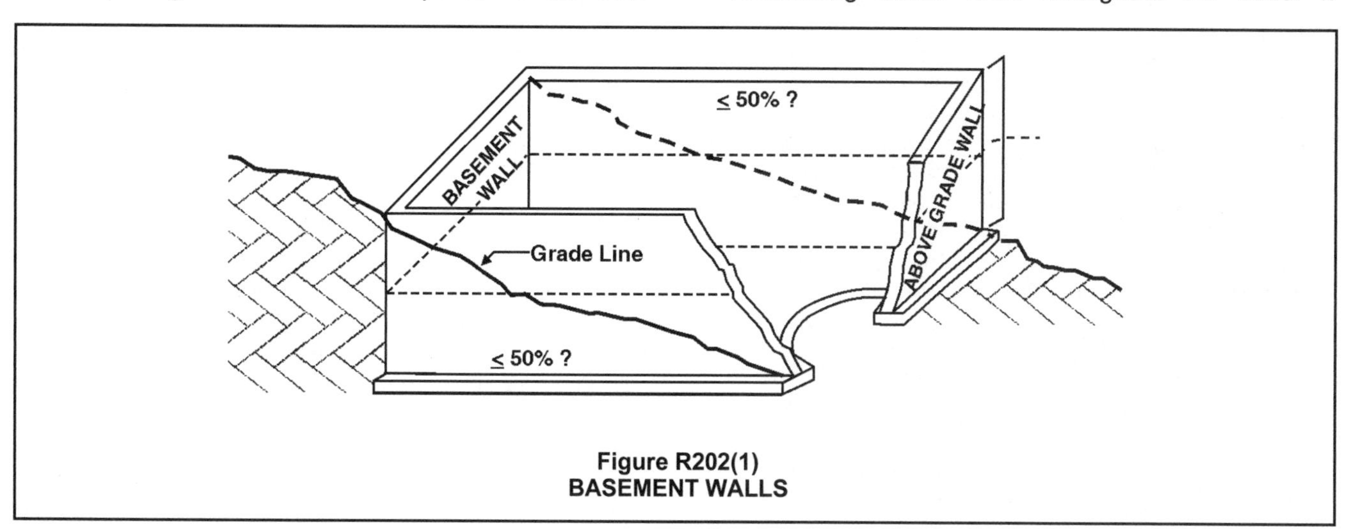

Figure R202(1)
BASEMENT WALLS

defines what portions of the building form a structurally bound conditioned space and are thereby covered by the insulation and infiltration (air leakage) requirements of the code. The building thermal envelope includes all building components separating conditioned spaces (see commentary, "Conditioned space") from unconditioned spaces or outside ambient conditions and through which heat is transferred. For example, the walls and doors separating an unheated garage (unconditioned space) from a living area (conditioned space) are part of the building thermal envelope. The walls and doors separating an unheated garage from the outdoors are not part of the building thermal envelope. Walls, floors and other building components separating two conditioned spaces are not part of the building thermal envelope. For example, interior partition walls, the common or party walls separating dwelling units in multiple-family buildings and the wall between a new conditioned addition and the existing conditioned space are not considered part of the building thermal envelope. Unconditioned spaces (areas having no heating or cooling sources) are placed outside the building envelope. A space is conditioned if it is heated or cooled directly or where a space is indirectly supplied with heating or cooling through uninsulated walls, floors or uninsulated ducts or heating, ventilating and air-conditioning (HVAC) piping. Boundaries that define the building envelope include the following:

- Building assemblies separating a conditioned space from outdoor ambient weather conditions.
- Building assemblies separating a conditioned space from the ground under or around that space, such as the ground around the perimeter of a slab or the soil at the exterior of a conditioned basement wall. Note that the code does not specify requirements for insulating basement floors or underneath slab floors (except at the perimeter edges).
- Building assemblies separating a conditioned space from an unconditioned garage, unconditioned sunroom or similar unheated/cooled area.

The code specifies requirements for ceiling, wall, floor, basement wall, slab-edge and crawl space wall components of the building envelope. In some cases, it may be unclear how to classify a particular part of a building. For example, skylight shafts have properties of a wall assembly but are located in the ceiling assembly. Because many of these items are not addressed specifically in the code, the code official should make the determination as to the appropriate classification and construction. When no distinction exists between roof and wall, such as in an A-frame structure, the code official should determine the appropriate classification. Historically, some codes have designated a wall as having a slope of 60 degrees or greater from the horizontal plane. In such situations, if the wall slope is less than 60 degrees, then classification as a "roof" is appropriate. Because the code is silent on this issue, other options such as stating that the roof could be considered to begin at a point 8 feet (2439 mm) above the floor surface of the uppermost story could be used. The phrase "exempt space" in the definition of "Building thermal envelope" refers to spaces identified as exempt from this scope of the code (see commentary, Section R101.4).

_C_-FACTOR (THERMAL CONDUCTANCE). The coefficient of heat transmission (surface to surface) through a building component or assembly, equal to the time rate of heat flow per unit area and the unit temperature difference between the warm side and cold side surfaces (Btu/h · ft^2 · °F) [W/(m^2 · K)].

❖ This definition addresses a term needed in the provisions of the code for the *U*-factor alternative given in Section R402.1.4.

CIRCULATING HOT WATER SYSTEM. A specifically designed water distribution system where one or more pumps are operated in the service hot water piping to circulate heated water from the water-heating equipment to fixtures and back to the water-heating equipment.

❖ In order to ensure hot water is available immediately at a faucet, circulating hot water systems can be installed to circulate hot water when the hot water supply system is not in use.

CLIMATE ZONE. A geographical region based on climatic criteria as specified in this code.

❖ See the commentary to Section R301.1.

CODE OFFICIAL. The officer or other designated authority charged with the administration and enforcement of this code, or a duly authorized representative.

❖ The statutory power to enforce the code is usually vested in a building department of a state, county or municipality with a designated enforcement officer who is termed the "code official."

COMMERCIAL BUILDING. For this code, all buildings that are not included in the definition of "Residential building."

❖ Commercial buildings include, among others, occupancies for assembly, educational, business, institutional, mercantile, factory/industrial, hazardous storage and utility occupancies (see definition and commentary, "Residential building").

One item that may easily be overlooked is that a Group R-1 occupancy building (a hotel or motel) would be classified as a "commercial building" because it is not included in the definition of "Residential building." Although classified as a residential occupancy by the IBC, hotels and motels tend to be more closely associated with commercial buildings as far as energy usage and operation.

CONDITIONED FLOOR AREA. The horizontal projection of the floors associated with the *conditioned space*.

❖ The conditioned floor area is the total area of all floors in the conditioned space of the building.

CONDITIONED SPACE. An area, room or space that is enclosed within the building thermal envelope and that is directly or indirectly heated or cooled. Spaces are indirectly heated or cooled where they communicate through openings with conditioned spaces, where they are separated from conditioned spaces by uninsulated walls, floors or ceilings, or where they contain uninsulated ducts, piping or other sources of heating or cooling.

❖ A conditioned space is typically any space that does not communicate directly to the outside; that is, a space not directly ventilated to the outdoors and meets one of the following criteria:

1. The space has a heating or cooling supply register.
2. The space has heating or cooling equipment designed to heat or cool the space, or both, such as a radiant heater built into the ceiling, a baseboard heater or a wall-mounted gas heater.
3. The space contains uninsulated ducts or uninsulated hydronic heating surfaces.
4. The space is inside the building thermal envelope. For example:
   - A basement with insulated walls but without insulation on the basement ceiling.
   - A closet on a home's exterior wall that is insulated on the exterior surface of the closet wall.
   - A space adjacent to and not physically separated from a conditioned space (such as a room adjacent to another room with a heating duct but without a door that can be closed between the rooms).
   - A room completely surrounded by conditioned spaces.

The builder/designer has some flexibility in defining the bounds of the conditioned space as long as the building envelope requirements are met. Spaces that are not conditioned directly but have uninsulated surfaces separating them from conditioned spaces are included within the insulated envelope of the building. For example, an unventilated crawl space below an uninsulated floor is considered part of the conditioned space, even where no heat is directly supplied to the crawl space area. Where the crawl space is included as a conditioned space, the builder must insulate the exterior crawl space walls instead of the floor above. The task of defining the building envelope is left to the permit applicant.

Examples of unconditioned spaces include garages and basements that are neither heated nor cooled if all duct surfaces running through these spaces are insulated; attached sunrooms that are neither heated nor cooled and have insulated/weatherstripped doors to separate the sun space from the conditioned space; attics; and ventilated crawl spaces. Note that the boundary between the conditioned and unconditioned space is subject to the infiltration control requirements of the code (see commentary, Section R402.4).

Historically, the code tied this definition to "conditioning for human comfort" or by specifying that the conditioning fall within a specific range. Starting with the 2006 edition, the code considers any type of conditioning as creating a "conditioned space." Therefore, providing heating within a storage building to keep the stock from freezing would still be considered as creating a conditioned space.

CONTINUOUS AIR BARRIER. A combination of materials and assemblies that restrict or prevent the passage of air through the building thermal envelope.

❖ This term is used throughout the code. This definition is necessary to clarify the options for meeting continuous air barrier requirements.

CONTINUOUS INSULATION (ci). Insulating material that is continuous across all structural members without thermal bridges other than fasteners and service openings. It is installed on the interior or exterior, or is integral to any opaque surface, of the building envelope.

❖ The need for more insulation on exterior walls has created the need to specify continuous insulation on the exterior side of the wall for certain climate zones in Chapter 4. Thus the need for this definition.

CRAWL SPACE WALL. The opaque portion of a wall that encloses a crawl space and is partially or totally below grade.

❖ Because exterior crawl space walls may be in contact with the ground, they exhibit different heat transfer properties than those of exterior above-grade walls or even basement walls. Thus, the code includes different thermal requirements for exterior crawl space walls.

This definition and distinction is also to help coordinate with Section R402.2.11. While a basement is defined as a "conditioned space," crawl spaces may be designed to be either conditioned or unconditioned spaces.

CURTAIN WALL. Fenestration products used to create an external nonload-bearing wall that is designed to separate the exterior and interior environments.

❖ This definition is included to help classify and properly apply the code requirements to these products. The industry uses the term to help establish the requirements and separate the products into different groupings. Without defining the curtain wall as a fenestration product, it would be difficult to determine how to properly apply the building envelope requirements.

DEMAND RECIRCULATION WATER SYSTEM. A water distribution system where pump(s) prime the service hot water piping with heated water upon demand for hot water.

❖ This is essential in the code to define a water distribution system in order to establish requirements for insulation to retain the heated water upon demand for hot water and conserve energy usage.

DUCT. A tube or conduit utilized for conveying air. The air passages of self-contained systems are not to be construed as air ducts.

❖ Ducts can be factory manufactured or field constructed of sheet metal, gypsum board, fibrous glass board or other approved materials. Ducts are used in air distribution systems, exhaust systems, smoke control systems and combustion air-supply systems. Air passageways that are integral parts of an air handler, packaged air-conditioning unit or similar piece of self-contained, factory-built equipment are not considered ducts in the context of the code.

DUCT SYSTEM. A continuous passageway for the transmission of air that, in addition to ducts, includes duct fittings, dampers, plenums, fans and accessory air-handling equipment and appliances.

❖ Duct systems are part of an air distribution system and include supply, return, transfer and relief/exhaust air systems.

DWELLING UNIT. A single unit providing complete independent living facilities for one or more persons, including permanent provisions for living, sleeping, eating, cooking and sanitation.

❖ A dwelling unit contains elements necessary for independent living (living spaces such as family rooms, dining rooms, living rooms, dens, etc.; sleeping quarters; food preparation and eating spaces; and personal hygiene, cleanliness and sanitation facilities). A dwelling unit is typically owner occupied, rented or leased. The code requirements are applied consistently to all dwellings whether owner occupied, rented or leased. A dwelling unit can exist singularly as a one-family dwelling or in combination with other dwelling units. When two dwelling units are grouped together in the same structure, the structure is considered a two-family dwelling. Three or more dwelling units in the same structure are considered a multiple-family dwelling. Note that under this definition, rooms in hotels and motels are not considered dwelling units but rather guestrooms or a sleeping unit, because they usually lack complete living, sanitation and eating facilities and are generally characterized as having transient occupancy patterns.

ENERGY ANALYSIS. A method for estimating the annual energy use of the *proposed design* and *standard reference design* based on estimates of energy use.

❖ Designs founded on simulated performance alternative (performance) (see Section R405) use energy analysis when the design professional requires more flexibility for a sophisticated or innovative design. Using the total-building-performance design methodology, the proposed design is evaluated based on the cost of various types of energy used rather than the units of energy used (Btu, kWh). That cost must be established using an hour-by-hour, full-year (8,760 hours) simulation tool capable of simulating the performance of both the proposed and standard designs (see commentary, "Energy simulation tool"). The simulation must be capable of converting calculated energy demand and consumption into utility costs using the actual utility rate schedules rather than the average cost of electricity or gas.

ENERGY COST. The total estimated annual cost for purchased energy for the building functions regulated by this code, including applicable demand charges.

❖ The total annual cost for purchased energy includes demand, power and fuel adjustment charges and the impact of special rate programs for large volume customers. A thorough evaluation of existing tariffs and fee schedules may uncover substantial savings opportunities. In some states, for example, manufacturing customers are exempt from sales taxes for energy. In other states, utilities may have multiple tariff options.

ENERGY SIMULATION TOOL. An *approved* software program or calculation-based methodology that projects the annual energy use of a building.

❖ An energy simulation tool is typically a software package incorporating, among other features, an hour-by-hour, full-year (8,760 hours), multiple-zone program to simulate the performance of both proposed and standard design buildings. It is possible to use other types of simulation tools to approximate the dynamics of hourly energy programs and that can be shown to produce equivalent results for the type of building and HVAC systems under consideration. However, the simulation must be capable of converting calculated energy demand and consumption into utility costs using the actual utility rate schedules (rather than average cost of electricity or gas). Some examples of when an hour-by-hour, full-year type of program is required are:

- When the features intended to reduce energy consumption require time-of-day interactions between weather, loads and operating criteria. Examples include: night ventilation or building thermal storage; chilled water or ice storage; heat recovery; daylighting; and water economizer cooling.
- When utility rates are time-of-day sensitive, and the proposed design uses time-of-day load shifting between different types of mechanical plant components.

Another distinguishing feature among simulation tools is their sophistication in modeling HVAC systems and plant equipment. Basically, three levels of complexity are used: constant efficiency models; models with simple part-load efficiency adjustment; and models with complex part-load efficiency adjustment. Simulation tools in the first category simply calculate hourly equipment input power requirements at part load by applying the full-load efficiency at any given hour. These programs should be avoided for all but constant load applications. Simulated tools with simple part-load efficiency adjustments use a profile of percent-rated input power versus percent-rated load. At each hour, these programs calculate input power to each piece of equipment. These tools are far more accurate than the constant-efficiency models, but still lack accurate compensation for environmental variables.

Building energy simulation tools, available for analyzing daylighting, passive solar design and solar systems, are described in more detail at www.eere.energy.gov/buildings/tools_directory/. This is a portion of the U.S. Department of Energy's Energy Efficiency and Renewable Energy website.

The information includes program uses, computer hardware required, price and contact information. The most sophisticated simulation tools incorporate a number of profiles for each and every piece of equipment. For variable flow fans, this might be as simple as a single profile of percent-rated input power versus percent-rated airflow. For more complex equipment, such as a cooling tower, the program considers such variables as the wet-bulb temperature, the approach (difference between the condenser water supply temperature and the wet-bulb temperature) and the range (difference between the condenser water entering and leaving temperatures). Each of these variables is used to adjust both the hourly capacity of the tower and the hourly operation (one fan, two fans, no fans). For all but the simplest systems, programs of this category must be used to obtain accurate results. Questions regarding a particular tool's ability to model the building on an hour-by-hour, full-year basis should be addressed to the proprietor or distributor.

ERI REFERENCE DESIGN. A version of the rated design that meets the minimum requirements of the 2006 *International Energy Conservation Code.*

❖ The Energy Rating Index compliance alternative utilizes the ERI Reference Design (see Section R406).

EXTERIOR WALL. Walls including both above-grade walls and basement walls.

❖ Wall insulation requirements defined in the code include almost all opaque exterior construction bounding conditioned space. Depending on the compliance method, the wall type, glazing percentage and whether the wall is on the exterior or just separating conditioned from unconditioned space can affect the wall insulation requirement. Note also that doors (both glazed and opaque) are considered fenestration (see commentary, "Fenestration").

In earlier editions, the code included the limitation that an exterior wall is vertical or sloped at an angle of 60 degrees (1.1 rad) or greater from the horizontal. This limitation may still be helpful to consider if dealing with unusual situations such as an A-frame building. Where a determination is needed to decide whether the roof or wall provisions are appropriate, this limitation could be helpful.

FENESTRATION. Products classified as either *vertical fenestration* or *skylights.*

❖ The term "fenestration" refers both to opaque and glazed doors and the light-transmitting areas of a wall or roof, but primarily windows and skylights. The code sets performance requirements for fenestration by establishing requirements that differ from the wall and roof requirements based on the type of fenestration and in the case of the prescriptive commercial requirements by limiting the fenestration area. In some of the compliance options the fenestration type and area allowed depend on the shading coefficient, the size of overhangs, the thermal performance (*U*-factor) and whether daylighting controls are installed.

FENESTRATION PRODUCT, SITE-BUILT. A fenestration designed to be made up of field-glazed or field-assembled units using specific factory cut or otherwise factory-formed framing and glazing units. Examples of site-built fenestration include storefront systems, curtain walls and atrium roof systems.

❖ Fenestration products can be site-built, not to be confused with field-fabricated fenestration products. This definition is installed to make that distinction.

***F*-FACTOR.** The perimeter heat loss factor for slab-on-grade floors (Btu/h·ft·°F) [W/(m·K)].

❖ As defined, *F*-factor is needed in the provisions of the code for the *U*-factor alternative given in Section R402.1.4.

HEATED SLAB. Slab-on-grade construction in which the heating elements, hydronic tubing, or hot air distribution system is in contact with, or placed within or under, the slab.

❖ The space above a heated slab is always conditioned (heated). The definition clarifies that certain slabs are the heating source and, as covered in the code, may require more insulation than unheated slabs. The installation of a radiant heat source in a space does not, in and of itself, qualify a slab as heated.

HIGH-EFFICACY LAMPS. Compact fluorescent lamps, T-8 or smaller diameter linear fluorescent lamps, or lamps with a minimum efficacy of:

1. 60 lumens per watt for lamps over 40 watts;
2. 50 lumens per watt for lamps over 15 watts to 40 watts; and
3. 40 lumens per watt for lamps 15 watts or less.

❖ The code requires that a percentage of permanent lighting fixtures have high-efficacy lamps. Therefore, this definition is necessary to specify the required performance to qualify as a high-efficacy lamp.

HISTORIC BUILDING. Buildings that are listed in or eligible for listing in the National Register of Historic Places, or designated as historic under an appropriate state or local law.

❖ Historic buildings are given some relief from the requirements of the code (see Section R501.6).

INFILTRATION. The uncontrolled inward air leakage into a building caused by the pressure effects of wind or the effect of differences in the indoor and outdoor air density or both.

❖ Air leakage is random movement of air into and out of a building through cracks and holes in the building envelope. In technical terms, air leakage is called "infiltration" (air moving into a building) or "exfiltration" (air moving out of a building). In nontechnical terms, air leaks are often referred to as "drafts." Infiltration may be reduced by either reducing the sources of air leakage (joints, penetrations and holes in the building envelope) or by reducing the pressures driving the airflow.

INSULATED SIDING. A type of continuous insulation with manufacturer-installed insulating material as an integral part of the cladding product having a minimum *R*-value of R-2.

INSULATING SHEATHING. An insulating board with a core material having a minimum *R*-value of R-2.

❖ Some exterior hardboard and vinyl siding products are not recommended for use in direct contact with aluminum-foil-faced sheathing products (check with the product manufacturer). Other sheathing products are uniquely designed for use in direct contact with wood, brick, vinyl, aluminum and hardboard-based sidings. To be considered insulating sheathing the product must have an *R*-value no less than R-2.

LABELED. Equipment, materials or products to which have been affixed a label, seal, symbol or other identifying mark of a nationally recognized testing laboratory, inspection agency or other organization concerned with product evaluation that maintains periodic inspection of the production of the above-labeled items and where labeling indicates either that the equipment, material or product meets identified standards or has been tested and found suitable for a specified purpose.

❖ When a product is labeled, the label indicates that the equipment or material has been tested for conformance to an applicable standard and that the component is subject to third-party inspection, which verifies that the minimum level of quality required by the applicable standard is maintained. Labeling is a readily available source of information that is useful for field inspection of installed products. The label identifies the product or material and provides other information that can be further investigated if there is a question concerning the suitability of the product or material for the specific installation. The labeling agency performing the third-party inspection must be approved by the code official and the basis for this approval may include, but is not necessarily limited to, the capacity and capability of the agency to perform the specific testing and inspection. The applicable referenced standard often states the minimum identifying information that must be on a label. The data contained on a label typically includes, but is not necessarily limited to, the name of the manufacturer, the product name or serial number, installation specifications, applicable tests and standards, the testing approved and the labeling agency.

LISTED. Equipment, materials, products or services included in a list published by an organization acceptable to the *code official* and concerned with evaluation of products or services that maintains periodic inspection of production of *listed* equipment or materials or periodic evaluation of services and where the listing states either that the equipment, material, product or service meets identified standards or has been tested and found suitable for a specified purpose.

❖ Not all testing laboratories, inspection agencies and other organizations concerned with product evaluation use the same means for identifying listed equipment, appliances or materials. Some do not recognize equipment, appliances or materials as listed unless they are also labeled. The authority having jurisdiction must use the same system used by the listing organization to identify a listed product. In general, all equipment and appliances regulated by the I-Codes are to be listed and labeled unless otherwise approved in accordance with Section R102.1, which allows use of alternative materials, methods and equipment. As stated in the definition, the listing states either that the equipment or material meets nationally recognized standards or it has been found suitable for use in a specified manner. The listing becomes part of the documentation that the code official can use to approve or disapprove the equipment or appliance.

LOW-VOLTAGE LIGHTING. Lighting equipment powered through a transformer such as a cable conductor, a rail conductor and track lighting.

❖ Track lighting is a form of low-voltage lighting featuring a continuous-powered track that accepts fixtures (called heads) anywhere along its length. The heads and the wide variety of lamps (bulbs) they accept suit virtually every lighting situation, and most heads are also adjustable. One-, two-, three- and four-circuit tracks let the lighting design professional mix standard and low-voltage heads or have multiple switching options. Track systems are easily modified or added to as lighting needs change. Although tracks can be recessed flush with a ceiling surface or suspended from posts, they are typically fastened to the surface either directly or with mounting clips. Controls ranging from a simple on/off switch to elaborate programmable dimming devices are widely available, making the system extremely flexible.

MANUAL. Capable of being operated by personal intervention (see "Automatic").

❖ Devices, systems or equipment having manual controls or overrides are designed to operate safely with only human intervention instead of having an automatic operation or control system (see commentary, "Automatic").

A typical light switch would be an example of a manual device. The lights that the switch controls stay either off or on until a person flips the switch to the appropriate position.

PROPOSED DESIGN. A description of the proposed *building* used to estimate annual energy use for determining compliance based on total building performance.

❖ The proposed design is simply a description of the proposed building, used to estimate annual energy costs for determining compliance based on total building performance. The proposed design is effectively the subject building intended to be built. The performance of the proposed design (the building exactly as it is anticipated to be constructed) is then compared to the standard reference design (a similar building that is assumed to be built to a prescriptive set of minimum code requirements). Although the method permits trade-offs in energy use between different systems, the proposed design must still comply with the general requirements. General "mandatory" requirements are separately specified in Chapter 4 [RE]. For instance, the air-leakage requirements are general requirements that apply, even if the project uses any one of the two performance methods referenced above. The general requirements still apply for several reasons, including:

- Many basic requirements cannot be accurately modeled (such as subdivision of feeders).
- Many requirements are inherently cost effective (such as variable speed drives or large motors and minimizing air leakage).
- Some basic requirements are calculation methodologies that establish a fair basis for comparing component performance (such as *U*-factor calculations).
- Some basic requirements are not intended for trade-offs (such as exterior lighting).

RATED DESIGN. A description of the proposed *building* used to determine the energy rating index.

❖ See Section 406 for application.

READILY ACCESSIBLE. Capable of being reached quickly for operation, renewal or inspection without requiring those to whom ready access is requisite to climb over or remove obstacles or to resort to portable ladders or access equipment (see "*Accessible*").

❖ Readily accessible can be described as able to be quickly reached or approached for operation, inspection, observation or emergency action. Ready access does not require the removal or movement of any door panel or similar obstruction, or overcoming physical obstructions or obstacles, including differential elevation.

REPAIR. The reconstruction or renewal of any part of an existing building for the purpose of its maintenance or to correct damage.

❖ The repair of an item, appliance, energy-using subsystem or other piece of equipment typically does not require a permit. This definition makes it clear that a repair is limited to work on the item, and does not include its replacement or other new work.

This definition is important when applying the provisions of Section R504 and determining how the code will apply to existing buildings and installations.

REROOFING. The process of recovering or replacing an existing *roof covering*. See "Roof recover" and "Roof replacement."

RESIDENTIAL BUILDING. For this code, includes detached one- and two-family dwellings and multiple single-family dwellings (townhouses) as well as Group R-2, R-3 and R-4 buildings three stories or less in height above grade plane.

❖ The definition of a residential building is important not only for what it does include, but also for what it does not. One of the primary limitations of this definition is the fact that this term will only include Group R-2 and R-4 occupancies when they are three stories or less in height. Therefore, if a Group R-2 or R-4 occupancy is over three stories in height, it would be defined as a "Commercial building" (see definition, "Commercial building") and be required to comply with the requirements of Chapter 4[CE] instead of the residential provisions of Chapter 4 [RE]. Buildings that are classified as an R-3 are not affected by the three-story limitation. Because of this, a one- or two-family dwelling would always be considered as a residential building regardless of the number of stories.

It should be noted that a Group R-1 building is not included in this definition. Therefore, any hotel, motel or similar use that is classified as an R-1 must comply with commercial building requirements. This applies even if the hotel or motel is three stories or less in height.

Though not specifically addressed within this definition, any building built under the provisions of the *International Residential Code®* (IRC®) would also be considered a "residential building."

The Group R-2 classification of residential buildings includes apartments and condominiums where three or more units are physically attached. In addition, it includes certain boarding houses, convents, dormitories, fraternities or other such facilities. The code defines a dwelling unit as a single housekeeping unit of one or more rooms providing complete independent living facilities, including permanent provisions for living, sleeping, cooking and sanitation (see commentary, "Dwelling unit"). Many of the code provisions address dwelling units specifically. All of the elements listed in the definition must be present for a dwelling unit to exist. Guestrooms (sleeping units) in hotels, motels, nursing homes and larger group care facilities do not meet this criteria. Therefore, these buildings are considered as commercial uses, even if they are three stories or less in height (see the IBC for additional information on dwelling units). Although this definition closely follows the Group R-2 occupancy classification of the IBC, it is important to note there are differences. The main distinction is the three-story limitation, which is found only in the code. It is also important to note that townhouses are not considered as being a part of this definition.

The R-4 classification is included in order to coordinate the provisions of the code with the IBC. These small residential care/assisted-living facilities are considered residential buildings based on Sections 310.1 and 310.2 of the IBC. This distinction is important because larger facilities are classified by the IBC as institutional occupancies and, therefore, would be considered commercial buildings by the code (see "Commercial building").

As mentioned earlier, it is important to note the application of the three-story limitation, which is found only in the code. This issue of defining a residential building over three stories in height as a commercial building coordinates the code with ANSI/ASHRAE/IESNA 90.1.

ROOF ASSEMBLY. A system designed to provide weather protection and resistance to design loads. The system consists of a roof covering and roof deck or a single component serving as both the roof covering and the roof deck. A roof assembly includes the roof covering, underlayment, roof deck, insulation, vapor retarder and interior finish.

❖ This section basically defines the cover or protection over a building but it will be important to the energy requirements in several sections. The roof assembly generally serves as the building thermal envelope for the top of the building. By defining the various elements that are included in the definition, the phrase can be applied to situations where the insulation is installed above the roof deck, beneath the roof deck or above a ceiling that occurs below.

The "Roof assembly" definition will be important not only when applying the prescriptive insulation requirements of the building thermal envelope but also when applying the performance requirements where the type of roof covering and ventilation can affect the energy efficiency.

Numerous horizontal and sloped surfaces may be associated with the roof or roof/ceiling assembly, including flat and cathedral ceilings, dormer roofs, bay-window roofs, overhead portions of an interior stairway to an attic, or other unconditioned space, attic hatches and skylights. When determining the area of the assembly under the performance options, ceiling assembly areas should be measured on the slope of the finished interior surface.

In earlier editions, the code included the limitation that a roof is generally horizontal or sloped at an angle of less than 60 degrees (1.1 rad) from the horizontal. This limitation may still be helpful to consider if dealing with unusual situations such as an A-frame building. Where a determination is needed to decide whether the roof or wall provisions are appropriate, this limitation could be helpful (see commentary, "Exterior wall" and "Skylight," as well as the definition in IBC Section 202).

ROOF RECOVER. The process of installing an additional *roof covering* over a prepared existing *roof covering* without removing the existing *roof covering*.

❖ Roof recover is considered an alteration to the building. In the application of Section R503, roof recover is exempt from the code requirements.

ROOF REPAIR. Reconstruction or renewal of any part of an existing roof for the purposes of its maintenance.

❖ See Section R504 for application.

ROOF REPLACEMENT. The process of removing the existing *roof covering*, repairing any damaged substrate and installing a new *roof covering*.

❖ This is a form of alteration to a building and may or may not require additional energy conservation measures, such as insulation in the roof cavity, depending upon the extent of work (see Section R503).

***R*-VALUE (THERMAL RESISTANCE).** The inverse of the time rate of heat flow through a body from one of its bounding surfaces to the other surface for a unit temperature difference between the two surfaces, under steady state conditions, per unit area ($h \cdot ft^2 \cdot °F/Btu$) [$(m^2 \cdot K)/W$].

❖ Thermal resistance measures how well a material or series of materials retards heat flow. Insulation thermal resistance is rated using *R*-values. As the *R*-value of an element or assembly increases, the heat loss or gain through that element or assembly decreases. Thus, a higher *R*-value is considered better than a lower *R*-value.

SERVICE WATER HEATING. Supply of hot water for purposes other than comfort heating.

❖ Although the definition makes it clear that the code requirement applies to equipment used to produce and distribute hot water for purposes other than comfort heating, the definition also applies to energy-efficient process water heating systems and equipment. Equipment providing or distributing hot water for uses such as restrooms, showers, laundries, kitchens, pools and spas, defrosting of sidewalks and driveways, carwashes, beauty salons and other commercial enterprises are included. Space-conditioning boilers and distribution systems are not considered service water heating components.

SKYLIGHT. Glass or other transparent or translucent glazing material installed at a slope of less than 60 degrees (1.05 rad) from horizontal.

❖ A skylight is a glazed opening in a roof to admit daylight. Skylights are often the only method of bringing natural light into an interior, enclosed area. Unfortunately, these fixtures often do their job too well. Installing too many skylights or ones too large for the room can lead to overheating during warm-weather months. Also, choosing the most energy-efficient models can compromise light transmission—the reason people buy skylights in the first place.

Skylights are available in a variety of sizes and shapes, though rectangular units are the most common. Although most skylights are fixed or inoperable, others can be opened and shut like a window or have hidden ventilating systems. Large operable skylights designed for the sloping ceilings of attic rooms are even marketed as "roof windows." These approaches help cool the room in warm weather by venting hot air. The IBC requires either tempered or laminated glass in skylights. Both types are designed to stand up to snow loads and provide protection against falling objects. Tempered glass breaks into small pieces, rather than large shards, if damaged. Laminated glass, which is fused with a thin layer of plastic, stays in place for added safety if broken. Laminated glass is also better at keeping out sound and is slightly more energy efficient, though also slightly more expensive. Skylights are not energy efficient. They collect little heat during the winter, which is when it is needed most, when the sun is low in the sky. Worse yet, because they're located where the pressure difference between the inside and outside of the house is greatest, skylights are an easy escape route for heated air. Also, in the summer they can heat up the home quickly—when it is not needed.

Like windows, skylights offer a variety of energy-efficient glazing options, including low-e and tinted glass. Green tints are better than bronze tints for reducing solar heat gain while letting in plenty of visible light. And because skylights usually are not visible from the street, tinted glazings are less likely to affect aesthetics.

It is important to note that the term "skylight" can also include glazed roofs and sloped walls. This distinction can affect the proper application of the code requirements. To help determine the appropriate *U*-factor, this definition includes the slope limitation, which is shown. If the slope of the roof or sloped wall is 15 degrees (0.26 rad) or more from the vertical, then the skylight *U*-factor is appropriate. If the slope is less than 15 degrees (0.26 rad), then the glazing would be considered a fenestration in a wall. As an example using the requirements of Table R402.1.2 for Climate Zone 1, a skylight would require a *U*-factor of 0.75, while the fenestration in a wall would require a *U*-factor of 1.20. Both the skylight and the more vertical wall fenestration would require a solar heat gain coefficient (SHGC) of 0.30 (see Table R402.1.2, Note b).

SOLAR HEAT GAIN COEFFICIENT (SHGC). The ratio of the solar heat gain entering the space through the fenestration assembly to the incident solar radiation. Solar heat gain includes directly transmitted solar heat and absorbed solar radiation that is then reradiated, conducted or convected into the space.

❖ The SHGC is the fraction of incident solar radiation admitted through a window or skylight. This includes the solar radiation that is directly transmitted, and that which is absorbed and subsequently released inward. Therefore, the SHGC measures how well a window blocks the heat from sunlight. The SHGC is the fraction of the heat from the sun that enters through a window. SHGC is expressed as a number between 0 and 1. The lower a window's SHGC, the less solar heat it transmits.

In the warmer climate zones, where cooling is the dominant requirement, the code will generally impose a limitation on the amount of solar heat gain permitted. In colder climate zones, the code will generally not place any SHGC requirement on the fenestration. Thus, it will depend on the climate zone as to whether a higher or lower SHGC is best. While windows with lower SHGC values reduce summer cooling and overheating, they also reduce free winter solar heat gain.

STANDARD REFERENCE DESIGN. A version of the *proposed design* that meets the minimum requirements of this code and is used to determine the maximum annual energy use requirement for compliance based on total building performance.

❖ The standard reference design is simply the same building design as that intended to be built (see commentary, "Proposed design"), except the energy conservation features required by the code (insulation, windows, infiltration, mechanical, lighting and service water-heating systems) are modified to meet the minimum prescriptive requirements, as applicable. Note that the standard reference design is not truly a separate building, and it is never actually built. It is the baseline against which the proposed design is measured.

The performance of the proposed design (the building exactly as it is anticipated to be constructed) is then compared to the standard reference design (a similar building that is assumed to be built to a prescriptive set of minimum code requirements). Under the performance paths, if the energy efficiency of the proposed design is equal to or better than that of the standard reference design, then the proposed design is acceptable and in compliance with the code requirements.

SUNROOM. A one-story structure attached to a dwelling with a glazing area in excess of 40 percent of the gross area of the structure's exterior walls and roof.

❖ Sunrooms are unique rooms that create a space differing in character from that provided by conventional portions of a dwelling. Sunrooms and other highly glazed structures are sometimes called conservatories or solariums. These sunrooms are often added onto an existing building but there is nothing in the code that would prohibit them from being constructed as a part of a new dwelling.

This definition distinguishes sunrooms from other conventional spaces because they are limited to one story in height and the total glazing area needs to be at least 40 percent of the exterior wall and roof area of the sunroom. This definition and its use in Sections R402.2.2 and R402.3.5 are important since most dwellings do not have such large amounts of glazing.

Many of these sunrooms are constructed to be unconditioned spaces and will not have any type of heating or cooling. Other sunrooms may be conditioned indirectly by openings from the adjacent conditioned space of the dwelling, provided with a separate space conditioning system or designed as a separate zone from the dwelling's space conditioning system. Due to the large amount of glazing and, therefore, the probability that the sunroom will either be hot in the summer or cold in the winter and may be unused during such times, the code provisions address the thermal isolation of these spaces (see definition, "Thermal isolation"). When addressing the thermal isolation of sunrooms, the provisions of Sections R402.2.13 and R402.3.5 need to be reviewed and code users need to realize how those sections cordinate with Section R501.1.1. Section R402.3.5 requires that "new" walls separating a sunroom from a conditioned space must comply with the building thermal envelope requirements. Therefore, if a sunroom is built as an addition to an existing dwelling, the existing wall between the sunroom and any conditioned space is not regulated.

THERMAL ISOLATION. Physical and space conditioning separation from *conditioned space(s)*. The *conditioned space*(s) shall be controlled as separate zones for heating and cooling or conditioned by separate equipment.

❖ This term is conceptually similar to the separation provided by the building envelope, but instead of being between conditioned space and the exterior or conditioned space and unconditioned space, this separation occurs between two conditioned spaces. In this situation it is the separation between the dwelling and the sunroom. Therefore, where the sunroom is an addition, the existing exterior wall of the dwelling will generally provide the thermal isolation between the dwelling and the sunroom at the point where the sunroom addition is attached. If a new door or window opening is added to permit passage between or to provide a connection with the dwelling and the sunroom, that new door or window is required to comply with the fenestration *U*-factor specified in Table R402.1.2. By requiring the maximum *U*-factor, it ensures a reasonable level of energy conservation between the dwelling and the sunroom, so they would not need to be brought into compliance with the *U*-factor specified in the table.

This exemption for the existing doors and windows is based not only on the language in this definition, but also on the scoping requirements found in Section R402.3.5. Since these existing doors and windows would serve as a part of the building envelope for the dwelling prior to the addition of the sunroom, they may be accepted as a part of the thermal isolation between the dwelling and any new sunroom addition (see commentary, "Sunroom" and Section R402.3.5).

This definition is also important due to the restriction that it places on the heating and cooling system for the various spaces. In order to be thermally isolated, the two spaces are either served by separate systems or as separate zones on a common system.

THERMOSTAT. An automatic control device used to maintain temperature at a fixed or adjustable set point.

❖ Thermostats combine control and sensing functions in a single device. Thermostats may signal other control devices to trigger an action when certain temperatures are reached or surpassed. Because thermostats are so prevalent, the various types and their operating characteristics are described here.

The occupied-unoccupied or dual-temperature room thermostat reduces temperature at night. It may be indexed (changed from occupied to unoccupied operation or vice versa) individually from a remote point or in a group by a manual or time switch. Some types have an individual clock and switch built in.

The pneumatic day-night thermostat uses a two-pressure air supply system, where changing the pressure at a central point from one value to the other actuates switching devices in the thermostat and indexes it.

The heating-cooling or summer-winter thermostat can have its action reversed and its set point changed in response to outdoor and comfort conditions. It is used to actuate controlled devices, such as valves or dampers that regulate a heating source at one time and a cooling source at another. The pneumatic heating-cooling thermostat uses a two-pressure air supply similar to that described for occupied-unoccupied thermostats.

Multistage thermostats are arranged to operate two or more successive steps in sequence.

A submaster thermostat has its set point raised or lowered over a predetermined range, in accordance with variations in output from a master controller. The master controller can be a thermostat, manual switch, pressure controller or similar device. For example, a master thermostat measuring outdoor air temperature can be used to readjust the set point of a submaster thermostat that controls the water temperature in a heating system.

A wet-bulb thermostat is often used (in combination with a dry-bulb thermostat) for humidity control. Using a wick or another means of keeping the bulb wet with pure (distilled) water and rapid air motion to ensure a true wet-bulb measurement is essential. Wet-bulb thermostats are seldom used.

A dew-point thermostat is designed to control dew-point temperatures.

A dead-band thermostat has a wide differential over which the thermostat remains neutral, requiring neither heating nor cooling. This differential may be adjustable up to 10°F (-12°C) (see Section C403.2.4.1.2).

***U*-FACTOR (THERMAL TRANSMITTANCE).** The coefficient of heat transmission (air to air) through a building component or assembly, equal to the time rate of heat flow per unit area and unit temperature difference between the warm side and cold side air films (Btu/h · ft^2 · °F) [W/(m^2 · K)].

❖ Thermal transmittance (*U*-factor) is a measure of how well a material or series of materials conducts heat. *U*-factors for window and door assemblies are the reciprocal of the assembly *R*-value:

$$U\text{-factor} = \frac{1}{R\text{-value}}$$

For other building assemblies, such as a wall or roof/ceiling, the *R*-value used in the above equation is the *R*-value of the entire assembly, not just the insulation. This distinction is important and reflects the provisions of Sections R402.1.4 and R402.1.5. It also explains why there are differences between the comparable values of Tables R402.1.2 and R402.1.4.

The *U*-factors will be used in a number of locations of the code. When using the performance options or the "total UA alternative" (see Section R402.1.5), the individual thermal transmittance (*U*-factor) of each element is multiplied by the area of each envelope component (walls, floors, and ceilings) and the area of each fenestration element (doors, windows and skylights). Therefore, the UA is simply the *U*-factor times the area. For example, a 400-square-foot (37 m^2) wall with a *U*-factor of 0.082 would result in a UA of 32.8 (400 × 0.082 = 32.8). The total UA would simply be the sum of all of the individual UAs for each building element (walls, floors, ceilings, doors, windows and skylights).

VENTILATION. The natural or mechanical process of supplying conditioned or unconditioned air to, or removing such air from, any space.

❖ Ventilation is provided by the wind, naturally occurring pressure and temperature differences or mechanical systems typically consisting of fans and blowers.

VENTILATION AIR. That portion of supply air that comes from outside (outdoors) plus any recirculated air that has been treated to maintain the desired quality of air within a designated space.

❖ Ventilation air can be used for comfort cooling, control of air contaminants, equipment cooling and replenishing oxygen levels (see commentary, "Ventilation").

VERTICAL FENESTRATION. Windows (fixed or moveable), opaque doors, glazed doors, glazed block and combination opaque/glazed doors composed of glass or other transparent or translucent glazing materials and installed at a slope of a least 60 degrees (1.05 rad) from horizontal.

❖ Vertical fenestration is that which is typically installed on exterior walls. The required thermal efficiency of fenestration varies by climate zone (see Section R402.3).

VISIBLE TRANSMITTANCE [VT]. The ratio of visible light entering the space through the fenestration product assembly to the incident visible light, Visible Transmittance, includes the effects of glazing material and frame and is expressed as a number between 0 and 1.

❖ Use of daylight for energy conservation is now a fundamental part of the code. However, requirements for lower SHGC values for glazing can work against this because glazing with lower SHGC values generally will allow less visible light through the glazing. Therefore, the code now must include a limit on VT.

WHOLE HOUSE MECHANICAL VENTILATION SYSTEM. An exhaust system, supply system, or combination thereof that is designed to mechanically exchange indoor air with outdoor air when operating continuously or through a programmed intermittent schedule to satisfy the whole house ventilation rates.

❖ It is beneficial to have "Whole house mechanical ventilation system" defined in the code, as findings from a recent study commissioned by the DOE and the California Energy Commission identified that energy consumption of whole house mechanical ventilation systems is significant. Further, the study revealed that large disparities exist in the energy consumption and associated operating costs of whole house mechanical ventilation systems in cold, mild and hot, dry climates. Within the study, exhaust only systems balanced heat recovery systems, supply only systems, and central fan integrated systems were all modeled to assess resultant energy use and associated costs.

ZONE. A space or group of spaces within a building with heating or cooling requirements that are sufficiently similar so that desired conditions can be maintained throughout using a single controlling device.

❖ The simplest all-air system is a supply unit serving a single-temperature control zone. Ideally, this system responds completely to the space needs, and well-designed control systems maintain temperature and humidity closely and efficiently. Single-zone systems can be shut down when not required without affecting the operation of adjacent areas. Thus, in a thorough discussion of the term "zone," the concept of "zoning" for temperature control requires consideration.

Exterior zoning. Exterior zones are affected by varying weather conditions—wind, temperature and sun—and, depending on the geographic area and season, require both heating and cooling. This variation gives the designer considerable flexibility in choosing a system and results in the greatest advantages. The need for separate perimeter zone heating is determined by:

- Severity of the heating load (i.e., geographic location).
- Nature and orientation of the building envelope.
- Effects of downdraft at windows and the radiant effect of the cold glass surface (type of glass, area, height and *U*-factor).
- Type of occupancy (sedentary versus transient).
- Operating costs (in buildings such as offices and schools that are occupied for considerable periods). Fan operating costs can be reduced by heating with perimeter radiation during unoccupied periods rather than operating the main supply fans or local unit fans.

Interior zoning. Conditions in interior spaces are relatively constant because they are isolated from external influences. Usually, interior spaces require cooling throughout the year. Interior spaces with a roof exposure, however, may require similar treatment to perimeter spaces requiring heat. To summarize, zone control is required when the conditions at the thermostat are not representative of all the rooms or the entire exposure. This situation will almost certainly occur if any of the following conditions exist:

- The building has more than one level.
- One or more spaces are used for entertaining large groups.
- One or more spaces have large glass areas.
- The building has an indoor swimming pool or hot tub.
- The building has a solarium or atrium. In addition, zoning may be required when several rooms or spaces are isolated from each other and from the thermostat.
- The building spreads out in many directions (wings).
- Some spaces are distinctly isolated from the rest of the building.
- The envelope only has one or two exposures.
- The building has a room or rooms in a basement.
- The building has a room or rooms in an attic space.
- The building has one or more rooms with slab or exposed floor.
- Zone control can be achieved by insulation.
- There are discrete heating/cooling duct systems for each zone requiring control.
- There are automatic zone damper systems in a single heating/cooling duct system.

Chapter 3 [RE]: General Requirements

General Comments

Chapter 3 [RE] specifies the climate zones establishing exterior design conditions and provides general requirements for interior design conditions, and materials, systems and equipment. In general, the climate zone provisions are determined simply by referring to the map (see Figure R301.1) or by looking at the tables [see Tables R301.1, R301.3(1) and R301.3(2)]. In addition, Section R302 details the interior design conditions that are used for heating and cooling load calculations. Section R303 provides requirements for fenestration, identification of insulation and other basic general requirements for insulation materials.

Purpose

Climate has a major impact on the energy use of most commercial and residential buildings. The code establishes many requirements, such as wall and roof insulation *R*-values, window and door thermal transmittance requirements (*U*-factors), as well as provisions that affect the mechanical systems based on the climate where the building is located. This chapter contains the information used to properly assign the building location into the correct climate zone, which will then be used as the basis for establishing or eliminating requirements.

Materials and systems used to provide insulation and fenestration values, including *U*-factor and solar heat gain coefficient (SHGC) ratings, must be based on data used by appropriate tests. This establishes a level playing field for manufacturers of products.

Discussion and Development of the Climate Zone Map

The 2006 code made a dramatic shift in the classification of climate zones. While this change in the climate zone map was a part of the major revision to help simplify the code and make both compliance and enforcement easier, the climate zone revisions were a lengthy, very detailed and complicated process. Much of the new climate zone development was based on a paper titled "Climate Classification for Building Energy Codes and Standards." This paper was written by Robert S. Briggs, Robert G. Lucas and Z. Todd Taylor of the U.S. Department of Energy's Pacific Northwest National Laboratory (PNNL). Some aspects of this paper may help users better understand the climate zones and also help people feel comfortable with these new classifications.

Climate zones were developed based on the following criteria:

1. Offer consistent climate materials for all compliance methods and code sections (including both commercial and residential).
2. Enable the code to be self-contained with respect to climate data.
3. Be technically sound.
4. Map to political boundaries.
5. Provide a long-term climate classification solution.
6. Be generic and neutral (i.e., not overly tailored to current code requirements).
7. Be useful in beyond-code and future-code contexts.
8. Offer a more concise set of climate zones and presentation formats.
9. Be acceptable to the American Society of Heating, Refrigerating and Air-Conditioning Engineers (ASHRAE), and usable in ASHRAE standards and guidelines.
10. Provide a basis for use outside of the United States.

The reasons that the authors cited for some of the less obvious items include:

Item 4 – Mapping climate zones to easily recognizable political boundaries instead of to abstract climatic parameters facilitates code implementation. Users and jurisdictions are able to easily tell what requirements apply, which is not the case in some locations when climate parameters are used.

Item 7 – "Useful in future-code and beyond-code contexts" reflects the view that minimum acceptable practice codes and standards can provide an effective platform on which to build other efficiency programs. Beyond-code programs are likely to encourage features and technologies not included in current codes, many of which are likely to be more climate-sensitive than current requirements.

Item 9 – "Usable in ASHRAE standards and guidelines" is important because effective coordination of both content and formats used in the code and ASHRAE standards offers the potential to facilitate rapid migration of ASHRAE standards into model codes. Previous efforts to translate ASHRAE criteria into the simpler and more prescriptive forms most desired by the code enforcement community has, in some cases, added years to the adoption process and widespread implementation of updated criteria.

The belief in developing the climate zones was that any new system needed to show substantial improvement over the previously existing systems. In addition, any new classification must be at least roughly compatible with the previous climate-dependent requirements in order to allow for the conversion and inclusion of existing, generally accepted requirements. The intent was to develop a set of climate classifications that could support simple, approximate ways of prescribing energy-efficiency measures for buildings. It was not intended to develop a set of categories that could be used for all purposes.

The new climate zones were developed in an open process involving several standards committees of ASHRAE, the U.S. Department of Energy (DOE) staff and other interested parties.

Given the interest of the International Code Council (ICC®) and ASHRAE in producing documents that are capable of being used internationally, an effort was made to develop a system and climate zones that could work outside of the United States. The new climate definitions were developed using the International System of Units, abbreviated SI, from the French Le Systéme International d'Unités. By using the SI units and climate indices, which are widely available internationally, the climate zones and the development of building energy-efficiency provisions can be applied anywhere in the world. The boundaries between the various climate zones in Table R301.3(2) occur in multiples of 900°F days, which converts to 500°C days. Distinguishing the climate zones with these numbers results in a clean and understandable division between the climate zones in either system of temperature measurement.

The developers of the climate zone map selected bands of 1000 HDD18°C (1800 HDD65°F) because they resulted in boundaries that align with boundaries established in ANSI/ASHRAE/IESNA 90.1, plus they facilitate the use of both SI and inch-pound (I-P) units and were able to affect a significant reduction in the number of climate zones.

An objective for any effective classification is to maximize the differences between the selected criteria for each climate zone, while minimizing the variations that occur within the group. A large variation between the groups enables generalizations embodied in the code requirements to be better tailored to each climate zone. A small variation in each climate zone will ensure that the generalizations better fit the climate zone. It was the developers' feeling that the new classification better represents the climatic diversity, while defining more coherent climate zones than what the code previously used. It should be noted that mountainous regions defy clean geographic separation of clusters.

SECTION R301
CLIMATE ZONES

R301.1 General. *Climate zones* from Figure R301.1 or Table R301.1 shall be used in determining the applicable requirements from Chapter 4. Locations not in Table R301.1 (outside the United States) shall be assigned a *climate zone* based on Section R301.3.

❖ Climate involves temperature, moisture, wind and sun and also includes both daily and seasonal patterns of variation of the parameters. To account for these variations, the code establishes climate zones that serve as the basis for code provisions.

This section serves as the starting point for determining virtually all of the code requirements, especially under the prescriptive compliance paths. Because of their easy-to-understand graphic nature, maps have proven useful over the years as an effective way to help code users determine climate-dependent requirements. Therefore, for the United States, the climate zones are shown in the map in Figure R301.1. Because of the limited size of the map, the code also includes a listing of the climate zones by states and counties in Table R301.1. Table R301.1 will allow users to positively identify climate-zone assignments in those few locations for which the map interpretation may be difficult. Whether the map or the county list is used, the climate classification for each area will be the same.

When dealing with prescriptive compliance paths, the code user would simply look at the map or listing and select the proper climate zone based on the location of the building. When using a performance approach, additional climatic data may be needed.

Virtually every building energy code that has been developed for use in the United States has included a performance-based compliance path, which allows users to perform an energy analysis and demonstrate compliance based on equivalence with the prescriptive requirements. To perform these analyses, users must select appropriate weather data for their given project's location. The selection of appropriate weather data is straightforward for any project located in or around one of the various weather stations in the United States. For other locations, selecting the most appropriate weather site can be problematic. The codes themselves provide little help with this selection process. During the development of the new climate zones, the developers mapped every county in the United States to the most appropriate SAMSON station (National Climatic Data Center "Solar and Meteorological Surface Observation Network" station) for each county as a whole. This mapping is not included in the code but may be used in some compliance software. Designating an appropriate SAMSON station should not be considered to be the only climate data permitted for a given county. It could, however, be used in the absence of better information. Where local data better reflects regional or microclimatic conditions of an area, they would be appropriate to use. For example, elevation has a large impact on climate and can vary dramatically within individual counties, especially in the western United States. Where elevation differences are significant, code officials may require use of sites that differ from the sites designated as being the most appropriate for the county. For additional information on this topic, review the paper "Climate Classification for Building Energy Codes and Standards," which is referenced in the commentary text that precedes Section R301.

The new climate classifications do not attempt to resolve the issue of what the appropriate treatment for elevation differences is. This aspect is left in the hands of the local code official.

R301.2 Warm humid counties. Warm humid counties are identified in Table R301.1 by an asterisk.

❖ Table R301.1 lists the counties in the southeastern United States that fall below the white-dashed line appearing on the map in Figure R301.1. The warm-humid climate designation includes parts of eight states and also covers all of Florida, Hawaii and the U.S. territories. Table R301.3(1) provides the details that were used to determine the classification of the warm-humid designation for the counties.

There currently are very few requirements in the code specifically tied to the warm-humid climate criteria. Although not tied directly to the warm-humid designation, many other code sections, such as those addressing moisture control and energy recovery ventilation systems, do take these climatic features into account.

R301.3 International climate zones. The *climate zone* for any location outside the United States shall be determined by applying Table R301.3(1) and then Table R301.3(2).

❖ Although the code and the climate zone classifications it includes are predominately used in the United States, they can be used in any location. Because the mapping and decisions that were made during the development of the climate zones focused primarily on the United States, this section details how to properly classify the climate zones based on thermal criteria [Table R301.3(2)], the major climate types [Table R301.3(1)] and the warm-humid criteria for locations outside of the United States (see Commentary Figure R301.3).

In developing the climate zone designations, two climate zones were defined in the classification, but not thoroughly evaluated or actively applied because no sites in the United States or its territories required their use. The two climate zones are 1B [dry and greater than 5,000 CDD10°C (9,000 CDD50°F)], characterized as "very hot-dry," and 5C [marine and 3,000 < HDD18°C ≤ 400 (5,400, HDD65°F ≤ 7,200)], characterized as "cool marine." The marine (C) designation was not used for climate zones colder than Climate Zone 5 or hotter than Cimate Zone 3, as marine climates are inherently neither very cold nor very hot. In addition, the humid (A) and dry (B) divisions were dropped for climate zones colder than Climate Zone 6 because they did not appear to be warranted based on differences in appropriate building design requirements. Reevaluation of these decisions might be warranted before applying the new climate classifications to locations outside of the United States.

R301.4 Tropical climate zone. The tropical *climate zone* shall be defined as:

1. Hawaii, Puerto Rico, Guam, American Samoa, U.S. Virgin Islands, Commonwealth of Northern Mariana Islands; and
2. Islands in the area between the Tropic of Cancer and the Tropic of Capricorn.

❖ Tropical areas are quite different from the U.S. mainland in climate, construction techniques, traditional construction and energy prices. Prior to the installation of these new provisions, the IECC treated tropical climates as if they were simply a southern extension of the U.S. mainland. Traditional residences, especially less-expensive residences, have evolved inexpensive ways to work with the tropical climates to provide comfortable interior spaces without the need for substantial space conditioning. Tropical electrical prices, usually over 20 cents per kwh, provide a substantial incentive for energy conservation. Solar water heating works particularly well in tropical climates.

This section provides a simple option for a newly defined climate zone, the "tropical zone." The area between the Tropic of Cancer and the Tropic of Capricorn is the area between 23.5 degrees northern and southern latitude of the equator.

Traditional construction, especially with solar water heating, is usually more energy efficient than the construction style typically assumed in the IECC, as is shown by an analysis done for Puerto Rico. Using energy-efficient versions of traditional construction saves more energy and is much more cost-effective than pushing those in tropical climates to adopt mainland construction practices. Traditional tropical construction focuses on greatly reducing or eliminating the need for space conditioning by making a living space that is comfortable without space conditioning.

These requirements are based on informal conversations with those who live in tropical regions.

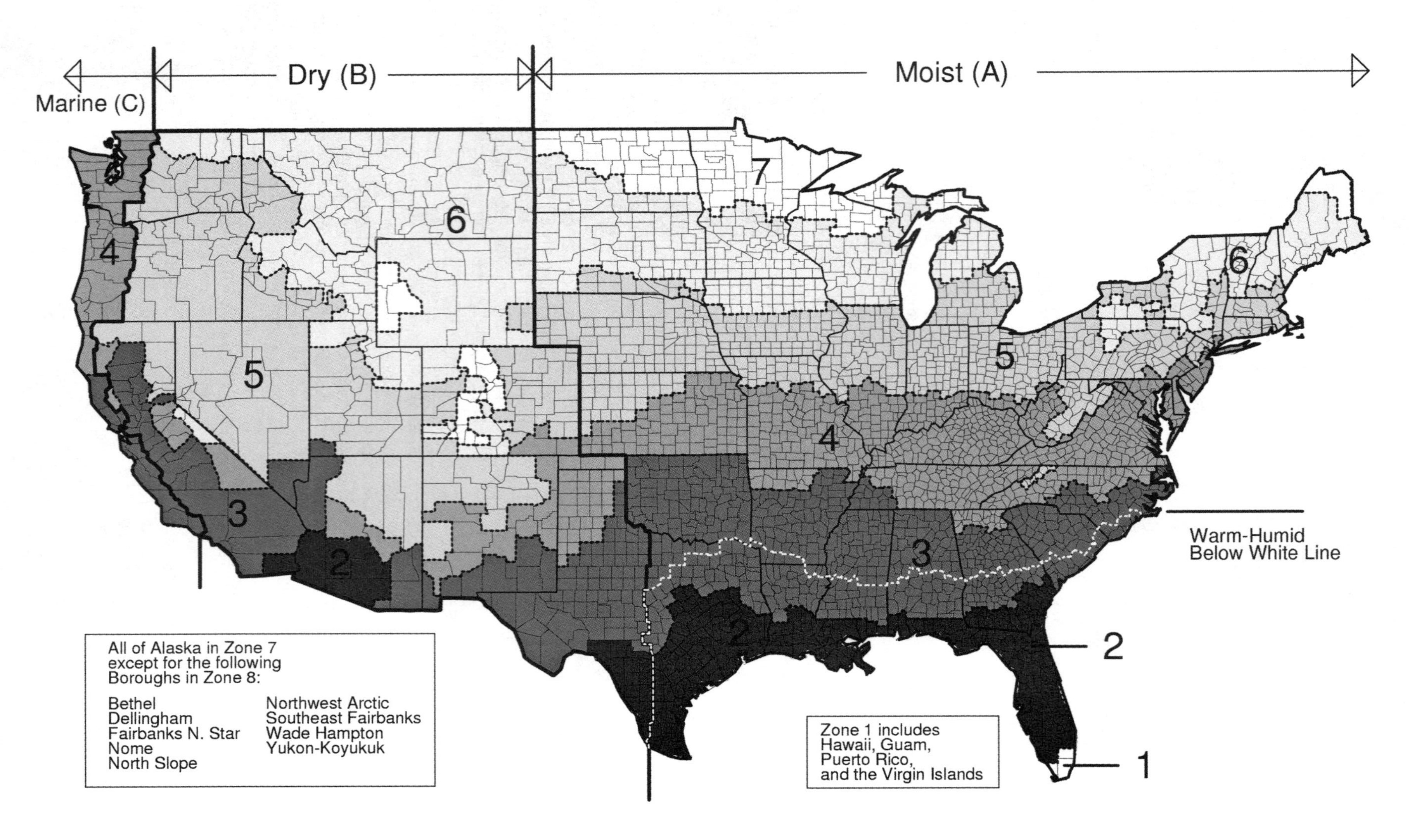

FIGURE R301.1
CLIMATE ZONES

ZONE NUMBER	CLIMATE ZONE NAME AND TYPE[2]	THERMAL CRITERIA [1, 3, 6]	REPRESENTATIVE U.S. CITY[4]
1A	Very Hot-Humid	9000 < CDD50°F	Miami, FL
1B[5]	Very Hot-Dry	9000 < CDD50°F	—
2A	Hot-Humid	6300 < CDD50°F ≤ 9000	Houston, TX
2B	Hot-Dry	6300 < CDD50°F ≤ 9000	Phoenix, AZ
3A	Warm-Humid	4500 < CDD50°F ≤ 6300	Memphis, TN
3B	Warm-Dry	4500 < CDD50°F ≤ 6300	El Paso, TX
3C	Warm-Marine	HDD65°F ≤ 3600	San Francisco, CA
4A	Mixed-Humid	CDD50°F ≤ 4500 AND HDD65°F ≤ 5400	Baltimore, MD
4B	Mixed-Dry	CDD50°F ≤ 4500 AND HDD65°F ≤ 5400	Albuquerque, NM
4C	Mixed-Marine	3600 < HDD65°F ≤ 5400	Salem, OR
5A	Cool-Humid	5400 < HDD65°F ≤ 7200	Chicago, IL
5B	Cool-Dry	5400 < HDD65°F ≤ 7200	Boise, ID
5C[5]	Cool-Marine	5400 < HDD65°F ≤ 7200	—
6A	Cool-Humid	7200 < HDD65°F ≤ 9000	Burlington, VT
6B	Cool-Dry	7200 < HDD65°F ≤ 9000	Helena, MT
7	Very Cold	9000 < HDD65°F ≤ 12600	Duluth, MN
8	Sub Arctic	12600 < HDD65°F	Fairbanks, AK

Notes:

1. Column 1 contains alphanumeric designations for each climate zone. These designations are intended for use when the climate zones are referenced in the code. The numeric part of the designation relates to the thermal properties of the climate zone. The letter part indicates the major climatic group to which the climate zone belongs; A indicates humid, B indicates dry, and C indicates marine. The climatic group designation was dropped for Climate Zones 7 and 8 because the developers of the new climate zone classifications did not anticipate any building design criteria sensitive to the humid/dry/marine distinction in very cold climates. Climate Zones 1B and 5C have been defined but are not used for the United States. Zone 6C [Marine and HDD18°C > 4000 (HDD65°F > 7200)] might appear to be necessary for consistency. However, very few locations in the world are both as mild as is required by the marine zone definition and as cold as necessary to accumulate that many heating degree days. In addition, such sites do not appear climatically very different from sites in Zone 6A, which is where they are assigned in the absence of a Zone 6C.
2. Column 2 contains a descriptive name for each climate zone and the major climate type. The names can be used in place of the alphanumeric designations wherever a more descriptive designation is appropriate.
3. Column 3 contains definitions for the climate zone divisions based on degree day cooling and/or heating criteria. The humid/dry/marine divisions must be determined first before these criteria are applied. The definitions in Tables R301.3(1) and R301.3(2) contain logic capable of assigning a climate zone designation to any location with the necessary climate data anywhere in the world. However, the work to develop this classification focused on the 50 United States. Application of the classification to locations outside of the United States is untested.
4. Column 4 contains the name of a SAMSON station (National Climatic Data Center "Solar and Meteorological Surface Observation Network" station) found to best represent the climate zone as a whole. See the discussions at the beginning of this chapter regarding the development of the new climate zones for an explanation of how the representative cities were selected.
5. Climate Zones 1B and 5C do not occur in the United States, and no representative cities were selected for these climate zones due to data limitations. Climates meeting the listed criteria do exist in such locations as Saudi Arabia; British Columbia, Canada; and Northern Europe.
6. SI to I-P Conversions:
   2500 CDD10°C = 4500 CDD50°F
   3000 HDD18°C = 5400 HDD65°F
   3500 CDD10°C = 6300 CDD50°F
   4000 HDD18°C = 7200 HDD65°F
   5000 CDD10°C = 9000 CDD50°F
   5000 HDD18°C = 9000 HDD65°F
   2000 HDD18°C = 3600 HDD65°F
   7000 HDD18°C = 12600 HDD65°F

Figure R301.3
CLIMATE ZONE DEFINITIONS

TABLE R301.3(1)
INTERNATIONAL CLIMATE ZONE DEFINITIONS

MAJOR CLIMATE TYPE DEFINITIONS
Marine (C) Definition—Locations meeting all four criteria: 1. Mean temperature of coldest month between -3°C (27°F) and 18°C (65°F). 2. Warmest month mean < 22°C (72°F). 3. At least four months with mean temperatures over 10°C (50°F). 4. Dry season in summer. The month with the heaviest precipitation in the cold season has at least three times as much precipitation as the month with the least precipitation in the rest of the year. The cold season is October through March in the Northern Hemisphere and April through September in the Southern Hemisphere.
Dry (B) Definition—Locations meeting the following criteria: Not marine and $P_{in} < 0.44 \times (TF - 19.5)$ [$P_{cm} < 2.0 \times (TC + 7)$ in SI units] where: P_{in} = Annual precipitation in inches (cm) T = Annual mean temperature in °F (°C)
Moist (A) Definition—Locations that are not marine and not dry.
Warm-humid Definition—Moist (A) locations where either of the following wet-bulb temperature conditions shall occur during the warmest six consecutive months of the year: 1. 67°F (19.4°C) or higher for 3,000 or more hours; or 2. 73°F (22.8°C) or higher for 1,500 or more hours.

For SI: °C = [(°F)-32]/1.8, 1 inch = 2.54 cm.

TABLE R301.3(2)
INTERNATIONAL CLIMATE ZONE DEFINITIONS

ZONE NUMBER	THERMAL CRITERIA	
	IP Units	SI Units
1	9000 < CDD50°F	5000 < CDD10°C
2	6300 < CDD50°F ≤ 9000	3500 < CDD10°C ≤ 5000
3A and 3B	4500 < CDD50°F ≤ 6300 AND HDD65°F ≤ 5400	2500 < CDD10°C ≤ 3500 AND HDD18°C ≤ 3000
4A and 4B	CDD50°F ≤ 4500 AND HDD65°F ≤ 5400	CDD10°C ≤ 2500 AND HDD18°C ≤ 3000
3C	HDD65°F ≤ 3600	HDD18°C ≤ 2000
4C	3600 < HDD65°F ≤ 5400	2000 < HDD18°C ≤ 3000
5	5400 < HDD65°F ≤ 7200	3000 < HDD18°C ≤ 4000
6	7200 < HDD65°F ≤ 9000	4000 < HDD18°C ≤ 5000
7	9000 < HDD65°F ≤ 12600	5000 < HDD18°C ≤ 7000
8	12600 < HDD65°F	7000 < HDD18°C

For SI: °C = [(°F)-32]/1.8.

TABLE R301.1
CLIMATE ZONES, MOISTURE REGIMES, AND WARM-HUMID DESIGNATIONS BY STATE, COUNTY AND TERRITORY

Key: A – Moist, B – Dry, C – Marine. Absence of moisture designation indicates moisture regime is irrelevant. Asterisk (*) indicates a warm-humid location.

US STATES

ALABAMA

3A Autauga*
2A Baldwin*
3A Barbour*
3A Bibb
3A Blount
3A Bullock*
3A Butler*
3A Calhoun
3A Chambers
3A Cherokee
3A Chilton
3A Choctaw*
3A Clarke*
3A Clay
3A Cleburne
3A Coffee*
3A Colbert
3A Conecuh*
3A Coosa
3A Covington*
3A Crenshaw*
3A Cullman
3A Dale*
3A Dallas*
3A DeKalb
3A Elmore*
3A Escambia*
3A Etowah
3A Fayette
3A Franklin
3A Geneva*
3A Greene
3A Hale
3A Henry*
3A Houston*
3A Jackson
3A Jefferson
3A Lamar
3A Lauderdale
3A Lawrence
3A Lee
3A Limestone
3A Lowndes*
3A Macon*
3A Madison
3A Marengo*
3A Marion
3A Marshall
2A Mobile*
3A Monroe*
3A Montgomery*
3A Morgan
3A Perry*
3A Pickens
3A Pike*
3A Randolph
3A Russell*
3A Shelby
3A St. Clair
3A Sumter
3A Talladega
3A Tallapoosa
3A Tuscaloosa
3A Walker
3A Washington*
3A Wilcox*
3A Winston

ALASKA

7 Aleutians East
7 Aleutians West
7 Anchorage
8 Bethel
7 Bristol Bay
7 Denali
8 Dillingham
8 Fairbanks North Star
7 Haines
7 Juneau
7 Kenai Peninsula
7 Ketchikan Gateway
7 Kodiak Island
7 Lake and Peninsula
7 Matanuska-Susitna
8 Nome
8 North Slope
8 Northwest Arctic
7 Prince of Wales-Outer Ketchikan
7 Sitka
7 Skagway-Hoonah-Angoon
8 Southeast Fairbanks
7 Valdez-Cordova
8 Wade Hampton
7 Wrangell-Petersburg
7 Yakutat
8 Yukon-Koyukuk

ARIZONA

5B Apache
3B Cochise
5B Coconino
4B Gila
3B Graham
3B Greenlee
2B La Paz
2B Maricopa
3B Mohave
5B Navajo
2B Pima
2B Pinal
3B Santa Cruz
4B Yavapai
2B Yuma

ARKANSAS

3A Arkansas
3A Ashley
4A Baxter
4A Benton
4A Boone
3A Bradley
3A Calhoun
4A Carroll
3A Chicot
3A Clark
3A Clay
3A Cleburne
3A Cleveland
3A Columbia*
3A Conway
3A Craighead
3A Crawford
3A Crittenden
3A Cross
3A Dallas
3A Desha
3A Drew
3A Faulkner
3A Franklin
4A Fulton
3A Garland
3A Grant
3A Greene
3A Hempstead*
3A Hot Spring
3A Howard
3A Independence
4A Izard
3A Jackson
3A Jefferson
3A Johnson
3A Lafayette*
3A Lawrence
3A Lee
3A Lincoln
3A Little River*
3A Logan
3A Lonoke
4A Madison
4A Marion
3A Miller*
3A Mississippi
3A Monroe
3A Montgomery
3A Nevada
4A Newton
3A Ouachita
3A Perry
3A Phillips
3A Pike
3A Poinsett
3A Polk
3A Pope
3A Prairie
3A Pulaski
3A Randolph
3A Saline
3A Scott
4A Searcy
3A Sebastian
3A Sevier*
3A Sharp
3A St. Francis
4A Stone
3A Union*
3A Van Buren
4A Washington
3A White
3A Woodruff
3A Yell

CALIFORNIA

3C Alameda
6B Alpine
4B Amador
3B Butte
4B Calaveras
3B Colusa
3B Contra Costa
4C Del Norte
4B El Dorado
3B Fresno
3B Glenn

(continued)

TABLE R301.1—continued
CLIMATE ZONES, MOISTURE REGIMES, AND WARM-HUMID DESIGNATIONS BY STATE, COUNTY AND TERRITORY

4C Humboldt
2B Imperial
4B Inyo
3B Kern
3B Kings
4B Lake
5B Lassen
3B Los Angeles
3B Madera
3C Marin
4B Mariposa
3C Mendocino
3B Merced
5B Modoc
6B Mono
3C Monterey
3C Napa
5B Nevada
3B Orange
3B Placer
5B Plumas
3B Riverside
3B Sacramento
3C San Benito
3B San Bernardino
3B San Diego
3C San Francisco
3B San Joaquin
3C San Luis Obispo
3C San Mateo
3C Santa Barbara
3C Santa Clara
3C Santa Cruz
3B Shasta
5B Sierra
5B Siskiyou
3B Solano
3C Sonoma
3B Stanislaus
3B Sutter
3B Tehama
4B Trinity
3B Tulare
4B Tuolumne
3C Ventura
3B Yolo
3B Yuba

COLORADO

5B Adams
6B Alamosa
5B Arapahoe
6B Archuleta
4B Baca
5B Bent
5B Boulder
5B Broomfield
6B Chaffee
5B Cheyenne
7 Clear Creek
6B Conejos
6B Costilla
5B Crowley
6B Custer
5B Delta
5B Denver
6B Dolores
5B Douglas
6B Eagle
5B Elbert
5B El Paso
5B Fremont
5B Garfield
5B Gilpin
7 Grand
7 Gunnison
7 Hinsdale
5B Huerfano
7 Jackson
5B Jefferson
5B Kiowa
5B Kit Carson
7 Lake
5B La Plata
5B Larimer
4B Las Animas
5B Lincoln
5B Logan
5B Mesa
7 Mineral
6B Moffat
5B Montezuma
5B Montrose
5B Morgan
4B Otero
6B Ouray
7 Park
5B Phillips
7 Pitkin
5B Prowers
5B Pueblo
6B Rio Blanco
7 Rio Grande
7 Routt
6B Saguache
7 San Juan
6B San Miguel
5B Sedgwick
7 Summit
5B Teller
5B Washington
5B Weld
5B Yuma

CONNECTICUT

5A (all)

DELAWARE

4A (all)

DISTRICT OF COLUMBIA

4A (all)

FLORIDA

2A Alachua*
2A Baker*
2A Bay*
2A Bradford*
2A Brevard*
1A Broward*
2A Calhoun*
2A Charlotte*
2A Citrus*
2A Clay*
2A Collier*
2A Columbia*
2A DeSoto*
2A Dixie*
2A Duval*
2A Escambia*
2A Flagler*
2A Franklin*
2A Gadsden*
2A Gilchrist*
2A Glades*
2A Gulf*
2A Hamilton*
2A Hardee*
2A Hendry*
2A Hernando*
2A Highlands*
2A Hillsborough*
2A Holmes*
2A Indian River*
2A Jackson*
2A Jefferson*
2A Lafayette*
2A Lake*
2A Lee*
2A Leon*
2A Levy*
2A Liberty*
2A Madison*
2A Manatee*
2A Marion*
2A Martin*
1A Miami-Dade*
1A Monroe*
2A Nassau*
2A Okaloosa*
2A Okeechobee*
2A Orange*
2A Osceola*
2A Palm Beach*
2A Pasco*
2A Pinellas*
2A Polk*
2A Putnam*
2A Santa Rosa*
2A Sarasota*
2A Seminole*
2A St. Johns*
2A St. Lucie*
2A Sumter*
2A Suwannee*
2A Taylor*
2A Union*
2A Volusia*
2A Wakulla*
2A Walton*
2A Washington*

GEORGIA

2A Appling*
2A Atkinson*
2A Bacon*
2A Baker*
3A Baldwin
4A Banks
3A Barrow
3A Bartow
3A Ben Hill*
2A Berrien*
3A Bibb
3A Bleckley*
2A Brantley*
2A Brooks*
2A Bryan*
3A Bulloch*
3A Burke
3A Butts
3A Calhoun*
2A Camden*
3A Candler*
3A Carroll
4A Catoosa
2A Charlton*
2A Chatham*
3A Chattahoochee*
4A Chattooga
3A Cherokee
3A Clarke
3A Clay*
3A Clayton
2A Clinch*
3A Cobb
3A Coffee*
2A Colquitt*
3A Columbia
2A Cook*
3A Coweta

(continued)

TABLE R301.1—continued
CLIMATE ZONES, MOISTURE REGIMES, AND WARM-HUMID DESIGNATIONS BY STATE, COUNTY AND TERRITORY

3A Crawford
3A Crisp*
4A Dade
4A Dawson
2A Decatur*
3A DeKalb
3A Dodge*
3A Dooly*
3A Dougherty*
3A Douglas
3A Early*
2A Echols*
2A Effingham*
3A Elbert
3A Emanuel*
2A Evans*
4A Fannin
3A Fayette
4A Floyd
3A Forsyth
4A Franklin
3A Fulton
4A Gilmer
3A Glascock
2A Glynn*
4A Gordon
2A Grady*
3A Greene
3A Gwinnett
4A Habersham
4A Hall
3A Hancock
3A Haralson
3A Harris
3A Hart
3A Heard
3A Henry
3A Houston*
3A Irwin*
3A Jackson
3A Jasper
2A Jeff Davis*
3A Jefferson
3A Jenkins*
3A Johnson*
3A Jones
3A Lamar
2A Lanier*
3A Laurens*
3A Lee*
2A Liberty*
3A Lincoln
2A Long*
2A Lowndes*
4A Lumpkin
3A Macon*
3A Madison
3A Marion*
3A McDuffie
2A McIntosh*
3A Meriwether
2A Miller*
2A Mitchell*
3A Monroe
3A Montgomery*
3A Morgan
4A Murray
3A Muscogee
3A Newton
3A Oconee
3A Oglethorpe
3A Paulding
3A Peach*
4A Pickens
2A Pierce*
3A Pike
3A Polk
3A Pulaski*
3A Putnam
3A Quitman*
4A Rabun
3A Randolph*
3A Richmond
3A Rockdale
3A Schley*
3A Screven*
2A Seminole*
3A Spalding
4A Stephens
3A Stewart*
3A Sumter*
3A Talbot
3A Taliaferro
2A Tattnall*
3A Taylor*
3A Telfair*
3A Terrell*
2A Thomas*
3A Tift*
2A Toombs*
4A Towns
3A Treutlen*
3A Troup
3A Turner*
3A Twiggs*
4A Union
3A Upson
4A Walker
3A Walton
2A Ware*
3A Warren
3A Washington
2A Wayne*
3A Webster*
3A Wheeler*
4A White
4A Whitfield
3A Wilcox*
3A Wilkes
3A Wilkinson
3A Worth*

HAWAII

1A (all)*

IDAHO

5B Ada
6B Adams
6B Bannock
6B Bear Lake
5B Benewah
6B Bingham
6B Blaine
6B Boise
6B Bonner
6B Bonneville
6B Boundary
6B Butte
6B Camas
5B Canyon
6B Caribou
5B Cassia
6B Clark
5B Clearwater
6B Custer
5B Elmore
6B Franklin
6B Fremont
5B Gem
5B Gooding
5B Idaho
6B Jefferson
5B Jerome
5B Kootenai
5B Latah
6B Lemhi
5B Lewis
5B Lincoln
6B Madison
5B Minidoka
5B Nez Perce
6B Oneida
5B Owyhee
5B Payette
5B Power
5B Shoshone
6B Teton
5B Twin Falls
6B Valley
5B Washington

ILLINOIS

5A Adams
4A Alexander
4A Bond
5A Boone
5A Brown
5A Bureau
5A Calhoun
5A Carroll
5A Cass
5A Champaign
4A Christian
5A Clark
4A Clay
4A Clinton
5A Coles
5A Cook
4A Crawford
5A Cumberland
5A DeKalb
5A De Witt
5A Douglas
5A DuPage
5A Edgar
4A Edwards
4A Effingham
4A Fayette
5A Ford
4A Franklin
5A Fulton
4A Gallatin
5A Greene
5A Grundy
4A Hamilton
5A Hancock
4A Hardin
5A Henderson
5A Henry
5A Iroquois
4A Jackson
4A Jasper
4A Jefferson
5A Jersey
5A Jo Daviess
4A Johnson
5A Kane
5A Kankakee
5A Kendall
5A Knox
5A Lake
5A La Salle
4A Lawrence
5A Lee
5A Livingston
5A Logan
5A Macon
4A Macoupin
4A Madison
4A Marion
5A Marshall
5A Mason
4A Massac
5A McDonough
5A McHenry

(continued)

TABLE R301.1—continued
CLIMATE ZONES, MOISTURE REGIMES, AND WARM-HUMID DESIGNATIONS BY STATE, COUNTY AND TERRITORY

5A McLean
5A Menard
5A Mercer
4A Monroe
4A Montgomery
5A Morgan
5A Moultrie
5A Ogle
5A Peoria
4A Perry
5A Piatt
5A Pike
4A Pope
4A Pulaski
5A Putnam
4A Randolph
4A Richland
5A Rock Island
4A Saline
5A Sangamon
5A Schuyler
5A Scott
4A Shelby
5A Stark
4A St. Clair
5A Stephenson
5A Tazewell
4A Union
5A Vermilion
4A Wabash
5A Warren
4A Washington
4A Wayne
4A White
5A Whiteside
5A Will
4A Williamson
5A Winnebago
5A Woodford

INDIANA

5A Adams
5A Allen
5A Bartholomew
5A Benton
5A Blackford
5A Boone
4A Brown
5A Carroll
5A Cass
4A Clark
5A Clay
5A Clinton
4A Crawford
4A Daviess
4A Dearborn
5A Decatur
5A De Kalb
5A Delaware
4A Dubois
5A Elkhart
5A Fayette
4A Floyd
5A Fountain
5A Franklin
5A Fulton
4A Gibson
5A Grant
4A Greene
5A Hamilton
5A Hancock
4A Harrison
5A Hendricks
5A Henry
5A Howard
5A Huntington
4A Jackson
5A Jasper
5A Jay
4A Jefferson
4A Jennings
5A Johnson
4A Knox
5A Kosciusko
5A LaGrange
5A Lake
5A LaPorte
4A Lawrence
5A Madison
5A Marion
5A Marshall
4A Martin
5A Miami
4A Monroe
5A Montgomery
5A Morgan
5A Newton
5A Noble
4A Ohio
4A Orange
5A Owen
5A Parke
4A Perry
4A Pike
5A Porter
4A Posey
5A Pulaski
5A Putnam
5A Randolph
4A Ripley
5A Rush
4A Scott
5A Shelby
4A Spencer
5A Starke
5A Steuben
5A St. Joseph
4A Sullivan
4A Switzerland
5A Tippecanoe
5A Tipton
5A Union
4A Vanderburgh
5A Vermillion
5A Vigo
5A Wabash
5A Warren
4A Warrick
4A Washington
5A Wayne
5A Wells
5A White
5A Whitley

IOWA

5A Adair
5A Adams
6A Allamakee
5A Appanoose
5A Audubon
5A Benton
6A Black Hawk
5A Boone
6A Bremer
6A Buchanan
6A Buena Vista
6A Butler
6A Calhoun
5A Carroll
5A Cass
5A Cedar
6A Cerro Gordo
6A Cherokee
6A Chickasaw
5A Clarke
6A Clay
6A Clayton
5A Clinton
5A Crawford
5A Dallas
5A Davis
5A Decatur
6A Delaware
5A Des Moines
6A Dickinson
5A Dubuque
6A Emmet
6A Fayette
6A Floyd
6A Franklin
5A Fremont
5A Greene
6A Grundy
5A Guthrie
6A Hamilton
6A Hancock
6A Hardin
5A Harrison
5A Henry
6A Howard
6A Humboldt
6A Ida
5A Iowa
5A Jackson
5A Jasper
5A Jefferson
5A Johnson
5A Jones
5A Keokuk
6A Kossuth
5A Lee
5A Linn
5A Louisa
5A Lucas
6A Lyon
5A Madison
5A Mahaska
5A Marion
5A Marshall
5A Mills
6A Mitchell
5A Monona
5A Monroe
5A Montgomery
5A Muscatine
6A O'Brien
6A Osceola
5A Page
6A Palo Alto
6A Plymouth
6A Pocahontas
5A Polk
5A Pottawattamie
5A Poweshiek
5A Ringgold
6A Sac
5A Scott
5A Shelby
6A Sioux
5A Story
5A Tama
5A Taylor
5A Union
5A Van Buren
5A Wapello
5A Warren
5A Washington
5A Wayne
6A Webster
6A Winnebago

(continued)

TABLE R301.1—continued
CLIMATE ZONES, MOISTURE REGIMES, AND WARM-HUMID DESIGNATIONS BY STATE, COUNTY AND TERRITORY

6A Winneshiek
5A Woodbury
6A Worth
6A Wright

KANSAS

4A Allen
4A Anderson
4A Atchison
4A Barber
4A Barton
4A Bourbon
4A Brown
4A Butler
4A Chase
4A Chautauqua
4A Cherokee
5A Cheyenne
4A Clark
4A Clay
5A Cloud
4A Coffey
4A Comanche
4A Cowley
4A Crawford
5A Decatur
4A Dickinson
4A Doniphan
4A Douglas
4A Edwards
4A Elk
5A Ellis
4A Ellsworth
4A Finney
4A Ford
4A Franklin
4A Geary
5A Gove
5A Graham
4A Grant
4A Gray
5A Greeley
4A Greenwood
5A Hamilton
4A Harper
4A Harvey
4A Haskell
4A Hodgeman
4A Jackson
4A Jefferson
5A Jewell
4A Johnson
4A Kearny
4A Kingman
4A Kiowa
4A Labette
5A Lane
4A Leavenworth
4A Lincoln
4A Linn
5A Logan
4A Lyon
4A Marion
4A Marshall
4A McPherson
4A Meade
4A Miami
5A Mitchell
4A Montgomery
4A Morris
4A Morton
4A Nemaha
4A Neosho
5A Ness
5A Norton
4A Osage
5A Osborne
4A Ottawa
4A Pawnee
5A Phillips
4A Pottawatomie
4A Pratt
5A Rawlins
4A Reno
5A Republic
4A Rice
4A Riley
5A Rooks
4A Rush
4A Russell
4A Saline
5A Scott
4A Sedgwick
4A Seward
4A Shawnee
5A Sheridan
5A Sherman
5A Smith
4A Stafford
4A Stanton
4A Stevens
4A Sumner
5A Thomas
5A Trego
4A Wabaunsee
5A Wallace
4A Washington
5A Wichita
4A Wilson
4A Woodson
4A Wyandotte

KENTUCKY

4A (all)

LOUISIANA

2A Acadia*
2A Allen*
2A Ascension*
2A Assumption*
2A Avoyelles*
2A Beauregard*
3A Bienville*
3A Bossier*
3A Caddo*
2A Calcasieu*
3A Caldwell*
2A Cameron*
3A Catahoula*
3A Claiborne*
3A Concordia*
3A De Soto*
2A East Baton Rouge*
3A East Carroll
2A East Feliciana*
2A Evangeline*
3A Franklin*
3A Grant*
2A Iberia*
2A Iberville*
3A Jackson*
2A Jefferson*
2A Jefferson Davis*
2A Lafayette*
2A Lafourche*
3A La Salle*
3A Lincoln*
2A Livingston*
3A Madison*
3A Morehouse
3A Natchitoches*
2A Orleans*
3A Ouachita*
2A Plaquemines*
2A Pointe Coupee*
2A Rapides*
3A Red River*
3A Richland*
3A Sabine*
2A St. Bernard*
2A St. Charles*
2A St. Helena*
2A St. James*
2A St. John the Baptist*
2A St. Landry*
2A St. Martin*
2A St. Mary*
2A St. Tammany*
2A Tangipahoa*
3A Tensas*
2A Terrebonne*
3A Union*
2A Vermilion*
3A Vernon*
2A Washington*
3A Webster*
2A West Baton Rouge*
3A West Carroll
2A West Feliciana*
3A Winn*

MAINE

6A Androscoggin
7 Aroostook
6A Cumberland
6A Franklin
6A Hancock
6A Kennebec
6A Knox
6A Lincoln
6A Oxford
6A Penobscot
6A Piscataquis
6A Sagadahoc
6A Somerset
6A Waldo
6A Washington
6A York

MARYLAND

4A Allegany
4A Anne Arundel
4A Baltimore
4A Baltimore (city)
4A Calvert
4A Caroline
4A Carroll
4A Cecil
4A Charles
4A Dorchester
4A Frederick
5A Garrett
4A Harford
4A Howard
4A Kent
4A Montgomery
4A Prince George's
4A Queen Anne's
4A Somerset
4A St. Mary's
4A Talbot
4A Washington
4A Wicomico
4A Worcester

MASSACHSETTS

5A (all)

MICHIGAN

6A Alcona
6A Alger

(continued)

TABLE R301.1—continued
CLIMATE ZONES, MOISTURE REGIMES, AND WARM-HUMID DESIGNATIONS BY STATE, COUNTY AND TERRITORY

5A Allegan
6A Alpena
6A Antrim
6A Arenac
7 Baraga
5A Barry
5A Bay
6A Benzie
5A Berrien
5A Branch
5A Calhoun
5A Cass
6A Charlevoix
6A Cheboygan
7 Chippewa
6A Clare
5A Clinton
6A Crawford
6A Delta
6A Dickinson
5A Eaton
6A Emmet
5A Genesee
6A Gladwin
7 Gogebic
6A Grand Traverse
5A Gratiot
5A Hillsdale
7 Houghton
6A Huron
5A Ingham
5A Ionia
6A Iosco
7 Iron
6A Isabella
5A Jackson
5A Kalamazoo
6A Kalkaska
5A Kent
7 Keweenaw
6A Lake
5A Lapeer
6A Leelanau
5A Lenawee
5A Livingston
7 Luce
7 Mackinac
5A Macomb
6A Manistee
6A Marquette
6A Mason
6A Mecosta
6A Menominee
5A Midland
6A Missaukee
5A Monroe
5A Montcalm
6A Montmorency
5A Muskegon
6A Newaygo
5A Oakland
6A Oceana
6A Ogemaw
7 Ontonagon
6A Osceola
6A Oscoda
6A Otsego
5A Ottawa
6A Presque Isle
6A Roscommon
5A Saginaw
6A Sanilac
7 Schoolcraft
5A Shiawassee
5A St. Clair
5A St. Joseph
5A Tuscola
5A Van Buren
5A Washtenaw
5A Wayne
6A Wexford

MINNESOTA

7 Aitkin
6A Anoka
7 Becker
7 Beltrami
6A Benton
6A Big Stone
6A Blue Earth
6A Brown
7 Carlton
6A Carver
7 Cass
6A Chippewa
6A Chisago
7 Clay
7 Clearwater
7 Cook
6A Cottonwood
7 Crow Wing
6A Dakota
6A Dodge
6A Douglas
6A Faribault
6A Fillmore
6A Freeborn
6A Goodhue
7 Grant
6A Hennepin
6A Houston
7 Hubbard
6A Isanti
7 Itasca
6A Jackson
7 Kanabec
6A Kandiyohi
7 Kittson
7 Koochiching
6A Lac qui Parle
7 Lake
7 Lake of the Woods
6A Le Sueur
6A Lincoln
6A Lyon
7 Mahnomen
7 Marshall
6A Martin
6A McLeod
6A Meeker
7 Mille Lacs
6A Morrison
6A Mower
6A Murray
6A Nicollet
6A Nobles
7 Norman
6A Olmsted
7 Otter Tail
7 Pennington
7 Pine
6A Pipestone
7 Polk
6A Pope
6A Ramsey
7 Red Lake
6A Redwood
6A Renville
6A Rice
6A Rock
7 Roseau
6A Scott
6A Sherburne
6A Sibley
6A Stearns
6A Steele
6A Stevens
7 St. Louis
6A Swift
6A Todd
6A Traverse
6A Wabasha
7 Wadena
6A Waseca
6A Washington
6A Watonwan
7 Wilkin
6A Winona
6A Wright
6A Yellow Medicine

MISSISSIPPI

3A Adams*
3A Alcorn
3A Amite*
3A Attala
3A Benton
3A Bolivar
3A Calhoun
3A Carroll
3A Chickasaw
3A Choctaw
3A Claiborne*
3A Clarke
3A Clay
3A Coahoma
3A Copiah*
3A Covington*
3A DeSoto
3A Forrest*
3A Franklin*
3A George*
3A Greene*
3A Grenada
2A Hancock*
2A Harrison*
3A Hinds*
3A Holmes
3A Humphreys
3A Issaquena
3A Itawamba
2A Jackson*
3A Jasper
3A Jefferson*
3A Jefferson Davis*
3A Jones*
3A Kemper
3A Lafayette
3A Lamar*
3A Lauderdale
3A Lawrence*
3A Leake
3A Lee
3A Leflore
3A Lincoln*
3A Lowndes
3A Madison
3A Marion*
3A Marshall
3A Monroe
3A Montgomery
3A Neshoba
3A Newton
3A Noxubee
3A Oktibbeha
3A Panola
2A Pearl River*
3A Perry*
3A Pike*

(continued)

TABLE R301.1—continued
CLIMATE ZONES, MOISTURE REGIMES, AND WARM-HUMID DESIGNATIONS BY STATE, COUNTY AND TERRITORY

3A Pontotoc
3A Prentiss
3A Quitman
3A Rankin*
3A Scott
3A Sharkey
3A Simpson*
3A Smith*
2A Stone*
3A Sunflower
3A Tallahatchie
3A Tate
3A Tippah
3A Tishomingo
3A Tunica
3A Union
3A Walthall*
3A Warren*
3A Washington
3A Wayne*
3A Webster
3A Wilkinson*
3A Winston
3A Yalobusha
3A Yazoo

MISSOURI

5A Adair
5A Andrew
5A Atchison
4A Audrain
4A Barry
4A Barton
4A Bates
4A Benton
4A Bollinger
4A Boone
5A Buchanan
4A Butler
5A Caldwell
4A Callaway
4A Camden
4A Cape Girardeau
4A Carroll
4A Carter
4A Cass
4A Cedar
5A Chariton
4A Christian
5A Clark
4A Clay
5A Clinton
4A Cole
4A Cooper
4A Crawford
4A Dade
4A Dallas
5A Daviess
5A DeKalb
4A Dent
4A Douglas
4A Dunklin
4A Franklin
4A Gasconade
5A Gentry
4A Greene
5A Grundy
5A Harrison
4A Henry
4A Hickory
5A Holt
4A Howard
4A Howell
4A Iron
4A Jackson
4A Jasper
4A Jefferson
4A Johnson
5A Knox
4A Laclede
4A Lafayette
4A Lawrence
5A Lewis
4A Lincoln
5A Linn
5A Livingston
5A Macon
4A Madison
4A Maries
5A Marion
4A McDonald
5A Mercer
4A Miller
4A Mississippi
4A Moniteau
4A Monroe
4A Montgomery
4A Morgan
4A New Madrid
4A Newton
5A Nodaway
4A Oregon
4A Osage
4A Ozark
4A Pemiscot
4A Perry
4A Pettis
4A Phelps
5A Pike
4A Platte
4A Polk
4A Pulaski
5A Putnam
5A Ralls
4A Randolph
4A Ray
4A Reynolds
4A Ripley
4A Saline
5A Schuyler
5A Scotland
4A Scott
4A Shannon
5A Shelby
4A St. Charles
4A St. Clair
4A St. Francois
4A St. Louis
4A St. Louis (city)
4A Ste. Genevieve
4A Stoddard
4A Stone
5A Sullivan
4A Taney
4A Texas
4A Vernon
4A Warren
4A Washington
4A Wayne
4A Webster
5A Worth
4A Wright

MONTANA

6B (all)

NEBRASKA

5A (all)

NEVADA

5B Carson City (city)
5B Churchill
3B Clark
5B Douglas
5B Elko
5B Esmeralda
5B Eureka
5B Humboldt
5B Lander
5B Lincoln
5B Lyon
5B Mineral
5B Nye
5B Pershing
5B Storey
5B Washoe
5B White Pine

NEW HAMPSHIRE

6A Belknap
6A Carroll
5A Cheshire
6A Coos
6A Grafton
5A Hillsborough
6A Merrimack
5A Rockingham
5A Strafford
6A Sullivan

NEW JERSEY

4A Atlantic
5A Bergen
4A Burlington
4A Camden
4A Cape May
4A Cumberland
4A Essex
4A Gloucester
4A Hudson
5A Hunterdon
5A Mercer
4A Middlesex
4A Monmouth
5A Morris
4A Ocean
5A Passaic
4A Salem
5A Somerset
5A Sussex
4A Union
5A Warren

NEW MEXICO

4B Bernalillo
5B Catron
3B Chaves
4B Cibola
5B Colfax
4B Curry
4B DeBaca
3B Dona Ana
3B Eddy
4B Grant
4B Guadalupe
5B Harding
3B Hidalgo
3B Lea
4B Lincoln
5B Los Alamos
3B Luna
5B McKinley
5B Mora
3B Otero
4B Quay
5B Rio Arriba
4B Roosevelt
5B Sandoval
5B San Juan
5B San Miguel
5B Santa Fe
4B Sierra
4B Socorro

(continued)

TABLE R301.1—continued
CLIMATE ZONES, MOISTURE REGIMES, AND WARM-HUMID DESIGNATIONS BY STATE, COUNTY AND TERRITORY

5B Taos
5B Torrance
4B Union
4B Valencia

NEW YORK

5A Albany
6A Allegany
4A Bronx
6A Broome
6A Cattaraugus
5A Cayuga
5A Chautauqua
5A Chemung
6A Chenango
6A Clinton
5A Columbia
5A Cortland
6A Delaware
5A Dutchess
5A Erie
6A Essex
6A Franklin
6A Fulton
5A Genesee
5A Greene
6A Hamilton
6A Herkimer
6A Jefferson
4A Kings
6A Lewis
5A Livingston
6A Madison
5A Monroe
6A Montgomery
4A Nassau
4A New York
5A Niagara
6A Oneida
5A Onondaga
5A Ontario
5A Orange
5A Orleans
5A Oswego
6A Otsego
5A Putnam
4A Queens
5A Rensselaer
4A Richmond
5A Rockland
5A Saratoga
5A Schenectady
6A Schoharie
6A Schuyler
5A Seneca
6A Steuben
6A St. Lawrence
4A Suffolk
6A Sullivan
5A Tioga
6A Tompkins
6A Ulster
6A Warren
5A Washington
5A Wayne
4A Westchester
6A Wyoming
5A Yates

NORTH CAROLINA

4A Alamance
4A Alexander
5A Alleghany
3A Anson
5A Ashe
5A Avery
3A Beaufort
4A Bertie
3A Bladen
3A Brunswick*
4A Buncombe
4A Burke
3A Cabarrus
4A Caldwell
3A Camden
3A Carteret*
4A Caswell
4A Catawba
4A Chatham
4A Cherokee
3A Chowan
4A Clay
4A Cleveland
3A Columbus*
3A Craven
3A Cumberland
3A Currituck
3A Dare
3A Davidson
4A Davie
3A Duplin
4A Durham
3A Edgecombe
4A Forsyth
4A Franklin
3A Gaston
4A Gates
4A Graham
4A Granville
3A Greene
4A Guilford
4A Halifax
4A Harnett
4A Haywood
4A Henderson
4A Hertford
3A Hoke
3A Hyde
4A Iredell
4A Jackson
3A Johnston
3A Jones
4A Lee
3A Lenoir
4A Lincoln
4A Macon
4A Madison
3A Martin
4A McDowell
3A Mecklenburg
5A Mitchell
3A Montgomery
3A Moore
4A Nash
3A New Hanover*
4A Northampton
3A Onslow*
4A Orange
3A Pamlico
3A Pasquotank
3A Pender*
3A Perquimans
4A Person
3A Pitt
4A Polk
3A Randolph
3A Richmond
3A Robeson
4A Rockingham
3A Rowan
4A Rutherford
3A Sampson
3A Scotland
3A Stanly
4A Stokes
4A Surry
4A Swain
4A Transylvania
3A Tyrrell
3A Union
4A Vance
4A Wake
4A Warren
3A Washington
5A Watauga
3A Wayne
4A Wilkes
3A Wilson
4A Yadkin
5A Yancey

NORTH DAKOTA

6A Adams
7 Barnes
7 Benson
6A Billings
7 Bottineau
6A Bowman
7 Burke
6A Burleigh
7 Cass
7 Cavalier
6A Dickey
7 Divide
6A Dunn
7 Eddy
6A Emmons
7 Foster
6A Golden Valley
7 Grand Forks
6A Grant
7 Griggs
6A Hettinger
7 Kidder
6A LaMoure
6A Logan
7 McHenry
6A McIntosh
6A McKenzie
7 McLean
6A Mercer
6A Morton
7 Mountrail
7 Nelson
6A Oliver
7 Pembina
7 Pierce
7 Ramsey
6A Ransom
7 Renville
6A Richland
7 Rolette
6A Sargent
7 Sheridan
6A Sioux
6A Slope
6A Stark
7 Steele
7 Stutsman
7 Towner
7 Traill
7 Walsh
7 Ward
7 Wells
7 Williams

OHIO

4A Adams
5A Allen

(continued)

TABLE R301.1—continued
CLIMATE ZONES, MOISTURE REGIMES, AND WARM-HUMID DESIGNATIONS BY STATE, COUNTY AND TERRITORY

5A Ashland
5A Ashtabula
5A Athens
5A Auglaize
5A Belmont
4A Brown
5A Butler
5A Carroll
5A Champaign
5A Clark
4A Clermont
5A Clinton
5A Columbiana
5A Coshocton
5A Crawford
5A Cuyahoga
5A Darke
5A Defiance
5A Delaware
5A Erie
5A Fairfield
5A Fayette
5A Franklin
5A Fulton
4A Gallia
5A Geauga
5A Greene
5A Guernsey
4A Hamilton
5A Hancock
5A Hardin
5A Harrison
5A Henry
5A Highland
5A Hocking
5A Holmes
5A Huron
5A Jackson
5A Jefferson
5A Knox
5A Lake
4A Lawrence
5A Licking
5A Logan
5A Lorain
5A Lucas
5A Madison
5A Mahoning
5A Marion
5A Medina
5A Meigs
5A Mercer
5A Miami
5A Monroe
5A Montgomery
5A Morgan
5A Morrow
5A Muskingum
5A Noble
5A Ottawa
5A Paulding
5A Perry
5A Pickaway
4A Pike
5A Portage
5A Preble
5A Putnam
5A Richland
5A Ross
5A Sandusky
4A Scioto
5A Seneca
5A Shelby
5A Stark
5A Summit
5A Trumbull
5A Tuscarawas
5A Union
5A Van Wert
5A Vinton
5A Warren
4A Washington
5A Wayne
5A Williams
5A Wood
5A Wyandot

OKLAHOMA

3A Adair
3A Alfalfa
3A Atoka
4B Beaver
3A Beckham
3A Blaine
3A Bryan
3A Caddo
3A Canadian
3A Carter
3A Cherokee
3A Choctaw
4B Cimarron
3A Cleveland
3A Coal
3A Comanche
3A Cotton
3A Craig
3A Creek
3A Custer
3A Delaware
3A Dewey
3A Ellis
3A Garfield
3A Garvin
3A Grady
3A Grant
3A Greer
3A Harmon
3A Harper
3A Haskell
3A Hughes
3A Jackson
3A Jefferson
3A Johnston
3A Kay
3A Kingfisher
3A Kiowa
3A Latimer
3A Le Flore
3A Lincoln
3A Logan
3A Love
3A Major
3A Marshall
3A Mayes
3A McClain
3A McCurtain
3A McIntosh
3A Murray
3A Muskogee
3A Noble
3A Nowata
3A Okfuskee
3A Oklahoma
3A Okmulgee
3A Osage
3A Ottawa
3A Pawnee
3A Payne
3A Pittsburg
3A Pontotoc
3A Pottawatomie
3A Pushmataha
3A Roger Mills
3A Rogers
3A Seminole
3A Sequoyah
3A Stephens
4B Texas
3A Tillman
3A Tulsa
3A Wagoner
3A Washington
3A Washita
3A Woods
3A Woodward

OREGON

5B Baker
4C Benton
4C Clackamas
4C Clatsop
4C Columbia
4C Coos
5B Crook
4C Curry
5B Deschutes
4C Douglas
5B Gilliam
5B Grant
5B Harney
5B Hood River
4C Jackson
5B Jefferson
4C Josephine
5B Klamath
5B Lake
4C Lane
4C Lincoln
4C Linn
5B Malheur
4C Marion
5B Morrow
4C Multnomah
4C Polk
5B Sherman
4C Tillamook
5B Umatilla
5B Union
5B Wallowa
5B Wasco
4C Washington
5B Wheeler
4C Yamhill

PENNSYLVANIA

5A Adams
5A Allegheny
5A Armstrong
5A Beaver
5A Bedford
5A Berks
5A Blair
5A Bradford
4A Bucks
5A Butler
5A Cambria
6A Cameron
5A Carbon
5A Centre
4A Chester
5A Clarion
6A Clearfield
5A Clinton
5A Columbia
5A Crawford
5A Cumberland
5A Dauphin
4A Delaware
6A Elk
5A Erie
5A Fayette
5A Forest
5A Franklin
5A Fulton
5A Greene

(continued)

TABLE R301.1—continued
CLIMATE ZONES, MOISTURE REGIMES, AND WARM-HUMID DESIGNATIONS BY STATE, COUNTY AND TERRITORY

5A Huntingdon
5A Indiana
5A Jefferson
5A Juniata
5A Lackawanna
5A Lancaster
5A Lawrence
5A Lebanon
5A Lehigh
5A Luzerne
5A Lycoming
6A McKean
5A Mercer
5A Mifflin
5A Monroe
4A Montgomery
5A Montour
5A Northampton
5A Northumberland
5A Perry
4A Philadelphia
5A Pike
6A Potter
5A Schuylkill
5A Snyder
5A Somerset
5A Sullivan
6A Susquehanna
6A Tioga
5A Union
5A Venango
5A Warren
5A Washington
6A Wayne
5A Westmoreland
5A Wyoming
4A York

RHODE ISLAND

5A (all)

SOUTH CAROLINA

3A Abbeville
3A Aiken
3A Allendale*
3A Anderson
3A Bamberg*
3A Barnwell*
3A Beaufort*
3A Berkeley*
3A Calhoun
3A Charleston*
3A Cherokee
3A Chester
3A Chesterfield
3A Clarendon
3A Colleton*
3A Darlington
3A Dillon
3A Dorchester*
3A Edgefield
3A Fairfield
3A Florence
3A Georgetown*
3A Greenville
3A Greenwood
3A Hampton*
3A Horry*
3A Jasper*
3A Kershaw
3A Lancaster
3A Laurens
3A Lee
3A Lexington
3A Marion
3A Marlboro
3A McCormick
3A Newberry
3A Oconee
3A Orangeburg
3A Pickens
3A Richland
3A Saluda
3A Spartanburg
3A Sumter
3A Union
3A Williamsburg
3A York

SOUTH DAKOTA

6A Aurora
6A Beadle
5A Bennett
5A Bon Homme
6A Brookings
6A Brown
6A Brule
6A Buffalo
6A Butte
6A Campbell
5A Charles Mix
6A Clark
5A Clay
6A Codington
6A Corson
6A Custer
6A Davison
6A Day
6A Deuel
6A Dewey
5A Douglas
6A Edmunds
6A Fall River
6A Faulk
6A Grant
5A Gregory
6A Haakon
6A Hamlin
6A Hand
6A Hanson
6A Harding
6A Hughes
5A Hutchinson
6A Hyde
5A Jackson
6A Jerauld
6A Jones
6A Kingsbury
6A Lake
6A Lawrence
6A Lincoln
6A Lyman
6A Marshall
6A McCook
6A McPherson
6A Meade
5A Mellette
6A Miner
6A Minnehaha
6A Moody
6A Pennington
6A Perkins
6A Potter
6A Roberts
6A Sanborn
6A Shannon
6A Spink
6A Stanley
6A Sully
5A Todd
5A Tripp
6A Turner
5A Union
6A Walworth
5A Yankton
6A Ziebach

TENNESSEE

4A Anderson
4A Bedford
4A Benton
4A Bledsoe
4A Blount
4A Bradley
4A Campbell
4A Cannon
4A Carroll
4A Carter
4A Cheatham
3A Chester
4A Claiborne
4A Clay
4A Cocke
4A Coffee
3A Crockett
4A Cumberland
4A Davidson
4A Decatur
4A DeKalb
4A Dickson
3A Dyer
3A Fayette
4A Fentress
4A Franklin
4A Gibson
4A Giles
4A Grainger
4A Greene
4A Grundy
4A Hamblen
4A Hamilton
4A Hancock
3A Hardeman
3A Hardin
4A Hawkins
3A Haywood
3A Henderson
4A Henry
4A Hickman
4A Houston
4A Humphreys
4A Jackson
4A Jefferson
4A Johnson
4A Knox
3A Lake
3A Lauderdale
4A Lawrence
4A Lewis
4A Lincoln
4A Loudon
4A Macon
3A Madison
4A Marion
4A Marshall
4A Maury
4A McMinn
3A McNairy
4A Meigs
4A Monroe
4A Montgomery
4A Moore
4A Morgan
4A Obion
4A Overton
4A Perry
4A Pickett
4A Polk
4A Putnam
4A Rhea

(continued)

TABLE R301.1—continued
CLIMATE ZONES, MOISTURE REGIMES, AND WARM-HUMID DESIGNATIONS BY STATE, COUNTY AND TERRITORY

4A Roane
4A Robertson
4A Rutherford
4A Scott
4A Sequatchie
4A Sevier
3A Shelby
4A Smith
4A Stewart
4A Sullivan
4A Sumner
3A Tipton
4A Trousdale
4A Unicoi
4A Union
4A Van Buren
4A Warren
4A Washington
4A Wayne
4A Weakley
4A White
4A Williamson
4A Wilson

TEXAS

2A Anderson*
3B Andrews
2A Angelina*
2A Aransas*
3A Archer
4B Armstrong
2A Atascosa*
2A Austin*
4B Bailey
2B Bandera
2A Bastrop*
3B Baylor
2A Bee*
2A Bell*
2A Bexar*
3A Blanco*
3B Borden
2A Bosque*
3A Bowie*
2A Brazoria*
2A Brazos*
3B Brewster
4B Briscoe
2A Brooks*
3A Brown*
2A Burleson*
3A Burnet*
2A Caldwell*
2A Calhoun*
3B Callahan
2A Cameron*
3A Camp*
4B Carson
3A Cass*
4B Castro
2A Chambers*
2A Cherokee*
3B Childress
3A Clay
4B Cochran
3B Coke
3B Coleman
3A Collin*
3B Collingsworth
2A Colorado*
2A Comal*
3A Comanche*
3B Concho
3A Cooke
2A Coryell*
3B Cottle
3B Crane
3B Crockett
3B Crosby
3B Culberson
4B Dallam
3A Dallas*
3B Dawson
4B Deaf Smith
3A Delta
3A Denton*
2A DeWitt*
3B Dickens
2B Dimmit
4B Donley
2A Duval*
3A Eastland
3B Ector
2B Edwards
3A Ellis*
3B El Paso
3A Erath*
2A Falls*
3A Fannin
2A Fayette*
3B Fisher
4B Floyd
3B Foard
2A Fort Bend*
3A Franklin*
2A Freestone*
2B Frio
3B Gaines
2A Galveston*
3B Garza
3A Gillespie*
3B Glasscock
2A Goliad*
2A Gonzales*
4B Gray
3A Grayson
3A Gregg*
2A Grimes*
2A Guadalupe*
4B Hale
3B Hall
3A Hamilton*
4B Hansford
3B Hardeman
2A Hardin*
2A Harris*
3A Harrison*
4B Hartley
3B Haskell
2A Hays*
3B Hemphill
3A Henderson*
2A Hidalgo*
2A Hill*
4B Hockley
3A Hood*
3A Hopkins*
2A Houston*
3B Howard
3B Hudspeth
3A Hunt*
4B Hutchinson
3B Irion
3A Jack
2A Jackson*
2A Jasper*
3B Jeff Davis
2A Jefferson*
2A Jim Hogg*
2A Jim Wells*
3A Johnson*
3B Jones
2A Karnes*
3A Kaufman*
3A Kendall*
2A Kenedy*
3B Kent
3B Kerr
3B Kimble
3B King
2B Kinney
2A Kleberg*
3B Knox
3A Lamar*
4B Lamb
3A Lampasas*
2B La Salle
2A Lavaca*
2A Lee*
2A Leon*
2A Liberty*
2A Limestone*
4B Lipscomb
2A Live Oak*
3A Llano*
3B Loving
3B Lubbock
3B Lynn
2A Madison*
3A Marion*
3B Martin
3B Mason
2A Matagorda*
2B Maverick
3B McCulloch
2A McLennan*
2A McMullen*
2B Medina
3B Menard
3B Midland
2A Milam*
3A Mills*
3B Mitchell
3A Montague
2A Montgomery*
4B Moore
3A Morris*
3B Motley
3A Nacogdoches*
3A Navarro*
2A Newton*
3B Nolan
2A Nueces*
4B Ochiltree
4B Oldham
2A Orange*
3A Palo Pinto*
3A Panola*
3A Parker*
4B Parmer
3B Pecos
2A Polk*
4B Potter
3B Presidio
3A Rains*
4B Randall
3B Reagan
2B Real
3A Red River*
3B Reeves
2A Refugio*
4B Roberts
2A Robertson*
3A Rockwall*
3B Runnels
3A Rusk*
3A Sabine*
3A San Augustine*
2A San Jacinto*
2A San Patricio*

(continued)

TABLE R301.1—continued
CLIMATE ZONES, MOISTURE REGIMES, AND WARM-HUMID DESIGNATIONS BY STATE, COUNTY AND TERRITORY

3A San Saba*
3B Schleicher
3B Scurry
3B Shackelford
3A Shelby*
4B Sherman
3A Smith*
3A Somervell*
2A Starr*
3A Stephens
3B Sterling
3B Stonewall
3B Sutton
4B Swisher
3A Tarrant*
3B Taylor
3B Terrell
3B Terry
3B Throckmorton
3A Titus*
3B Tom Green
2A Travis*
2A Trinity*
2A Tyler*
3A Upshur*
3B Upton
2B Uvalde
2B Val Verde
3A Van Zandt*
2A Victoria*
2A Walker*
2A Waller*
3B Ward
2A Washington*
2B Webb
2A Wharton*
3B Wheeler
3A Wichita
3B Wilbarger
2A Willacy*
2A Williamson*
2A Wilson*
3B Winkler
3A Wise
3A Wood*
4B Yoakum
3A Young
2B Zapata
2B Zavala

UTAH

5B Beaver
6B Box Elder
6B Cache
6B Carbon
6B Daggett
5B Davis
6B Duchesne
5B Emery
5B Garfield
5B Grand
5B Iron
5B Juab
5B Kane
5B Millard
6B Morgan
5B Piute
6B Rich
5B Salt Lake
5B San Juan
5B Sanpete
5B Sevier
6B Summit
5B Tooele
6B Uintah
5B Utah
6B Wasatch
3B Washington
5B Wayne
5B Weber

VERMONT

6A (all)

VIRGINIA

4A (all)

WASHINGTON

5B Adams
5B Asotin
5B Benton
5B Chelan
4C Clallam
4C Clark
5B Columbia
4C Cowlitz
5B Douglas
6B Ferry
5B Franklin
5B Garfield
5B Grant
4C Grays Harbor
4C Island
4C Jefferson
4C King
4C Kitsap
5B Kittitas
5B Klickitat
4C Lewis
5B Lincoln
4C Mason
6B Okanogan
4C Pacific
6B Pend Oreille
4C Pierce
4C San Juan
4C Skagit
5B Skamania
4C Snohomish
5B Spokane
6B Stevens
4C Thurston
4C Wahkiakum
5B Walla Walla
4C Whatcom
5B Whitman
5B Yakima

WEST VIRGINIA

5A Barbour
4A Berkeley
4A Boone
4A Braxton
5A Brooke
4A Cabell
4A Calhoun
4A Clay
5A Doddridge
5A Fayette
4A Gilmer
5A Grant
5A Greenbrier
5A Hampshire
5A Hancock
5A Hardy
5A Harrison
4A Jackson
4A Jefferson
4A Kanawha
5A Lewis
4A Lincoln
4A Logan
5A Marion
5A Marshall
4A Mason
4A McDowell
4A Mercer
5A Mineral
4A Mingo
5A Monongalia
4A Monroe
4A Morgan
5A Nicholas
5A Ohio
5A Pendleton
4A Pleasants
5A Pocahontas
5A Preston
4A Putnam
5A Raleigh
5A Randolph
4A Ritchie
4A Roane
5A Summers
5A Taylor
5A Tucker
4A Tyler
5A Upshur
4A Wayne
5A Webster
5A Wetzel
4A Wirt
4A Wood
4A Wyoming

WISCONSIN

6A Adams
7 Ashland
6A Barron
7 Bayfield
6A Brown
6A Buffalo
7 Burnett
6A Calumet
6A Chippewa
6A Clark
6A Columbia
6A Crawford
6A Dane
6A Dodge
6A Door
7 Douglas
6A Dunn
6A Eau Claire
7 Florence
6A Fond du Lac
7 Forest
6A Grant
6A Green
6A Green Lake
6A Iowa
7 Iron
6A Jackson
6A Jefferson
6A Juneau
6A Kenosha
6A Kewaunee
6A La Crosse
6A Lafayette
7 Langlade
7 Lincoln
6A Manitowoc
6A Marathon
6A Marinette
6A Marquette
6A Menominee
6A Milwaukee
6A Monroe
6A Oconto
7 Oneida
6A Outagamie

(continued)

TABLE R301.1—continued
CLIMATE ZONES, MOISTURE REGIMES, AND WARM-HUMID DESIGNATIONS BY STATE, COUNTY AND TERRITORY

6A Ozaukee
6A Pepin
6A Pierce
6A Polk
6A Portage
7 Price
6A Racine
6A Richland
6A Rock
6A Rusk
6A Sauk
7 Sawyer
6A Shawano
6A Sheboygan
6A St. Croix
7 Taylor
6A Trempealeau
6A Vernon
7 Vilas
6A Walworth
7 Washburn
6A Washington
6A Waukesha
6A Waupaca
6A Waushara
6A Winnebago
6A Wood

WYOMING

6B Albany
6B Big Horn
6B Campbell
6B Carbon
6B Converse
6B Crook
6B Fremont
5B Goshen
6B Hot Springs
6B Johnson
6B Laramie
7 Lincoln
6B Natrona
6B Niobrara
6B Park
5B Platte
6B Sheridan
7 Sublette
6B Sweetwater
7 Teton
6B Uinta
6B Washakie
6B Weston

US TERRITORIES

AMERICAN SAMOA

1A (all)*

GUAM

1A (all)*

NORTHERN MARIANA ISLANDS

1A (all)*

PUERTO RICO

1A (all)*

VIRGIN ISLANDS

1A (all)*

SECTION R302 DESIGN CONDITIONS

R302.1 Interior design conditions. The interior design temperatures used for heating and cooling load calculations shall be a maximum of 72°F (22°C) for heating and minimum of 75°F (24°C) for cooling.

❖ While the previous sections of IECC Chapter 3 address outdoor design conditions, this section provides the interior conditions that will be used for properly sizing the mechanical equipment. The proper sizing of mechanical equipment (see Section R403.7) can vary depending on the selected design conditions. While the code does address oversizing equipment, it is not enforceable without establishing the exact design parameters. This section is included in the code only for system sizing. It does not affect the interior design temperatures required by other codes such as Section 1204 of the IBC or Section 602.2 of the *International Property Maintenance Code*® (IPMC®).

The 75°F (24°C) design temperature for cooling was used in the code so that it coordinated with both the ASHRAE *Handbook of Fundamentals* and the Air Conditioning Contractors of America (ACCA) Manuals S and J, established standards that deal with equipment sizing.

SECTION R303 MATERIALS, SYSTEMS AND EQUIPMENT

R303.1 Identification. Materials, systems and equipment shall be identified in a manner that will allow a determination of compliance with the applicable provisions of this code.

❖ This section is intended to make certain that sufficient information exists to determine compliance with the code during the plan review and field inspection phases. The permittee can submit the required equipment and materials information on the building plans, specification sheets or schedules, or in any other way that allows the code official to clearly identify which specifications apply to which portions of the building (i.e., which parts of the building are insulated to the levels listed). Materials information includes envelope insulation levels, glazing assembly *U*-factors and duct and piping insulation levels. Equipment information includes heating and cooling equipment and appliance efficiencies where high-efficiency equipment is claimed to meet code requirements.

This section contains specific material, equipment and system identification requirements for the approval and installation of the items required by the code. Although the means for permanent marking (tag, stencil, label, stamp, sticker, bar code, etc.) is often determined and applied by the manufacturer, if there is any uncertainty about the product, the mark is subject to the approval of the code official.

R303.1.1 Building thermal envelope insulation. An *R*-value identification mark shall be applied by the manufacturer to each piece of *building thermal envelope* insulation 12 inches (305 mm) or greater in width. Alternately, the insulation installers shall provide a certification listing the type, manufacturer and *R*-value of insulation installed in each element of the *building thermal envelope*. For blown or sprayed insulation (fiberglass and cellulose), the initial installed thickness, settled thickness, settled *R*-value, installed density, coverage area and number of bags installed shall be *listed* on the certification. For sprayed polyurethane foam (SPF) insulation, the installed thickness of the areas covered and *R*-value of installed thickness shall be *listed* on the certification. For insulated siding, the *R*-value shall be labeled on the product's package and shall be *listed* on the certification. The insulation installer shall sign, date and post the certification in a conspicuous location on the job site.

❖ The thermal performance of insulation is rated in terms of *R*-value. For products lacking an *R*-value identification, the installer (or builder) must provide the insulation performance data. For example, some insulation materials, such as foamed-in-place urethane, can be installed in wall, floor and cathedral ceiling cavities. These products are not labeled, as is batt insulation, nor is it appropriate for them to be evaluated as required in the code for blown or sprayed insulation; however, the installer must certify the type, thickness and *R*-value of these materials.

The *R*-value of loose-fill insulation (blown or sprayed) is dependent on both the installed thickness and density (number of bags used). Therefore, loose-fill insulation cannot be directly labeled by the manufacturer. Many blown insulation products carry a manufacturer's *R*-value guarantee when installed to a designated thickness, "inches = *R*-value." Blown insulation products lacking this manufacturer's guarantee can be subjected to special inspection and testing; what is referred to as "cookie cutting." Cookie cutting involves extracting a column of insulation with a cylinder to determine its density. The insulation depth and density must yield the specified *R*-value according to the manufacturer's bag label specification.

The code and Federal Trade Commission Rule 460 require that installers of insulation in homes, apartments and manufactured housing units report this information to the authority having jurisdiction in the form of a certification posted in a conspicuous location (see Commentary Figure R303.1.1).

R303.1.1.1 Blown or sprayed roof/ceiling insulation. The thickness of blown-in or sprayed roof/ceiling insulation (fiberglass or cellulose) shall be written in inches (mm) on markers that are installed at least one for every 300 square feet (28 m^2) throughout the attic space. The markers shall be affixed to the trusses or joists and marked with the minimum initial installed thickness with numbers not less than 1 inch (25 mm) in height. Each marker shall face the attic access opening. Spray polyurethane foam thickness and installed *R*-value shall be *listed* on certification provided by the insulation installer.

❖ To help verify the installed *R*-value of blown-in or spray-applied insulation, the installer must certify the following information in a signed statement posted in a conspicuous place (see Section R303.1.1):

- The type of insulation used and manufacturer.
- The insulation's coverage per bag (the number of bags required to result in a given *R*-value for a given area), as well as the settled *R*-value.
- The initial and settled thickness.
- The number of bags installed.

Under circumstances where the insulation *R*-value is guaranteed, only the initial thickness is required on the certification.

This section helps demonstrate compliance and enforcement of the provisions found in Section R303.1.1. To assist with application and enforcement, loose-fill ceiling insulation also requires thickness markers that are attached to the framing and face the attic access. In a large space, markers placed evenly about every 17 feet (5182 mm) (with some markers at the edge of the space) will meet this requirement. For sprayed polyurethane, such markers are not effective. When using this product, the code requires that the measured thickness and *R*-value be recorded on the certificate.

R303.1.2 Insulation mark installation. Insulating materials shall be installed such that the manufacturer's *R*-value mark is readily observable upon inspection.

❖ For batt insulation, manufacturers' *R*-value designations and stripe codes are often printed directly on the insulation. Where possible, the insulation must be installed so these designations are readable. Backed floor batts can be installed with the designation

This Attic Has Been Insulated To

INSULATION CONTRACTORS ICAA ASSOCIATION OF AMERICA

R- ☐

By A Professional Insulation Contractor

The insulation in this attic was installed by a qualified professional Contractor to the R-value stated above

CIMA

INSULATE TODAY SAVE TOMORROW

NAIMA
NORTH AMERICAN INSULATION MANUFACTURERS ASSOCIATION

Certificate of Insulation

BUILDING ADDRESS: ______________________

Installation Date ______________

CONTRACTOR: ______________________

License# ______________

Area Insulated	R-Value	Installed Thickness	Settled Thickness	Installed Density	No. Bags	Sq. Ft.
Attic						
Walls						
Floors						

I, ______________________________, (print name) certify that this residence/building has been insulated to the stated R-value and that the installation is in conformance with all applicable codes, standards, regulations and specifications.

Authorized Signature ______________________ Date __________

Figure R303.1.1
SAMPLE CERTIFICATE OF INSULATION

(Logos courtesy of Cellulose Insulation Manufacturers Association, http://cellulose.org, Insulation Contractors Association of America, www.insulate.org, and North American Insulation Manufacturers Association, www.NAIMA.org)

against the underfloor, which means it would not be visible. In those cases, the *R*-value must be certified by the installer or be validated by some other means (see commentary, Section R303.1.1).

R303.1.3 Fenestration product rating. *U*-factors of fenestration products (windows, doors and skylights) shall be determined in accordance with NFRC 100.

Exception: Where required, garage door *U*-factors shall be determined in accordance with either NFRC 100 or ANSI/DASMA 105.

U-factors shall be determined by an accredited, independent laboratory, and *labeled* and certified by the manufacturer.

Products lacking such a *labeled U*-factor shall be assigned a default *U*-factor from Table R303.1.3(1) or R303.1.3(2). The solar heat gain coefficient (SHGC) and *visible transmittance* (VT) of glazed fenestration products (windows, glazed doors and skylights) shall be determined in accordance with NFRC 200 by an accredited, independent laboratory, and *labeled* and certified by the manufacturer. Products lacking such a *labeled* SHGC or VT shall be assigned a default SHGC or VT from Table R303.1.3(3).

❖ Until recently, the buyers of fenestration products received energy performance information in a variety of ways. Some manufacturers described performance by showing *R*-values of the glass. While the glass might have been a good performer, the rating did not include the effects of the frame. Other manufacturers touted the insulating value of different window components, but these, too, did not reflect the total window system performance. When manufacturers rated the entire product, some used test laboratory measurements and others used computer calculations. Even among those using test laboratory reports, the test laboratories often tested the products under different procedures, making an "apples-to-apples" comparison difficult. The different rating methods confused builders and consumers. They also created headaches for manufacturers trying to differentiate the performance of their products from the performance of their competitors' products.

The National Fenestration Rating Council (NFRC) has developed a fenestration energy rating system based on whole-product performance. This accurately accounts for the energy-related effects of all the product's component parts and prevents information about a single component from being compared in a misleading way to other whole-product properties. With energy ratings based on whole-product performance, the NFRC helps builders, designers and consumers directly compare products with different construction details and attributes.

Products that have been rated by NFRC-approved testing laboratories and certified by NFRC-accredited independent certification and inspection agencies carry a temporary and permanent label featuring the "NFRC-certified" mark. With this mark, the manufacturer stipulates that the energy performance of the product was determined according to NFRC rules and procedures.

By certifying and labeling their products, manufacturers are demonstrating their commitment to providing accurate energy and energy-related performance information. The code purposely sets the default values fairly high. This helps to encourage the use of products that have been tested and also ensures that products that have little energy-saving values are not used inappropriately in the various climate zones. By setting the default value so high, it will also prevent someone from removing the label from a tested window and then using the default values. Therefore, the default values are most representative of the lower end of the energy-efficient products.

A product that is not NFRC certified and does not exactly match the specifications in Tables R303.1.3(1) and R303.1.3(2) must use the tabular specification for the product it most closely resembles. In the absence of tested *U*-factors, the default *U*-factor for doors containing glazing can be a combination of the glazing and door *U*-factor as described in the definition for "*U*-factor" (see commentary, Section R202, "*U*-factor"). NFRC procedures determine *U*-factor and SHGC ratings based on the whole fenestration assembly [untested fenestration products have default *U*-factors and SHGCs assigned as described in the commentary to Tables R303.1.3(1) through R303.1.3(3)]. During construction inspection, the label on each glazing assembly should be checked for conformance to the *U*-factor specified on the approved plans. These labels must be left on the glazing until after the building has been inspected for compliance. A sample NFRC label is shown in Commentary Figure R303.1.3(1).

Products certified according to NFRC procedures are listed in the *Certified Products Directory*. The directory is published annually and contains energy performance information for over 1.4 million fenestration product options listed by over 450 manufacturers. When using the directory or shopping for NFRC-certified products, it is important to note:

1. A product is considered to be NFRC certified only if it carries the NFRC label. Simply being listed in this directory is not enough.

2. The NFRC-certified mark does not signify that the product meets any energy-efficiency standards or criteria.

3. The NFRC neither sets minimum performance standards nor mandates specific performance levels. Rather, NFRC ratings can be used to determine whether a product meets a state or local code or other performance requirement and to compare the energy performance of different products during plan review. For questions about the NFRC and its rating and labeling system, more information is available on the organization's website at www.nfrc.org. The NFRC adopted a new energy performance

label in 2005. It lists the manufacturer, describes the product, provides a source for additional information and includes ratings for one or more energy performance characteristics.

The IECC offers an alternative to NFRC-certified glazed fenestration product *U*-factor ratings. In the absence of *U*-factors based on NFRC test procedures, the default *U*-factors in Table R303.1.3(1) must be used. When a composite of materials from two different product types is used, the code official should be consulted regarding how the product will be rated. Generally, the product must be assigned the higher *U*-factor, although an average based on the *U*-factors and areas may be acceptable in some cases.

The product cannot receive credit for a feature that cannot be seen. Because performance features such as argon fill and low-emissivity coatings for glass are not visually verifiable, they do not receive credit in the default tables. Tested *U*-factors for these windows are often lower, so using tested *U*-factors is to the applicant's advantage. Commentary Figure R303.1.3(2) illustrates visually verifiable window characteristics, among other various window performance, function and cost considerations.

A single-glazed window with an installed storm window may be considered a double-glazed assembly and use the corresponding *U*-factor from the default table. For example, the *U*-factor 0.80 in Table R303.1.3(1) applies to a single-glazed, metal window without a thermal break (but with an installed storm window). If the storm window was not installed, the *U*-factor would be 1.20.

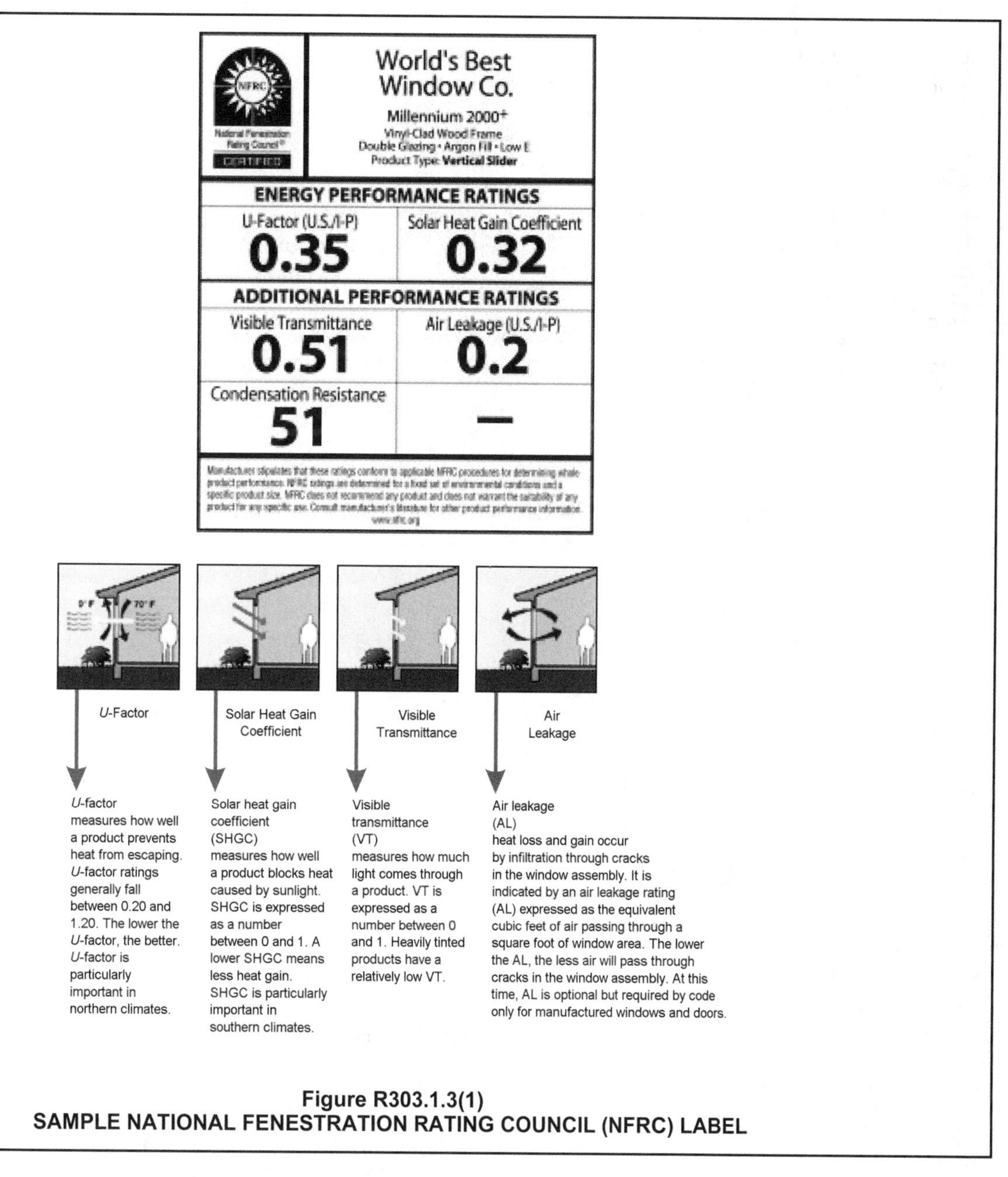

Figure R303.1.3(1)
SAMPLE NATIONAL FENESTRATION RATING COUNCIL (NFRC) LABEL

TABLE R303.1.3(1)
DEFAULT GLAZED FENESTRATION *U*-FACTORS

FRAME TYPE	SINGLE PANE	DOUBLE PANE	SKYLIGHT	
			Single	Double
Metal	1.20	0.80	2.00	1.30
Metal with Thermal Break	1.10	0.65	1.90	1.10
Nonmetal or Metal Clad	0.95	0.55	1.75	1.05
Glazed Block	0.60			

TABLE R303.1.3(2)
DEFAULT DOOR *U*-FACTORS

DOOR TYPE	*U*-FACTOR
Uninsulated Metal	1.20
Insulated Metal	0.60
Wood	0.50
Insulated, nonmetal edge, max 45% glazing, any glazing double pane	0.35

❖ Door *U*-factors in Table R303.1.3(2) should be used wherever NFRC-certified ratings are not available. There are a few other aspects to note about doors. Opaque door *U*-factors must include the effects of the door edge and the frame. Calculating *U*-factors based on a cross section through the insulated portion is not acceptable. To take credit for a thermal break, the door must have a thermal break in both the door slab and in the frame. The values in the table are founded on principles established in the ASHRAE *Handbook of Fundamentals*.

TABLE R303.1.3(3)
DEFAULT GLAZED FENESTRATION SHGC AND VT

	SINGLE GLAZED		DOUBLE GLAZED		GLAZED BLOCK
	Clear	Tinted	Clear	Tinted	
SHGC	0.8	0.7	0.7	0.6	0.6
VT	0.6	0.3	0.6	0.3	0.6

❖ This table offers an alternative to NFRC-certified SHGC and visible transmittance (VT) values based on visually verifiable characteristics of the fenestration product. The SHGC is the fraction of incident solar radiation absorbed and directly transmitted by the window area, then subsequently reradiated, conducted or convected inward. SHGC is a ratio, expressed as a number between 0 and 1. The lower a window's SHGC, the less solar heat it transmits. The VT is the ratio of visible light entering the space through the fenestration product assembly to the incident visible light. VT includes the effects of glazing material and frame and is expressed as a number between 0 and 1.

An SHGC of 0.40 or less is recommended in cooling-dominated climates (Climate Zones 1–3). In heating-dominated climates, a high SHGC increases passive solar gain for the heating but reduces cooling season performance. A low SHGC improves cooling season performance but reduces passive solar gains for heating.

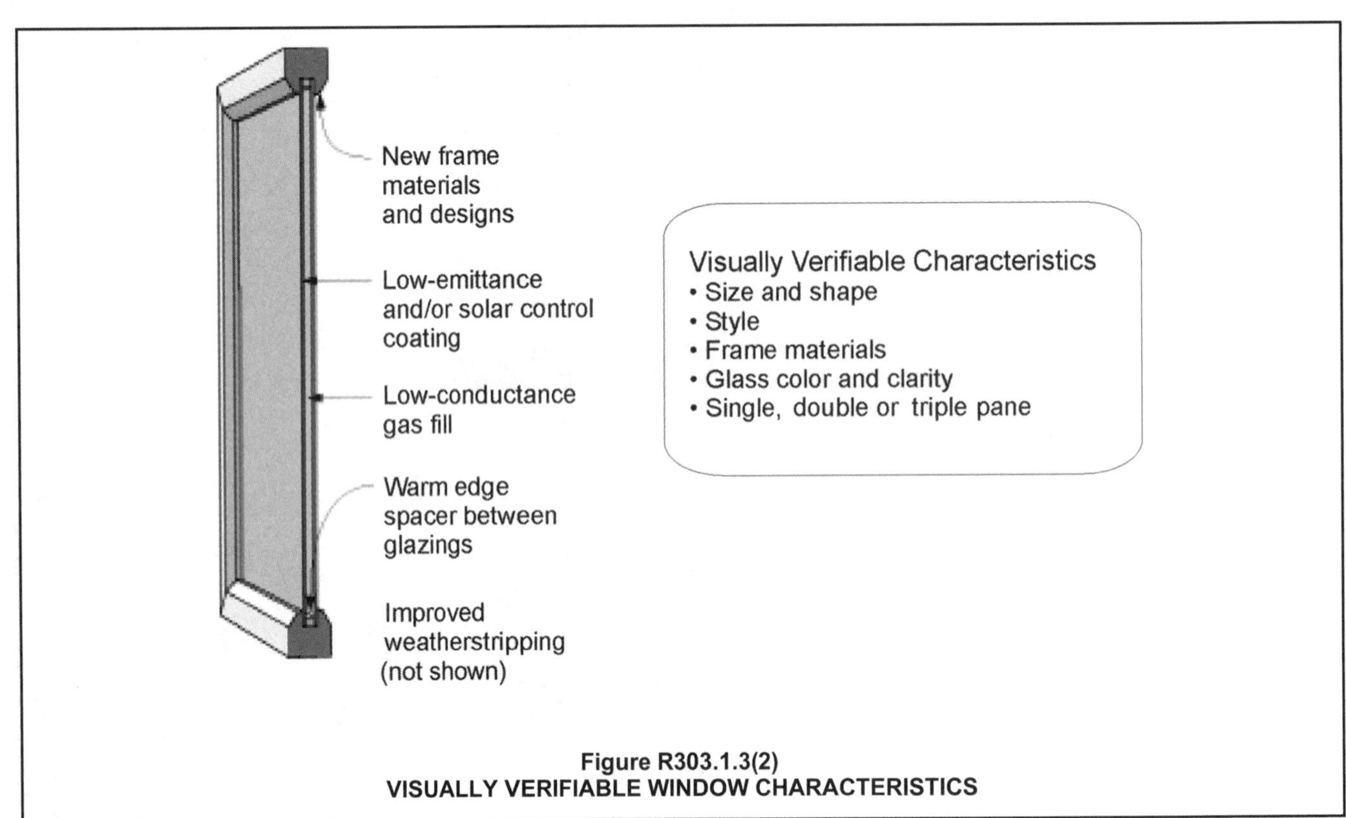

Figure R303.1.3(2)
VISUALLY VERIFIABLE WINDOW CHARACTERISTICS

R303.1.4 Insulation product rating. The thermal resistance (*R*-value) of insulation shall be determined in accordance with the U.S. Federal Trade Commission *R*-value rule (CFR Title 16, Part 460) in units of h • ft^2 • °F/Btu at a mean temperature of 75°F (24°C).

❖ This section brings two important requirements to the code.

First, the Federal Trade Commission *R*-value rule details test standards for insulation. The test standards are specific to the type of insulation and intended use. This clarifies any questions on the rating conditions to be used for insulation materials.

Second, the text above specifies the rating temperature to be used when evaluating the *R*-value of the product, providing consistency not currently in the code. Insulation products sometimes list several *R*-values based on different test temperatures. This eliminates any question as to which *R*-value to use. The temperature selected is a standard rating condition.

R303.1.4.1 Insulated siding. The thermal resistance (*R*-value) of insulated siding shall be determined in accordance with ASTM C 1363. Installation for testing shall be in accordance with the manufacturer's instructions.

❖ Insulated siding is a unique product that requires a special test method for determination of *R*-value. This test method is ASTM C 1363, *Standard Test Method for Thermal Performance of Building Materials and Envelope Assemblies by Means of a Hot Box Apparatus*.

R303.2 Installation. Materials, systems and equipment shall be installed in accordance with the manufacturer's instructions and the *International Building Code* or *International Residential Code*, as applicable.

❖ Manufacturers' installation instructions are thoroughly evaluated by the listing agency, verifying that a safe installation is prescribed. When an appliance is tested to obtain a listing and label, the approval agency installs the appliance in accordance with the manufacturer's instructions. The appliance is tested under these conditions; thus, the installation instructions become an integral part of the labeling process. The listing agency can require that the manufacturer alter, delete or add information to the instructions as necessary to achieve compliance with applicable standards and code requirements.

Manufacturers' installation instructions are enforceable extensions of the code and must be in the hands of the code official when an inspection takes place. Inspectors must carefully and completely read and comprehend the manufacturer's instructions in order to properly perform an installation inspection. In some cases, the code will specifically address an installation requirement that is also addressed in the manufacturer's installation instructions. The code requirement may be the same or may exceed the requirement in the manufacturer's installation instructions. The manufacturer's installation instructions could contain requirements that exceed those in the code. In such cases, the more restrictive requirements would apply (see commentary, Section 106).

Even if an installation appears to be in compliance with the manufacturer's instructions, the installation cannot be completed or approved until all associated components, connections and systems that serve the appliance or equipment are also in compliance with the requirements of the applicable *International Codes*® (I-Codes®) of reference. For example, a gas-fired boiler installation must not be approved if the boiler is connected to a deteriorated, undersized or otherwise unsafe chimney or vent. Likewise, the same installation must not be approved if the existing gas piping has insufficient capacity to supply the boiler load or if the electrical supply circuit is inadequate or unsafe.

Manufacturers' installation instructions are often updated and changed for various reasons, such as changes in the appliance, equipment or material design; revisions to the product standards; or as a result of field experiences related to existing installations. The code official should stay abreast of any changes by reviewing the manufacturer's instructions for every installation.

R303.2.1 Protection of exposed foundation insulation. Insulation applied to the exterior of basement walls, crawlspace walls and the perimeter of slab-on-grade floors shall have a rigid, opaque and weather-resistant protective covering to prevent the degradation of the insulation's thermal performance. The protective covering shall cover the exposed exterior insulation and extend not less than 6 inches (153 mm) below grade.

❖ The ultimate performance of insulation material is directly proportional to the workmanship involved in the material's initial installation, as well as the material's integrity over the life of the structure. Accordingly, foundation wall and slab-edge insulation materials installed in the vicinity of the exterior grade line require protection from damage that could occur from contact by lawn-mowing and maintenance equipment, garden hoses, garden tools, perimeter landscape materials, etc. In addition, the long-term thermal performance of foam-plastic insulation materials is adversely affected by direct exposure to the sun. To protect the insulation from sunlight and physical damage, it must have a protective covering that is inflexible, puncture resistant, opaque and weather resistant.

R303.3 Maintenance information. Maintenance instructions shall be furnished for equipment and systems that require preventive maintenance. Required regular maintenance actions shall be clearly stated and incorporated on a readily accessible label. The label shall include the title or publication number for the operation and maintenance manual for that particular model and type of product.

❖ This section establishes an owner's responsibility for maintaining the building in accordance with the requirements of the code and other referenced standards. This section requires, among others, that mechanical and service water heating equipment and appliance maintenance information be made available to the owner/operator. This section does not require that labels be added to existing equipment; having the manufacturer's maintenance literature is usually sufficient to meet this requirement. During final occupancy inspection, the mechanical equipment and water heater should be inspected to verify that the information is taped to each unit or referenced on a label mounted in a conspicuous location on the units.

The code official has the authority to rule on the performance of maintenance work when equipment functions would be affected by such work. He or she also has the authority to require a building and its energy-using systems to be maintained in compliance with the public health and safety provisions required by other I-Codes.

Bibliography

The following resource materials were used in the preparation of the commentary for this chapter of the code.

Morales, C.G.; and A. J. Malavé, A.J. *Energy Modeling of Low Income Residencies*. http://library.witpress.com/pages/Papelnfo.asp?PaperID=22547. (The paper above is not free. The proponents will send a Puerto Rico Energy Center presentation done for DOE that summarizes that work to anyone who requests this by email.)

Chapter 4 [RE]: Residential Energy Efficiency

General Comments

Chapter 4 [RE] contains the energy-efficiency-related requirements for the design and construction of residential buildings regulated under the code. The applicable portions of the building must comply with the provisions within this chapter for energy efficiency.

Section R401 contains the scope and application of the chapter and also regulates a certificate that must be left with the building. In addition, Section R401 provides an alternative for residential buildings in the Tropical Zone. Section R402 contains the insulation *R*-value requirements for the building envelope, which includes the roof/ceiling assembly, wall assembly and floor assembly, as well as fenestration requirements. Section R403 contains the requirements for heating and cooling systems and includes requirements for equipment sizing, duct installation, piping insulation and the requirements for controls. Section R404 contains electrical power and lighting systems information. Section R405 provides a performance option that will not only provide an additional means of demonstrating compliance with the code but also allows trade-offs between the various systems.

Purpose

This chapter details requirements for the portions of the building and building systems that impact energy use in new residential construction and promotes the effective use of energy. The provisions in the chapter promote energy efficiency in the building envelope, the heating and cooling system, the service water-heating system, and the electrical power and lighting system of the building. Compliance with this chapter will provide a minimum level of energy efficiency for new construction. Greater levels of efficiency can be installed to decrease the energy use of new construction.

SECTION R401 GENERAL

R401.1 Scope. This chapter applies to residential buildings.

❖ This chapter covers "residential" buildings as they are defined in Chapter 2 [RE]. A review of the definition is important because it does not include all buildings that are classified as "residential" by the *International Building Code®* (IBC®). Hotels, motels and other transient occupancies that are classified as a Group R-1 occupancy by the IBC are not included in the definition of "Residential" and would, therefore, need to comply with the "Commercial" provisions that are found in Chapter 4[CE].

Chapter 4 [RE] applies to portions of the building thermal envelope that enclose conditioned space as shown in Commentary Figure R401.1(1). Conditioned space is the area provided with heating and/or cooling either directly through a positive heating/cooling supply system, such as registers located in the space, or indirectly through an opening that allows heated or cooled air to communicate directly with the space. For example, a walk-in closet connected to a master bedroom suite may not contain a positive heating supply through a register, but it would be conditioned indirectly by the free passage of heated or cooled air into the spaces from the bedroom.

The code through Section R402.1 exempts areas that do not contain conditioned space and are separated from the conditioned spaces of the building by the building envelope from the building thermal envelope requirements. A good example of this would be an unconditioned garage or attic space. In the case of a garage, if the unconditioned garage area is separated from the conditioned portions of the residence by an assembly that meets the building thermal envelope criteria (meaning that the wall between them is insulated), the exterior walls of the garage would not need to be insulated to separate the garage from the exterior climate.

The building thermal envelope consists of the wall, roof/ceiling and floor assemblies that surround the conditioned space. Raised floors over a crawl space or garage or directly exposed to the outside air are considered to be part of the floor assembly. Walls surrounding a conditioned basement (in addition to surrounding conditioned spaces above grade) are part of the building envelope. The code defines "Above-grade walls" surrounding conditioned spaces as exterior walls. This definition includes walls between the conditioned space and unconditioned garage, roof and basement knee walls, dormer walls, gable end walls, walls enclosing a mansard roof and basement walls with an average below-grade wall area that is less than 50 percent of the total basement gross wall area. This definition would not include walls separating an unconditioned garage from the outdoors. The code's definition of "Exterior walls" would also include basement walls. The roof/ceiling assembly is the surface where insulation will be installed, typically on top of the gypsum board [see Commentary Figure R401.1(2)].

R401.2 Compliance. Projects shall comply with one of the following:

1. Sections R401 through R404.
2. Section R405 and the provisions of Sections R401 through R404 labeled "Mandatory."
3. An energy rating index (ERI) approach in Section R406.

❖ This section allows residential buildings to comply with either the prescriptive requirements of Sections R402 through R404, the performance options that are provided in Section R405, or the energy rating index option in Section R406. Under all options, the building must comply with the mandatory requirements that are found in Sections R402.4, R402.5, R403.1, R403.1.2, R403.3.2, R403.3.3, R403.3.5, R403.4, R403.6, R403.7, R403.8, R403.9, R404 and R406.3. A code user may evaluate both options and use the one that fits the project best, as these two differing methods can result in different requirements. Most requirements are given prescriptively. Alternative trade-offs are specified for many requirements, such as for the building thermal envelope requirements. For requirements specified by *U*-factors, an overall

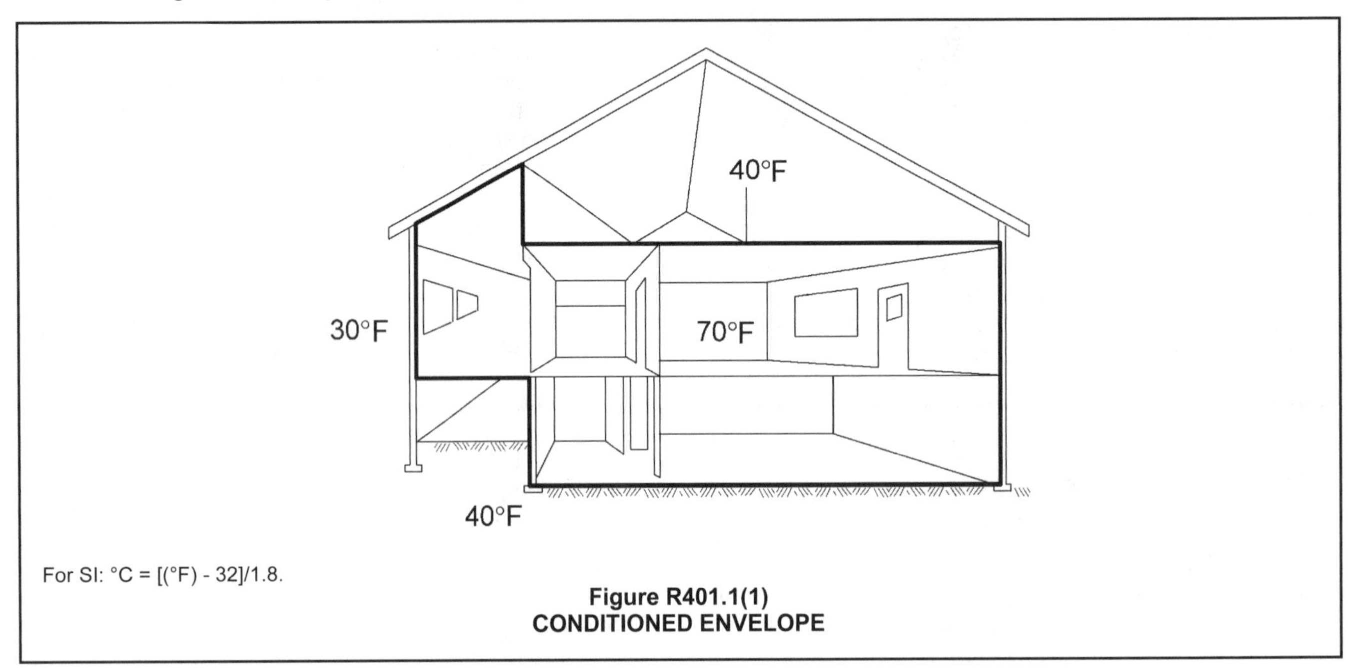

Figure R401.1(1)
CONDITIONED ENVELOPE

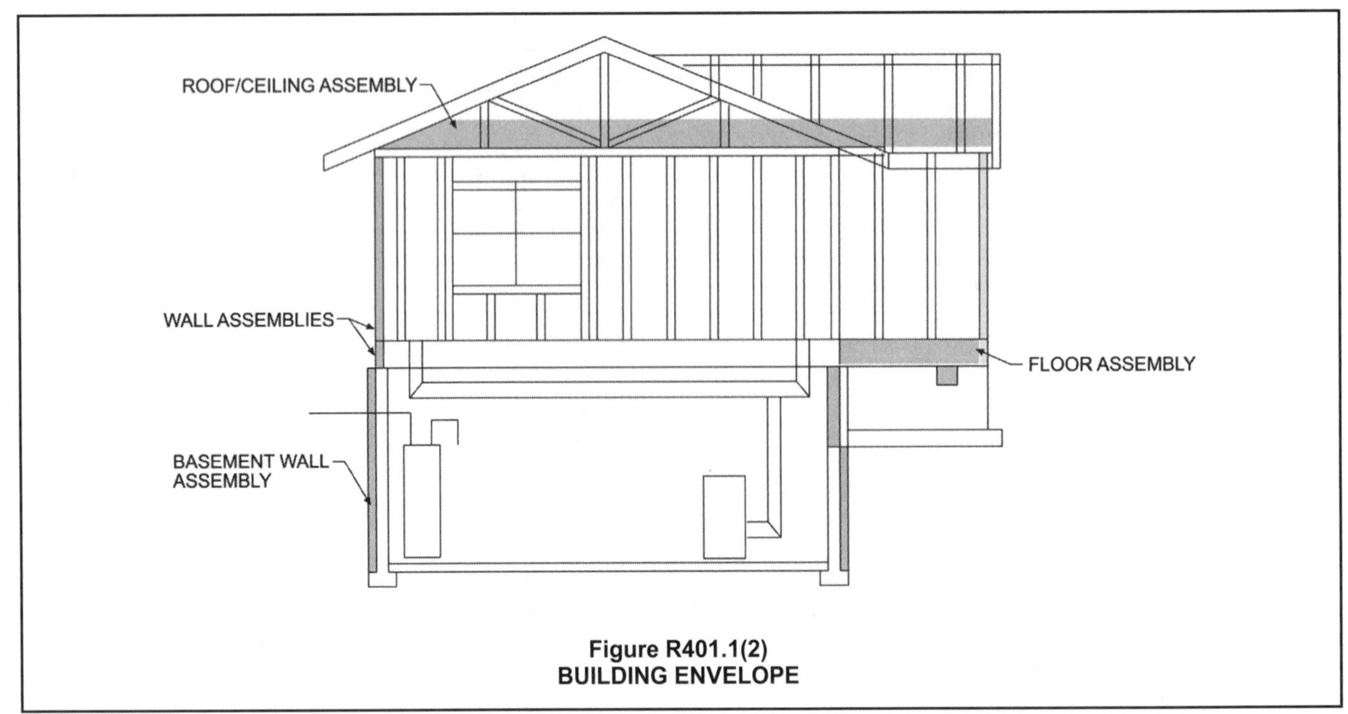

Figure R401.1(2)
BUILDING ENVELOPE

UA (*U*-factor times the area) can be used to show equivalence. A performance-based annual energy calculation also can be met by showing overall energy equivalence.

The majority of the requirements of this chapter are based on the climate zone where the project is being built. The appropriate climate zone can be found in Section R301.1 of the code. Climate Zones 1 through 7 apply to various parts of the continental United States and are defined by county lines. Climate Zones 7 and 8 apply to various parts of Alaska, and Hawaii is classified as Climate Zone 1. The climate zones have been divided into marine, dry and moist to deal with levels of humidity. For more details and background on the development of the new climate zones, see the commentary to Section R301.1.

R401.2.1 Tropical zone. *Residential buildings* in the tropical zone at elevations below 2,400 feet (731.5 m) above sea level shall be deemed to comply with this chapter where the following conditions are met:

1. Not more than one-half of the *occupied* space is air conditioned.
2. The *occupied* space is not heated.
3. Solar, wind or other renewable energy source supplies not less than 80 percent of the energy for service water heating.
4. Glazing in *conditioned* space has a *solar heat gain coefficient* of less than or equal to 0.40, or has an overhang with a projection factor equal to or greater than 0.30.
5. Permanently installed lighting is in accordance with Section R404.
6. The exterior roof surface complies with one of the options in Table C402.3 or the roof/ceiling has insulation with an *R-value* of R-15 or greater. If present, attics above the insulation are vented and attics below the insulation are unvented.
7. Roof surfaces have a minimum slope of $^1/_4$ inch per foot of run. The finished roof does not have water accumulation areas.
8. Operable fenestration provides ventilation area equal to not less than 14 percent of the floor area in each room. Alternatively, equivalent ventilation is provided by a ventilation fan.
9. Bedrooms with exterior walls facing two different directions have operable fenestration.
10. Interior doors to bedrooms are capable of being secured in the open position.
11. A ceiling fan or ceiling fan rough-in is provided for bedrooms and the largest space that is not used as a bedroom.

❖ This section provides an alternative for residences in the tropical climates, a simple option for a newly defined climate zone, the "tropical zone." The area between the Tropic of Cancer and the Tropic of Capricorn is the area between 23.5 degrees northern and southern latitude of the equator. A zone that recognizes the unusually constant and unique climate of this region would help make the *International Codes®* (I-Codes®) more of an "international code."

Tropical areas are quite different from the US mainland in climate, construction techniques, traditional construction and energy prices. This chapter, and the IECC heretofore, has treated tropical climates as if they were simply a southern extension of the US mainland. Traditional residences, especially the less expensive residences, have evolved inexpensive ways to work with the tropical climates to provide comfortable interior spaces without the need for substantial space conditioning. Tropical electrical prices, usually over 20 cents per hWh, provide a substantial incentive for energy conservation. Solar water heating works particularly well in tropical climates.

Traditional construction, especially with solar water heating, is usually more energy efficient than the construction style assumed in the IECC, as is shown by an analysis done for Puerto Rico. Using energy efficient versions of traditional construction saves more energy and is much more cost-effective than pushing those in tropical climates to adopt mainland construction practices. Traditional tropical construction focuses on greatly reducing or eliminating the need for space conditioning by making a living space that is comfortable without space conditioning.

The following are explanations, by item:

1. Air conditioning only a portion of the residence is common in some residences and saves energy compared to air conditioning the whole occupied space.
2. Heating is seldom needed.
3. Consistently warm temperatures and high power costs make solar water heating very attractive. Solar water heating is widely used. Water heating is often 35 percent or more of the residential energy use. Substantial energy savings come from solar water heating.
4. Limiting solar gains and providing ventilation is the energy focus for windows. Window *U*-factor has little impact. Window air tightness is of little value when the important feature of the windows is their ability to be operable and provide ventilation.
5. High efficiency lighting makes sense with tropical energy prices.
6. This references the "cool roof" provisions. This is similar to an option in Hawaii's code and the Puerto Rico Energy Center's analysis. Insulation is less valuable in mild climates where the outside temperature is often comfortable as an inside temperature.
7. Even flat roofs need to drain.

8. Ventilation provided by tropical winds makes occupied spaces more comfortable. 14 percent of the floor area of the room is an option for unconditioned residences in Hawaii's new energy code.
9. When bedroom walls facing two directions are available, ventilation on both walls will be more effective.
10. Interior doors should not block bedroom ventilation. This is similar to Hawaii's new energy code and recommended by the Puerto Rico Energy Center.
11. Ceiling fans increase comfort without conditioning the air. This is similar to Hawaii's new energy code and recommended by the Puerto Rico Energy Center.

R401.3 Certificate (Mandatory). A permanent certificate shall be completed by the builder or registered design professional and posted on a wall in the space where the furnace is located, a utility room or an approved location inside the building. Where located on an electrical panel, the certificate shall not cover or obstruct the visibility of the circuit directory label, service disconnect label or other required labels. The certificate shall list the predominant *R*-values of insulation installed in or on ceiling/roof, walls, foundation (slab, basement wall, crawlspace wall and floor) and ducts outside conditioned spaces; *U*-factors for fenestration and the solar heat gain coefficient (SHGC) of fenestration, and the results from any required duct system and building envelope air leakage testing done on the building. Where there is more than one value for each component, the certificate shall list the value covering the largest area. The certificate shall list the types and efficiencies of heating, cooling and service water heating equipment. Where a gas-fired unvented room heater, electric furnace or baseboard electric heater is installed in the residence, the certificate shall list "gas-fired unvented room heater," "electric furnace" or "baseboard electric heater," as appropriate. An efficiency shall not be *listed* for gas-fired unvented room heaters, electric furnaces or electric baseboard heaters.

❖ This section is intended to increase consumer awareness of the energy-efficiency ratings for various building elements in the home. The builder or registered design professional has to complete the certificate and place it on or inside the electrical panel (see Commentary Figure R401.3). The permanent certificate shall not cover or obstruct the visibility of the circuit directory label, service disconnect label or other required labels.

The certificate must disclose the building's *R*-values, fenestration *U*-factors and fenestration SHGC, HVAC equipment types and efficiencies. The energy efficiency of a building as a system is a function of many elements considered as separate parts of the whole. It is difficult to have a proper identification and analysis of a building's energy efficiency once the building is completed because many of the elements may not be readily accessible.

This information is also valuable for existing structures undergoing alterations and additions to help determine the appropriate sizing for the mechanical systems. This is meant to be a simple certificate that is easy to read. The certificate does not contain all the information required for compliance and cannot be substituted for information on the required construction documents. Instead, the certificate is meant to provide the housing owner, occupant or buyer with a simple-to-understand overview of the home's energy efficiency. Where there is a mixture of insulation and fenestration values, the value applying to the largest area is specified. For example, if most of the wall insulation was R-19, but a limited area bordering the garage was R-13, the certificate would specify R-19 for the walls. (In contrast, plans and overall compliance would need to account for both *R*-values.)

The code specifies the minimum information on the certificate, but does not prohibit additional information being added so long as the required information is clearly visible. For example, a builder might choose to list energy-efficiency features beyond those required by the code.

SECTION R402
BUILDING THERMAL ENVELOPE

R402.1 General (Prescriptive). The *building thermal envelope* shall meet the requirements of Sections R402.1.1 through R402.1.5.

Exception: The following low-energy buildings, or portions thereof, separated from the remainder of the building by *building thermal envelope* assemblies complying with

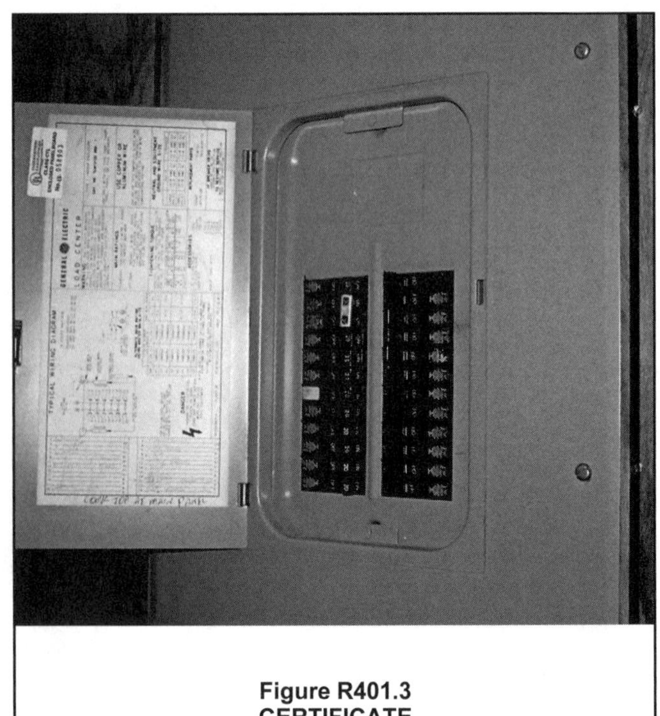

Figure R401.3
CERTIFICATE

this section shall be exempt from the *building thermal envelope* provisions of Section R402.

1. Those with a peak design rate of energy usage less than 3.4 Btu/h • ft^2 (10.7 W/m^2) or 1.0 watt/ft^2 of floor area for space-conditioning purposes.
2. Those that do not contain *conditioned space*.

❖ The provisions of Section R402 are detailed requirements of insulation levels, the performance of openings (fenestrations) and air-leakage and moisture-control provisions that serve to establish the building's energy efficiency. When combined with the "systems" requirements (Section R403) and the electrical power requirements (Section R404), these three sections provide the total package of energy conservation that the code requires.

The term "building thermal envelope" is defined in Section R202 as being "the basement walls, exterior walls, floor, roof and any other building elements that enclose conditioned spaces." This boundary also includes the boundary between conditioned space or provides a boundary between conditioned space and exempt or unconditioned space. Therefore, when combined with the definition of "Conditioned space," the code has defined the boundaries of the building that will be regulated by this section. The building thermal envelope is a key term and resounding theme used throughout the energy requirements. It defines what portions of the building structure bound conditioned space and are thereby covered by the insulation and infiltration (air leakage) requirements of the code. The building thermal envelope includes all building components separating conditioned spaces (see commentary, "Conditioned space") from unconditioned spaces or outside ambient conditions and through which heat is transferred. For example, the walls and doors separating an unheated garage (unconditioned space) from a living area (conditioned space) are part of the building envelope. The walls and doors separating an unheated garage from the outdoors are not part of the building thermal envelope. Walls, floors and other building components separating two conditioned spaces are not part of the building envelope. For example, interior partition walls, the common or party walls separating dwelling units in multiple-family buildings and the wall between a new conditioned addition and the existing conditioned space are not considered part of the building envelope.

Unconditioned spaces (areas having no heating or cooling sources) are considered outside the building thermal envelope and are exempt from these requirements (see Section R202). A space is conditioned if it is heated or cooled directly; communicates directly with a conditioned space; or is indirectly supplied with heating, cooling or both through uninsulated walls, floors, uninsulated ducts or HVAC piping. Boundaries that define the building envelope include the following:

- Building assemblies separating a conditioned space from outdoor ambient weather conditions.
- Building assemblies separating a conditioned space from the ground under or around that space, such as the ground around the perimeter of a slab or the soil at the exterior of a conditioned basement wall. Note that the code does not specify requirements for insulating basement floors or underneath slab floors (except at the perimeter edges).
- Building assemblies separating a conditioned space from an unconditioned garage, unconditioned sunroom or similar unheated/cooled area.

The code specifies requirements for ceiling, wall, floor, basement wall, slab-edge and crawl space wall components of the building envelope. In some cases, it may be unclear how to classify a particular part of a building. For example, skylight shafts have properties of a wall assembly but are located in the ceiling assembly. In these situations, a determination needs to be made and approved by the code official prior to construction so that the proper level of insulation can be installed to complete the building thermal envelope. Generally, skylight shafts and other items that are vertical or at an angle of greater than 60 degrees (1.1 rad) from the horizontal would typically use the wall insulation value.

The exception exempts certain buildings from the thermal envelope requirements of Section R402. The obvious one is any building that does not contain a conditioned space. The other type of building is a low-energy building (design rate of energy usage less than 1 watt/ft^2). For these spaces, insulation and rated fenestration are not cost effective.

R402.1.2 Insulation and fenestration criteria. The *building thermal envelope* shall meet the requirements of Table R402.1.2, based on the climate zone specified in Chapter 3.

❖ This section serves as the basis for the code's general insulation and fenestration requirements. Therefore, this is the first place to determine what the requirements for the building thermal envelope will be. There are specific requirements for certain assemblies and locations that are addressed in Sections R402.2 and R402.3. First, this section mandates compliance with the proper component insulation and fenestration requirements of Table R402.1.2. However, once that general requirement is established, Sections R402.1.3, R402.1.4 and R402.1.5 provide three possible means of showing that the building thermal envelope will comply. Any of the three methods may be used at the discretion of the designer. The three options and their advantages are discussed in the commentary with the subsections. In general, the later subsections will provide the designer with more options and flexibility, but they are also more complex than using Table R402.1.2 on an individual component basis.

Table R402.1.2 lists the minimum *R*-value and maximum *U*-factor and SHGC requirements for different portions of the building thermal envelope, including basement and exterior walls, floor, ceiling and any other building elements that enclose conditioned space. Using the table begins with determining the climate zone for the proposed location from Table R301.1 or Figure R301.1. Once the climate zone has been determined, each of the *R*-value, *U*-factor or SHGC requirements must be met for the applicable component (e.g., ceilings, walls, floors, etc.).

Maximum fenestration *U*-factor is the first column in Table R402.1.2 that must be complied with (see definition for "Fenestration" in Chapter 2). Except as modified or exempted by Section R402.3, each fenestration product in the proposed building must not exceed the maximum *U*-factor requirement presented in the table for a particular climate zone. For example, a single-family residence located in Climate Zone 5 would require installation of glazed fenestration products with a maximum *U*-factor of 0.32. This would include all glazing in the walls of the building thermal envelope (e.g., vertical windows); skylights in the roof would be limited to a maximum *U*-factor of 0.55. The proposed glazing *U*-factor should be called out in the building plans either on the floor plan or in a window schedule. This will provide the necessary information to the field inspector, who will then need to verify that what is on the plans is installed in the field.

Fenestration products that do not have labels on them must use the default *U*-factors contained in Table R303.1.3(1) or R303.1.3(3) (see commentary, Section R303.1.3). Note that the lowest default *U*-factor included in the table for glazed fenestration is listed at 0.55 for a "nonmetal or metal-clad double-pane window." This *U*-factor will not meet the requirements of the code in Climate Zones 2 and higher.

TABLE R402.1.2. See below.

❖ Table R402.1.2 serves as the basis for establishing the building thermal envelope requirements based on the text of Section R402.1.2 and sets the performance level for each of the individual components listed. See the commentary for Sections R402.1.2, R402.2 and R402.3 for additional discussion related to the components in the table. The simplest compliance approach is to meet these requirements directly. Note that the requirements do not change based on the area of the components of the residence. These same requirements apply to changes in existing buildings; for example, additions.

A few specifics of Table R402.1.2 may benefit from clarification.

When applying the fenestration requirements of this table, it is important to remember the definition of

TABLE R402.1.2
INSULATION AND FENESTRATION REQUIREMENTS BY COMPONENT[a]

CLIMATE ZONE	FENESTRATION *U*-FACTOR[b]	SKYLIGHT[b] *U*-FACTOR	GLAZED FENESTRATION SHGC[b, e]	CEILING *R*-VALUE	WOOD FRAME WALL *R*-VALUE	MASS WALL *R*-VALUE[i]	FLOOR *R*-VALUE	BASEMENT[c] WALL *R*-VALUE	SLAB[d] *R*-VALUE & DEPTH	CRAWL SPACE[c] WALL *R*-VALUE
1	NR	0.75	0.25	30	13	3/4	13	0	0	0
2	0.40	0.65	0.25	38	13	4/6	13	0	0	0
3	0.35	0.55	0.25	38	20 or 13+5[h]	8/13	19	5/13[f]	0	5/13
4 except Marine	0.35	0.55	0.40	49	20 or 13+5[h]	8/13	19	10 /13	10, 2 ft	10/13
5 and Marine 4	0.32	0.55	NR	49	20 or 13+5[h]	13/17	30[g]	15/19	10, 2 ft	15/19
6	0.32	0.55	NR	49	20+5 or 13+10[h]	15/20	30[g]	15/19	10, 4 ft	15/19
7 and 8	0.32	0.55	NR	49	20+5 or 13+10[h]	19/21	38[g]	15/19	10, 4 ft	15/19

For SI: 1 foot = 304.8 mm.

a. R-values are minimums. *U*-factors and SHGC are maximums. When insulation is installed in a cavity which is less than the label or design thickness of the insulation, the installed *R*-value of the insulation shall not be less than the *R*-value specified in the table.

b. The fenestration *U*-factor column excludes skylights. The SHGC column applies to all glazed fenestration. Exception: Skylights may be excluded from glazed fenestration SHGC requirements in climate zones 1 through 3 where the SHGC for such skylights does not exceed 0.30.

c. "15/19" means R-15 continuous insulation on the interior or exterior of the home or R-19 cavity insulation at the interior of the basement wall. "15/19" shall be permitted to be met with R-13 cavity insulation on the interior of the basement wall plus R-5 continuous insulation on the interior or exterior of the home. "10/13" means R-10 continuous insulation on the interior or exterior of the home or R-13 cavity insulation at the interior of the basement wall.

d. R-5 shall be added to the required slab edge *R*-values for heated slabs. Insulation depth shall be the depth of the footing or 2 feet, whichever is less in Climate Zones 1 through 3 for heated slabs.

e. There are no SHGC requirements in the Marine Zone.

f. Basement wall insulation is not required in warm-humid locations as defined by Figure R301.1 and Table R301.1.

g. Or insulation sufficient to fill the framing cavity, R-19 minimum.

h. The first value is cavity insulation, the second value is continuous insulation, so "13+5" means R-13 cavity insulation plus R-5 continuous insulation.

i. The second *R*-value applies when more than half the insulation is on the interior of the mass wall.

"Fenestration" and that it includes items such as doors, glass blocks and other items as well as windows. Therefore, any door located in the building thermal envelope would still be subject to these limitations. Although vertical fenestration (vertical windows and doors) and skylights have a separate column for *U*-factor, the SHGC applies to both. This is reinforced by the provisions of Note b.

The ceiling *R*-value requirements are precalculated for insulation only and already assume a credible *R*-value for other building materials such as air films, interior sheathing and exterior sheathing. The only *R*-value for ceiling insulation that may be used to meet the requirements is that installed between the conditioned space and the vented airspace in the roof/ceiling assembly. This is typically not an issue because most insulation is installed directly on top of the gypsum board ceiling and the ceiling location represents the building thermal envelope. Insulation installed in the ceiling must meet or exceed the required insulation level. The "conditioned attic" requirements of the code may be viewed as an acceptable alternative if approved by the code official. These minimum ceiling *R*-values would still be applicable where the provisions of Section R806.4 of the IRC were used to create a conditioned attic assembly. In those cases, the location of the insulation and air barrier (building thermal envelope) are simply located at the roof instead of at the ceiling line. See the commentary to Sections R402.2.1 and R402.2.2 for additional information regarding ceiling insulation requirements.

The *R*-values presented under the "wall" columns represent the sum of the insulation materials installed between the framing cavity and, if used, the insulating sheathing. See Section R402.1.2 regarding how to compute the *R*-value. Insulating sheathing must have an *R*-value of at least R-2 to be considered. The R-2 limitation comes from the definition of "Insulating sheathing." The *R*-value of noninsulative interior finishes, such as sheet rock, or exterior coverings (e.g., wood structural panel siding) is not considered when determining whether the proposed wall assembly meets the requirements. For example, in Commentary Figure R402.1.2, the *R*-value of the cavity insulation installed between framing (R-13) is added to the insulating sheathing installed on the outside of the studs (R-6) resulting in an R-19 wall. The R-19 total insulation value can then be compared to the *R*-value requirement for the specific climate zone in Table R402.1.2 to determine compliance.

The insulation *R*-value requirement for exterior walls assumes wood framing. Walls framed using steel studs or constructed of materials such as a concrete masonry unit (CMU) are addressed in Sections R402.2.5 and R402.2.6. In residences with more than one type of wall (frame or mass) or more than one type of below-grade wall (conditioned basement or crawl space), the requirement for each component is taken from the appropriate column in Table R402.1.2.

Mass walls are defined and have additional requirements in Section R402.2.5. Mass walls are intended to be above-grade walls and do not include basement walls, which have a separate entry in the table.

Note a reminds the code user which level of performance is required. Therefore, when dealing with *R*-values, a higher number would be better. When dealing with *U*-factors, the lower the number, the better the performance.

In accordance with Note c, for basement walls and crawl space walls, the two numbers separated by a "/" represent the values for continuous and cavity insulation; either will meet the code's requirements. For example, in Climate Zone 6, the wall can either be covered with continuous insulation to a minimum level of R-15 or, if some type of framing is used (such as a wood-frame wall used to finish out a basement), R-19 insulation must be installed in the cavity. This higher level of cavity insulation adjusts for the bridging or reduction in energy efficiency that the framing elements would create.

In accordance with Note d, heated slabs require R-5 insulation in Climate Zones 1, 2 and 3 and R-15 slab edge insulation in Climate Zones 4 and above. This R-15 insulation is the result of R-5 being "added" to the R-10 insulation level specified in the table for Climate Zones 4 through 8.

In accordance with Note g, where R-30 under-floor insulation is required, less insulation may be used if the framing cavity is filled, down to a minimum of R-19. This recognizes that extending the framing solely to hold more insulation can cost more than it is worth.

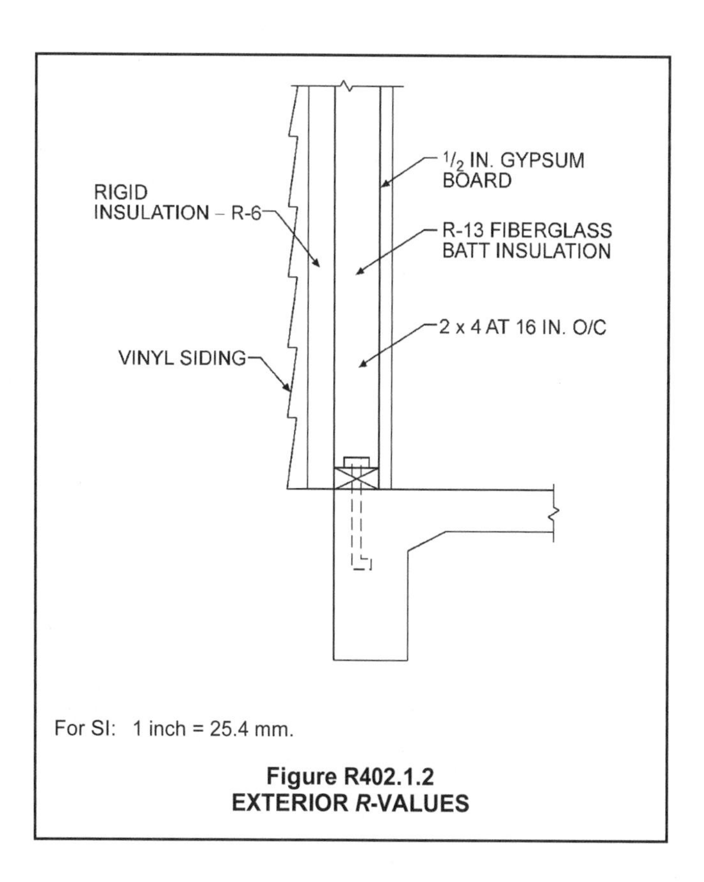

For SI: 1 inch = 25.4 mm.

Figure R402.1.2
EXTERIOR *R*-VALUES

R402.1.3 *R*-value computation. Insulation material used in layers, such as framing cavity insulation, or continuous insulation shall be summed to compute the corresponding component *R*-value. The manufacturer's settled *R*-value shall be used for blown insulation. Computed *R*-values shall not include an *R*-value for other building materials or air films. Where insulated siding is used for the purpose of complying with the continuous insulation requirements of Table R402.1.2, the manufacturer's labeled *R*-value for insulated siding shall be reduced by R-0.6.

❖ This section indicates how the *R*-value in Table R402.1.2 is to be determined. Table R402.1.2 specifies the required *R*-values for the insulation products, the nominal *R*-value. This is the *R*-value of the insulation products only. Although other products and features such as finish materials, air films and airspaces may contribute to overall energy efficiency, when determining the *R*-value in the code, these additional items are not considered and do not contribute to the nominal *R*-value. For example, if a wall had R-13 cavity insulation, gypsum board with an *R*-value of almost R-1 and exterior siding that has an *R*-value of R-1, the overall wall *R*-value is simply R-13 because the gypsum board and the exterior siding do not contribute to the *R*-value for purposes of determining code compliance. Where there is more than one layer of insulation, the *R*-values for the layers are summed. For example, a wall with R-13 batts within the framing cavity and R-4 insulated sheathing would be treated as an R-17 wall (13 + 4 = 17). It is only insulation materials that may be summed to determine the component's *R*-value.

R402.1.4 *U*-factor alternative. An assembly with a *U*-factor equal to or less than that specified in Table R402.1.4 shall be permitted as an alternative to the *R*-value in Table R402.1.2.

❖ For residences built with common insulation products, the most direct method of compliance is often via the *R*-values in Table R402.1.2. As an alternative, compliance can be demonstrated by calculating the *U*-factor for a component. Table R402.1.4 gives *U*-factors that are deemed to be equivalent to the *R*-values in the prescriptive tables. Unlike the *R*-values in Table R402.1.2, which consider only the insulation, *U*-factors consider all the parts of the construction. *U*-factors for a wall might include exterior siding, gypsum board and air films, all of which would be excluded from the *R*-value computation by Section R402.1.2; for example, whether wall framing is 16 or 24 inches (406 or 610 mm) on center matters in computing the *U*-factor. Whether framing is metal or wood can also have a significant impact on *U*-factor.

U-factors are well suited to several applications. Construction types that limit the amount of framing or include thermal breaks as part of their design may benefit from *U*-factor calculations. Components with complex or nonuniform geometries can use testing to establish *U*-factors. Compliance with the "total UA alternative" or tradeoff approach in Section R402.1.4 requires the use of the *U*-factor tables.

Example of *U*-factor calculation in Table R402.1.4:

WALL COMPONENT	CAVITY	STUDS, PLATES	HEADERS
	R-value	*R*-value	*R*-value
Outside air film	0.25	0.25	0.25
Plywood siding	0.59	0.59	0.59
Continuous insulation	5	5	5
Plywood sheathing	0.83	0.83	0.83
Wood studs	—	4.38	4.38
Cavity insulation	13	—	—
$^1/_2$" Gypsum board	0.45	0.45	0.45
Inside air film	0.68	0.68	0.68
Sum of thermal resistance	20.8	12.18	12.18

The above table includes the *R*-values for the cavity and framing (studs, plates and headers). The *U*-factor is 1/*R*-value. To calculate the *U*-factor for the combination of the cavity and framing at 16 inches on-center spacing, calculate the *U*-factor based on the weighting factors of 75-percent cavity, 21-percent studs and plates, and 4-percent headers (78-percent cavity, 18-percent studs and plates and 4-percent headers for 24 inches on-center spacing). An assembly with a *U*-factor equal to or less than that calculated in this table must be permitted as an alternative to an assembly in Table R402.1.2.

The *U*-factor for the above assembly may be calculated as shown below:

$$U = \frac{0.75}{20.8} + \frac{0.21}{12.18} + \frac{0.04}{12.18} = 0.057$$

Example of *U*-factor alternative:

WALL COMPONENT	CAVITY	STUDS, PLATES	HEADERS
	R-value	*R*-value	*R*-value
Outside air film	0.25	0.25	0.25
Plywood siding	0.59	0.59	0.59
Continuous insulation	13.1	13.1	31.1
Plywood sheathing	0.83	0.83	0.83
Wood studs	—	4.38	4.38
Cavity insulation	0.91	—	—
$^1/_2$" Gypsum board	0.45	0.45	0.45
Inside air film	0.68	0.68	0.68
Sum of thermal resistance	16.81	20.28	20.28

In the above assembly, the R-13 cavity insulation is removed and a value of 0.91 is assumed for the airspace in the stud cavity. The continuous insulation is increased to 13.1. The framing spacing is increased to 24 inches (610 mm) on center.

The U-factor for the above assembly may be calculated as shown below:

$$U = \frac{0.78}{16.81} + \frac{0.18}{20.28} + \frac{0.04}{20.28} = 0.057$$

The U-factor from the above assembly is equal to or less than that specified in Table R402.1.4.

It should be noted that the U-factor alternative could be used for all or part of a wall. Thus, the prescriptive R-values in Section R402.1.2 can be applied to one segment of the wall, and the U-factors in this section can be used to determine the construction of another segment of the wall. One possible combination would be to apply the prescriptive requirements of Table R402.1.2 for the portion of the exterior wall thermal envelope with structural sheathing, and the U-factor alternative (Table R402.1.4) for the portion of the exterior wall thermal envelope that does not have structural sheathing. Note h of Table R402.1.2 could be applied to 40 percent of the wall when structural sheathing is present, allowing a reduction in the R-value of R-3 for this segment.

Where approved by the code official in accordance with Section R102, alternative methods of construction, materials and insulation systems may also be used instead of the prescriptive R-values of Table R402.1.2. For example, designers may substitute greater stud spacing, insulated plates and insulated headers in walls as an alternative to the prescribed cavity or continuous insulation.

R402.1.5 Total UA alternative. If the total *building thermal envelope* UA (sum of U-factor times assembly area) is less than or equal to the total UA resulting from using the U-factors in Table R402.1.4 (multiplied by the same assembly area as in the proposed building), the building shall be considered in compliance with Table R402.1.2. The UA calculation shall be done using a method consistent with the ASHRAE *Handbook of Fundamentals* and shall include the thermal bridging effects of framing materials. The SHGC requirements shall be met in addition to UA compliance.

❖ This alternative allows one portion of the building to make up for another. It recognizes that there may be reasons for less insulation in some parts of the building, which can be compensated for by more insulation in other parts of the residence. The key concept is that the overall building thermal flow (UA) meets the code. This concept could allow a ceiling to make up for a wall or vice versa. As a practical matter, whether a building will comply by this method can sometimes be estimated quickly. A large area that is significantly over the required R-value will make up for a small area only mildly under the required R-value. Likewise, it will sometimes be obvious that a small area that mildly exceeds the requirement will not make up for a large area well below the requirement.

This section allows for such tradeoffs but only if the total UA for the proposed building is below the aggregate UA calculation using the required values in Table R402.1.4 and the same assembly areas as the actual building. In other words, under this alternative, components with varying insulating values can be "traded off" with one another as the builder sees fit, as long as the total UA calculation for the entire building equals or is less than a calculation for that same house using the same assembly areas and the maximum UA values from Table R402.1.4.

The UA is the sum of the component U-factors times each assembly area. The maximum allowable UA is the UA for a proposed design as if it was insulated to meet exactly the individual component U-factor requirements. This tradeoff provision allows the type of insulation and installed fenestration to vary, which permits significant design flexibility. The desire for tradeoffs in construction is common because of unexpected problems or design conflicts, and a UA trade-off analysis is usually calculated with the assistance of electronic compliance tools, depending on

TABLE R402.1.4
EQUIVALENT *U*-FACTORS[a]

CLIMATE ZONE	FENESTRATION *U*-FACTOR	SKYLIGHT *U*-FACTOR	CEILING *U*-FACTOR	FRAME WALL *U*-FACTOR	MASS WALL *U*-FACTOR[b]	FLOOR *U*-FACTOR	BASEMENT WALL *U*-FACTOR	CRAWL SPACE WALL *U*-FACTOR
1	0.50	0.75	0.035	0.084	0.197	0.064	0.360	0.477
2	0.40	0.65	0.030	0.084	0.165	0.064	0.360	0.477
3	0.35	0.55	0.030	0.060	0.098	0.047	0.091[c]	0.136
4 except Marine	0.35	0.55	0.026	0.060	0.098	0.047	0.059	0.065
5 and Marine 4	0.32	0.55	0.026	0.060	0.082	0.033	0.050	0.055
6	0.32	0.55	0.026	0.045	0.060	0.033	0.050	0.055
7 and 8	0.32	0.55	0.026	0.045	0.057	0.028	0.050	0.055

a. Nonfenestration U-factors shall be obtained from measurement, calculation or an approved source.

b. When more than half the insulation is on the interior, the mass wall U-factors shall be a maximum of 0.17 in Climate Zone 1, 0.14 in Climate Zone 2, 0.12 in Climate Zone 3, 0.087 in Climate Zone 4 except Marine, 0.065 in Climate Zone 5 and Marine 4, and 0.057 in Climate Zones 6 through 8.

c. Basement wall U-factor of 0.360 in warm-humid locations as defined by Figure R301.1 and Table R301.1.

the jurisdiction. For example, the Department of Energy (DOE) has online compliance software for the code called REScheck, which can be downloaded from the DOE website at www.energycodes.gov. REScheck, if approved by the jurisdiction as compliant with the code, can be used to perform a UA tradeoff analysis.

This section explicitly prohibits the tradeoff of SHGC requirements, requiring that the "SHGC requirements be met in addition to UA compliance." As a result, glazed fenestration must comply with the SHGC values shown in Table R402.1.2 even if the *U*-factor is modified by trading off against some other component.

The requirements of this section establish specific additional requirements for any tradeoff. First, the baseline house must have the same assembly areas as the proposed house (e.g., the same area of each assembly—fenestration, skylights, ceiling, wall and floor). Second, the calculation should be done consistent with the *ASHRAE Handbook of Fundamentals*. Third, the calculation must include the thermal bridging effects of framing materials. To meet these requirements, the calculation method must either specifically combine the actual framing and insulation paths (with their specific areas and *U*-values) or use framing factors such as those found in the *ASHRAE Handbook of Fundamentals* for all framed building components (Note: this is not necessary for fenestration, which is a whole product value). To illustrate this approach, assume 1,000 square feet (93 m^2) of wall, of which 250 square feet (23 m^2) is framing (assuming 0.81 *U*-factor) and 750 square feet (70 m^2) is cavity (R-13 insulation). The baseline (general code requirement) and proposed opaque wall UAs are computed as follows:

- Baseline Opaque Wall for Climate Zone 2 = (0.082 x 1000) = 82. (The 0.082 value used in the calculation is taken from Table R402.1.4.)
- Proposed Opaque Wall UA = (0.81 x 250) + (0.077 x 750) = 78. (The 0.81 value used in the calculation was given above as an assumption. The 0.077 value is determined based on the *R*-value of 13; that is, 1 ÷ 13 = 0.077.)

Similar computations would be done for each assembly (such as fenestration or ceilings) and the baseline and proposed values then totaled and compared. If the baseline is greater than or equal to the proposed values, the house satisfies the UA alternative. The home still must meet all other prescriptive requirements, including the fenestration SHGC in Table R402.1.2, the air leakage requirements of Section R402.4 and the moisture-control requirements of Section R402.5.

It also should be noted that this alternative is limited to UA tradeoffs for the building's thermal envelope. The 2015 code does not authorize or establish any basis for HVAC tradeoffs associated with its UA tradeoff option. HVAC performance is simply not addressed under the code prescriptive or UA tradeoff paths. It is addressed only in Section R405 under the "Simulated Performance Alternative." As a result, the code limits these simplified tradeoffs to UA envelope tradeoffs and defers any more complex tradeoffs exclusively to the "Simulated Performance Alternative" in Section R405. If the builder wishes to factor in HVAC performance for tradeoffs, Section R405 can be used and is permitted based on Section R101.3, accepting Section R405 as a compliance option.

Documentation acceptable to the local building department generally must be submitted to the appropriate authority to certify acceptable component UA tradeoffs.

UA alternative example (for calculating standard and proposed designs):

Building areas and *U*-factors (for single-family, Zone 5 from Table R402.1.4)

Standard (Code) Design	**Proposed Design**
Exterior Wall 1050 ft^2 U_w = -0.057	Exterior Wall 1050 ft^2 U = 0.0451
Glazed doors and windows 192 ft^2 U_g = 0.32	Glazed doors and windows 192 ft^2 U = 0.40
Opaque exterior door 38 ft^2 U_g = 0.32	Opaque exterior door 38 ft^2 U = 0.35
Roof 1500 ft^2 U_r = 0.026	Roof 1500 ft^2 U_r = 0.026
Floor (slab) 1500 ft^2 U = 0.033	Floor (slab) 1500 ft^2 U_{fe} = 0.033

UA Standard = (0.057 × 1050) + (0.32 × 192) + (0.32 × 38) + (0.026 × 1500) + (0.033 × 1500) = 222

UA Proposed = (0.0451 × 1050) + (0.40 × 192) + (0.35 × 38) + (0.026 × 1500) + (0.033 × 1500) = 225

UA Proposed > UA Standard; therefore design fails to meet code requirements.

R402.2 Specific insulation requirements (Prescriptive). In addition to the requirements of Section R402.1, insulation shall meet the specific requirements of Sections R402.2.1 through R402.2.13.

❖ This section contains specific requirements to be followed for the individual items listed in the subsections. Although Section R402.1 and Tables R402.1.2 and R402.1.4 provide the general basis for complying with the energy requirements of this chapter, Section R402.2 provides additional details regarding the actual construction of the assemblies or modifications that may affect the general requirements. Relying on the principle that specific requirements apply over general requirements will help ensure that these specific provisions are properly followed.

R402.2.1 Ceilings with attic spaces. Where Section R402.1.2 would require R-38 insulation in the ceiling, installing R-30 over 100 percent of the ceiling area requiring insulation shall be deemed to satisfy the requirement for R-38 wherever the full height of uncompressed R-30 insulation extends over the wall top plate at the eaves. Similarly, where Section R402.1.2 would require R-49 insulation in the ceiling, installing R-38 over 100 percent of the ceiling area requiring insulation shall be deemed to satisfy the requirement for R-49 insulation wherever the full height of uncompressed R-38 insulation extends over the wall top plate at the eaves. This reduction shall not apply to the *U*-factor alternative approach in Section R402.1.4 and the total UA alternative in Section R402.1.5.

❖ The required ceiling *R*-value found in the code is based on the assumption that standard truss or rafter construction was being used. Where raised-heel trusses or other methods of framing that do not permit the ceiling insulation to be installed to its full depth over the entire area are used, the code permits the installation of a lower *R*-value insulation. The general assumption is that ceiling insulation will be compressed at the edges and, if special construction techniques are used, the level of insulation required can be reduced.

Insulation installed in a typical roof assembly will be full height throughout the center portions of the assembly and will taper (be compressed) at the edges as the roof nears the top plate of the exterior wall system [see Commentary Figure R402.2.1(1)]. The slope of the roof causes this tapering, which is further amplified by any baffling installed to direct ventilation air from the eave vents up and over the insulation. Because of this tapering, the installed *R*-value near the plate lines will be less than the rated *R*-value for the insulation. This is caused by compression (compressed insulation has a lower *R*-value than insulation installed to its full thickness) and the limited space between the floor of the attic and the roof sheathing near the exterior plate line. Thus, a typical installation, on average, will have a lower *R*-value than that of the rated insulation. Because of this, the code will allow installation of a lower insulation value if it can be installed full thickness, to its rated *R*-value, over the plate line of the exterior wall. This allowance recognizes that a partial thermal "bypass" has been made more efficient by using insulation with the full *R*-value at the eaves. The full insulation *R*-value is sometimes achieved by what is termed an "energy truss" or "advanced framing." This can be achieved by using an oversized truss or raised-heel truss as shown in Commentary Figure R402.2.1(2). Another way to achieve the full *R*-value would be by use of insulation with a higher *R*-value per inch at the eaves. The use of the options permitted by this section allows substituting R-30 for R-38 insulation, and R-38 may be substituted for R-49 insulation to meet the requirements of the code. When using the conditioned attic requirements of the code, the same option of using a reduced *R*-value would apply if the insulation was installed directly under the roof deck, rather than on the attic floor. The insulation would be permitted to meet the lesser *R*-value, presuming the full *R*-value was met over the eaves. Of course, this situation would also presume that the attic space beneath the insulation was not vented.

Note that this text applies only to the *R*-value portion of the code; there is no reduction in requirements if the *U*-factor alternative or the total UA alternative is used. In those cases, the reduced thickness must be

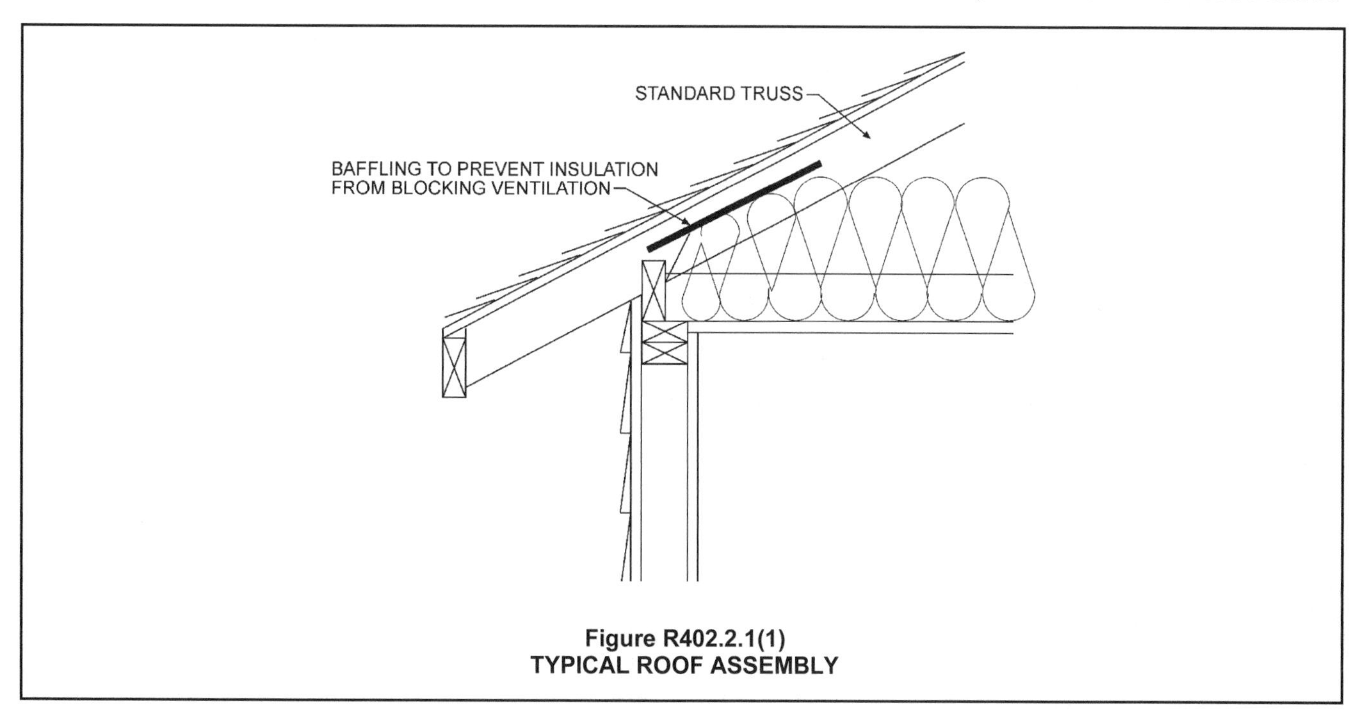

Figure R402.2.1(1)
TYPICAL ROOF ASSEMBLY

accounted for in the calculations. In addition, if the residence had more than one separate attic space, it is possible this section could apply to one attic space, but not another.

R402.2.2 Ceilings without attic spaces. Where Section R402.1.2 would require insulation levels above R-30 and the design of the roof/ceiling assembly does not allow sufficient space for the required insulation, the minimum required insulation for such roof/ceiling assemblies shall be R-30. This reduction of insulation from the requirements of Section R402.1.2 shall be limited to 500 square feet (46 m^2) or 20 percent of the total insulated ceiling area, whichever is less. This reduction shall not apply to the *U*-factor alternative approach in Section R402.1.4 and the total UA alternative in Section R402.1.5.

❖ In situations where the ceiling is installed directly onto the roof rafters and no attic space is created, this section will allow a reduced level of ceiling insulation for a limited area. This section addresses the construction of what are typically called "cathedral" or "vaulted" ceilings, which result in that portion of the home not having an attic area above the ceiling. See the definition of "Attic" in Chapter 2 of the code. Based on the use of solid sawn lumber (2 × 8, 2 × 10 or 2 × 12) in conventional construction, as the ceiling *R*-value requirement increases, it may be impossible to install the required ceiling insulation from Table R402.1.2 in the available cavity depth. In addition, the ventilation requirements of Section R806.3 of the IRC or Section 1203.2 of the IBC further reduce the available space by requiring a minimum space of 1 inch (25 mm) between the insulation and the roof sheathing. Therefore, when the depth of the cavity will not permit the required insulation level, this section permits a reduction to R-30 ceiling insulation instead of the normally required R-38 or R-49 requirement from Table R402.1.2. This will generally result in reducing the required insulation level instead of having to increase the depth of the framing members. This section in the code recognizes that increases in framing size done only to accommodate higher *R*-values are an expensive way to achieve a limited increase in *R*-value.

It is important to note that this section applies only to areas that have a required insulation level above R-30 (Climate Zones 4 through 8). Further, the reduction is limited to portions of ceiling assemblies that do not exceed 500 square feet (46 m^2) or 20 percent of the ceiling area. The intent is that the 500-square-foot (46 m^2) or 20-percent limitation be the total aggregate exempted amount of the building's thermal envelope (ceiling) that can use this reduction. It is not the intent that a home could have multiple areas that were each under the 500-square-foot (46 m^2) or 20-percent limit but would aggregate to more than that amount. In situations that cannot meet these limitations [homes in Climate Zones 1 through 3 or homes needing more than 500 square feet (46 m^2) of reduced ceiling insulation], the depth of the rafters would have to be increased to meet the table's required insulation level, or other design changes or means of compliance would be necessary. This provision does not apply if the *U*-factor alternative or the total UA alternative is being used. In those cases, the reduced insulation level might be accomplished, but it would be accounted for with additional insulating components in the case of the *U*-factor alternative or with added insulation elsewhere in the case of the total UA alternative.

R402.2.3 Eave baffle. For air-permeable insulations in vented attics, a baffle shall be installed adjacent to soffit and eave vents. Baffles shall maintain an opening equal or greater than the size of the vent. The baffle shall extend over

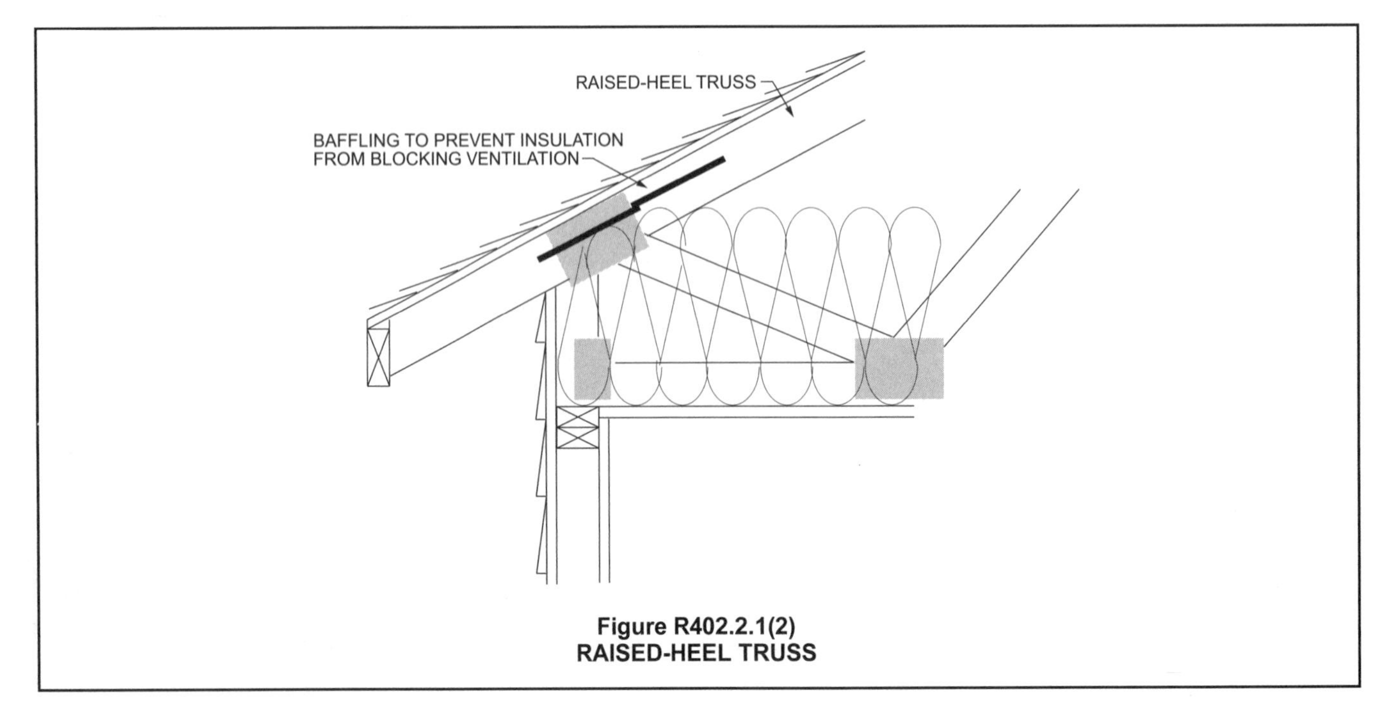

Figure R402.2.1(2)
RAISED-HEEL TRUSS

the top of the attic insulation. The baffle shall be permitted to be any solid material.

❖ For air-permeable insulations in vented attics, a baffle shall be installed adjacent to soffit and eave vents. This will help prevent wind from degrading the attic insulation performance. Baffles serve to keep vents open, insulation in place and keep the wind from blowing through the insulation and reducing its effectiveness.

R402.2.4 Access hatches and doors. Access doors from conditioned spaces to unconditioned spaces such as attics and crawl spaces shall be weatherstripped and insulated to a level equivalent to the insulation on the surrounding surfaces. Access shall be provided to all equipment that prevents damaging or compressing the insulation. A wood- framed or equivalent baffle or retainer is required to be provided when loose-fill insulation is installed, the purpose of which is to prevent the loose-fill insulation from spilling into the living space when the attic access is opened, and to provide a permanent means of maintaining the installed *R*-value of the loose-fill insulation.

> **Exception:** Vertical doors that provide access from conditioned to unconditioned spaces shall be permitted to meet the fenestration requirements of Table R402.1.2 based on the applicable climate zone specified in Chapter 3.

❖ The portion of a ceiling used for an attic access door is a weak part of the building thermal envelope. The purpose of this provision is to ensure that measures are taken to prevent large loss of energy through this opening.

R402.2.5 Mass walls. Mass walls for the purposes of this chapter shall be considered above-grade walls of concrete block, concrete, insulated concrete form (ICF), masonry cavity, brick (other than brick veneer), earth (adobe, compressed earth block, rammed earth) and solid timber/logs, or any other walls having a heat capacity greater than or equal to 6 Btu/ft^2 × °F (123 kJ/m^2 × K).

❖ The code uses a simple definition for mass walls. Walls made of the specified materials and mass are mass walls. Mass walls may meet the lower mass wall *R*-value (as compared to frame wall values) specified in their respective climate zones because of the energy-conserving characteristics of mass walls. Note that the difference between the wood-frame *R*-value is greatest in southern climates. This recognizes that the thermal "averaging" provided by mass walls is most effective in warmer climates. In the very northern climates where there is almost continual heating during parts of the year, the thermal mass is of limited value. The code simply lists the types of walls that are considered as being "mass" walls. There is no additional limitation or characteristic specified for the walls that would be applicable when using the code.

In general terms, the heat capacity is a measure of how well a material stores heat. The higher the heat capacity, the greater the amount of heat stored in the material. For example, a 6-inch (152 mm) heavyweight concrete wall has a heat capacity of 14 Btu/ft^2 × °F (620 J/m^2 × K) compared to a conventional 2-inch by 4-inch wood-framed wall with a heat capacity of approximately 3 Btu/ft^2 × °F (138 J/m^2 × K). Tables R402.1.2 and R402.1.4 are used to determine the insulation or equivalent *U*-factor requirements for mass walls. To use the tables, first consult either Table R301.1 or Figure R301.1 to determine the climate zone of the proposed project. The *R*-value for the assembly is then given for the mass wall in Table R402.1.2 or the *U*-factor is given in Table R402.1.4.

When dealing with the mass walls, the insulation location is important. Note i of Table R402.1.2 indicates that the second *R*-value, which is the higher *R*-value for mass walls in each climate zone, is applicable when more than half of the insulation is on the interior of the mass wall. The first values listed in Table R402.1.2 are, therefore, based on the installation of the majority of the insulation ("at least 50 percent") being located on the exterior side of the wall or being integral to the wall. Likewise, Note b of Table R402.1.4 provides lower maximum *U*-factors for mass walls with more than 50 percent of the insulation on the interior of the wall.

For an example of a wall assembly that has insulation on the exterior of the wall (between the mass wall and the exterior), see Commentary Figure R402.2.5(1). Concrete masonry units with insulated cores or masonry cavity walls are examples of integral insulation. For mass walls with the insulation installed on the interior (an insulated furred wall located between the conditioned space and the mass wall), see Commentary Figure R402.2.5(2). As shown, these two figures would be examples of mass walls that meet or exceed the general requirements of this section and Table R402.1.2 for Climate Zones 1 through 4, except for those in Marine Zone 4.

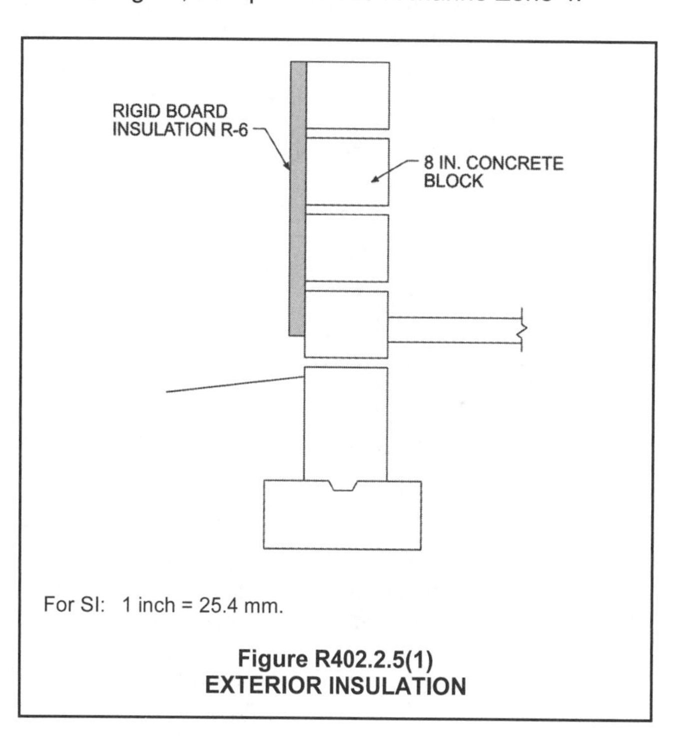

For SI: 1 inch = 25.4 mm.

Figure R402.2.5(1)
EXTERIOR INSULATION

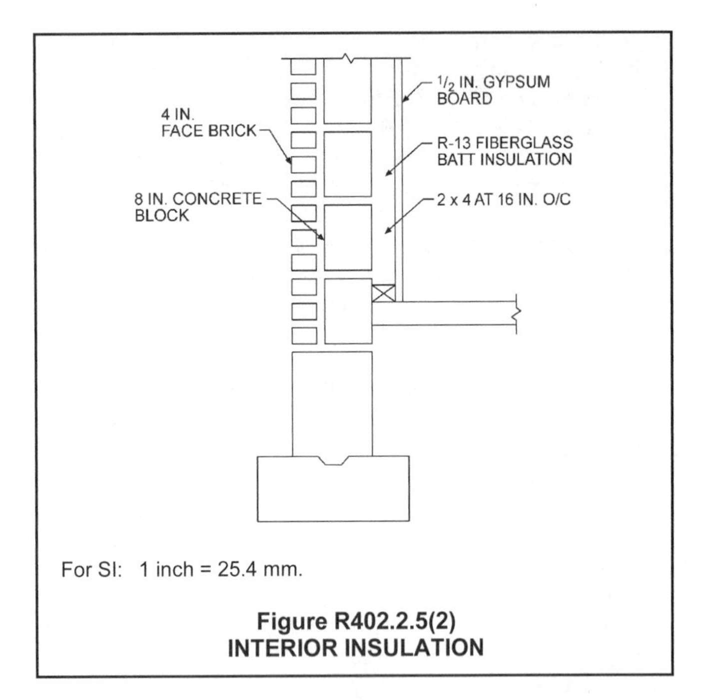

For SI: 1 inch = 25.4 mm.

Figure R402.2.5(2)
INTERIOR INSULATION

R402.2.6 Steel-frame ceilings, walls and floors. Steel-frame ceilings, walls, and floors shall meet the insulation requirements of Table R402.2.6 or shall meet the *U*-factor requirements of Table R402.1.4. The calculation of the *U*-factor for a steel-frame envelope assembly shall use a series-parallel path calculation method.

❖ The insulation requirements of Table R402.1.2 are based on conventional wood-frame construction methods. Because the code also includes provisions applicable to steel-framing methods, this chapter has been written to include this material. Table R402.2.6 specifies combinations of cavity and continuous insulation for steel framing that are equivalent to the specified wood-frame component *R*-values in Table R402.1.2.

Table R402.1.2 cannot be used directly for steel-frame components. When using Table R402.2.6, all listed options comply; the code user can choose any option on the list that corresponds to the correct *R*-value in the left-hand column. Steel has a much higher thermal conductivity (ability to transfer heat) than wood. Therefore, steel-frame cavities require a higher *R*-value or include a requirement for insulated sheathing that acts as a thermal break or both.

Instead of using this table, the code user could choose to calculate or measure a *U*-factor for a building component and show compliance based on meeting the *U*-factor requirement in Table R402.1.4. The code user could use the total UA tradeoff in Section R402.1.4 or even the performance-based approach in Section R405 to show compliance based on the overall building, even if the steel-frame wall did not meet the code requirements directly. Other combinations of cavity insulation and continuous insulation not shown in Table R402.2.6 would also be allowed; however, if a combination from the table is used, no additional calculation is necessary.

R402.2.7 Walls with partial structural sheathing. Where Section R402.1.2 would require continuous insulation on exterior walls and structural sheathing covers 40 percent or less of the gross area of all exterior walls, the continuous insulation *R*-value shall be permitted to be reduced by an amount necessary to result in a consistent total sheathing thickness, but not more than R-3, on areas of the walls covered by structural sheathing. This reduction shall not apply to the *U*-factor alternative approach in Section R402.1.4 and the total UA alternative in Section R402.1.5.

❖ The provisions of this section are intended to deal with a practical situation that occurs when structural sheathing is used over a portion of the outside wall, such as at the corners for horizontal bracing. In those

TABLE R402.2.6
STEEL-FRAME CEILING, WALL AND FLOOR INSULATION (*R*-VALUE)

WOOD FRAME *R*-VALUE REQUIREMENT	COLD-FORMED STEEL EQUIVALENT *R*-VALUE[a]
Steel Truss Ceilings[b]	
R-30	R-38 or R-30 + 3 or R-26 + 5
R-38	R-49 or R-38 + 3
R-49	R-38 + 5
Steel Joist Ceilings[b]	
R-30	R-38 in 2 × 4 or 2 × 6 or 2 × 8 R-49 in any framing
R-38	R-49 in 2 × 4 or 2 × 6 or 2 × 8 or 2 × 10
Steel-framed Wall, 16″ on center	
R-13	R-13 + 4.2 or R-19 + 2.1 or R-21 + 2.8 or R-0 + 9.3 or R-15 + 3.8 or R-21 + 3.1
R-13 + 3	R-0 + 11.2 or R-13 + 6.1 or R-15 + 5.7 or R-19 + 5.0 or R-21 + 4.7
R-20	R-0 + 14.0 or R-13 + 8.9 or R-15 + 8.5 or R-19 + 7.8 or R-19 + 6.2 or R-21 + 7.5
R-20 + 5	R-13 + 12.7 or R-15 + 12.3 or R-19 + 11.6 or R-21 + 11.3 or R-25 + 10.9
R-21	R-0 + 14.6 or R-13 + 9.5 or R-15 + 9.1 or R-19 + 8.4 or R-21 + 8.1 or R-25 + 7.7
Steel-framed Wall, 24″ on center	
R-13	R-0 + 9.3 or R-13 + 3.0 or R-15 + 2.4
R-13 + 3	R-0 + 11.2 or R-13 + 4.9 or R-15 + 4.3 or R-19 + 3.5 or R-21 + 3.1
R-20	R-0 + 14.0 or R-13 + 7.7 or R-15 + 7.1 or R-19 + 6.3 or R-21 + 5.9
R-20 + 5	R-13 + 11.5 or R-15 + 10.9 or R-19 + 10.1 or R-21 + 9.7 or R-25 + 9.1
R-21	R-0 + 14.6 or R-13 + 8.3 or R-15 + 7.7 or R-19 + 6.9 or R-21 + 6.5 or R-25 + 5.9
Steel Joist Floor	
R-13	R-19 in 2 × 6, or R-19 + 6 in 2 × 8 or 2 × 10
R-19	R-19 + 6 in 2 × 6, or R-19 + 12 in 2 × 8 or 2 × 10

a Cavity insulation *R*-value is listed first, followed by continuous insulation *R*-value.

b. Insulation exceeding the height of the framing shall cover the framing.

cases, the thickness of the portion of the wall with structural sheathing could be different than the portion of the wall without sheathing. The code therefore allows a reduction of R-3 in the required insulation of the sheathed portion to facilitate some flexibility in the choice of insulation. This option is only allowed in the application of the prescriptive table, Table R402.1.2, and not when utilizing the *U*-factor alternative or the UA alternative.

R402.2.8 Floors. Floor framing-cavity insulation shall be installed to maintain permanent contact with the underside of the subfloor decking.

> **Exception:** The floor framing-cavity insulation shall be permitted to be in contact with the topside of sheathing or continuous insulation installed on the bottom side of floor framing where combined with insulation that meets or exceeds the minimum wood frame wall *R*-value in Table 402.1.2 and that extends from the bottom to the top of all perimeter floor framing members.

❖ Floors that are a part of the building thermal envelope, such as those over a crawl space or an unconditioned garage, are required to meet or exceed the floor *R*-value requirements listed in Table R402.1.2. The insulation *R*-value requirements range from R-13 in warm climates to R-38 in extremely cold climates. Insulation must be installed between the floor joists and must be well supported with netting, wire, wood strips or another method of support so that the insulation does not droop or fall out of the joist cavities over time. Some floor insulation has a tendency to sag or drop with time. This sag or drop exposes the subfloor directly to the temperature beneath the floor. Sagging also has a tendency to open airflow paths to parts of the floor, producing cold spots and negating the value of the floor insulation in the effected section of the floor. Even small areas that lack insulation or allow air circulation between the floor insulation and the subfloor can have a marked effect on the energy efficiency of the floor. This section specifies that floor insulation must be installed so it will maintain "permanent contact" with the subfloor (meaning over the useful life of the residence).

If a floor of a building extends over outside air, such as over an open carport, the floor is still insulated to the "floor" requirement of Table R402.1.2 rather than being treated as part of the exterior envelope card meeting the same *R*-value as for ceilings.

The exception provides a specific configuration for insulating a floor. Requiring insulation in floors to be in direct contact with the underside of subfloor decking is one insulating option. Another option is to have an airspace between the floor sheathing and the top-of-cavity insulation where this cavity insulation is in direct contact with the top side of sheathing or continuous insulation installed on the underside of the floor framing and is combined with perimeter insulation that meets or exceeds the *R*-value requirements for walls. This second option leads to fewer cold spots yet does not change the heat loss as long as the cavity insulation is in direct contact with sheathing or continuous insulation below it. It also facilitates services to be enclosed in the thermal envelope. An example of this configuration is illustrated in Commentary Figure R402.2.8.

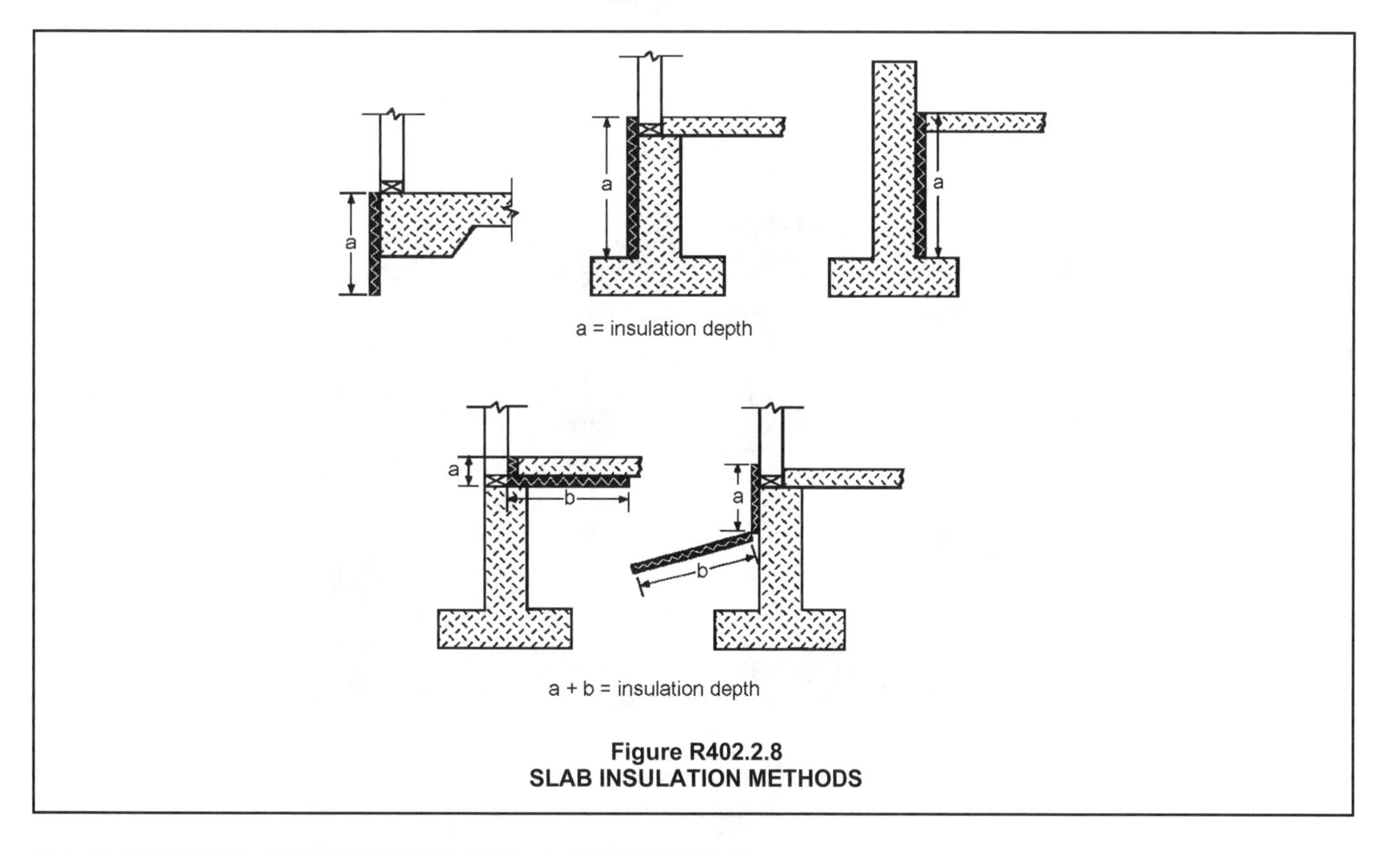

Figure R402.2.8
SLAB INSULATION METHODS

R402.2.9 Basement walls. Walls associated with conditioned basements shall be insulated from the top of the *basement wall* down to 10 feet (3048 mm) below grade or to the basement floor, whichever is less. Walls associated with unconditioned basements shall meet this requirement unless the floor overhead is insulated in accordance with Sections R402.1.2 and R402.2.8.

❖ The walls of conditioned basements must be insulated to meet the requirements of Table R402.1.2. Each wall of a basement must be considered separately to determine whether it is a basement wall or an exterior wall. It is a basement wall if it has an average below-grade wall area of 50 percent or more. A wall that is less than 50 percent below grade is an exterior wall and must meet the insulation requirements for walls. Most walls associated with basements will be at least 50 percent below grade and, therefore, must meet basement wall requirements.

Walkout basements offer a challenge in determining compliance with the code. A walkout basement [see Commentary Figures R402.2.9(1) and (2)] may have a back wall that is entirely below grade, a front wall or the walkout portion that is entirely above grade and two sidewalls with a grade line running diagonally. In this case, the back wall must meet the requirements for basement walls, the front wall would need to meet the requirements for walls (either framed wall or mass wall, as applicable) and the side walls would need to be evaluated to determine whether they were 50 percent or more below grade and, therefore, basement walls.

Basement insulation must extend up to 10 feet (3048 mm) under the ground, or at least as far as the basement wall extends under the ground. Heat flow into the ground occurs all along the buried portion of the wall, as well as along the above-ground portion of

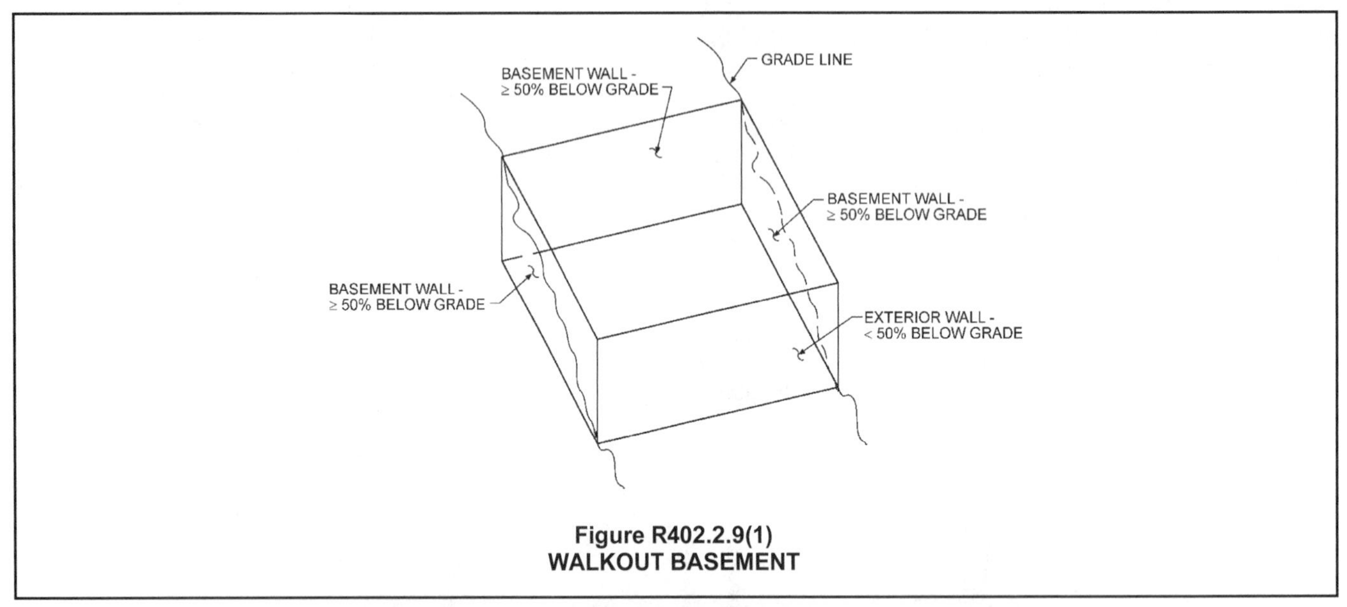

Figure R402.2.9(1)
WALKOUT BASEMENT

Figure R402.2.9(2)
EXTERIOR BASEMENT INSULATION

the wall. Heat flow below 10 feet (3048 mm) is greatly diminished so the code requires basement insulation only down to 10 feet (3048 mm), or the depth of the basement wall.

The code does not specify whether the insulation is to be placed on the inside or outside of a basement wall. In some localities moisture considerations may suggest the type and location for insulation.

The last part of this section allows insulating unconditioned basement walls as an alternative to insulating the floor above the unconditioned basement. Therefore, it essentially shifts the location of the building thermal envelope from the floor to the basement walls. Although not required, insulating the unconditioned basement walls makes a good deal of sense if a basement is likely to be conditioned at some time after construction.

Because the rim joist between floors is a part of the building envelope, this must be insulated also if the basement is conditioned.

When applying the provisions of this section, it is important to review the definitions for both "Above grade wall" and "Basement wall" in Section R202. For code users who also deal with commercial buildings, remember that the rule for residential buildings is 50 percent above or below grade and not the 85-percent minimum below grade with 15-percent maximum above-grade area that is applicable under Section C402.2.2 for commercial construction.

R402.2.10 Slab-on-grade floors. Slab-on-grade floors with a floor surface less than 12 inches (305 mm) below grade shall be insulated in accordance with Table R402.1.2. The insulation shall extend downward from the top of the slab on the outside or inside of the foundation wall. Insulation located below grade shall be extended the distance provided in Table R402.1.2 by any combination of vertical insulation, insulation extending under the slab or insulation extending out from the building. Insulation extending away from the building shall be protected by pavement or by not less than 10 inches (254 mm) of soil. The top edge of the insulation installed between the *exterior wall* and the edge of the interior slab shall be permitted to be cut at a 45-degree (0.79 rad) angle away from the *exterior wall.* Slab-edge insulation is not required in jurisdictions designated by the *code official* as having a very heavy termite infestation.

❖ The perimeter edges of slab-on-grade floors must be insulated to the *R*-values listed in Table R402.1.2. These requirements apply only to slabs 12 inches (305 mm) or less below grade. The listed *R*-value requirements in the table are for unheated slabs. A heated slab must add another R-5 to the required insulation levels based on Note d of Table R402.1.2.

The insulation must extend downward from the top of the slab or downward to the bottom of the slab and then horizontally in either direction until the distance listed in Table R402.1.2 is reached. See Commentary Figure R402.2.10 for examples on how the distance is measured. Most of the heat loss from a slab will occur in the edge that is exposed directly to the outside air. The insulation must be installed to the top of the slab edge to prevent this heat loss. Slab insulation may be installed on the exterior of the slab edge or between the interior wall and the edge of the interior slab as in a nonmonolithic slab. In this type of installation, the exposed insulation could cause problems with tack strips for carpeting. Therefore, the

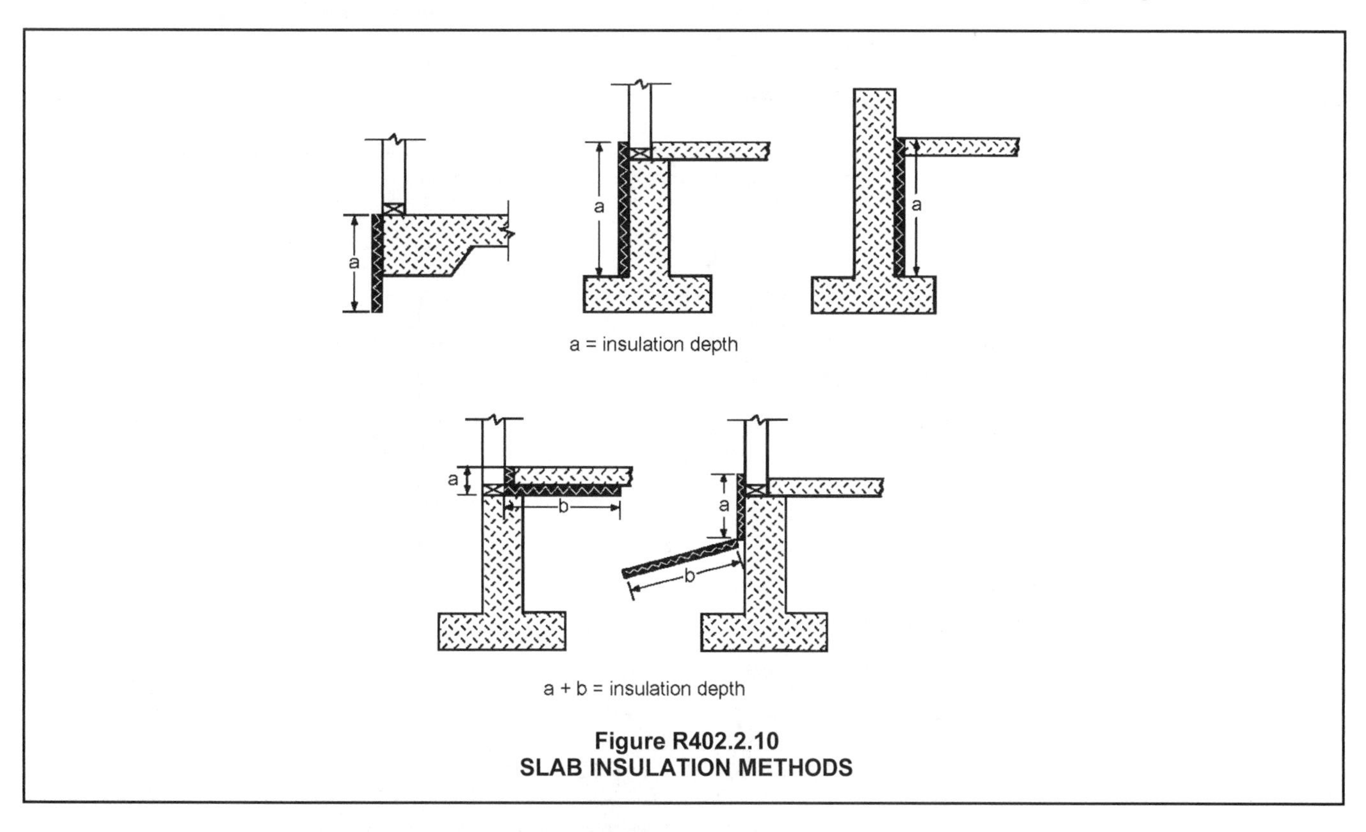

Figure R402.2.10
SLAB INSULATION METHODS

insulation is allowed to be cut at a 45-degree angle away from the exterior of the wall. If a monolithic slab and foundation is being used, the required insulation would obviously need to be installed on the exterior and then either extended to the required depth or turned out to the exterior and protected by either some type of pavement or a minimum of 10 inches (254 mm) of soil. Insulation that is exposed on or near the surface is easily damaged. This protection method ensures that the insulation remains in place and provides the intended energy savings.

In areas with very heavy termite infestation, slab perimeter insulation need not be installed in accordance with Table R402.1.2. These areas are identified in Figure 301.2(6) of the IRC or Figure 2603.9 of the IBC or by the jurisdiction, based on the local history and situation. It is important to understand that this exemption from the slab insulation provisions are for any area with heavy termite infestations. The fact that this is an exemption and does not contain any requirement for a compensating increase of insulation at other locations is important.

The requirements of Section R402.1.4 could still be used in areas that do not have a heavy termite infestation to eliminate the slab edge insulation if desired. Typically, slab perimeter insulation can be traded off entirely in these climates by increasing the ceiling or wall insulation *R*-values or by using glazing with a lower *U*-factor.

R402.2.11 Crawl space walls. As an alternative to insulating floors over crawl spaces, crawl space walls shall be permitted to be insulated when the crawl space is not vented to the outside. Crawl space wall insulation shall be permanently fastened to the wall and extend downward from the floor to the finished grade level and then vertically and/or horizontally for at least an additional 24 inches (610 mm). Exposed earth in unvented crawl space foundations shall be covered with a continuous Class I vapor retarder in accordance with the *International Building Code* or *International Residential Code*, as applicable. All joints of the vapor retarder shall overlap by 6 inches (153 mm) and be sealed or taped. The edges of the vapor retarder shall extend not less than 6 inches (153 mm) up the stem wall and shall be attached to the stem wall.

❖ The code allows for the insulation of crawl space walls instead of insulating the floor between the crawl space and the conditioned space. In essence, the code user is defining the thermal boundary as either the floor or the crawl space wall. Because the ground under the crawl space is tempered by the thermal mass of the dirt, the temperature of the crawl space is usually more favorable than the outside temperature. This is a popular practice for freeze protection of plumbing pipes in colder climates because it is common to install plumbing in the crawl space. The heat transferred through the uninsulated floor to the crawl space helps keep the crawl space temperature above freezing when the outside air temperature drops below freezing. To comply with this provision, the crawl space must be mechanically vented or supplied with conditioned air from the living space. IBC Section 1203.3 and its subsections address this crawl space ventilation requirement.

The code also requires installation of insulation from the sill plate downward to the exterior finished grade level and then an additional 24 inches (610 mm) either vertically or horizontally (see Commentary Figure R402.2.11). Under this insulation scenario, the rim joist is considered part of the conditioned envelope and must be insulated to the same level as the exterior wall. The insulation must be attached securely to the crawl space wall so that it does not fall off. The code also requires installing a continuous Class I vapor retarder on the floor of the crawl space

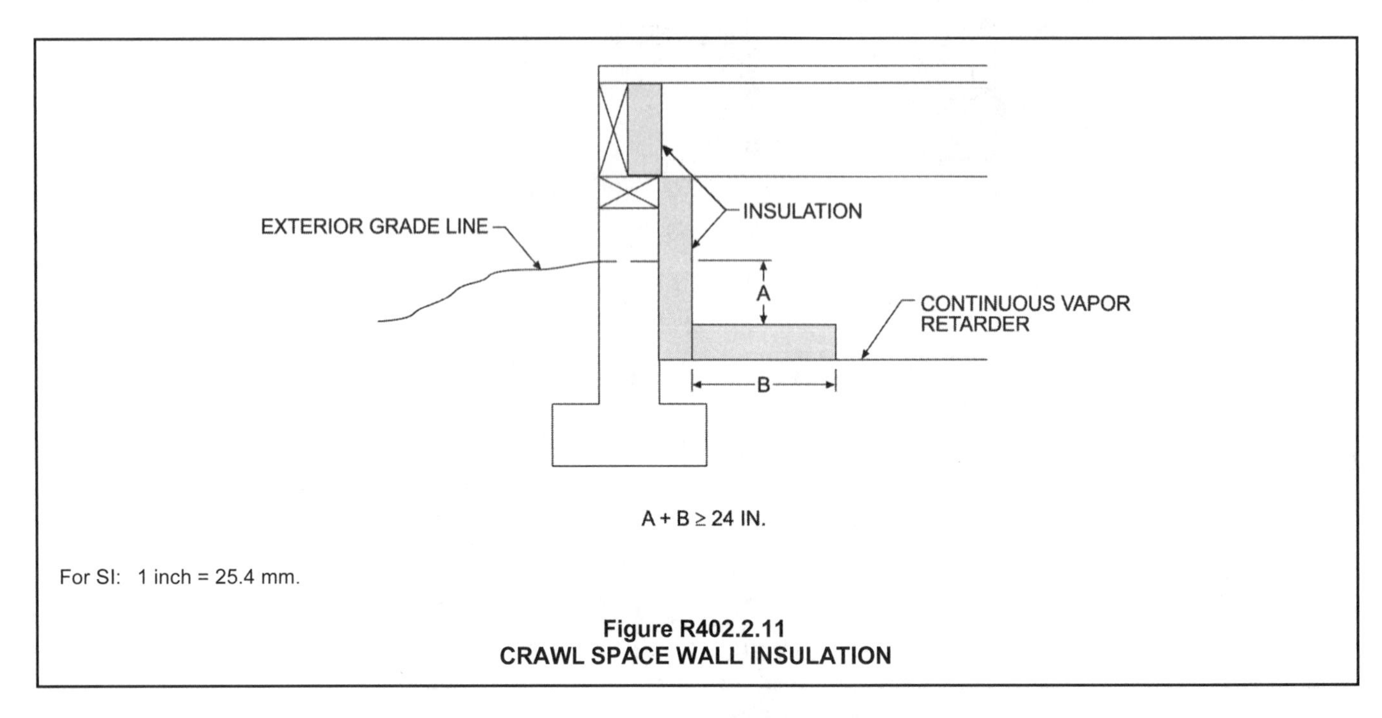

Figure R402.2.11
CRAWL SPACE WALL INSULATION

to prevent ground-water vapor from entering the crawl space. A Class I vapor retarder is defined in the IBC as a material with a permeance of 0.1 perm or less (see the definition of "Vapor retarder class" in the IBC). The vapor retarder is to be installed with all joints overlapped and sealed or taped to provide continuity. Also, the vapor retarder must extend up the crawl space wall and be secured to the wall with an appropriate attachment, such as an approved mastic or a treated wood nailer.

R402.2.12 Masonry veneer. Insulation shall not be required on the horizontal portion of the foundation that supports a masonry veneer.

❖ For exterior foundation insulation, the horizontal portion of the foundation that supports a masonry veneer need not be insulated. For slab-edge insulation installed on the exterior of the slab, the code allows the insulation to start at the bottom of the masonry veneer and extend downward. This is essentially a matter of practicality and accommodates the construction of a "brick ledge" without the need for insulating the foundation at the point where the masonry would bear on that foundation.

R402.2.13 Sunroom insulation. *Sunrooms* enclosing conditioned space shall meet the insulation requirements of this code.

Exception: For *sunrooms* with *thermal isolation*, and enclosing conditioned space, the following exceptions to the insulation requirements of this code shall apply:

1. The minimum ceiling insulation *R*-values shall be R-19 in *Climate Zones* 1 through 4 and R-24 in *Climate Zones* 5 through 8.
2. The minimum wall *R*-value shall be R-13 in all *climate zones*. Walls separating a *sunroom* with a *thermal isolation* from *conditioned space* shall meet the *building thermal envelope* requirements of this code.

❖ With the amount of glass in sunrooms, it is imperative the insulation requirements of the code be complied with to ensure minimal current requirements. The exceptions also tighten the requirements per zone to minimize energy usage.

R402.3 Fenestration (Prescriptive). In addition to the requirements of Section R402, fenestration shall comply with Sections R402.3.1 through R402.3.6.

❖ This section contains specific requirements that affect the requirements for individual items listed in the subsections. Although Section R402.1 and Tables R402.1.2 and R402.1.4 provide the general basis for complying with the energy requirements of this chapter, Section R402.3 provides additional details regarding the application of the provisions or modifications that may affect the general fenestration requirements. Relying on the general code policy that specific requirements apply over general requirements will help ensure that these specific provisions are properly applied.

The term "fenestration" in this section refers to opaque doors and the light-transmitting areas of a residential building's wall, floor or roof, generally window, skylight and nonopaque door products (see the definition of "Fenestration" in Chapter 2). Prior to the 2006 edition, the residential energy provisions of the code applied only to buildings with 15 percent or less of glazing areas (fenestration). The code's prescriptive requirements varied depending on the type of residential occupancy, but did include limitations on the amount of fenestration (maximum of 25 percent for one- and two-family dwellings and 30 percent for R-2, R-4 or townhouses). These earlier versions of the code established whole building performance requirements, with fenestration performance requirements as a derivative value dependent on window area, overall envelope area and the performance of other assemblies (e.g., walls, ceilings, floors). The code establishes specific simplified prescriptive requirements (without area considerations) for fenestration products in Table R402.1.2—specifically, fenestration *U*-factors, skylight *U*-factors and glazed fenestration SHGCs. The elimination of these fenestration area limitations helped to greatly simplify the application of the code's envelope requirements.

The fenestration requirements of the code are critical to the overall energy efficiency of the residence. First, unlike opaque assemblies, glazed fenestration can transmit a substantial amount of heat through the glazing into the living space in both the summer and winter, resulting in a unique concern about solar heat gain (and, as a result, necessitating SHGC requirements). Second, the insulating value (*U*-factor) of typical fenestration is much higher than that of a typical wall. For example, a good low emissivity (low-E), insulated glass wood or vinyl fenestration product will have less than one-fourth the insulating value of an equivalent area of R-13 insulated opaque wall. These issues have an effect not only on energy use, but overall occupant comfort, condensation and other issues.

In accordance with Section R303.1.3, the *U*-factor and SHGC for each fenestration product must be obtained from a label attached to the product certifying that the values were determined in accordance with NFRC procedures by an accredited, independent lab or from a limited default table.

R402.3.1 *U*-factor. An area-weighted average of fenestration products shall be permitted to satisfy the *U*-factor requirements.

❖ Section R402.3.1 permits using the calculated area-weighted average *U*-factor of all fenestration products in the building to satisfy the fenestration *U*-factor requirements set by Table R402.1.2 or R402.1.4. As a result, if all fenestration products (window, door or skylight) do not meet the specific value, the user can still achieve compliance if the weighted average of all products is equal to or less than the specified value.

This option permits the use of some windows that have values lower than the prescriptive general requirement, as long as these poorly performing windows are offset by windows with values better than the requirement.

When applying this area-weighted option, it is important to remember that the term "fenestration" includes windows, skylights, doors with glazing and opaque doors, all of which would be included in the average calculation.

Using the *U*-factor requirement of 0.35 for Climate Zone 6 as an illustration, this section provides two options for compliance. The simplest option is to ensure that all windows and doors have labeled NFRC values of 0.35 or less. This approach is also more likely to ensure adequate performance and comfort throughout the home. Alternatively, a weighted average may be taken of the values from all windows and doors to see if the weighted average is less than or equal to 0.35.

As a simple example, assume 100 square feet (9.3 m^2) of 0.32 windows, 100 square feet (9.3 m^2) of 0.36 windows and one 20-square-foot (1.8 m^2) 0.40 *U*-factor door $[(100 \times 0.32) + (100 \times 0.36) + (20 \times 0.40)]/(100 + 100 + 20) = 0.345$ (weighted average *U*-factor). Therefore, because the weighted average *U*-factor is less than the required 0.35, the fenestration in this example would be in compliance with the code.

In accordance with Section R303.1.2, the *U*-factor for each fenestration product must be obtained from a label attached to the product certifying that the *U*-factor was determined in accordance with NFRC procedures by an accredited, independent lab. In the absence of an NFRC-labeled *U*-factor, a value from the limited default tables [Tables R303.1.3(1) and R303.1.3(2)] must be used. The provisions of Section R402.4.5 should be reviewed when using the area-weighted average in the tradeoff options (see commentary, Section R402.5).

R402.3.2 Glazed fenestration SHGC. An area-weighted average of fenestration products more than 50-percent glazed shall be permitted to satisfy the SHGC requirements.

Dynamic glazing shall be permitted to satisfy the SHGC requirements of Table R402.1.2 provided the ratio of the higher to lower labeled SHGC is greater than or equal to 2.4, and the *dynamic glazing* is automatically controlled to modulate the amount of solar gain into the space in multiple steps. *Dynamic glazing* shall be considered separately from other fenestration, and area-weighted averaging with other fenestration that is not dynamic glazing shall not be permitted.

> **Exception:** *Dynamic glazing* is not required to comply with this section when both the lower and higher labeled SHGC already comply with the requirements of Table R402.1.2.

❖ Under Table R402.1.2, all glazed fenestration products in Climate Zones 1 through 3 must have an SHGC equal to or less than 0.30 (there is no requirement in Climate Zones 4 through 8). This requirement is intended to control unwanted solar gain in cooling-dominated climates to increase comfort, reduce air-conditioning energy and peaks, reduce HVAC sizing and reduce energy costs.

Similar to Section R402.3.1, Section R402.3.2 allows some latitude for individual product variability by permitting this performance requirement to be met using an area-weighted average of all of the fenestration products that are more than 50-percent glazed. The 50-percent glazing threshold is established to exclude doors or other fenestration products that are either completely or largely opaque from the equation. The reason for this exclusion is that opaque elements do not allow solar heat gain as glazing does. The area-weighted calculation approach is explained in an example with Section R402.3.1. An additional example for SHGC is as follows:

Window 1	SHGC - 0.24	200 ft^2
Window 2	SHGC - 0.32	100 ft^2
Window 3	SHGC - 0.30	100 ft^2
Sliding glass door	SHGC - 0.40	40 ft^2

$[(200\ ft^2 \times 0.24) + (100\ ft^2 \times 0.32) + (100\ ft^2 \times 0.30) + (40\ ft^2 \times 0.40)]/440\ ft^2$

440 ft^2 = SHGC - 0.29 Average

Using the figures in the example, even though Window 2 and the sliding glass door both have SHGC values that exceed the 0.25 limitation of Table R402.1.2 for Climate Zones 1 through 3, this design will comply because the weighted average is 0.34.

In accordance with Section R303.1.3, the SHGC for each glazed fenestration product must be obtained from a label attached to the product certifying that the SHGC was determined in accordance with NFRC procedures by an accredited, independent lab. In the absence of an NFRC-labeled SHGC, a value from the limited default table [Table R303.1.3(1)] must be used.

It is important to note that the SHGC requirement must be met by the fenestration product on a stand-alone basis. The code does not permit the alternative of a "permanent solar shading device" such as eave overhangs or awnings, as was permitted by other energy codes and previous versions of the code, to assist in code compliance.

R402.3.3 Glazed fenestration exemption. Up to 15 square feet (1.4 m^2) of glazed fenestration per dwelling unit shall be permitted to be exempt from *U*-factor and SHGC requirements in Section R402.1.2. This exemption shall not apply to the *U*-factor alternative approach in Section R402.1.4 and the Total UA alternative in Section R402.1.5.

❖ In addition to using the area-weighted average approach (Sections R402.3.1 and R402.3.2) to allow maximum compliance flexibility for builders, the code allows up to 15 square feet (1.4 m^2) of the building's total glazed fenestration area to be exempt from the *U*-factor and SHGC requirements listed in Table

R402.1.2. All other glazing must meet or exceed the designated *U*-factor and SHGC requirements. The exempted glazing area should be designated on the building plan, either on the floor plan or in a window schedule. This will give the necessary information to the field inspector who will then need to verify that what is on the plans is installed in the field. This exemption allows the use of ornate or unique window, skylight or glazed door assemblies in a building without going to another compliance approach. The area, the *U*-factor and SHGC of the exempt product(s) should be excluded from the area-weighted calculations that may be performed under Sections R402.3.1 and R402.3.2. In addition, the exception provided by this section would also allow this 15 square feet (1.4 m^2) of glazing to be excluded from the limits of Section R402.5. Note that this exemption does not apply where the basis for the thermal envelope was the *U*-factor alternative in Section R402.1.4 or the total UA alternative in Section R402.1.4. In these cases, the amount of insulation does not need to be exempted.

R402.3.4 Opaque door exemption. One side-hinged opaque door assembly up to 24 square feet (2.22 m^2) in area is exempted from the *U*-factor requirement in Section R402.1.4. This exemption shall not apply to the *U*-factor alternative approach in Section R402.1.4 and the total UA alternative in Section R402.1.5.

❖ Similar to the exemption provided in Section R402.3.3 to enhance design flexibility, the code allows one side-hinged opaque door to be exempt from fenestration *U*-factor requirements as contained in Table R402.1.2, as well as the limitations of Section R402.5. Although the code does not define it, an opaque door is generally considered to be a fenestration product with an overall glazing area of less than 50 percent. The opaque door exemption allows builders to use an ornate or otherwise *U*-factor noncompliant entrance door assembly in a building without going to another compliance approach. The area and the *U*-factor of the exempt product should be excluded from the area-weighted calculations under Section R402.4.1.

R402.3.5 Sunroom fenestration. *Sunrooms* enclosing *conditioned space* shall meet the fenestration requirements of this code.

Exception: For *sunrooms* with *thermal isolation* and enclosing *conditioned space* in *Climate Zones* 2 through 8, the maximum fenestration *U*-factor shall be 0.45 and the maximum skylight *U*-factor shall be 0.70.

New fenestration separating the *sunroom* with *thermal isolation* from *conditioned space* shall meet the *building thermal envelope* requirements of this code.

❖ This section simply reminds the user that sunrooms enclosing conditioned spaces are subject to code requirements. The exception tightens the maximum fenestration *U*-factor for Climate Zones 4 through 8. The maximum skylight *U*-factor shall be 0.70. Given the amount of glass in a sunroom, this section of the code tightens up the fenestration requirements and minimizes energy usage.

R402.4 Air leakage (Mandatory). The *building thermal envelope* shall be constructed to limit air leakage in accordance with the requirements of Sections R402.4.1 through R402.4.4.

❖ Sealing the building envelope is critical to good thermal performance of the building. The seal will prevent warm, conditioned air from leaking out around doors, windows and other cracks during the heating season, thereby reducing the cost of heating the residence. During hot summer months, a proper seal will stop hot air from entering the residence, helping to reduce the air-conditioning load on the building. Any penetration in the building envelope must be thoroughly sealed during the construction process, including holes made for the installation of plumbing, electrical and heating and cooling systems (see Commentary Figure R402.4). The code lists several areas that must be caulked, gasketed, weatherstripped, wrapped or otherwise sealed to limit uncontrolled air movement. Most of the air sealing will be done prior to the installation of an interior wall covering because any penetration will be noticeable and accessible at this time. The code allows the use of airflow retarders (house wraps) or other solid materials as an acceptable method to meet this requirement. To be effective, the building thermal envelope seal must be:

- Impermeable to airflow.
- Continuous over the entire building envelope.
- Able to withstand the forces that may act on it during and after construction.
- Durable over the expected lifetime of the building.

It is unlikely that the same type of barrier will be used on all portions of the building's thermal envelope. Therefore, joints between the various elements, as well as joints or splices within products (such as the overlap in separate pieces of house wrap), must be effectively addressed to provide the continuity needed to perform as desired.

R402.4.1 Building thermal envelope. The *building thermal envelope* shall comply with Sections R402.4.1.1 and R402.4.1.2. The sealing methods between dissimilar materials shall allow for differential expansion and contraction.

❖ Air infiltration is a major source of energy use because the incoming air usually requires conditioning. The uncontrolled introduction of outside air (infiltration) creates a load that varies with time. Ventilation, the controlled introduction of fresh air, is more manageable and provides a more controlled air quality. Uncontrolled infiltration also has a tendency to create or aggravate moisture problems, providing an additional reason to limit infiltration.

R402.4.1.1 Installation. The components of the *building thermal envelope* as listed in Table R402.4.1.1 shall be installed in accordance with the manufacturer's instructions and the criteria listed in Table R402.4.1.1, as applicable to the method of construction. Where required by the *code official*, an *approved* third party shall inspect all components and verify compliance.

❖ The provisions of this section are intended to reduce the energy loss to infiltration and to improve insulation installation.

The code allows all materials that are commonly used as sheathing to be part of the thermal envelope, including interior drywall. By the definition of an air barrier, gypsum board should be considered, as should exterior sheathing. The solid sheathing is not enough; for the air barrier to be effective the joints and openings must be sealed. The code does not require the air barrier on the inside of the air-permeable insulation.

In some cases, inspection of the air barrier is better performed by individuals with more expertise than the staff of the authority having jurisdiction might have. Therefore, the code gives the code official authority to require third-party inspectors.

Note that the line item in the table regarding fireplaces requires gaskets on fireplace doors. All wood-burning fireplaces are a source of air leakage. Loss of energy through these units can be reduced with gasketed doors and a requirement that combustion air be brought directly from the outdoors to the firebox. Gaskets on the fireplace doors will help to minimize air leakage into the firebox. Air that leaks past a poorly gasketed fireplace door, or a door that is simply left open, will flow up the chimney aided by the chimney draft. During the majority of the year the fireplace will not be operating. The combination of well-gasketed doors and a well-sealed flue damper will prevent air leakage through what is effectively an enormous hole in the thermal envelope of the building.

However, it should be noted that the difficulty with this requirement is that most factory-built fireplaces tested in accordance with the code-required test standard for factory-built fireplaces, UL127, are not tested and listed with doors. Therefore, there is a significant safety hazard in adding gasketed doors to a factory-built fireplace tested and manufactured in accordance with UL127, without doors. That hazard would be overheating of the unit and high likelihood of causing fires. This provision in Table R402.4.1.1 for gasketed fireplace doors is a general provision to prevent air leakage and was not intended to prohibit the use of factory-built fireplaces listed for use without doors, in violation of that listing. The specific requirements of the test standard UL127, the product listing and the manufacturer's instructions would prevail for factory-built fireplaces. Keep in mind that Section R402.4.3 requires tight fitting dampers for all fireplaces as well, which will prevent air leakage.

TABLE R402.4.1.1. See page R4-23.

❖ This table contains the list of items that are required to be inspected if the visual inspection option for demonstrating building air tightness given in Section R402.4.1.1 is chosen.

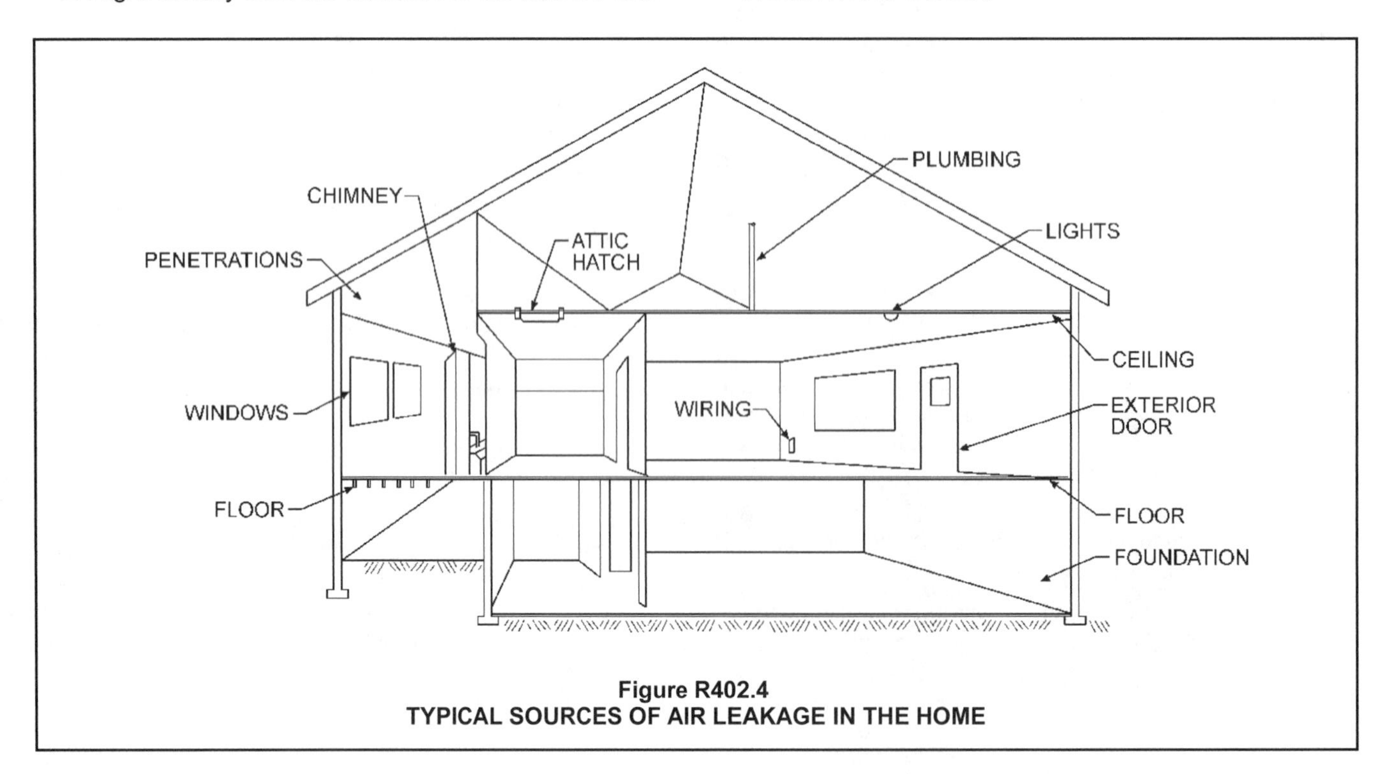

Figure R402.4
TYPICAL SOURCES OF AIR LEAKAGE IN THE HOME

TABLE R402.4.1.1
AIR BARRIER AND INSULATION INSTALLATION

COMPONENT	AIR BARRIER CRITERIA[a]	INSULATION INSTALLATION CRITERIA
General requirements	A continuous air barrier shall be installed in the building envelope. The exterior thermal envelope contains a continuous air barrier. Breaks or joints in the air barrier shall be sealed.	Air-permeable insulation shall not be used as a sealing material.
Ceiling/attic	The air barrier in any dropped ceiling/soffit shall be aligned with the insulation and any gaps in the air barrier shall be sealed. Access openings, drop down stairs or knee wall doors to unconditioned attic spaces shall be sealed.	The insulation in any dropped ceiling/soffit shall be aligned with the air barrier.
Walls	The junction of the foundation and sill plate shall be sealed. The junction of the top plate and the top of exterior walls shall be sealed. Knee walls shall be sealed.	Cavities within corners and headers of frame walls shall be insulated by completely filling the cavity with a material having a thermal resistance of R-3 per inch minimum. Exterior thermal envelope insulation for framed walls shall be installed in substantial contact and continuous alignment with the air barrier.
Windows, skylights and doors	The space between window/door jambs and framing, and skylights and framing shall be sealed.	
Rim joists	Rim joists shall include the air barrier.	Rim joists shall be insulated.
Floors (including above garage and cantilevered floors)	The air barrier shall be installed at any exposed edge of insulation.	Floor framing cavity insulation shall be installed to maintain permanent contact with the underside of subfloor decking, or floor framing cavity insulation shall be permitted to be in contact with the top side of sheathing, or continuous insulation installed on the underside of floor framing and extends from the bottom to the top of all perimeter floor framing members.
Crawl space walls	Exposed earth in unvented crawl spaces shall be covered with a Class I vapor retarder with overlapping joints taped.	Where provided instead of floor insulation, insulation shall be permanently attached to the crawlspace walls.
Shafts, penetrations	Duct shafts, utility penetrations, and flue shafts opening to exterior or unconditioned space shall be sealed.	
Narrow cavities		Batts in narrow cavities shall be cut to fit, or narrow cavities shall be filled by insulation that on installation readily conforms to the available cavity space.
Garage separation	Air sealing shall be provided between the garage and conditioned spaces.	
Recessed lighting	Recessed light fixtures installed in the building thermal envelope shall be sealed to the drywall.	Recessed light fixtures installed in the building thermal envelope shall be air tight and IC rated.
Plumbing and wiring		Batt insulation shall be cut neatly to fit around wiring and plumbing in exterior walls, or insulation that on installation readily conforms to available space shall extend behind piping and wiring.
Shower/tub on exterior wall	The air barrier installed at exterior walls adjacent to showers and tubs shall separate them from the showers and tubs.	Exterior walls adjacent to showers and tubs shall be insulated.
Electrical/phone box on exterior walls	The air barrier shall be installed behind electrical or communication boxes or air-sealed boxes shall be installed.	
HVAC register boots	HVAC register boots that penetrate building thermal envelope shall be sealed to the subfloor or drywall.	
Concealed sprinklers	When required to be sealed, concealed fire sprinklers shall only be sealed in a manner that is recommended by the manufacturer. Caulking or other adhesive sealants shall not be used to fill voids between fire sprinkler cover plates and walls or ceilings.	

a. In addition, inspection of log walls shall be in accordance with the provisions of ICC-400.

R402.4.1.2 Testing. The building or dwelling unit shall be tested and verified as having an air leakage rate not exceeding five air changes per hour in Climate Zones 1 and 2, and three air changes per hour in Climate Zones 3 through 8. Testing shall be conducted in accordance with ASTM E 779 or ASTM E 1827 and reported at a pressure of 0.2 inch w.g. (50 Pascals). Where required by the *code official*, testing shall be conducted by an *approved* third party. A written report of the results of the test shall be signed by the party conducting the test and provided to the *code official*. Testing shall be performed at any time after creation of all penetrations of the *building thermal envelope*.

During testing:

1. Exterior windows and doors, fireplace and stove doors shall be closed, but not sealed, beyond the intended weatherstripping or other infiltration control measures.
2. Dampers including exhaust, intake, makeup air, backdraft and flue dampers shall be closed, but not sealed beyond intended infiltration control measures.
3. Interior doors, if installed at the time of the test, shall be open.
4. Exterior doors for continuous ventilation systems and heat recovery ventilators shall be closed and sealed.
5. Heating and cooling systems, if installed at the time of the test, shall be turned off.
6. Supply and return registers, if installed at the time of the test, shall be fully open.

❖ The purpose of this code section is to test the building or dwelling unit to demonstrate the building's air tightness. A blower door test, which is a house pressurization test, should be done with a blower door at a pressure of 0.2 inches water gauge [50 Pascals (1psf)]. The building or dwelling unit shall be tested and verified as having an air leakage rate of not exceeding 5 ACH50, or five air changes per hour at 50 Pascals (1 psf) in Climate Zones 1 and 2, and 3 ACH50, or three air changes per hour at 50 Pascals (1 psf), in Climate Zones 3 through 8. The ACH50 is a common measurement made when doing air infiltration tests and, therefore, a reasonable metric for use in the code. Testing can be conducted by an approved third party, if allowed by the code official. This requires that HVAC ducts not be sealed during the test. In this context, "sealed" is intended to mean sealed off from the interior of the house. The maximum is 5ACH50, or five air changes per hour at 50 Pascals (1 psf).

R402.4.2 Fireplaces. New wood-burning fireplaces shall have tight-fitting flue dampers or doors, and outdoor combustion air. Where using tight-fitting doors on factory-built fireplaces listed and labeled in accordance with UL 127, the doors shall be tested and listed for the fireplace. Where using tight-fitting doors on masonry fireplaces, the doors shall be listed and labeled in accordance with UL 907.

❖ All wood-burning fireplaces are a source of air leakage. Loss of energy through these units can be reduced with gasketed doors and a requirement that combustion air be brought directly from the outdoors to the firebox. Gaskets on the fireplace doors will help to minimize air leakage into the firebox. Air that leaks past a poorly fitted fireplace door, or a door that is simply left open, will flow up the chimney aided by the chimney draft. Throughout the majority of the year, the fireplace will not be in use. The combination of tight-fitting doors and a tight flue damper will prevent air leakage through what is effectively an enormous hole in the thermal envelope of the building. However, not all factory-built fireplaces are tested for use with doors; therefore, in some cases installing doors on the fireplace would violate the unit's listing and could create a fire hazard. Doors on factory-built fireplaces are required to be tested and listed for the fireplace in accordance with UL 127. The manufacturers' instructions for factory-built fireplaces should be followed and such instructions could prohibit the installation of doors. Doors on masonry fireplaces are required to be listed and labeled in accordance with UL 907. The code also allows tight-fitting flue dampers for fireplaces to deal with the issue of air loss.

Lastly, this section and Section R1006 of the IRC require factory-built and masonry fireplaces to be equipped with a direct outdoor combustion supply. The provisions of Section R1006.2 of the IRC would prevent the installation of a fireplace with the firebox floor below finished grade.

R402.4.3 Fenestration air leakage. Windows, skylights and sliding glass doors shall have an air infiltration rate of no more than 0.3 cfm per square foot (1.5 $L/s/m^2$), and swinging doors no more than 0.5 cfm per square foot (2.6 $L/s/m^2$), when tested according to NFRC 400 or AAMA/WDMA/CSA 101/I.S.2/A440 by an accredited, independent laboratory and *listed* and *labeled* by the manufacturer.

Exception: Site-built windows, skylights and doors.

❖ Windows, skylights and doors should be tested and labeled by the manufacturer as meeting the air infiltration requirements. The intent of this section is to effectively complete the sealing of the building's thermal envelope by providing specific testing and performance criteria for windows, skylights and doors. This testing and labeling requirement provides an easy method for both the builder and the inspector to demonstrate compliance with the code. While "site built" fenestration is exempted from these requirements, units would have to be "durably sealed" to limit infiltration according to the requirements in Section R402.4.1.

R402.4.4 Rooms containing fuel-burning appliances. In Climate Zones 3 through 8, where open combustion air ducts provide combustion air to open combustion fuel burning appliances, the appliances and combustion air opening shall be located outside the building thermal envelope or enclosed in a room, isolated from inside the thermal envelope. Such rooms shall be sealed and insulated in accordance with the envelope requirements of Table R402.1.2, where the walls, floors and ceilings shall meet not less than the basement wall *R*-value requirement. The door into the room shall be fully gasketed and any water lines and ducts in the room insulated in accordance with Section R403. The combustion air duct shall be insulated where it passes through conditioned space to a minimum of R-8.

Exceptions:

1. Direct vent appliances with both intake and exhaust pipes installed continuous to the outside.
2. Fireplaces and stoves complying with Section R402.4.2 and Section R1006 of the *International Residential Code*.

❖ The entire set of provisions in the code for air leakage is of little value when a combustion air duct is installed, open to the conditioned space, virtually placing a large hole through the thermal envelope. Blower door testing as now required by the code cannot be accomplished with a combustion air opening inside the thermal envelope. Testers regularly block such openings as this is the only way they can pressurize the home; only to be opened after the test is completed. Ideally, direct vent, sealed combustion appliances solve the problem. Where less efficient, open combustion fuel burning appliances are used, it is reasonable and proper to isolate the appliances and the required combustion air from inside the thermal envelope.

R402.4.5 Recessed lighting. Recessed luminaires installed in the *building thermal envelope* shall be sealed to limit air leakage between conditioned and unconditioned spaces. All recessed luminaires shall be IC-rated and *labeled* as having an air leakage rate not more than 2.0 cfm (0.944 L/s) when tested in accordance with ASTM E 283 at a 1.57 psf (75 Pa) pressure differential. All recessed luminaires shall be sealed with a gasket or caulk between the housing and the interior wall or ceiling covering.

❖ To correctly apply this provision, it is important to realize that it deals only with recessed lights that are installed in the building thermal envelope. Therefore, lights that are located so that all sides of the luminaire are surrounded by conditioned space would not fall under this section's requirements. For example, a light located in a soffit would not be regulated if the soffit was below a ceiling that served as the building's thermal envelope. Additionally, a light installed in the floor/ceiling assembly between a first-floor living room and a bedroom above it would be exempted.

Because of their typical location of installation, recessed lighting fixtures pose a potential fire hazard if incorrectly covered with insulation. In addition, the ceiling or a barrier directly above it often serves as the air leakage or moisture barrier for the home.

Therefore, holes cut through the ceiling to install these fixtures also act as chimneys that transfer heat loss and moisture through the building envelope into attic spaces. The heat loss resulting from improperly insulated recessed lighting fixtures can be significant.

Recessed lighting fixtures must be insulation contact (IC) rated lights, which are typically double-can fixtures, with one can inside another (see Commentary Figure R402.4.5). The outer can (in contact with the insulation) is tested to make sure it remains cool enough to avoid a fire hazard. An IC-rated fixture should have the IC rating stamped on the fixture or printed on an attached label.

Recessed lights must also be tightly sealed or gasketed to prevent air leakage through the fixture into the ceiling cavity.

R402.5 Maximum fenestration *U*-factor and SHGC (Mandatory). The area-weighted average maximum fenestration *U*-factor permitted using tradeoffs from Section R402.1.5 or R405 shall be 0.48 in Climate Zones 4 and 5 and 0.40 in Climate Zones 6 through 8 for vertical fenestration, and 0.75 in Climate Zones 4 through 8 for skylights. The area-weighted average maximum fenestration SHGC permitted using tradeoffs from Section R405 in Climate Zones 1 through 3 shall be 0.50.

❖ This section is intended to clarify the application of the fenestration performance maximums and to set reasonable performance levels for these products when using the tradeoffs of Section R402.1.5 or R405.

This section does not define the minimum code requirements for fenestration *U*-factors that are set in Table R402.1.2. Rather, this section sets limits on the tradeoffs allowed based on the total UA calculations

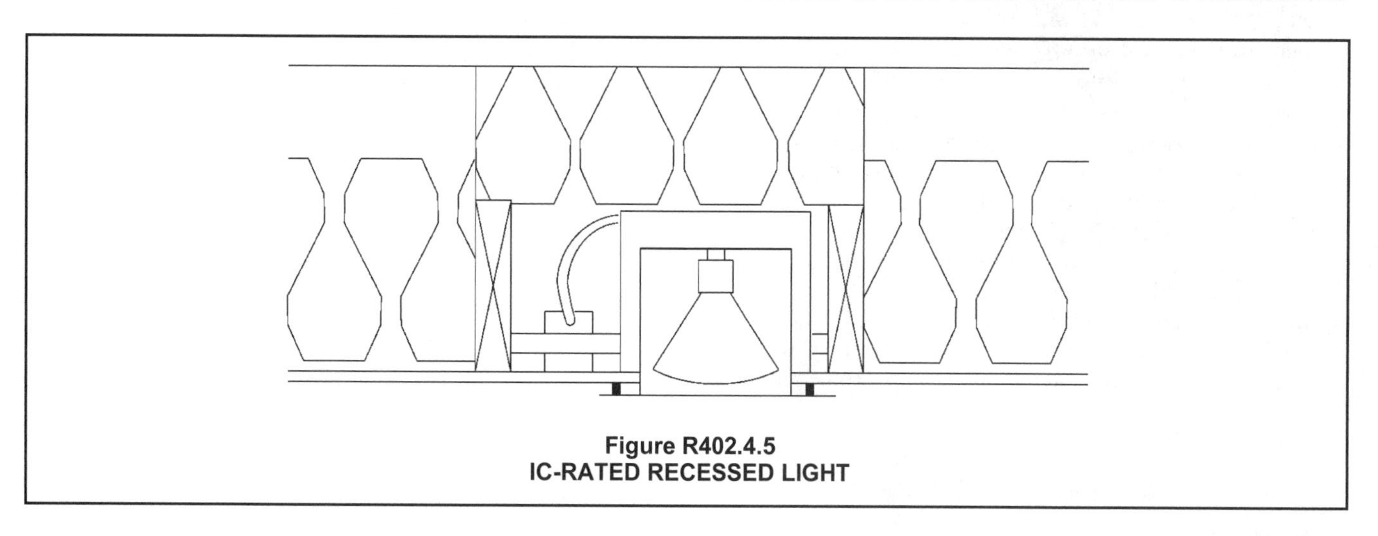

Figure R402.4.5
IC-RATED RECESSED LIGHT

(see Section R402.1.4) or the simulated performance option found in Section R405. This section is in contrast to the basic principle used in Section R402.1.5. Although that section allows a *U*-factor that does not meet the general code requirements to be offset by the increased efficiency of another part of the residence, this section establishes a limitation on the level of efficiency that can be compensated for.

Note that this section does not set a limit on individual products because Sections R402.3.1 and R402.3.2 are not affected. Rather, this section sets an overall limit on the weighted average of those products. These limits are often called "hard limits" or "maximum tradeoff" limits.

In situations where the code is applied to only a portion of the residence, such as only to the addition but not the existing residence, the weighted average could be calculated for just the addition (see commentary, Sections R402.3.1 and R402.3.2).

The thermal properties (*U*-factor) of skylights are different than those of window products. After installation, they perform differently than windows and are, therefore, rated differently. As a result, even the highest rated skylights cannot achieve the same level of *U*-factor performance as windows and glass doors. Therefore, the code provides different values for these "maximum" limits in the tradeoff.

The SHGC provisions are limited in the cooling-dominated climate zones. Therefore, where solar heat gain through the fenestration can affect the cooling load, comfort and efficiency, a "hard limit" of 0.5 would be applied when using Section R405.

SECTION R403
SYSTEMS

R403.1 Controls (Mandatory). At least one thermostat shall be provided for each separate heating and cooling system.

❖ This provision ensures that a separate thermostat is installed for each system. As an example, if separate systems are installed so that one serves the downstairs and one serves the upstairs of a two-story residence, two separate thermostats would be required, one regulating each level. This allows for greater flexibility, control and energy savings than would be possible if both systems were controlled by a single thermostat.

R403.1.1 Programmable thermostat. The thermostat controlling the primary heating or cooling system of the dwelling unit shall be capable of controlling the heating and cooling system on a daily schedule to maintain different temperature set points at different times of the day. This thermostat shall include the capability to set back or temporarily operate the system to maintain *zone* temperatures down to 55°F (13°C) or up to 85°F (29°C). The thermostat shall initially be programmed by the manufacturer with a heating temperature set point no higher than 70°F (21°C) and a cooling temperature set point no lower than 78°F (26°C).

❖ This code section provides each household an opportunity for energy savings by requiring a programmable thermostat that allows changing the temperature setpoints automatically throughout the day.

R403.1.2 Heat pump supplementary heat (Mandatory). Heat pumps having supplementary electric-resistance heat shall have controls that, except during defrost, prevent supplemental heat operation when the heat pump compressor can meet the heating load.

❖ Heat pump systems must have controls that prevent supplementary electric resistance heater operation when the heating load can be met by the heat pump alone. Typically, these controls will be thermostats that will have a "heat pump" designation on them. The make and model of the thermostat should be called out on the building plans so the inspector can verify that what is installed in the field matches the plans. Because change-outs in the field are common, the instructions that come with the thermostat can also be checked to verify that the thermostat is designed for use with a particular heat pump. To limit the hours of use of the electric-resistance heating unit and provide the most cost-efficient operation of the equipment, the specific control language is included in this section.

R403.2 Hot water boiler outdoor temperature setback. Hot water boilers that supply heat to the building through one- or two-pipe heating systems shall have an outdoor setback control that lowers the boiler water temperature based on the outdoor temperature.

❖ This section provides a requirement that gives each household with a hot water boiler an opportunity for energy savings by requiring a setback that will lower the temperature of the water based on ambient conditions.

R403.3 Ducts. Ducts and air handlers shall be in accordance with Sections R403.3.1 through R403.3.5.

❖ The five subsections of Section R403.3 provide the requirements of this chapter, which apply to the ducts and air handlers used in residential buildings.

R403.3.1 Insulation (Prescriptive). Supply and return ducts in attics shall be insulated to a minimum of R-8 where 3 inches (76 mm) in diameter and greater and R-6 where less than 3 inches (76 mm) in diameter. Supply and return ducts in other portions of the building shall be insulated to a minimum of R-6 where 3 inches (76 mm) in diameter or greater and R-4.2 where less than 3 inches (76 mm) in diameter.

Exception: Ducts or portions thereof located completely inside the *building thermal envelope.*

❖ HVAC ductwork located outside of the conditioned space must have insulation with minimum *R*-values of R-8 or R-6, depending on the size of the duct. This includes both supply and return ducts. For areas other than attics, the values are R-6 and R-4.2. Energy losses are less in smaller ducts, therefore the *R*-value of the insulation is lower for ducts less than 3 inches in diameter.

The exception addresses the fact that ductwork in the thermal envelope is already protected from energy loss by virtue of being in the thermal envelope.

R403.3.2 Sealing (Mandatory). Ducts, air handlers and filter boxes shall be sealed. Joints and seams shall comply with either the *International Mechanical Code* or *International Residential Code*, as applicable.

Exceptions:

1. Air-impermeable spray foam products shall be permitted to be applied without additional joint seals.
2. For ducts having a static pressure classification of less than 2 inches of water column (500 Pa), additional closure systems shall not be required for continuously welded joints and seams, and locking-type joints and seams of other than the snap-lock and button-lock types.

❖ Ducts must be sealed in accordance with the duct sealing requirements of either the IMC or the code. Joints and seams that fail in a duct system can result in increased energy use because the conditioned air will be delivered to an unconditioned space or to a building space where it is not needed, such as a wall or floor assembly, crawl space or attic, instead of to a room or area inside the conditioned envelope. Return-air ductwork installed in basements or concealed building spaces may conduct chemicals or other products that produce potentially harmful fumes. Because the return air operates under negative pressure, any leaks could draw fumes, moisture, soil gases or odors from the surrounding area and direct them into the house. Therefore, sealing of return-air ductwork is also a requirement.

Duct tightness must be checked by leakage testing of the ducts at the end of construction or at the time of rough in. The pass/fail criteria for leakage rate is more stringent if the rough-in test time is chosen. This is simply because it is probably that some of the ductwork will be disturbed during subsequent construction and, therefore, have a higher leakage rate than what was tested.

R403.3.2.1 Sealed air handler. Air handlers shall have a manufacturer's designation for an air leakage of no more than 2 percent of the design air flow rate when tested in accordance with ASHRAE 193.

❖ This section of the code requires that air handlers have a manufacturer's designation for an air leakage of not more than 2 percent of the design flow rate when tested in accordance with ASHRAE 193. Energy conservation measures in the air-conditioning industry have driven the manufacturers of systems and components to establish compliance with leakage limits in ducts and air-handling units. The standards set by the American Society of Heating, Refrigerating and Air-conditioning Engineers (ASHRAE) form the basis for testing. Establishing an air-handler leakage rate, given the availability of a uniform test procedure, is prudent since any leakage in the air-handling unit contributes to waste of energy. The magnitude of leakage has a direct bearing on energy use and indoor air quality.

R403.3.3 Duct testing (Mandatory). Ducts shall be pressure tested to determine air leakage by one of the following methods:

1. Rough-in test: Total leakage shall be measured with a pressure differential of 0.1 inch w.g. (25 Pa) across the system, including the manufacturer's air handler enclosure if installed at the time of the test. All registers shall be taped or otherwise sealed during the test.
2. Postconstruction test: Total leakage shall be measured with a pressure differential of 0.1 inch w.g. (25 Pa) across the entire system, including the manufacturer's air handler enclosure. Registers shall be taped or otherwise sealed during the test.

Exception: A duct air leakage test shall not be required where the ducts and air handlers are located entirely within the building thermal envelope.

A written report of the results of the test shall be signed by the party conducting the test and provided to the *code official*.

❖ Duct tightness must be checked by leakage testing at the end of construction or at the time of rough-in. This affords some flexibility to the builder. The test at the end of construction, if failed, could be more difficult to correct. However, the test after rough-in is more stringent as noted in the commentary to Section R403.3.4.

R403.3.4 Duct leakage (Prescriptive). The total leakage of the ducts, where measured in accordance with Section R403.3.3, shall be as follows:

1. Rough-in test: The total leakage shall be less than or equal to 4 cubic feet per minute (113.3 L/min) per 100 square feet (9.29 m^2) of conditioned floor area where the air handler is installed at the time of the test. Where the air handler is not installed at the time of the test, the total leakage shall be less than or equal to 3 cubic feet per minute (85 L/min) per 100 square feet (9.29 m^2) of conditioned floor area.
2. Postconstruction test: Total leakage shall be less than or equal to 4 cubic feet per minute (113.3 L/min) per 100 square feet (9.29 m^2) of conditioned floor area.

❖ The pass/fail criteria for the tested duct leakage rate is more stringent if the rough-in test time is chosen. This is simply because it is probable that some of the ductwork will be disturbed during subsequent construction and, therefore, have a higher leakage rate than what was tested.

R403.3.5 Building cavities (Mandatory). Building framing cavities shall not be used as ducts or plenums.

❖ Building framing cavities shall not be used as ducts or plenums. Stud bays and other building cavities that are exposed to the differing outside temperatures cannot be used as supply air ducts. In addition, these spaces are limited to use for return air only because the negative pressures in the return air plenum with respect to surrounding spaces will decrease the likelihood of spreading smoke to other spaces via the plenum. In addition, for many of the same reasons that sealing (Section R403.2.2) is required, building cavities are a very inefficient means of distributing air.

R403.4 Mechanical system piping insulation (Mandatory). Mechanical system piping capable of carrying fluids above 105°F (41°C) or below 55°F (13°C) shall be insulated to a minimum of R-3.

❖ Heat losses during mechanical fluid distribution impact building energy use both in the energy required to make up for the lost heat and in the additional load that can be placed on the space-cooling system if the heat is released to air-conditioned space. These losses can be effectively limited by insulating the mechanical system piping that conveys fluids at extreme temperatures.

R403.4.1 Protection of piping insulation. Piping insulation exposed to weather shall be protected from damage, including that caused by sunlight, moisture, equipment maintenance and wind, and shall provide shielding from solar radiation that can cause degradation of the material. Adhesive tape shall not be permitted.

❖ Proper protection of piping insulation exposed to the weather elements is necessary to prevent damage, including that due to sunlight, moisture, equipment maintenance and wind.

R403.5 Service hot water systems. Energy conservation measures for service hot water systems shall be in accordance with Sections R403.5.1 and R403.5.4.

❖ A review of the definition of "Service water heating" in Chapter 2 is appropriate before applying these requirements. Any time the system is set up to circulate the water through it, the piping is required to be insulated as listed in Section R403.5.2.

R403.5.1 Heated water circulation and temperature maintenance systems (Mandatory). Heated water circulation systems shall be in accordance with Section R403.5.1.1. Heat trace temperature maintenance systems shall be in accordance with Section R403.5.1.2. Automatic controls, temperature sensors and pumps shall be accessible. Manual controls shall be readily accessible.

❖ When the distribution piping is heated to maintain usage temperatures, such as in circulating hot water systems or systems using pipe heating cable, the system pump or heat trace cable must have conveniently located manual or automatic switches or other controls that can be set to optimize system operation or turn off the system during periods of reduced demand. The simplest of these devices is an automatic time clock. Notice, however, that the code will accept either a readily accessible manual switch or an automatic means to turn off the circulating pump.

R403.5.1.1 Circulation systems. Heated water circulation systems shall be provided with a circulation pump. The system return pipe shall be a dedicated return pipe or a cold water supply pipe. Gravity and thermo-syphon circulation systems shall be prohibited. Controls for circulating hot water system pumps shall start the pump based on the identification of a demand for hot water within the occupancy. The controls shall automatically turn off the pump when the water in the circulation loop is at the desired temperature and when there is no demand for hot water.

❖ Demand-activated circulation is an efficient energy conservation strategy, given that the code, the *International Plumbing Code®* (IPC®) and the IRC require that the hot water piping in automatic temperature maintenance systems in new buildings be insulated with pipe insulation. This means the water in the circulation loop will stay hot for a very long time—up to 45 minutes for $^3/_4$-inch nominal pipe and up to 2 hours for 2-inch nominal pipe—even if the circulating pump is shut off. If this is the case, there is no reason to run the pump when the water is still hot and no reason to run the pump when no one is in the building or when no one is demanding hot water. The only time it makes sense to run the pump is shortly before hot water is needed, hence the requirement that the pump be controlled on-demand.

R403.5.1.2 Heat trace systems. Electric heat trace systems shall comply with IEEE 515.1 or UL 515. Controls for such systems shall automatically adjust the energy input to the heat tracing to maintain the desired water temperature in the piping in accordance with the times when heated water is used in the occupancy.

❖ The requirements for heat trace are partly to ensure that the systems can be operated in the most energy-

TABLE R403.6.1
MECHANICAL VENTILATION SYSTEM FAN EFFICACY

FAN LOCATION	AIR FLOW RATE MINIMUM (CFM)	MINIMUM EFFICACY (CFM/WATT)	AIR FLOW RATE MAXIMUM (CFM)
Range hoods	Any	2.8 cfm/watt	Any
In-line fan	Any	2.8 cfm/watt	Any
Bathroom, utility room	10	1.4 cfm/watt	< 90
Bathroom, utility room	90	2.8 cfm/watt	Any

For SI: 1 cfm = 28.3 L/min.

efficient manner consistent with providing heated water to the occupancy. The referenced standards are included to ensure that installed systems are safe for the intended application. The energy consequences of using heat trace are very reasonable. An example is an analysis of a small (100-foot-long) loop. The energy requirements of keeping the trunk line hot, which is the same as keeping the supply portion of the loop hot in a circulating system, are 701 kWh per year, assuming 12 hours at high temp (115°F) and 12 hours at economy temp (105°F). This is equivalent to operating the loop about 3 hours per day, but with hot water available 24 hours a day/7 days a week in the supply trunk. This is a significant savings when water heating is done electrically or with a similarly expensive fuel. If the branches are also traced, heated water can be delivered even more quickly to the fixtures using only 1,682 kWh per year, which is the same energy as running the loop a little more than 6 hours a day.

R403.5.2 Demand recirculation systems. A water distribution system having one or more recirculation pumps that pump water from a heated water supply pipe back to the heated water source through a cold water supply pipe shall be a *demand recirculation water system*. Pumps shall have controls that comply with both of the following:

1. The control shall start the pump upon receiving a signal from the action of a user of a fixture or appliance, sensing the presence of a user of a fixture or sensing the flow of hot or tempered water to a fixture fitting or appliance.
2. The control shall limit the temperature of the water entering the cold water piping to 104°F (40°C).

❖ The purpose of this section is to clarify the requirements for installing circulation pumps in applications that use a cold water supply pipe to circulate the water back to the water heater. Demand recirculation water systems are significantly more energy efficient than other recirculation systems and are inherently safer when the cold water supply is used as the return. The greater safety comes from the fact that the controls specified for demand recirculation water systems limit the flow of water from the hot water supply into the cold water supply to only minutes a day and because they limit the temperature of the water that is allowed to go into the cold water supply.

R403.5.3 Hot water pipe insulation (Prescriptive). Insulation for hot water pipe with a minimum thermal resistance (*R*-value) of R-3 shall be applied to the following:

1. Piping $^3/_4$ inch (19.1 mm) and larger in nominal diameter.
2. Piping serving more than one dwelling unit.
3. Piping located outside the conditioned space.
4. Piping from the water heater to a distribution manifold.
5. Piping located under a floor slab.
6. Buried in piping.
7. Supply and return piping in recirculation systems other than demand recirculation systems.

❖ This section simply prescribes conditions where hot water piping needs to be provided with a minimum R-3 insulation.

R403.5.4 Drain water heat recovery units. Drain water heat recovery units shall comply with CSA B55.2. Drain water heat recovery units shall be tested in accordance with CSA B55.1. Potable water-side pressure loss of drain water heat recovery units shall be less than 3 psi (20.7 kPa) for individual units connected to one or two showers. Potable water-side pressure loss of drain water heat recovery units shall be less than 2 psi (13.8 kPa) for individual units connected to three or more showers.

❖ Drain water heat recovery is often a cost-effective way to add to energy efficiency by recapturing hot water energy that is literally "going down the drain." The proposed standards have already been in use by designers for 10 years and the resulting ratings are in use by a variety of energy efficiency programs. Commercial (i.e., nonmultiunit residential) applications are engineered systems while multiunit residential applications are nonengineered and straightforward.

CSA B55.2 standard is for fabrication and material quality of DWHR units. The CSA B55.1 standard is for testing and labeling of DWHR unit efficiency and pressure loss at 2.5 gpm (9.5 lpm).

R403.6 Mechanical ventilation (Mandatory). The building shall be provided with ventilation that meets the requirements of the *International Residential Code* or *International Mechanical Code*, as applicable, or with other approved means of ventilation. Outdoor air intakes and exhausts shall have automatic or gravity dampers that close when the ventilation system is not operating.

❖ Mechanical ventilation is the alternative to having natural ventilation. Both natural and mechanical ventilation can be provided to a space. Unlike natural ventilation, mechanical ventilation does not depend on unpredictable air pressure differentials between the indoors and outdoors to create airflow. The volume of air supplied to a space must be approximately equal to the volume of the air removed from the space. Otherwise, the space will be either positively or negatively pressurized and the actual ventilation flow rate will be equivalent to the lower rate of either the air supply or air exhaust.

The requirements of this section are intended to reduce infiltration into the building when ventilation systems are off. Infiltration speeds up natural cooling or warming of the space during off hours and can increase the energy use required to maintain normal temperatures.

Any outdoor air inlets or outlets serving fans, boilers and other HVAC systems equipped with an on/off switch that either introduces outside air into a building or exhaust air outside of a building must have dampers that automatically close when the fan is shut off. These dampers may be either gravity type or motorized, regardless of whether the fan is supplying or exhausting air. One of the most common such dampers that is seen on homes is the gravity-type backdraft damper on a clothes dryer exhaust duct.

R403.6.1 Whole-house mechanical ventilation system fan efficacy. Mechanical ventilation system fans shall meet the efficacy requirements of Table R403.6.1.

Exception: Where mechanical ventilation fans are integral to tested and listed HVAC equipment, they shall be powered by an electronically commutated motor.

❖ This section of the code is applicable to whole-house mechanical ventilation systems that meet the efficacy requirements of Table R403.6.1. To reduce the amount of energy consumed by residential mechanical ventilation systems is to address the power consumption of the fans that are powering the system. This is important because these fans will operate many hours per day. The table offers energy efficacy levels for exhaust fans that are the same levels as current ENERGY STAR ventilation fan specifications.

R403.7 Equipment sizing and efficiency rating (Mandatory). Heating and cooling equipment shall be sized in accordance with ACCA Manual S based on building loads calculated in accordance with ACCA Manual J or other *approved* heating and cooling calculation methodologies. New or replacement heating and cooling equipment shall have an efficiency rating equal to or greater than the minimum required by federal law for the geographic location where the equipment is installed.

❖ Once the building's thermal envelope is properly insulated and sealed, it will often allow for the reduction of equipment size from what has typically been installed using a "rule-of-thumb" method or other means of estimating. A properly sized system will operate more efficiently, help improve occupant comfort and extend the equipment's service life. Section M1401.3 of the IRC stipulates that the heating and cooling equipment must be sized based on the building loads calculated in accordance with the ACCA Manual J. This manual contains a simplified method of calculating heating and cooling loads. It includes a room-by-room calculation method that allows the designer to determine the required capacities of the heating and cooling equipment. In addition, it provides a means to estimate the airflow requirements for each of the areas in the house. This estimate can be used in sizing the duct system for the types of heating and cooling units that use air as the medium for heat transfer. Other approved methods may be used with the code official's approval.

Though not required by the code, the calculated airflows would provide a means to evaluate the installation of the mechanical system. By providing the proper airflow to the various portions of the building, the system can operate more efficiently and can help prevent spaces from being too hot or too cold.

R403.8 Systems serving multiple dwelling units (Mandatory). Systems serving multiple dwelling units shall comply with Sections C403 and C404 of the IECC—Commercial Provisions in lieu of Section R403.

❖ The criteria in Section R403 primarily address standalone mechanical systems in single-family houses. However, for buildings, many residential building projects such as townhouses will have more complicated mechanical systems that may consist of a single system serving multiple dwelling units.

R403.9 Snow melt and ice system controls (Mandatory). Snow- and ice-melting systems, supplied through energy service to the building, shall include automatic controls capable of shutting off the system when the pavement temperature is above 50°F (10°C), and no precipitation is falling and an automatic or manual control that will allow shutoff when the outdoor temperature is above 40°F (4.8°C).

❖ Snow melt equipment is being installed at a greater frequency in residential projects in communities with high snow accumulation. Previously, the code only required that the building be built to a certain level of efficiency; however, there was no limit placed on the energy use for snow melt, which can be twice the energy use per square foot of the building.

This section does not restrict the use or sizing of snow melt, but it does require that controls be installed on the equipment so that the system will operate more efficiently. The automatic controls provide efficient operation by keeping the system in an idle mode until light snow begins to fall and allowing adequate warm-up before a heavy snowfall. Systems that only use manual controls require the building owner to manually turn on the system when it starts to snow or to leave the system running in the snow-melting mode, using significantly more energy. Chapter 50, Snow Melting and Freeze Protection, of the 2003 *ASHRAE Applications Handbook* states that using a manual switch to operate snow melt equipment may not melt snow effectively; thus, snow will accumulate. This requirement is also referenced in ANSI/ASHRAE/IESNA Standard 90.1, Section 6.4.3.8, Freeze protection and Snow/Ice Melting Systems.

R403.10 Pools and permanent spa energy consumption (Mandatory). The energy consumption of pools and permanent spas shall be in accordance with Sections R403.10.1 through R403.10.3.

❖ Because of the heating and filtering operations involved with pools and inground permanently installed spas, they provide a good opportunity to save energy by limiting heat loss or pump operation. This section provides the scoping requirements for the pool and spa heaters, time switches and pool covers that can make the operation more energy efficient. These features would provide energy savings for residential pools if their owners use them. The requirements of this section were added to the code in order to help coordinate with requirements found in ASHRAE 90.1 and to help reduce the energy used by these pools and inground permanently installed spa systems.

R403.10.1 Heaters. The electric power to heaters shall be controlled by a readily *accessible* on-off switch that is an integral part of the heater mounted on the exterior of the heater, or external to and within 3 feet (914 mm) of the heater. Operation of such switch shall not change the setting of the heater thermostat. Such switches shall be in addition to a circuit breaker for the power to the heater. Gas-fired heaters shall not be equipped with continuously burning ignition pilots.

❖ An accessible on-off switch allows heaters to be turned off when heat is not needed or when the pool or spa may not be used for a period of time.

R403.10.2 Time switches. Time switches or other control methods that can automatically turn off and on according to a preset schedule shall be installed for heaters and pump motors. Heaters and pump motors that have built-in time switches shall be in compliance with this section.

Exceptions:

1. Where public health standards require 24-hour pump operation.
2. Pumps that operate solar- and waste-heat-recovery pool heating systems.

❖ The use of a time switch or other control method to control the heater and pumps provides an easy system for pool and spa operations and energy savings. The application of Exception 1 is dependent on the requirements of the health department in the jurisdiction. Because these are often public pools and spas, the health department may require continuous filtering or circulation. Exception 2 grants a credit for using other systems that help the pool and spa operate more efficiently. Therefore, when solar- and waste-heat-recovery systems are used to heat the pool, the exception eliminates the time-switch requirement.

R403.10.3 Covers. Outdoor heated pools and outdoor permanent spas shall be provided with a vapor-retardant cover or other *approved* vapor-retardant means.

Exception: Where more than 70 percent of the energy for heating, computed over an operation season, is from site-recovered energy, such as from a heat pump or solar energy source, covers or other vapor-retardant means shall not be required.

❖ Where energy is used to heat a pool or a spa, a cover is required to help hold in the heat and keep it from being lost to the surrounding air. The level of protection or insulation that the cover must provide depends on the temperature to which the pool is heated. Any time a pool or spa is heated, the code will require a vapor-retardant pool cover. This type of cover is not required to provide any minimum level of insulation value. It simply will help hold some of the heat in, much like placing a lid on a pot. In situations where the pool is heated above 90°F (32°C), the cover must be insulated to the specified R-12 level. The exception is similar to that found in Section R403.10.2.

R403.11 Portable spas (Mandatory). The energy consumption of electric-powered portable spas shall be controlled by the requirements of APSP-14.

❖ This section coordinates the code with the *International Swimming Pool and Spa Code®* (ISPSC®). Both codes rely on APSP-14 for energy conservation requirements for portable spas. The standard addresses efficiency requirements for water heating equipment and water circulation pumps.

R403.12 Residential pools and permanent residential spas. Residential swimming pools and permanent residential spas that are accessory to detached one- and two-family dwellings and townhouses three stories or less in height above grade plane and that are available only to the household and its guests shall be in accordance with APSP-15.

❖ This section coordinates the code with the ISPSC. Both codes rely on APSP-15 for energy conservation requirements for residential pools and permanent residential spas. The standard addresses efficiency requirements for water heating equipment and water circulation pumps.

SECTION R404 ELECTRICAL POWER AND LIGHTING SYSTEMS

R404.1 Lighting equipment (Mandatory). Not less than 75 percent of the lamps in permanently installed lighting fixtures shall be high-efficacy lamps or not less than 75 percent of the permanently installed lighting fixtures shall contain only high-efficacy lamps.

Exception: Low-voltage lighting.

❖ Lighting accounts for roughly 12 percent of the energy used in residences relying on incandescent bulbs. Thus, this requirement is a substantial energy saver. Incandescent lighting—still used in the vast majority of residences—is the least energy efficient of all light types.

One more efficient lighting option is the compact fluorescent light (CFL). CFLs use about 80 percent less energy than standard incandescent lighting. Limiting this requirement to 75 percent of the lamps in a residence ensures there will be plenty of exceptions for situations where a CFL might not work as well, such as dimmable fixtures.

R404.1.1 Lighting equipment (Mandatory). Fuel gas lighting systems shall not have continuously burning pilot lights.

❖ Continuously burning pilots waste energy; therefore, the code does not use them.

SECTION R405 SIMULATED PERFORMANCE ALTERNATIVE (PERFORMANCE)

❖ Section R405 describes an alternative way to meet the code's goal of effective energy use based on showing that the predicted annual energy use of a proposed design is less than or equal to that of the same home if it had been built to meet the prescriptive criteria in Sections R402 and R403. Section R406 does not prescribe a single set of requirements; rather, it provides a process to reach the energy-efficiency goal based on establishing equivalence with the intent of the code. Because of the level of detail required in the analysis, this method of design is not often used for residential buildings; however, with the changes that have been made in Section R402 and in newer editions of the DOE's REScheck software, Section R405 may become a more popular means of demonstrating code compliance. This section may allow designers to show that many of their current plans meet the overall code requirements even though individual components of the home fall below required code compliance levels.

When using this section and other performance options, the terms "standard reference design" and "proposed design" are used extensively. These terms can create confusion for some users, but they really are very easy to understand. The "proposed design" is exactly that. It is the building as it is proposed to be built, including building size and room configurations, window and wall assemblies, mechanical equipment, orientation, etc. It is essentially the building as it is shown on the construction documents (plans and specifications). The "standard reference design" is the same building configuration and orientation as the proposed design, but instead of matching the plans for the building as it will actually be built, the standard design is shown to only meet the minimum requirements of the code. For example, the standard reference design building in Climate Zone 4 would require that the walls and roof be insulated to a level of R-13 and R-38, respectively, based on the requirements in Table R402.1.2 (*U*-factors of 0.082 and 0.030 in accordance with Table R402.1.4). On the other hand, the proposed design may show that the intent is to install R-19 insulation in the walls and R-49 in the roof. Therefore, if everything else in the two buildings was identical, it would be easy to see in this example that the proposed design does exceed what is required by the code in the standard reference design.

There are two fundamental requirements for using the code. First, Section R405 compliance is based on total estimated annual energy usage across the major energy-using systems in a residential building: envelope, mechanical and service water heating. Note that Section R405 does not include lighting in the analysis process. Second, Section R405 compares the energy use of the proposed design to that of a standard reference design. As mentioned above, the standard reference design is the same building design as that proposed, except that the energy features required by the code (insulation, windows, HVAC, infiltration) are modified to meet the minimum prescriptive requirements in Sections R402 and R403. The standard reference design is used only for comparison and is never actually built.

Section R405 sets both general principles and specific guidelines for use in computing the estimated annual energy cost of the proposed and standard reference designs.

These guidelines constitute a large portion of Section R405 and are easily seen in Table R405.5.2(1). They are necessary to maintain fairness and consistency between the proposed and standard designs. Although the simulated performance alternative method is the most complex method, it gives the design professional the flexibility to introduce exterior walls, roof/ceiling components, etc., that do not meet the requirements of the prescriptive performance approach in Sections R402 and R403 but are considered acceptable where the annual energy use of the proposed building is equal to or less than that of the standard reference design building. Envelope features that lower energy consumption (window orientation, passive-solar features or the use of "cool" reflective roofing products in cooling-dominated climates) and mechanical and service water-heating systems that are more efficient than those required by the minimum prescriptive requirements in Sections R402 and R403 are used to offset the potentially high thermal transmittance of an innovative exterior envelope design in this instance.

The simulated performance alternative also allows energy supplied by renewable energy sources on the building site to be discounted from the total energy consumption of the proposed design building. Because renewable energy obtained on the site comes from nondepletable sources such as solar radiation, wind, plant byproducts and geothermal sources, its use is not counted as part of the proposed building's energy use. The definition of "Energy cost" in Section R202 and the provisions of Section R405.3 should be reviewed when dealing with renewable energy. Renewable energy that is purchased from an off-site source cannot be excluded and would be included within the definition of "energy cost."

R405.1 Scope. This section establishes criteria for compliance using simulated energy performance analysis. Such analysis shall include heating, cooling and service water heating energy only.

❖ This section simply indicates that the performance analysis can include not only building envelope performance (as is limited in Section R402.1.4) but that the tradeoffs can also include the energy used for heating, cooling and service water heating. The provisions of this section do not include an allowance for lighting energy to be included. If a designer wished to include lighting, it would be done under the alternative materials and methods provisions of Section R102 (see commentary, Section R102).

Section R405.1 establishes the terms of performance-based comparison for residential buildings. Under the Section R405 Simulated Performance approach, the candidate building (proposed design) is evaluated based on the cost of energy used. In simple terms, Section R406 states: build the residence any way as long as it is designed to use no more energy than a home built exactly to the minimum requirements in Sections R402 and R403.

This performance option can also be used to take advantage of energy-efficiency improvements that are only partly reflected in the total UA (sum of *U*-factor times the assembly area) of the residence, which is found in Section R402.1.4. For example, in a cooling climate, reduction in ceiling UA will save more energy than an equal reduction in slab UA. This energy savings occurs because cooling climates have ground temperatures that approach comfortable indoor temperatures, diminishing the value of slab insulation for such climates. In contrast, the ceiling insulation in cooling climates usually receives higher solar loads during the warmest part of the day, increasing the value of ceiling insulation. Therefore, in cooling climates, a change in ceiling UA (perhaps adding more ceiling insulation) may have a substantially greater impact on energy use than the same UA change in slab-edge insulation.

A variety of calculation methods can be used to show compliance under the simulated performance alternative approach. The calculations can be complex; for example, a detailed building energy simulation tool covering all aspects of building energy use. Much of Section R405 deals with establishing guidelines for defining aspects of the various sophisticated calculations that form the basis of the annual energy-use comparison. Alternatively, the analysis can be a simple calculation or correlation focused on one aspect of energy use. Where a method of analysis does not calculate, model or estimate a specific energy-using feature, it cannot be used for tradeoffs based on that specific feature. For example, software using degree-day-based climate and building envelope calculations does not specifically account for water heater energy use. Therefore, this type of software cannot be used to take credit for a high-efficiency water heater. Note, however, that software using degree-day-based climate and building envelope calculations accounts for changes in insulation levels; therefore, it could be used for an insulation tradeoff under the simulated performance approach.

The use of Section R405 to show compliance is not required by all homes with special features designed to save energy. Very energy-efficient designs may not give credit for all of the design's features to show compliance. For example, a highly energy-efficient, passive solar home may be so well insulated and have such low glazing *U*-factors that those items alone may be sufficient to show compliance with the energy-related requirements in the code by using only UA computations or by meeting the prescriptive requirements of Sections R402 and R403. Because of these very efficient elements, compliance may be shown by using the prescriptive provisions or the easier UA computations instead of a more complex evaluation, which would include the other energy-using subsystems in the building.

R405.2 Mandatory requirements. Compliance with this section requires that the mandatory provisions identified in Section R401.2 be met. All supply and return ducts not completely inside the *building thermal envelope* shall be insulated to a minimum of R-6.

❖ This section serves as a reminder of the requirements found in Section R101.3. When using Section R405, compliance with the various mandatory provisions in this chapter is still required. The sections that must be complied with separately include items that often cannot be effectively modeled. When using Section R405, it is important to review the provisions of Section R402.5, which place "hard limits" on the fenestration *U*-factors and SHGC that may be used for tradeoffs (see commentary, Section R402.5).

R405.3 Performance-based compliance. Compliance based on simulated energy performance requires that a proposed residence (*proposed design*) be shown to have an annual energy cost that is less than or equal to the annual energy cost of the *standard reference design*. Energy prices shall be taken from a source *approved* by the *code official*, such as the Department of Energy, Energy Information Administration's *State Energy Price and Expenditure Report*. *Code officials* shall be permitted to require time-of-use pricing in energy cost calculations.

Exception: The energy use based on source energy expressed in Btu or Btu per square foot of *conditioned floor area* shall be permitted to be substituted for the energy cost. The source energy multiplier for electricity shall be 3.16. The source energy multiplier for fuels other than electricity shall be 1.1.

❖ The general procedure is to show that the annual energy cost for the building is less than the annual energy cost of a building that just meets the prescriptive requirements. The applicant must estimate the annual energy cost for two buildings: the one to be built and the standard reference design building. Because the two are compared on the basis of annual energy costs, designs that have lower demand charges or use energy when rates are lower may be able to gain an advantage using Section R405. It is also common to equate energy sources on the basis of cost because energy bills are the principal motivator for energy efficiency. An annual energy analysis could be used to compare a wide variety of energy-using features and conservation opportunities for the candidate building. The annual energy use can be used to trade off insulation; window/door areas; *U*-factors; HVAC equipment efficiency; water heating efficiency; infiltration control measures; duct insulation and sealing; pipe insulation and renewable energy technologies; or new energy technologies. Appliances not regulated by the code (refrigerators, dishwashers, clothes-washing machines, residential lighting) are not eligible for tradeoffs under this chapter.

The approach in this section is targeted for use in residences with energy-saving features that are not reflected in the UA or overall U-factor. Examples of features that lower energy consumption but not U-factor include high-efficiency HVAC systems; windows predominantly oriented toward the south; passive solar features; highly reflective ("cool") roofing products in cooling-dominated climates; and renewable energy sources such as photovoltaic, geothermal heat pumps and wind farms.

A building will comply with the requirements of the code and be acceptable when the annual energy cost of the proposed design is less than or equal to the annual energy cost of the standard design. The following sections will provide additional information and guidance related to the standard reference design and the design to which it is compared.

The purpose of the exception is to allow the use of source energy as an alternative metric to energy cost for compliance with the performance provisions.

Adding source energy as an alternative to energy cost offers many benefits to compliance, as follows:

- Using cost will be a liability to the home builder if home buyers do not achieve the savings listed in the compliance documentation.
- Energy cost changes frequently. This means that a home that complies today may not comply a few months from now if costs change.
- Energy cost focuses attention on first-year energy costs, which misses the point of an energy code where features that are generally life-cycle cost effective to the homeowner are added to save energy and make homes more comfortable over the life of the home, not to reduce first cost.

The source multipliers of 3.16 and 1.1 are from the 2002 *DOE Core Databook*. One way to think of this is that electric energy utilized at the site requires 3.16 times the source energy to produce the electricity at power plants and distribute via power lines to homes. This is because the efficiency of power plants is much less than 100 percent and there are losses in transmission and distribution as well. Other fuels, such as natural gas and fuel oil, have less source energy losses and a lower source energy multiplier.

Standard Reference Design

This section reiterates the simulated performance approach's strong reliance on principles established in Sections R402 and R403 as the foundation for the baseline level of energy efficiency established in the code. Therefore, all tradeoffs are judged against the prescriptive requirements of the standard reference design. Accordingly, insulation levels for the standard design building are set by the prescriptive requirements in Section R402 and are generally based on the values in Table R402.1.4. To discourage designers from picking a combination for the standard design that results in an inflated energy budget, thus permitting easy compliance by the proposed design (i.e., gaming), general principles and specific guidelines are given for various energy-using systems of the building [see Table R405.5.2(1)].

The entries in Table R405.5.2(1) for the envelope and fenestration requirements assist the code user in defining the standard reference design without bias. This is especially true in southern climates, where air-conditioning requirements are significant. In these cooling-dominated climates, where *U*-factor requirements are relatively high, residences with very large window areas and virtually no wall insulation can comply with the code through the use of windows with very low *U*-factors. This results in excessive air-conditioning costs in these climates because unshaded windows, dominated by solar heat gain, are required to meet only the code thermal-resistance and solar heat gain requirements. As a result, it is possible to construct residences in these climates with limited wall insulation, as long as windows with high thermal resistance and low SHGC are used.

The Table R405.5.2(1) entries for the building's thermal envelope components and fenestration are necessary to pin down the overall energy consumption of the standard design as described in Section R402. Without the guidance provided by these entries, the areas and *U*-factors of the windows and building thermal envelope could assume a multitude of different values and still satisfy Section R402 requirements. Furthermore, different values for these components would result in different overall energy cost levels for the standard design. The annual energy cost of the standard design represents the maximum "budget" the proposed design must meet. Without specific details of the standard reference design's features and components, this budget becomes a moving target and will allow a designer to pick a combination for the standard design that results in a very large energy budget, thus permitting the proposed design to easily comply with Section R405 but not meet the energy conservation levels intended in the code.

Proposed Design

In general, the proposed design is the building that will be constructed exactly as it is shown in the plans and construction documents. Therefore, most of the entries for Table R405.5.2(1) will be "as proposed." There are, however, some sections that must match what is required or used for the standard reference design. The proposed design must be similar in many characteristics to the standard design. Otherwise, it may not provide an "apples-to-apples" comparison, thereby limiting the value of this method in demonstrating that the proposed design truly is more efficient. Some of the items that must be comparable include:

1. The standard and proposed designs must use the same energy sources (fuels) for the same function. For example, a gas-heated standard design cannot be compared to an electrically heated proposed design.
2. The areas of the building components (ceiling, wall and floor), including the building's shape, configuration and orientation, must all remain the same in both the standard and proposed designs.
3. Both the standard and proposed designs must assume the same outdoor and comfort indoor climate conditions.
4. Both the standard and proposed designs must assume the same occupant diversity and usage patterns; for example, the same thermostat set points, water usage, internal gains, etc.

The exception in Section R405.3 is important to understand because some jurisdictions may require the comparison to be done on the basis of "site energy" versus "annual energy cost." Because of the fact that utility charges for various types of energy can change over time, some code officials may prefer that the comparison be made based on the amount of energy delivered to the home instead of the cost of that energy.

The code establishes the effective use of site energy as opposed to primary (a.k.a., source) energy as a goal and then delineates specific criteria, both prescriptive and performance based, for meeting that goal. Primary energy is defined as the amount of energy delivered to a sector adjusted to account for the energy sources used to produce the energy (e.g., energy used to generate electricity). Included also is the energy lost in delivering the fuel to a customer; for instance, the electricity lost in the transmission and distribution of electricity. Site (delivered) energy is the amount of energy delivered to a household. Energy generation, transmission and distribution losses are not included. Primary energy is useful to show the ultimate resource impact of sectoral energy demand with respect to global particulate contributions of carbon dioxide, for example. Site energy is necessary if one wants to know what is going on inside the building. In this case, fuel choice does matter; oil-heated homes may consume more site energy than electrically heated homes, for instance.

From an economist's perspective, using energy cost instead of primary or site energy is preferable. Deregulation will affect the choice of fuels and households will decide what to consume. This view assumes that competitive pressures will ultimately equalize prices. This primary-versus-site debate centers on electricity, but losses are also associated with other energy sources. For example, one could take distillate fuel oil all the way back to the refinery, even factoring in energy losses in the fuel delivery trucks. So while Section R405 will generally require the comparison to be based on the annual energy cost for the homeowner, some jurisdictions prefer to know the amount of energy delivered and not the cost. Regardless of which comparison basis is used, it is clearly the intent of Section R405 to require the effective use of energy.

This exception permits equating different energy sources on the basis of the site energy. Site energy is energy use measured at the building boundary. For designs under Section R405, site energy is input to the heating, cooling-service and water-heating equipment.

As mentioned earlier, it is common to equate energy sources on the basis of cost because energy bills are the principal motivator for energy efficiency and, over time, the market will move to the more economical energy source. The results of using either energy cost or delivered "site energy" tend to be similar. The use of tradeoffs based on energy costs may be preferable in cooling-dominated climates. The cost of a unit of gas is commonly much less than the same unit of electricity; therefore, equating fossil fuels and electricity in terms of heat content may not reflect the interest of the building owner based primarily on the applicable utility rate structures and tariffs. In southern climates, the use of site-energy-based tradeoffs tends to overcredit changes in heating energy use and undercredit changes in cooling energy use. The intent of the code to "achieve the effective use of energy" (see Section R101.3) can be preserved through the use of an annual energy cost or site-energy-based comparison. The use of site energy, or delivered energy, as the basis for the comparison under Section R405 requires the approval of the code official.

R405.4 Documentation. Documentation of the software used for the performance design and the parameters for the building shall be in accordance with Sections R405.4.1 through R405.4.3.

❖ This section provides a list of the minimum information needed to demonstrate compliance with code requirements. The three subsections in Section R405.4 list the minimum level of information needed for this purpose. Because much of the information and software used for the performance option must be approved by the code official, the jurisdiction should be consulted to see if any additional information may be required (see commentary, Section R405.6.2).

A review of the code requirements of Section R104 should also be included when looking at Section R405.4. The construction documents must contain the data necessary to confirm that the results of the design have been incorporated into the construction documents and also to allow assessment of compliance. This is of particular importance when coupled with performing a comprehensive plan review given the sometimes very lengthy and sophisticated submittals required by the energy-efficiency method (see commentary, Section R104). Though not listed in the code, some state laws may require that the design of the building construction and, in this case, the development of a comparative building-energy analysis be done by a registered design professional in accordance with the licensing laws of the state where the work takes place.

In such states, the code official should consider and enforce the requirement to coordinate with state licensing laws that often establish thresholds for when the services of a registered design professional are required.

R405.4.1 Compliance software tools. Documentation verifying that the methods and accuracy of the compliance software tools conform to the provisions of this section shall be provided to the *code official.*

❖ This section is essentially a general requirement that where software is used to demonstrate compliance under Section R405, the software be shown to provide accurate comparisons and results. Many of the software systems that may be used will be familiar and readily acceptable to the code official. Where a less commonly used software is proposed for use, the code official will need to be shown that the software performs its intended function and is accurately comparing the standard reference and proposed designs.

R405.4.2 Compliance report. Compliance software tools shall generate a report that documents that the *proposed design* complies with Section R405.3. A compliance report on the *proposed design* shall be submitted with the application for the building permit. Upon completion of the building, a compliance report based on the as-built condition of the building shall be submitted to the *code official* before a certificate of occupancy is issued. Batch sampling of buildings to determine energy code compliance for all buildings in the batch shall be prohibited.

Compliance reports shall include information in accordance with Sections R405.4.2.1 and R405.4.2.2. Where the *proposed design* of a building could be built on different sites where the cardinal orientation of the building on each site is different, compliance of the *proposed design* for the purposes of the application for the building permit shall be based on the worst-case orientation, worst-case configuration, worst-case building air leakage and worst- case duct leakage. Such worst-case parameters shall be used as inputs to the compliance software for energy analysis.

❖ The approved software should be able to demonstrate that the proposed design has an annual energy cost that is either less than or equal to that of the standard reference design. The code requires that a compliance report be submitted on two occasions: at the time of permit application, to verify that the proposed design complies with the code; and at the end of the project before the certificate of occupancy is issued, to verify that the "as-built" building complies with the code.

R405.4.2.1 Compliance report for permit application. A compliance report submitted with the application for building permit shall include the following:

1. Building street address, or other building site identification.
2. A statement indicating that the *proposed design* complies with Section R405.3.
3. An inspection checklist documenting the building component characteristics of the *proposed design* as indicated in Table R405.5.2(1). The inspection checklist shall show results for both the *standard reference design* and the *proposed design* with user inputs to the compliance software to generate the results.
4. A site-specific energy analysis report that is in compliance with Section R405.3.
5. The name of the individual performing the analysis and generating the report.
6. The name and version of the compliance software tool.

❖ Providing the address of the project will not only help in the tracking of the project and various permits, but it also ensures that the calculations run were for the intended project and not based on a project that may not be applicable. While it may be possible to run a calculation based on a stock set of plans [see orientation provisions in Table R405.5.2(1)], the best way to provide a truly accurate comparison is to provide a site-specific evaluation, as required in Item 4.

Item 2 provides the primary information from which approval will be granted. A summary must be submitted showing how the annual energy cost of the proposed design compares to the annual energy cost of the standard design. The comparison summary must include, as a minimum, annual energy cost by design (standard versus proposed) and could include the fuel type (electric versus gas versus renewable sources) if it was a part of a trade-off.

Besides showing the actual comparison between the two designs, Item 3 requires an inspection checklist that can be used by the inspector to ensure that the proposed design matches what is actually constructed in the field. This checklist addresses the details of construction on which the comparison was conducted. See Section R405.4.3, Item 1, for the equivalent requirement for the standard reference design. These two sections not only ensure that the comparison is accurate but also provide information so that the comparison can be run again and verified if needed. While it would be best if the checklist included information for all of the items listed in Table R405.5.2(1), it is only necessary that the information is provided for items that are compared or for which trade-offs are based or taken.

Because various versions of software can include different inputs or evaluations, it is important that both the name and version of the software be provided. Changes to new or an updated version of existing software could often provide different results. Listing the software edition helps the code official evaluate the report and also provides the information to conduct a verification review if necessary.

R405.4.2.2 Compliance report for certificate of occupancy. A compliance report submitted for obtaining the certificate of occupancy shall include the following:

1. Building street address, or other building site identification.
2. A statement indicating that the as-built building complies with Section R405.3.
3. A certificate indicating that the building passes the performance matrix for code compliance and listing the energy saving features of the buildings.
4. A site-specific energy analysis report that is in compliance with Section R405.3.
5. The name of the individual performing the analysis and generating the report.
6. The name and version of the compliance software tool.

❖ The items required for the compliance report for obtaining the certificate of occupancy are basically the same as those required for the permit application, except that the report is based on the "as-built" condition of the building. During the course of construction, changes in the windows, type of insulation, equipment or building dimensions are all required to be documented and approved. The changes must be approved based on compliance with the code and the original design. At the end of the day, however, one must verify that the changes made did not have a negative impact on the performance of the building, as analyzed before the permit application.

R405.4.3 Additional documentation. The *code official* shall be permitted to require the following documents:

1. Documentation of the building component characteristics of the *standard reference design.*
2. A certification signed by the builder providing the building component characteristics of the *proposed design* as given in Table R405.5.2(1).
3. Documentation of the actual values used in the software calculations for the *proposed design.*

❖ The items in this section are not automatically required as those of Sections R405.4.1 and R405.4.2 but are instead only required when the code official wishes them to be. Item 1 provides information related to the standard reference design similar to what Section R405.4.2.2, Item 2, provides for the proposed design. Having this information allows the code official or designer to easily replicate the calculations if additional verification reviews or changes are needed.

Item 2 can be useful in a couple of ways. It can be used not only by the inspector, similar to the checklist in Section R405.4.3, Item 2, but it helps to ensure that items that may not be seen by or visible to the inspector have been installed as proposed. This certification can essentially be considered the same as the builder stating the as-built structure complies with the proposed and originally approved design plans.

Item 3 provides for a possible need to check input values for computer programs utilized. This enables the plan checker to readily verify compliance with the basic requirements of the code.

R405.5 Calculation procedure. Calculations of the performance design shall be in accordance with Sections R405.5.1 and R405.5.2.

❖ The provisions of this section simply ensure that comparisons between the standard reference design and the proposed design are accurate and reflect an "apples-to-apples" comparison.

R405.5.1 General. Except as specified by this section, the *standard reference design* and *proposed design* shall be configured and analyzed using identical methods and techniques.

❖ Items that are not involved in the tradeoffs must be comparable between the standard reference design and the proposed design. This helps to ensure that the energy savings are truly based on the differences from Table R405.5.2(1) that are being evaluated. Though it should go without saying, the same calculation method or software must be used to estimate the annual energy usage for space heating and cooling of the standard design and the proposed design. The calculation tool must be approved by the code official. A jurisdiction may want to make known the methods it prefers for comparing annual energy use. The code official retains the authority to determine whether a specific set of computations for a particular residence is acceptable. The use of Section R405 is optional; therefore, the applicant is free to choose the method for achieving compliance (see commentary, Sections R101.3 and R401.2). Regardless, the calculation procedure used to evaluate the standard design must also be used to evaluate the proposed design.

R405.5.2 Residence specifications. The *standard reference design* and *proposed design* shall be configured and analyzed as specified by Table R405.5.2(1). Table R405.5.2(1) shall include, by reference, all notes contained in Table R402.1.2.

❖ While the majority of Section R405 addresses the process to evaluate the standard design and proposed design, this section serves as the backbone of Section R405 and the information on which the comparisons are conducted. Tables R405.5.2(1) and R405.5.2(2) provide the list of items that are compared and establish not only the requirements for the standard design but also give a good view of what items may be included and evaluated for tradeoffs. The determination of requirements and application of the various items is fairly easy to follow due to the way the components are listed in separate rows of the table. Under Section R405, only these items may differ between the proposed and standard reference design. As discussed in the commentary to Section R405.1, if there are any additional components or features for which a designer wishes to make a tradeoff, the approval of such items would need to be based on Section R102.

The information that is determined for Tables R405.5.2(1) and R405.5.2(2) would be the type of information needed for both Sections R405.4.2.2, Item 2, and R405.4.3, Item 1.

TABLE R405.5.2(1). See page R4-41.

❖ As discussed in the commentary to Section R405.5.2, this table serves as the backbone of the requirements for Section R405. By comparing the annual energy cost of a standard reference design to that of a proposed design, designers may trade off the efficiency of one element or component for an increased efficiency in another.

The column dealing with "building components" helps to distinguish what building elements may be considered and included in the performance option of Section R405. The column labeled "Standard Reference Design" provides details used to determine the annual energy cost serving as the maximum energy cost that the proposed design could use. This column simply states how the standard design home is to be configured and treated in the simulations that will compare it to the proposed design home. This column simply ensures that comparisons between the two designs truly are based on a plan that meets the intended performance of the code. The "Proposed Design" column represents the home for which compliance is trying to be determined. As indicated, most of the items in a proposed design will be "as proposed." There are a few entries in this column where the proposed design must use the same values as the standard design. These items are restricted so that a fair comparison is provided and so that the design parameters may not simply be changed in an attempt to show improved efficiency. Additional comments about the table or specific component requirements are as follows:

Walls, roofs and fenestration: Similar to Sections R402.1 and R402.3, this table separates the impact of solar gains for fenestration such as windows or energy loss through windows and doors from the impact of wall *U*-factors. The provisions simply ensure that each component of the building's thermal envelope is evaluated separately.

Walls, floors, ceilings and roofs: To avoid gaming the comparison, the areas of these items must be the same in both designs. No valid comparison of the proposed and standard designs can be made if the size of the two designs is not equal.

Foundations: This section explicitly requires the foundation and on-grade floor type be the same for both designs. Without this provision, the possibility for the designer to select a less energy-efficient foundation or floor type (including modeling a floor system over a heated space) for the standard design and a more energy-efficient foundation type for the proposed design is left open. Under this circumstance, heat loss for the proposed design would be much less significant because of the lower heat loss through the foundation. The resulting loss "credit" results in less insulation being required for the proposed design. This type of credit is prohibited because the designer may not have intended to use the unfavorable foundation in the first place.

Doors: The amount of door area and the orientation for the door are specified. Whether the door is glazed or opaque is not a consideration for the standard design because the area of glazing is dealt with under a separate entry. By requiring the door to be on the north orientation, the effects of SHGC on a glazed door are minimized.

Glazing: This requirement exists in response to strong evidence showing that glazing area and geometry have some impact on the building's overall thermal transmittance, and that solar heat gain through windows constitutes a large portion of the air-conditioning load (typically 25 percent) in southern climates.

The area limitations help to ensure that the glazing areas are not manipulated to affect the efficiency of the various designs. The proposed glazing area limitation of 18 percent of the conditioned floor area of the residence is identical to the requirement of the Home Energy Rating System Council Guidelines (Version 2) for its standard design, and is equal to the maximum window area that was allowed by ASHRAE Standard 90.2-1993 for its standard design.

The code language, found in Note a, indicating that the glazing area includes the sash, framing and glazing is needed for completeness and clarification. The NFRC guidelines on window labeling require that window performance ratings be based on the full window system product. In other words, the *U*-factor, SHGC, visible light transmittance, air leakage and other properties of the window must be based on the combined impacts of both the glazing and its associated opaque constituents (defined here as "sash").

Orientation significantly affects the annual energy consumption on a building and is critical to achieving an optimum passive solar design. This section recognizes that building energy consumption is affected by orientation and that an orientation change of the glazing could be a way to manipulate the standard reference design so that it uses an increased amount of energy or gains an increased advantage by limiting solar heat gain.

Experimental evidence using eight cardinal exposures with equal glazing areas for the standard design as input for simulation programs shows that the procedure is inconvenient with very small increases in accuracy. Due to this, the four cardinal exposures are sufficient for achieving solar neutrality on the standard design. Because the percentage of glass-area facing in a particular orientation can vary significantly on a proposed design, the actual orientations are needed for accuracy. The difference in a large amount of glazing facing directly south, versus southwest, will significantly increase annual energy use, peak loads and the time the peak loads occur. The trend is that the more passive solar techniques are applied to the building, the more important the orientation becomes.

The section establishes an SHGC standard of 0.4 for glazed fenestration products in climates with significant cooling loads. Specifically, this requirement applies to warmer climates (Climate Zones 1 through 3). As a result, this requirement applies throughout much of the southern region of the United States in Oklahoma, Arkansas, Tennessee and North Carolina and extending as far north as the warmer parts of California and Nevada. Although Table R402.1.2 does not require an SHGC in Climate Zones 4 through 8, this table establishes a requirement of 0.40 for these climate zones. When applying the SHGC provisions of the standard design, remember the application of Note e in Table R402.1.2 for Climate Zone 3, Marine.

This section explicitly addresses interior shading and limits considering the benefits of it on the performance of windows in residences. The term "interior shading" hours and the values to be used are specified. Most homeowners use some form of interior window treatment, such as drapes, blinds or shades, on their windows. In addition to their decorative aspects, drapes and curtains have been traditionally used by homeowners to control privacy and daylight, provide protection from overheating and reduce the fading of fabrics. To most effectively reduce solar heat gain, the drapery used to block the sunlight should have high reflectance and low transmittance. A densely woven, light-colored fabric would achieve this objective. Drapes can reduce the SHGC of clear glass from 20 to 70 percent, depending on the color and openness of the drapery fabric. The impact of drapery on solar heat gain is proportionally lessened if the window glass is shaded or tinted. The main disadvantage of drapes and other interior devices as solar control measures is that once the solar energy has entered a window, a large proportion of the energy absorbed by the shading system will remain inside the house as heat gain. Interior devices are thus most effective when they are highly reflective, with minimum absorption of solar energy. Interior shading devices, such as blinds and shades, primarily provide light and privacy control but also can have an impact on controlling solar heat gain. They include horizontal Venetian blinds, mini blinds, vertical slatted blinds of various materials, a wide variety of pleated and honeycomb shades and roll-down shades. White- or silver-colored blinds, coupled with clear glass, have the greatest potential for reducing solar heat gains. Some manufacturers have offered window-unit options that include mini blinds mounted inside sealed or unsealed insulating glass. The blinds, in the sealed dust-free environment, can be operated with a magnetic lever without breaking the air seal. Blinds in the unsealed glazing unit are protected as well but can be easily removed for cleaning or repair. These between-glass shading devices have a lower shading coefficient than equivalent blinds mounted on the interior. They also provide additional insulating value to the double glass by reducing convective loops in the airspace.

Unlike the other strategies to reduce heat gain, interior shades generally require consistent and active intervention by the homeowner. It is unlikely that anyone would operate all shades in a consistent, optimal pattern, as an analysis assumes they are to be operated. A value of 0.7 is proposed for summer conditions to approximate the condition of predominantly closed, medium-colored interior draperies. The winter value is increased to 0.85 to not unduly penalize winter heating performance. Variance from these values is specifically not allowed. It is possible to install motorized and automated shading systems, but these are quite costly and not yet in common use. When contemplating the use of high-performance glazing for the necessary solar control as opposed to just using interior shading, there are two important benefits: there is less need for operating the shades and the window is rarely covered, resulting in a clear view and daylight at all times. Of course, shades also provide privacy and darkness when desired so they may be closed part of the day in any case, but the high-performance glazing means there is less need to operate them in a particular manner to achieve significant reductions in energy use. Because of the uncertainty of actual application and more limited benefits of the interior shades, the code will generally require the proposed design to use the same values as the standard reference design.

Permanent, exterior-mounted shading devices for the standard design are considered atypical and, therefore, are not required or allowed to be included. Where credit is taken in the proposed design for these devices, the code official must approve and confirm the installation of the actual shading device proposed.

Skylights: This is intended to reduce the complexity of the design requirements as applied to skylight areas in the standard design. If this section was not included, ceilings and skylights would also require detailed treatment as in Section R402.3 and the glazing section above to fully specify the standard design.

Air exchange rate: This section establishes conditions under which air-leakage reduction can be claimed reliably. Without using an actual measurement, it often is not practical to document infiltration reduction measures beyond those in the standard design because the code (see Section R402.4) already assumes a fairly comprehensive air-sealing regimen. Where the test reports that the anticipated building infiltration performance is not achieved as was assumed in the permit application, infiltration must be lowered or some other means used to make up the increased energy use attributable to the deficiency in infiltration rates. Note that "tight" buildings may require mechanical ventilation to improve occupant comfort. Controlling ventilation rates reduces cooking odors, damp musty smells, stale air and elevated levels of carbon dioxide. Ventilation also helps reduce concentrations of air-borne contaminants off-gassed from building materials and related to other household activities.

Internal gains: The calculations for both the standard and proposed designs must assume the same internal loads. Where the simulation tool allows specification of internal gains, the total heat gains (sensible + latent) are to be calculated based on this equation in both the standard and proposed designs (the code does not distinguish between sensible and latent gains). This equation makes internal gains a function of the conditioned area and the number of bedrooms.

Heating systems: This section simply requires that the heating and cooling equipment in the standard design meet the minimum efficiency requirements that are currently required at the time of the evaluation.

This section contains general guidelines for the proposed design for any annual energy analysis under Section R405. The standard and proposed designs must use the same energy sources (fuels) for the same function. For example, a gas-heated standard design cannot be compared to an electrically heated proposed design. For a fossil-fuel-heated building, the standard design must assume the applicable minimum efficiency as dictated by the code and the National Appliance Energy Conservation Act (NAECA).

Service water heating: Where the simulation tool or calculation models water heating energy, domestic service water-heating calculations must assume a set point of 120°F (49°C), with a daily usage of 30 gallons (113 L) per unit plus 10 gallons (37.8 L) per bedroom.

Thermal distribution systems [see Table R405.5.2(2)]: Research and practice in the past few years has shown that the major component of heating and cooling system efficiency stems from air leakage in hot and cold air distribution systems (primarily ducts, but also air-handler cabinets). The research has been conducted by ASHRAE; Lawrence-Berkley National Laboratory; Brookhaven National Laboratory; Electric Power Research Institute; Gas Research Institute; Florida Solar Energy Center; and numerous other research organizations and national utilities. Many, if not most, utilities in the nation now offer customers duct-leakage diagnosis and repair efforts as part of their energy conservation and demand side management programs. The Sheet Metal and Air-Conditioning Contractors National Association (SMACNA) has published test procedures in outlining the methods and air distribution systems through duct-system pressurization testing (see the SMACNA *HVAC Air Duct Leakage Test Manual*), and a number of calibrated pressurization test equipment manufacturers are now producing equipment for sale in the market.

Technical questions regarding the energy impacts and mechanisms of specific leak types in certain locations remain; however, there is no question the elimination of air leakage results in energy inefficiency in duct systems being limited to conduction gains and losses through the ducts themselves.

The virtual elimination of these leaks also results in much-improved heating- and cooling-system efficiencies (on the order of 95- to 99-percent efficient as opposed to 75- to 80-percent efficient). There is no technical disagreement that, once air leakage is no longer in question, air distribution system efficiency can be determined through straightforward engineering calculations using conduction heat-transfer equations, unencumbered by the complexities of air-leakage flows. A conservative assumption is that the ductwork in the standard design is 50-percent inside and 50-percent outside the conditioned space; however, the arrangement of ductwork in the standard design must be representative of that proposed.

Unless directly heated and cooled, attics and crawl spaces are unconditioned spaces. The specified distribution system efficiency (DSE) is to be used in the standard design where the designer wishes to take advantage of the impacts from improved DSEs in the proposed design and the entire distribution system is substantially leak free. Proposed designs use specified system efficiencies based on the proportion of ducts inside and outside a substantially leak-free conditioned system. It is acceptable for the proposed design to use system efficiencies other than those given where such factors result from a code-official-approved, post-construction duct system performance test. [Note that Table R405.5.2(2) is titled "Default Distribution System Efficiencies" and that Note a of that table indicates that the values are for untested distribution systems.] The provisions in this section are a mechanism that allows builders to take credit for substantially leak-free duct systems where they are installed.

R405.6 Calculation software tools. Calculation software, where used, shall be in accordance with Sections R405.6.1 through R405.6.3.

❖ See the commentary to Section R405.6.1.

TABLE R405.5.2(1)
SPECIFICATIONS FOR THE STANDARD REFERENCE AND PROPOSED DESIGNS

BUILDING COMPONENT	STANDARD REFERENCE DESIGN	PROPOSED DESIGN
Above-grade walls	Type: mass wall if proposed wall is mass; otherwise wood frame.	As proposed
	Gross area: same as proposed	As proposed
	U-factor: as specified in Table R402.1.4	As proposed
	Solar absorptance = 0.75	As proposed
	Emittance = 0.90	As proposed
Basement and crawl space walls	Type: same as proposed	As proposed
	Gross area: same as proposed	As proposed
	U-factor: from Table R402.1.4, with insulation layer on interior side of walls	As proposed
Above-grade floors	Type: wood frame	As proposed
	Gross area: same as proposed	As proposed
	U-factor: as specified in Table R402.1.4	As proposed
Ceilings	Type: wood frame	As proposed
	Gross area: same as proposed	As proposed
	U-factor: as specified in Table R402.1.4	As proposed
Roofs	Type: composition shingle on wood sheathing	As proposed
	Gross area: same as proposed	As proposed
	Solar absorptance = 0.75	As proposed
	Emittance = 0.90	As proposed
Attics	Type: vented with aperture = 1 ft^2 per 300 ft^2 ceiling area	As proposed
Foundations	Type: same as proposed	As proposed
	Foundation wall area above and below grade and soil characteristics: same as proposed	As proposed
Opaque doors	Area: 40 ft^2	As proposed
	Orientation: North	As proposed
	U-factor: same as fenestration from Table R402.1.4	As proposed
Vertical fenestration other than opaque doors	Total area[b] = (a) The proposed glazing area, where the proposed glazing area is less than 15 percent of the conditioned floor area (b) 15 percent of the conditioned floor area, where the proposed glazing area is 15 percent or more of the conditioned floor area	As proposed
	Orientation: equally distributed to four cardinal compass orientations (N, E, S & W).	As proposed
	U-factor: as specified in Table R402.1.4	As proposed
	SHGC: as specified in Table R402.1.2 except that for climates with no requirement (NR) SHGC = 0.40 shall be used.	As proposed
	Interior shade fraction: 0.92-(0.21 × SHGC for the standard reference design)	0.92-(0.21 × SHGC as proposed)
	External shading: none	As proposed
Skylights	None	As proposed
Thermally isolated sunrooms	None	As proposed
Air exchange rate	Air leakage rate of 5 air changes per hour in climate zones 1 and 2, and 3 air changes per hour in climate zones 3 through 8 at a pressure of 0.2 inches w.g (50 Pa). The mechanical ventilation rate shall be in addition to the air leakage rate and the same as in the proposed design, but no greater than $0.01 \times CFA + 7.5 \times (N_{br} + 1)$ where: CFA = conditioned floor area N_{br} = number of bedrooms Energy recovery shall not be assumed for mechanical ventilation.	For residences that are not tested, the same air leakage rate as the standard reference design. For tested residences, the measured air exchange rate[a]. The mechanical ventilation rate[b] shall be in addition to the air leakage rate and shall be as proposed.

(continued)

TABLE R405.5.2(1)—continued
SPECIFICATIONS FOR THE STANDARD REFERENCE AND PROPOSED DESIGNS

BUILDING COMPONENT	STANDARD REFERENCE DESIGN	PROPOSED DESIGN
Mechanical ventilation	None, except where mechanical ventilation is specified by the proposed design, in which case: Annual vent fan energy use: kWh/yr = 0.03942 × CFA + 29.565 × (N_{br} +1) where: CFA = conditioned floor area N_{br} = number of bedrooms	As proposed
Internal gains	IGain = 17,900 + 23.8 × CFA + 4104 × N_{br} (Btu/day per dwelling unit)	Same as standard reference design.
Internal mass	An internal mass for furniture and contents of 8 pounds per square foot of floor area.	Same as standard reference design, plus any additional mass specifically designed as a thermal storage element[c] but not integral to the building envelope or structure.
Structural mass	For masonry floor slabs, 80 percent of floor area covered by R-2 carpet and pad, and 20 percent of floor directly exposed to room air.	As proposed
	For masonry basement walls, as proposed, but with insulation required by Table R402.1.4 located on the interior side of the walls	As proposed
	For other walls, for ceilings, floors, and interior walls, wood frame construction	As proposed
Heating systems[d, e]	As proposed for other than electric heating without a heat pump, where the proposed design utilizes electric heating without a heat pump the standard reference design shall be an air source heat pump meeting the requirements of Section C403 of the IECC-Commercial Provisions. Capacity: sized in accordance with Section R403.7	As proposed
Cooling systems[d, f]	As proposed Capacity: sized in accordance with Section R403.7.	As proposed
Service water heating[d, e, f, g]	As proposed Use: same as proposed design	As proposed gal/day = 30 + (10 × N_{br})
Thermal distribution systems	Duct insulation: From Section R403.2.1 A thermal distribution system efficiency (DSE) of 0.88 shall be applied to both the heating and cooling system efficiencies for all systems other than tested duct systems. For tested duct systems, the leakage rate shall be 4 cfm (113.3 L/min) per 100 ft^2 (9.29 m^2) of *conditioned floor* area at a pressure of differential of 0.1 inches w.g. (25 Pa).	As tested or as specified in Table R405.5.2(2) if not tested. Duct insulation shall be as proposed.
Thermostat	Type: Manual, cooling temperature setpoint = 75°F; Heating temperature setpoint = 72°F	Same as standard reference

For SI: 1 square foot = 0.93 m^2, 1 British thermal unit = 1055 J, 1 pound per square foot = 4.88 kg/m^2, 1 gallon (US) = 3.785 L, °C = (°F-32)/1.8, 1 degree = 0.79 rad.

a. Where required by the *code official*, testing shall be conducted by an *approved* party. Hourly calculations as specified in the ASHRAE *Handbook of Fundamentals*, or the equivalent shall be used to determine the energy loads resulting from infiltration.

b. The combined air exchange rate for infiltration and mechanical ventilation shall be determined in accordance with Equation 43 of 2001 ASHRAE *Handbook of Fundamentals*, page 26.24 and the "Whole-house Ventilation" provisions of 2001 ASHRAE *Handbook of Fundamentals*, page 26.19 for intermittent mechanical ventilation.

c. Thermal storage element shall mean a component not part of the floors, walls or ceilings that is part of a passive solar system, and that provides thermal storage such as enclosed water columns, rock beds, or phase-change containers. A thermal storage element must be in the same room as fenestration that faces within 15 degrees (0.26 rad) of true south, or must be connected to such a room with pipes or ducts that allow the element to be actively charged.

d. For a proposed design with multiple heating, cooling or water heating systems using different fuel types, the applicable standard reference design system capacities and fuel types shall be weighted in accordance with their respective loads as calculated by accepted engineering practice for each equipment and fuel type present.

e. For a proposed design without a proposed heating system, a heating system with the prevailing federal minimum efficiency shall be assumed for both the standard reference design and proposed design.

f. For a proposed design home without a proposed cooling system, an electric air conditioner with the prevailing federal minimum efficiency shall be assumed for both the standard reference design and the proposed design.

g. For a proposed design with a nonstorage-type water heater, a 40-gallon storage-type water heater with the prevailing federal minimum energy factor for the same fuel as the predominant heating fuel type shall be assumed. For the case of a proposed design without a proposed water heater, a 40-gallon storage-type water heater with the prevailing federal minimum efficiency for the same fuel as the predominant heating fuel type shall be assumed for both the proposed design and standard reference design.

TABLE R405.5.2(2)
DEFAULT DISTRIBUTION SYSTEM EFFICIENCIES FOR PROPOSED DESIGNS[a]

DISTRIBUTION SYSTEM CONFIGURATION AND CONDITION	FORCED AIR SYSTEMS	HYDRONIC SYSTEMS[b]
Distribution system components located in unconditioned space	—	0.95
Untested distribution systems entirely located in conditioned space[c]	0.88	1
"Ductless" systems[d]	1	—

For SI: 1 cubic foot per minute = 0.47 L/s, 1 square foot = 0.093 m^2, 1 pound per square inch = 6895 Pa, 1 inch water gauge = 1250 Pa.

a. Default values given by this table are for untested distribution systems, which must still meet minimum requirements for duct system insulation.

b. Hydronic systems shall mean those systems that distribute heating and cooling energy directly to individual spaces using liquids pumped through closed-loop piping and that do not depend on ducted, forced airflow to maintain space temperatures.

c. Entire system in conditioned space shall mean that no component of the distribution system, including the air-handler unit, is located outside of the conditioned space.

d. Ductless systems shall be allowed to have forced airflow across a coil but shall not have any ducted airflow external to the manufacturer's air-handler enclosure.

R405.6.1 Minimum capabilities. Calculation procedures used to comply with this section shall be software tools capable of calculating the annual energy consumption of all building elements that differ between the *standard reference design* and the *proposed design* and shall include the following capabilities:

1. Computer generation of the *standard reference design* using only the input for the *proposed design.* The calculation procedure shall not allow the user to directly modify the building component characteristics of the *standard reference design.*
2. Calculation of whole-building (as a single *zone*) sizing for the heating and cooling equipment in the *standard reference design* residence in accordance with Section R403.6.
3. Calculations that account for the effects of indoor and outdoor temperatures and part-load ratios on the performance of heating, ventilating and air-conditioning equipment based on climate and equipment sizing.
4. Printed *code official* inspection checklist listing each of the *proposed design* component characteristics from Table R405.5.2(1) determined by the analysis to provide compliance, along with their respective performance ratings (*R*-value, *U*-factor, SHGC, HSPF, AFUE, SEER, EF are some examples).

❖ This section states the general capabilities for the calculation software and its ability to evaluate the effects of building parametrics, system design, climatic factors, operational characteristics and mechanical equipment on annual energy usage.

There are a number of different software programs available to perform these calculations. The complexity of the programs will often depend on the amount of parameters that may be varied between the designs. The phrase in the base section that the software must be capable of evaluating the consumption of "all building elements that differ" between the two designs is important because it is the differences between the standard design and the proposed design that determine whether the proposed design is acceptable.

To ensure that the comparisons are made only on the items that are being traded, Item 1 looks to keep the software from allowing any manipulation that would help to reduce the energy costs of the standard design.

R405.6.2 Specific approval. Performance analysis tools meeting the applicable provisions of Section R405 shall be permitted to be *approved.* Tools are permitted to be *approved* based on meeting a specified threshold for a jurisdiction. The *code official* shall be permitted to approve tools for a specified application or limited scope.

❖ Though this section may be viewed as accepting a reduced level of efficiency, the intent is really more in line with the provisions of Section R102. An alternative program for evaluating the energy efficiency of a building using performance criteria can be substituted here. However, that alternative must meet the performance criteria of Section R405.

R405.6.3 Input values. When calculations require input values not specified by Sections R402, R403, R404 and R405, those input values shall be taken from an *approved* source.

❖ This section simply requires that any additional input information required by the calculation software be approved by the code official. Through the use of the phrase "approved source," the code official would be able to review the source of the information that is being used for inputting information. This will help to ensure that the selected values are reasonable for the situation and not just an estimate.

SECTION R406 ENERGY RATING INDEX COMPLIANCE ALTERNATIVE

R406.1 Scope. This section establishes criteria for compliance using an Energy Rating Index (ERI) analysis.

❖ The Residential Provisions of the code allow for varying methods for demonstrating compliance. This includes a prescriptive and simulated performance option in addition to allowing efficiency programs designed to go above the minimum code levels, as "deemed to comply" programs. These above-code programs must be approved by the code official to be used in the jurisdiction. Alternative programs that depend on an Energy Rating Index (ERI) have been approved as an alternative code or above-code program in at least six states and in over 130 jurisdictions. These types of programs typically take the form of a Home Energy Rating System (HERS) program. Under the current code there is no guidance on setting ERI scores, which will lead to inconsistent application of these types of programs based on climate zones.

This section provides an ERI with established rating numbers to allow alternative programs using an ERI to be designed to meet these criteria. The section provides guidelines for the development of the index, requirements for documentation to be provided to ensure compliance and a requirement that an approved third party verify that the building complies with the applicable ERI.

R406.2 Mandatory requirements. Compliance with this section requires that the provisions identified in Sections R401 through R404 be labeled as mandatory and Section R4103.5.3 be met. The building thermal envelope shall be greater than or equal to levels of efficiency and Solar Heat Gain Coefficient in Table 402.1.1 or 402.1.3 of the 2009 *International Energy Conservation Code*.

Exception: Supply and return ducts not completely inside the building thermal envelope shall be insulated to a minimum of R-6.

❖ As with Section R405.2 for the performance-based alternative, minimum requirements have been established in the code for all buildings. The sections that must be complied with separately include items that often cannot be effectively modeled. When using Section R406, it is important to review the provisions of Section R402.5, which place "hard limits" on the fenestration *U*-factors and SHGC that may be used for trade-offs (see commentary, Section R402.5).

R406.3 Energy Rating Index. The Energy Rating Index (ERI) shall be a numerical integer value that is based on a linear scale constructed such that the *ERI reference design* has an Index value of 100 and a *residential building* that uses no net purchased energy has an Index value of 0. Each integer value on the scale shall represent a 1-percent change in the total energy use of the rated design relative to the total energy use of the *ERI reference design*. The ERI shall consider all energy used in the *residential building*.

❖ To avoid confusion in the application of different ERI programs, the parameters for the index are established here. The reason is simple and logical—to allow a clear understanding of the meaning of the index and to be able to uniformly apply the requirements of the code.

R406.3.1 ERI reference design. The *ERI reference design* shall be configured such that it meets the minimum requirements of the 2006 *International Energy Conservation Code* prescriptive requirements.

The proposed *residential building* shall be shown to have an annual total normalized modified load less than or equal to the annual total loads of the *ERI reference design*.

❖ The reference house is based on a home built to the 2006 IECC, which is consistent with ERI-based programs that are used today.

R406.4 ERI-based compliance. Compliance based on an ERI analysis requires that the *rated design* be shown to have an ERI less than or equal to the appropriate value listed in Table R406.4 when compared to the *ERI reference design*.

TABLE R406.4
MAXIMUM ENERGY RATING INDEX

CLIMATE ZONE	ENERGY RATING INDEX
1	52
2	52
3	51
4	54
5	55
6	54
7	53
8	53

❖ Based on the energy conservation levels set by the 2015 IECC, the ERI required for the rated design (i.e., the proposed building) are set as shown.

R406.5 Verification by approved agency. Verification of compliance with Section R406 shall be completed by an *approved* third party.

❖ In order to ensure proper application of the parameters of the ERI chosen and the proper use of computer software, this option for energy conservation requires a third party to review the proposed design to ensure compliance with the code. The approved third party should be knowledgeable in the energy code, the ERI system and the use of computer software for design. By stating the agency or individual must be a "third party," the code is requiring financial independence, with no possible conflicts of interest.

R406.6 Documentation. Documentation of the software used to determine the ERI and the parameters for the residential building shall be in accordance with Sections R406.6.1 through R406.6.3.

❖ See the commentary to Section R405.4.

R406.6.1 Compliance software tools. Documentation verifying that the methods and accuracy of the compliance software tools conform to the provisions of this section shall be provided to the *code official*.

❖ See the commentary to Section R405.4.1.

R406.6.2 Compliance report. Compliance software tools shall generate a report that documents that the ERI of the *rated design* complies with Sections R406.3 and R406.4. The compliance documentation shall include the following information:

1. Address or other identification of the residential building.
2. An inspection checklist documenting the building component characteristics of the *rated design*. The inspection checklist shall show results for both the *ERI reference design* and the *rated design*, and shall document all inputs entered by the user necessary to reproduce the results.
3. Name of individual completing the compliance report.
4. Name and version of the compliance software tool.

Exception: Multiple orientations. Where an otherwise identical building model is offered in multiple orientations, compliance for any orientation shall be permitted by documenting that the building meets the performance requirements in each of the four (north, east, south and west) cardinal orientations.

❖ See the commentary to Section R405.4.2.

R406.6.3 Additional documentation. The *code official* shall be permitted to require the following documents:

1. Documentation of the building component characteristics of the *ERI reference design*.
2. A certification signed by the builder providing the building component characteristics of the *rated design*.
3. Documentation of the actual values used in the software calculations for the *rated design*.

❖ See the commentary to Section R405.4.3.

R406.7 Calculation software tools. Calculation software, where used, shall be in accordance with Sections R406.7.1 through R406.7.3.

❖ See the commentary to Sections R405.6.1 through R405.6.3.

R406.7.1 Minimum capabilities. Calculation procedures used to comply with this section shall be software tools capable of calculating the ERI as described in Section R406.3, and shall include the following capabilities:

1. Computer generation of the *ERI reference design* using only the input for the *rated design*.

The calculation procedure shall not allow the user to directly modify the building component characteristics of the *ERI reference design*.

2. Calculation of whole building, as a single *zone*, sizing for the heating and cooling equipment in the *ERI reference design* residence in accordance with Section R403.7.
3. Calculations that account for the effects of indoor and outdoor temperatures and part-load ratios on the performance of heating, ventilating and air-conditioning equipment based on climate and equipment sizing.
4. Printed *code official* inspection checklist listing each of the *rated design* component characteristics determined by the analysis to provide compliance, along with their respective performance ratings.

❖ See the commentary to Sections R405.6.1 through R405.6.3.

R406.7.2 Specific approval. Performance analysis tools meeting the applicable sections of Section R406 shall be *approved*. Tools are permitted to be *approved* based on meeting a specified threshold for a jurisdiction. The *code official* shall approve tools for a specified application or limited scope.

❖ See the commentary to Sections R405.6.1 through R405.6.3.

R406.7.3 Input values. When calculations require input values not specified by Sections R402, R403, R404 and R405, those input values shall be taken from an approved source.

❖ See the commentary to Sections R405.6.1 through R405.6.3.

Chapter 5 [RE]: Existing Buildings

General Comments

New for the 2015 IECC Residential Provisions is a separate chapter dealing with alterations, repairs, additions and change of occupancy of existing buildings. These provisions, for the most part, originated in Chapter 1 of the 2012 IECC Residential Provisions; however, the 2015 edition contains options dealing with situations where compliance with the code for additions is difficult. The provisions now allow an energy-neutral method for difficult-to-comply projects, essentially mandating simply that the building with the addition uses no more energy than the existing building. This will allow projects to take advantage of energy-efficient alterations on the existing building to offset compliance difficulties with features on the addition. Requirements for alterations are provided, giving specific details and exceptions for the building envelope, heating and cooling systems, hot water systems, and lighting. Finally, the code allows that repairs be done without consideration of compliance with this code.

Purpose

The purpose of this chapter is to provide requirements for the unique circumstances involved with existing building additions, alterations and repairs.

SECTION R501 GENERAL

R501.1 Scope. The provisions of this chapter shall control the *alteration*, repair, addition and change of occupancy of existing buildings and structures.

❖ The scope of this chapter includes specific circumstances related to changes to existing buildings. This chapter provides a roadmap when dealing with different types of projects—alterations, repairs, additions and changes of occupancy.

R501.1.1 Additions, alterations, or repairs: General. Additions, alterations, or repairs to an existing building, building system or portion thereof shall comply with Section R502, R503 or R504. Unaltered portions of the existing building or building supply system shall not be required to comply with this code.

❖ For commentary on additions, see Section R502; for alterations, see Section R503; and for repairs, see Section R504.

R501.2 Existing buildings. Except as specified in this chapter, this code shall not be used to require the removal, *alteration* or abandonment of, nor prevent the continued use and maintenance of, an existing building or building system lawfully in existence at the time of adoption of this code.

❖ This section addresses the fact that, in general, the code does not affect existing buildings. It will permit an addition to be made to an existing building without requiring the existing building to conform to the code. In such a case, the addition is expected to comply with the current code, but it will not require changes for the existing portion. Therefore, the code does not apply retroactively to existing buildings. When an existing building is modified by an addition, alteration, renovation or repair, Section R502, R503 or R504 will provide the guidance and requirements for such changes.

R501.3 Maintenance. Buildings and structures, and parts thereof, shall be maintained in a safe and sanitary condition. Devices and systems that are required by this code shall be maintained in conformance to the code edition under which installed. The owner or the owner's authorized agent shall be responsible for the maintenance of buildings and structures. The requirements of this chapter shall not provide the basis for removal or abrogation of energy conservation, fire protection and safety systems and devices in existing structures.

❖ The code is designed to regulate new construction and new work and is not intended to be applied retroactively to existing buildings except where existing envelope, lighting, mechanical or service water heating systems are specifically affected by Section R502, R503 or R504. Maintenance of building systems and components in accordance with the effective code at the time of their installation is required, but such systems and components need not be upgraded or modified to meet the code.

R501.4 Compliance. *Alterations*, *repairs*, *additions* and changes of occupancy to, or relocation of, existing buildings and structures shall comply with the provisions for *alterations*, *repairs*, *additions* and changes of occupancy or relocation, respectively, in the *International Residential Code*, *International Building Code*, *International Fire Code*, *International*

Fuel Gas Code, International Mechanical Code, International Plumbing Code, International Property Maintenance Code, International Private Sewage Disposal Code and NFPA 70.

❖ Compliance with the provisions of other *International Codes*® (I-Codes®), such as the IBC, should not be ignored. The I-Codes are intended to be a coordinated set of construction codes. This section clarifies the relationship between this chapter of the code and the IRC, IBC, *International Fire Code*® (IFC®), *International Fuel Gas Code*® (IFGC®), *International Mechanical Code*® (IMC®), *International Plumbing Code*® (IPC®), *International Property Maintenance Code*® (IPMC®), *International Private Sewage Disposal Code*® (IPSDC®) and NFPA 70.

When alterations and repairs are made to existing mechanical and plumbing systems, the provisions of the I-Codes and NFPA 70 for alterations and repairs must be followed. Those codes indicate the extent to which existing systems must comply with the stated requirements. Where portions of existing building systems, such as plumbing, mechanical and electrical systems, are not being altered or repaired, those systems may continue to exist without being upgraded as long as they are not hazardous or unsafe to building occupants.

R501.5 New and replacement materials. Except as otherwise required or permitted by this code, materials permitted by the applicable code for new construction shall be used. Like materials shall be permitted for repairs, provided hazards to life, health or property are not created. Hazardous materials shall not be used where the code for new construction would not permit their use in buildings of similar occupancy, purpose and location.

❖ There are two options for materials used in repairs to an existing building. Generally, the materials used for repairs should be those that are presently required or permitted for new construction under the I-Codes. It is also acceptable to use materials consistent with those that are already present, except where those materials pose a hazard. This allowance follows the general concept that any repair should not make a building more hazardous than it was prior to the repair. It is generally possible to repair a structure, its components and its systems with materials consistent with those materials that were used previously. However, where materials now deemed hazardous are involved in the repair work, they may no longer be used. For example, the code identifies asbestos and lead-based paint as two common hazardous building materials that cannot be used in the repair process. Certain materials previously considered acceptable for building construction are now known to threaten the health of occupants.

R501.6 Historic buildings. No provision of this code relating to the construction, *repair, alteration,* restoration and movement of structures, and *change of occupancy* shall be mandatory for *historic buildings* provided a report has been submitted to the code official and signed by the owner, a registered *design professional*, or a representative of the State Historic Preservation Office or the historic preservation authority having jurisdiction, demonstrating that compliance with that provision would threaten, degrade or destroy the historic form, fabric or function of the *building*.

❖ In some aspects, this is a bit of a continuation of the "existing building" provisions, but it goes even further—historic buildings are exempt. In earlier editions of the code, this exemption applied only to the exterior envelope of such buildings and to the interior only in those cases where the ordinance explicitly designated elements of the interior. With the current text, historic buildings are exempt from all aspects of the code. This exemption, however, is not without conditions. The most important criterion for application of this section is that the building must be specifically classified as being of historic significance by a qualified party or agency. Usually this is done by a state or local authority after considerable scrutiny of the historical value of the building. Most states and many local jurisdictions have authorities such as a landmark commission.

Because of the unique issues involved, historic buildings are exempt from the requirements of the code. This exemption could be extended to include all parts that are "historic," including additions, alterations and repairs that would normally be addressed by Section R502 or R503. If the proposed addition, alteration or renovation is not "historic," then the provisions of Section R502 or R503 should be applied. Consideration of energy conservation and compliance with the code is still of value in historic buildings. In exempting historic buildings, the code is simply recognizing that energy efficiency may be difficult to accomplish while maintaining the "historic" nature of the building.

SECTION R502 ADDITIONS

R502.1 General. Additions to an existing building, building system or portion thereof shall conform to the provisions of this code as those provisions relate to new construction without requiring the unaltered portion of the existing building or building system to comply with this code. Additions shall not create an unsafe or hazardous condition or overload existing building systems. An addition shall be deemed to comply with this code where the addition alone complies, where the existing building and addition comply with this code as a single building, or where the building with the addition uses no more energy than the existing building. Additions shall be in accordance with Section R502.1.1 or R502.1.2.

❖ Simply stated, new work must comply with the current requirements for new work. Any addition to an existing system involving new work is subject to the requirements of the code. Additions can place additional loads or different demands on an existing system and those loads or demands could necessitate changing all or part of the existing system. Additions and alterations must not cause an existing system to be any less in compliance with the code than it was before the changes.

Additions to existing buildings must comply with the code when the addition is within the scope of the code and would not otherwise be exempted. Additions include new construction, such as a conditioned bedroom, sun space or enclosed porch added to an existing building. Additions also include existing spaces converted from unconditioned or exempt spaces to conditioned spaces. For example, a finished basement, an attic converted to a bedroom or a carport converted to a den are additions. The addition of an unconditioned garage would not be considered within the scope of the code because the code applies to heated or cooled (conditioned) spaces.

Although not specifically defined in the code, building codes typically define an "addition" as any increase in a building's habitable floor area (which can be interpreted as any increase in the conditioned floor area). For example, an unconditioned garage converted to a bedroom is an addition. If a conditioned floor area is expanded, such as a room made larger by moving out a wall, only the newly conditioned space must meet the code. A flat window added to a room does not increase the conditioned space and thus is not an addition by this definition. If several changes are made to a building at the same time, only the changes that expand the conditioned floor area are required to meet the code. The addition (the newly conditioned floor space) complies with the code if it complies with all of the applicable requirements in Chapter 4 [RE]. For example, requirements applicable to the addition of a new room would most likely include insulating the exterior walls, ceiling and floor to the levels specified in the code; sealing all joints and penetrations; installing a vapor retarder in unventilated frame walls, floors and ceilings; identifying installed insulation *R*-values and window *U*-factors; and insulating and sealing any ducts passing through unconditioned portions or within exterior envelope components (walls, ceilings or floors) of the new space.

R502.1.1 Prescriptive compliance. Additions shall comply with Sections R502.1.1.1 through R502.1.1.4.

❖ The basic premise of this section is that the addition alone, and not the remainder of the building, must comply with the code.

R502.1.1.1 Building envelope. New building envelope assemblies that are part of the addition shall comply with Sections R402.1, R402.2, R402.3.1 through R402.3.5, and R402.4.

Exception: Where nonconditioned space is changed to conditioned space, the building envelope of the addition shall comply where the UA, as determined in Section 402.1.4, of the existing building and the addition, and any alterations that are part of the project, is less than or equal to UA generated for the existing building.

❖ The exception is the key to this section. Again, the insulation, sealing and air infiltration barriers required in the code for new construction apply to the addition. However, the exception provides some guidance regarding the circumstance where nonconditioned space becomes conditioned space. A typical example of this would be enclosing an attached garage to become a family room. The baseline for the entire building including the existing building would be the UA of the existing building. That is the sum of the *U*-values for the walls, windows and doors multiplied by the respective areas of each of those components. Accordingly, the total sum of the UA for the completed building including the newly conditioned garage must be less than the UA for the existing building. This ensures that, regardless of how well insulated the existing building is, the addition will not make the building envelope less energy efficient.

Example: An existing house has a total UA of 222 (using the "Standard" Design as the existing house from the example in commentary Section R402.1.4). A 20-foot by 10-foot garage is proposed to be converted to a family room. For the example, assume the newly renovated garage has a 20-square-foot window.

The total UA of the existing building is 222 as calculated in Section R402.1.4. For the building with the new family room, we are using R-20 wall insulation in the cavity, with R-13 continuous insulation. Floor insulation is increased to R-38. The total UA for that combination of exterior wall values is 234, which is greater than 222. Therefore, some additional insulation will be required in order for the newly created conditioned space to meet the code.

EXISTING BUILDING	EXISTING BUILDING PLUS NEW FAMILY ROOM
Exterior Wall: 1,050 ft^2 U_w = 0.057 Glazed doors and windows: 192 ft^2 U_g = 0.32	Exterior Wall: 1,170 ft^2 850 ft^2 @ U_w = 0.057 320 ft^2 @ U_w = 0.032 Glazed doors and windows: 222 ft^2 @ U_g = 0.32
Opaque Exterior Door: 38 ft^2 U_g = 0.32 Roof: 1,500 ft^2 U_r = 0.026 Floor (slab): 1,500 ft^2 U_s = 0.033	Opaque Exterior Door: 19 ft^2 U_g = 0.32 Roof: 1,700 ft^2 1,500 ft^2 @ U_r = 0.026 200 ft^2 @ U_r = 0.022 Floor: 1,700 ft^2 1,500 @ U_s = 0.033 200 ft^2 @ U_s = 0.028

R502.1.1.2 Heating and cooling systems. New heating, cooling and duct systems that are part of the addition shall comply with Sections R403.1, R403.2, R403.3, R403.5 and R403.6.

> **Exception:** Where ducts from an existing heating and cooling system are extended to an addition, duct systems with less than 40 linear feet (12.19 m) in unconditioned spaces shall not be required to be tested in accordance with Section R403.3.3.

❖ Heating and cooling systems for additions are required to meet the current code for additions. The key to this section, again, is the exception that allows relaxation on duct-testing requirements. This accommodates a simple addition where there is a short reach with the ductwork in an attic or crawl space to the conditioned spaces in the addition.

R502.1.1.3 Service hot water systems. New service hot water systems that are part of the addition shall comply with Section R403.4.

❖ There is no relaxation of requirements for construction of new service hot water systems in additions.

R502.1.1.4 Lighting. New lighting systems that are part of the addition shall comply with Section R404.1.

❖ There is no relaxation of requirements for construction of new lighting systems in additions.

R502.1.2 Existing plus addition compliance (Simulated Performance Alternative). Where nonconditioned space is changed to conditioned space, the addition shall comply where the annual energy cost or energy use of the addition and the existing building, and any alterations that are part of the project, is less than or equal to the annual energy cost of the existing building when modeled in accordance with Section R405. The addition and any alterations that are part of the project shall comply with Section R405 in its entirety.

❖ Section R405 provides the performance alternative for compliance with the code. This section simply states that this performance alternative can be used in the design of an addition as well. The difference here is that the "standard design" is the existing building, rather than the standard building modeled from compliance with the code as used in Section R405.

SECTION R503 ALTERATIONS

R503.1 General. *Alterations* to any building or structure shall comply with the requirements of the code for new construction. *Alterations* shall be such that the existing building or structure is no less conforming to the provisions of this code than the existing building or structure was prior to the *alteration*.

Alterations to an existing building, building system or portion thereof shall conform to the provisions of this code as they relate to new construction without requiring the unaltered portions of the existing building or building system to comply with this code. Alterations shall not create an unsafe or hazardous condition or overload existing building systems. *Alterations* shall be such that the existing building or structure uses no more energy than the existing building or structure prior to the *alteration*. Alterations to existing buildings shall comply with Sections R503.1.1 through R503.2.

❖ Alterations include renovations, which implies that something is changed in the structure. For example, the removal, rearranagement or replacement of partition walls in an office building is an alteration because of, in part, possible impact on the means of egress, fire resistance or other life safety features of the building. Conversely, the replacement of damaged trim pieces on a door frame is considered a repair, not an alteration.

Basically, alterations are to conform to the requirements for a new structure, except as specifically stated in this section. For example, a new window installed where there was none will be required to have the *U*-factor rating and SHGC rating required for fenestration in the code, even if the insulation in the walls is not in conformance with the code. With this basic intent in mind, Sections R503.1.1, R503.1.2, R503.1.3, R503.1.4 and R503.2 address specific systems in the code that could be part of an alteration.

R503.1.1 Building envelope. Building envelope assemblies that are part of the alteration shall comply with Section R402.1.2 or R402.1.4, Sections R402.2.1 through R402.2.12, R402.3.1, R402.3.2, R402.4.3 and R402.4.4.

> **Exception:** The following alterations need not comply with the requirements for new construction provided the energy use of the building is not increased:
>
> 1. Storm windows installed over existing fenestration.
> 2. Existing ceiling, wall or floor cavities exposed during construction provided that these cavities are filled with insulation.
> 3. Construction where the existing roof, wall or floor cavity is not exposed.
> 4. Roof recover.
> 5. Roofs without insulation in the cavity and where the sheathing or insulation is exposed during reroofing shall be insulated either above or below the sheathing.
> 6. Surface-applied window film installed on existing single pane fenestration assemblies to reduce solar heat gain provided the code does not require the glazing or fenestration assembly to be replaced.

❖ The exceptions to this section are the key to these provisions. Alterations involving exterior walls and roofs are required to meet the requirements for a new building unless one of these exceptions apply.

The exceptions address situations where the alteration or repair of a structure or element is not required to comply with the provisions of the code. Typically, such situations would be a normal part of ongoing maintenance of the building, would improve the building's energy performance, or would not present an opportunity for improved energy savings. All of these exceptions are tied to the fact that they are permitted, provided "the energy use of the building is not increased."

Exception 1 is a fairly self-evident provision. Because of the limited nature of the work, there is little opportunity to make additional changes. This helps to reinforce the statement from the main paragraph that the intent is not to make "the unaltered portions(s) of the existing building or building system" comply with the code. In this case, the addition of a storm window over an existing window will only improve performance of the existing fenestration.

Exception 2 is important for a couple of the limitations that it contains. The provision only applies when the ceiling, wall or floor cavity is "exposed during construction." If the cavity is not opened up, then there is no requirement to do anything. If the cavity is exposed, the requirement will only be to "fill" it with insulation. Therefore, the level of insulation is not required to comply with the building thermal envelope requirements but is instead only required to be "filled" with any type of insulation and not to any specific *R*-value.

Exception 3 will exempt the need to make changes to the building thermal envelope because the building cavities are not exposed.

Exception 4 is straightforward. A roof recover is an alteration but does not impact the building envelope.

Exception 5 applies to roofs that are part of the building envelope and typically would have below-deck or above-deck insulation. The second sentence of Exception 4 permits the code-required insulation to be above or below the deck. For a typical single-family home (with nonconditioned space), the ceiling is the building thermal envelope and the roof is not; therefore, Exception 5 would not apply. However, if during reroofing the existing ceiling cavities are exposed, then Exception 3 would apply.

Exception 6 is straightforward as well. Surface film used to provide a lower SHGC rating is acceptable for windows that do not otherwise need to be replaced.

R503.1.1.1 Replacement fenestration. Where some or all of an existing fenestration unit is replaced with a new fenestration product, including sash and glazing, the replacement fenestration unit shall meet the applicable requirements for *U*-factor and SHGC as provided in Table R402.1.2.

❖ Replacement windows and doors are required to meet the requirements of the code for new construction, with no exceptions.

Replacing only a glass pane in an existing sash and frame would not fall under this provision if the *U*-factor and SHGC are equal to or lower than the values prior to the replacement. In situations where the existing values are not known or where the replacement is more than just replacing the glass pane, the provisions may be applicable. It is often common practice when fenestration is replaced to remove only the sash and glazing of an existing window and replace them with an entirely new fenestration product. Sometimes the existing frame also is removed, but many times the new fenestration product is custom made to fit in the existing space left after the sash and glazed portions are removed. In essence, the new fenestration is installed in or over the existing frame. Whether the existing frame is removed or not, these types of replacements are regulated by this section.

Section R503.1.1.1 requires that each fenestration unit replaced in a residence not exceed the maximum fenestration *U*-factor and SHGC for the applicable climate zone. This requirement applies to all replacement windows, even if the existing frame is not removed (e.g., the new window is placed inside the old frame), so long as the sash and glazing are replaced. In addition, remember that the definition of "Fenestration" includes doors, which must meet the same *U*-factor requirements as windows. Therefore, the replacement of a door also would have to meet these requirements.

When dealing with replacement fenestration, the code official should be consulted to explain how this requirement will be applied. For simple ease of application for both the code official and the installer, the *U*-factor and SHGC requirements could be applied to each fenestration unit. Therefore, the *U*-factor and SHGC required for each unit would be the values listed directly in Table R402.1.2. However, if acceptable to the code official and additional information is available regarding the performance of the remaining existing windows in the home, it may be reasonable to permit the use of the area-weighted values of Sections R402.3.1 and R402.3.2 or the exemptions of Sections R402.3.3 and R402.3.4 to the replaced fenestration unit.

In accordance with Section R102.1, the *U*-factor and SHGC for each replacement fenestration product must be obtained from a label attached to the product certifying that the values were determined in accordance with NFRC procedures by an accredited, independent laboratory. In the absence of an NFRC-labeled *U*-factor or SHGC, a value from the limited default tables [see Tables R303.1.3(1) through R303.1.3(3)] must be used. The NFRC procedures do include applicable methods to test various replacement products.

R503.1.2 Heating and cooling systems. New heating, cooling and duct systems that are part of the alteration shall comply with Sections R403.1, R403.2, R403.3 and R403.6.

> **Exception:** Where ducts from an existing heating and cooling system are extended, duct systems with less than 40 linear feet (12.19 m) in unconditioned spaces shall not be required to be tested in accordance with Section R403.3.3.

❖ Heating and cooling systems for additions are required to meet code requirements for new construction for alterations. The key to this section is the exception that allows relaxation on duct testing requirements. This accommodates a simple alteration where there is a short reach with the ductwork in an attic or crawl space to the conditioned spaces in the alteration.

R503.1.3 Service hot water systems. New service hot water systems that are part of the alteration shall comply with Section R403.4.

❖ Service hot water systems in an alteration must meet the requirements of the code for new construction, with no exceptions.

R503.1.4 Lighting. New lighting systems that are part of the alteration shall comply with Section 404.1.

Exception: Alterations that replace less than 50 percent of the luminaires in a space, provided that such alterations do not increase the installed interior lighting power.

❖ Alterations involving lighting systems can be relaxed somewhat when fewer than 50 percent of the luminaires are being replaced. However, in keeping with the primary intent of the code for alterations, the final product cannot use more energy than the building that is being renovated. The code does not allow an increase in lighting power.

R503.2 Change in space conditioning. Any nonconditioned or low-energy space that is altered to become *conditioned space* shall be required to be brought into full compliance with this code.

Exception: Where the simulated performance option in Section R405 is used to comply with this section, the annual energy cost of the proposed design is permitted to be 110 percent of the annual energy cost otherwise allowed by Section R405.3.

❖ When nonconditioned spaces are converted to conditioned space, the impact on the community energy resources is the same as that for new construction. As such, they should be required to meet the minimum standards set by the code for new construction.

The exception recognizes the fact that full compliance with this code for existing buildings is extremely difficult and costly. Conditions such as slab edges, structural thermal bridges, and window configurations cannot be practically remedied in many cases. Therefore, an alternate compliance path allows a 10-percent higher total building performance value. This could result in the preservation and adaptive reuse of more existing buildings, which itself is a significant energy conservation measure.

SECTION R504 REPAIRS

R504.1 General. Buildings, structures and parts thereof shall be repaired in compliance with Section R501.3 and this section. Work on nondamaged components necessary for the required *repair* of damaged components shall be considered part of the *repair* and shall not be subject to the requirements for *alterations* in this chapter. Routine maintenance required by Section R501.3, ordinary repairs exempt from *permit*, and abatement of wear due to normal service conditions shall not be subject to the requirements for *repairs* in this section.

❖ This section and Section R504.2 detail examples of what a repair is. A repair is work not required to comply with the alterations provisions of the code. Section R503.2 provides some specific work that is considered repairs. This is done simply to clarify the code on these specific issues; the list is not intended to be comprehensive—other work could also be considered repairs, as described in this section.

R504.2 Application. For the purposes of this code, the following shall be considered repairs:

1. Glass-only replacements in an existing sash and frame.
2. Roof repairs.
3. Repairs where only the bulb and/or ballast within the existing luminaires in a space are replaced provided that the replacement does not increase the installed interior lighting power.

❖ See the commentary to Section R504.1.

SECTION R505 CHANGE OF OCCUPANCY OR USE

R505.1 General. Spaces undergoing a change in occupancy that would result in an increase in demand for either fossil fuel or electrical energy shall comply with this code.

❖ When a building undergoes a change of occupancy, energy-using systems (envelope, mechanical, service water heating, electrical distribution or illumination) must be evaluated to determine the effect the change of occupancy has on system performance and energy use. For example, if a mercantile building were converted to a restaurant, additional ventilation would be required for the public based on the increased occupant load. If an existing system serves an occupancy that is different from the occupancy it served when the code went into effect, the mechanical system must comply with the applicable code requirements for a mechanical system serving the newer occupancy. Depending on the nature of the previous occupancy, changing a building's occupancy classification could result in a change to the mechanical, service water heating, electrical distribution or illumination systems or any combination of these.

Buildings undergoing a change of occupancy must meet the applicable requirements of the code when peak demand is increased. For example, if a hotel is converted to multiple-family residential use and the conversion results in an increase in the building's peak connected load (space conditioning, lighting or service water heating), the entire building must be brought into compliance.

When the occupancy changes in a portion of an existing building (residential or commercial) and the new occupancy results in an increase in the peak demand for either fossil fuel or electrical energy supply, the portion of the building associated with the new occupancy must meet the code.

When a permittee claims that a change in occupancy will not increase the peak design rate of energy use for the building, it is the applicant's responsibility to demonstrate that the peak load of the converted building will not exceed the peak load of the original building. Without supporting documentation, the peak load generally must be assumed to increase with a change in occupancy.

It is also important that users realize that under the code there can be a difference between the change of occupancy (the way a building is used) and what the IBC deals with when a change of occupancy classification occurs. Therefore, if a storage building that has no heating or cooling is modified so that the building is heated to prevent stock items from freezing, the IBC would not consider this as a change of occupancy because the occupancy classification would still be Group S-1. The code, however, would consider this a change in occupancy because the way the building is used would change and it would result in an increase in the demand for energy.

R505.2 General. Any space that is converted to a dwelling unit or portion thereof from another use or occupancy shall comply with this code.

Exception: Where the simulated performance option in Section R405 is used to comply with this section, the annual energy cost of the proposed design is permitted to be 110 percent of the annual energy cost otherwise allowed by Section R405.3.

❖ This is a specific example of a change in occupancy. See commentary to Section R505.1.

The exception recognizes the fact that full compliance with the code for existing buildings is extremely difficult and costly. Conditions such as slab edges, structural thermal bridges and window configurations cannot be practically remedied in many cases. Therefore, an alternate compliance path allows a 10-percent higher total building performance value. This could result in the preservation and adaptive reuse of more existing buildings, which itself is a significant energy conservation measure.

Chapter 6 [RE]: Referenced Standards

General Comments

Chapter 6 [RE] contains a comprehensive list of standards that are referenced in the Residential Provisions of this code. It is organized to make locating specific document references easy.

It is important to understand that not every document related to energy conservation is qualified to be a "referenced standard." The International Code Council (ICC)® has adopted a criterion that referenced standards in the *International Codes*® (I-Codes®) and standards intended for adoption into the I-Codes must meet to qualify as a referenced standard. The policy is summarized as follows:

- Code references: The scope and application of the standard must be clearly identified in the code text.
- Standard content: The standard must be written in mandatory language and be appropriate for the subject covered. The standard cannot have the effect of requiring proprietary materials or prescribing a proprietary testing agency.
- Standard promulgation: The standard must be readily available and developed and maintained in a consensus process such as those used by ASTM or ANSI.

The ICC Code Development Procedures CP #28, of which the standards policy is a part, are updated periodically. A copy of the latest version can be obtained from the ICC offices.

Once a standard is incorporated into the code through the code development process, it becomes an enforceable part of the code. When the code is adopted by a jurisdiction, the standard also is part of that jurisdiction's adopted code. It is for this reason that the criteria were developed. Compliance with this policy means that documents incorporated into the code are developed through the use of a consensus process, are written in mandatory language and do not mandate the use of proprietary materials or agencies. The requirement for a standard to be developed through a consensus process means that the standard is representative of the most current body of available knowledge on the subject as determined by a broad range of interested or affected parties without dominance by any single interest group. A true consensus process has many attributes, including but not limited to:

- An open process that has formal (published) procedures allowing for the consideration of all viewpoints.
- A definitive review period that allows for the standard to be updated and/or revised.
- A process of notification to all interested parties.
- An appeals process.

Many available documents related to mechanical system design and installation and construction, though useful, are not "standards" and are not appropriate for reference in the code. Often, these documents are developed or written with the intention of being used for regulatory purposes and are unsuitable for use as a standard because of extensive use of recommendations, advisory comments and nonmandatory terms. Typical examples include installation instructions, guidelines and practices.

The objective of ICC's standards policy is to provide regulations that are clear, concise and enforceable; thus the requirement for standards to be written in mandatory language. This requirement is not intended to mean that a standard cannot contain informational or explanatory material that will aid the user of the standard in its application. When the standard's promulgating agency wants such material to be included, however, the information must appear in a nonmandatory location, such as an annex or appendix, and be clearly identified as not being part of the standard.

Overall, standards referenced by the code must be authoritative, relevant, up to date and, most important, reasonable and enforceable. Standards that comply with the ICC's standards policy fulfill these expectations.

Purpose

As a performance-based code, the *International Energy Conservation Code*® (IECC®) contains numerous references to documents that are used to regulate materials and methods of construction. The references to these documents within the code text consist of the promulgating agency's acronym and its publication designation (for example, ACCA Manual J) and a further indication that the document being referenced is the one that is listed in this chapter. This chapter contains all of the information that is necessary to identify the specific referenced document. Included is the following information on a document's promulgating agency (see Commentary Figure 6 [RE]):

- The promulgating agency (the agency's title).
- The promulgating agency's acronym.
- The promulgating agency's address.

For example, a reference to ASTM E 283 indicates that this document can be found in Chapter 6 [RE] under the heading ASTM. The specific standards designation is E 283. For convenience, these designations are listed in alphanumeric order. This chapter identifies that ASTM E 283 is titled *Test Method for Determining the Rate of Air Leakage Through Exterior Windows, Curtain Walls and Doors Under Specified Pressure Differences Across the Specimen*, the applicable edition (i.e., its year of publication) is 2004 and it is referenced in one section of the code.

This chapter will also indicate when a document has been discontinued or replaced by its promulgating agency. When a document is replaced by a different one, a note will appear to tell the user the designation and title of the new document.

The key aspect of the manner in which standards are referenced by the code is that a specific edition of a specific standard is clearly identified. In this manner, the requirements necessary for compliance can be readily determined. The basis for code compliance is, therefore, established and available on an equal basis to the building official, contractor, designer and owner.

This chapter lists the standards that are referenced in various sections of this document. The standards are listed herein by the promulgating agency of the standard, the standard identification, the effective date and title, and the section or sections of this document that reference the standard. The application of the referenced standards shall be as specified in Section R106.

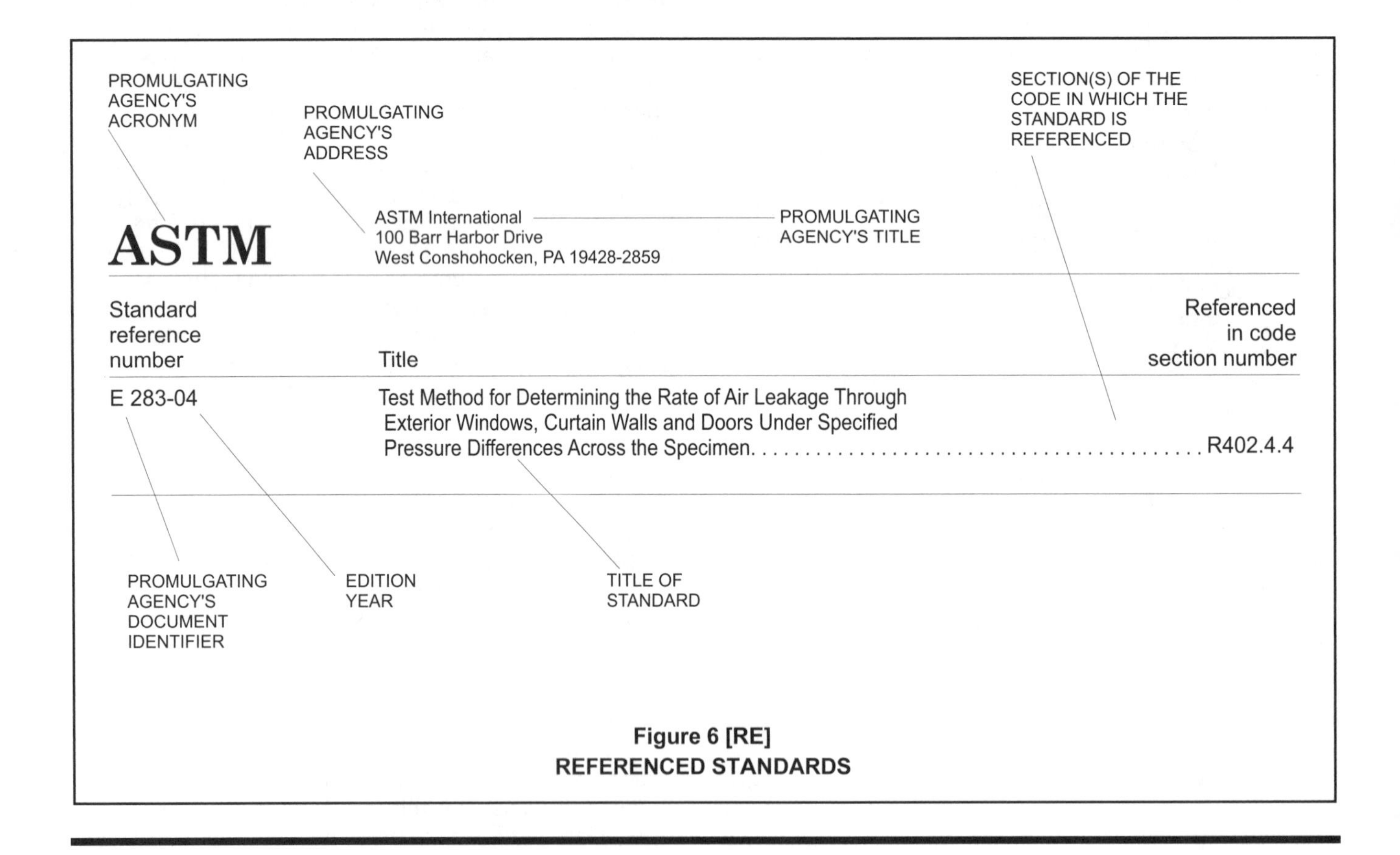

Figure 6 [RE]
REFERENCED STANDARDS

This chapter lists the standards that are referenced in various sections of this document. The standards are listed herein by the promulgating agency of the standard, the standard identification, the effective date and title, and the section or sections of this document that reference the standard. The application of the referenced standards shall be as specified in Section 106.

AAMA

American Architectural Manufacturers Association
1827 Walden Office Square
Suite 550
Schaumburg, IL 60173-4268

Standard reference number	Title	Referenced in code section number
AAMA/WDMA/CSA 101/I.S.2/A C440—11	North American Fenestration Standard/ Specifications for Windows, Doors and Unit Skylights	R402.4.3

ACCA

Air Conditioning Contractors of America
2800 Shirlington Road, Suite 300
Arlington, VA 22206

Standard reference number	Title	Referenced in code section number
Manual J—2011	Residential Load Calculation Eighth Edition	R403.7
Manual S—13	Residential Equipment Selection	R403.7

APSP

The Association of Pool and Spa Professionals
2111 Eisenhower Avenue
Alexandria, VA 22314

Standard reference number	Title	Referenced in code section number
APSP 14—11	American National Standard for Portable Electric Spa Energy Efficiency	R403.10.1, 403.11
APSP 15a—2013	American National Standard for Residential Swimming Pool and Spa Energy Efficiency	R403.12

ASHRAE

American Society of Heating, Refrigerating and Air-Conditioning Engineers, Inc.
1791 Tullie Circle, NE
Atlanta, GA 30329-2305

Standard reference number	Title	Referenced in code section number
ASHRAE—2013	ASHRAE Handbook of Fundamentals	R402.1.5, Table R405.5.2(1)
ASHRAE 193—2010	Method of Test for Determining the Airtightness of HVAC Equipment	R403.3.2.1

ASTM

ASTM International
100 Barr Harbor Drive
West Conshohocken, PA 19428-2859

Standard reference number	Title	Referenced in code section number
C 1363—11	Standard Test Method for Thermal Performance of Building Materials and Envelope Assemblies by Means of a Hot Box Apparatus	R303.1.4.1
E 283—04	Test Method for Determining the Rate of Air Leakage Through Exterior Windows, Curtain Walls and Doors Under Specified Pressure Differences Across the Specimen	R402.4.4

ASTM—continued

E 779—10	Standard Test Method for Determining Air Leakage Rate by Fan Pressurization	R402.4.1.2
E 1827—11	Standard Test Methods for Determining Airtightness of Building Using an Orifice Blower Door	R402.4.1.2

CSA

CSA Group
8501 East Pleasant Valley
Cleveland, OH 44131-5575

Standard reference number	Title	Referenced in code section number
AAMA/WDMA/CSA 101/I.S.2/A440—11	North American Fenestration Standard/Specification for Windows, Doors and Unit Skylights	R402.4.3
CSA 55.1—2012	Test Method for measuring efficiency and pressure loss of drain water heat recovery units	R403.5.4
CSA 55.2—2012	Drain water heat recover units	R403.5.4

DASMA

Door and Access Systems Manufacturers Association
1300 Sumner Avenue
Cleveland, OH 44115-2851

Standard reference number	Title	Referenced in code section number
105—92(R2004)—13	Test Method for Thermal Transmittance and Air Infiltration of Garage Doors	R303.1.3

ICC

International Code Council, Inc.
500 New Jersey Avenue, NW
6th Floor
Washington, DC 20001

Standard reference number	Title	Referenced in code section number
IBC—15	International Building Code®	R201.3, R303.2, R402.1.1, R501.4
ICC 400—12	Standard on the Design and Construction of Log Structures	Table R402.5.1.1
IECC—15	International Energy Conservation Code®	R101.4.1, 403.8
IECC—09	2009 International Energy Conservation Code®	R406.2
ECC—06	2006 International Energy Conservation Code®	R202, R406.3.1
IFC—15	International Fire Code®	R201.3,R501.4
IFGC—15	International Fuel Gas Code®	R201.3,R501.4
IMC—15	International Mechanical Code®	R201.3, R403.3.2, R403.6, R501.4
IPC—15	International Plumbing Code®	R201.3,R501.4
IPSDC—15	International Private Sewage Disposal Code®	501.4
IPMC—15	International Property Maintenance Code®	501.4
IRC—15	International Residential Code®	R201.3, R303.2, R402.1.1, R402.2.11, R403.3.2, R403.6, R501.4

IIEEE

The Institute of Electrical and Electronic Engineers, Inc.
3 Park Avenue
New York, NY 1016-5997

Standard reference number	Title	Referenced in code section number
515.1—2012	IEEE Standard for the Testing, Design, Installation, and Maintenance of Electrical Resistance Trace Heating for Commercial Applications	R403.5.1.2

NFPA

National Fire Protection Association.
1 Batterymarch Park
Quincy, MA 02169-7471

Standard reference number	Title	Referenced in code section number
70—14	National Electrical Code	R501.4

NFRC

National Fenestration Rating Council, Inc.
6305 Ivy Lane, Suite 140
Greenbelt, MD 20770

Standard reference number	Title	Referenced in code section number
100—2009	Procedure for Determining Fenestration Products *U*-factors—Second Edition	R303.1.3
200—2009	Procedure for Determining Fenestration Product Solar Heat Gain Coefficients and Visible Transmittance at Normal Incidence—Second Edition	R303.1.3
400—2009	Procedure for Determining Fenestration Product Air Leakage—Second Edition	R402.4.3

UL

UL LLC
333 Pfingsten Road
Northbrook, IL 60062

Standard reference number	Title	Referenced in code section number
127—11	Standard for Factory Built Fireplaces	R402.4.2
515—11	Electrical Resistance Heat Tracing for Commercial and Industrial Applications including revisions through November 30, 2011	R403.5.1.2

US-FTC

United States-Federal Trade Commission
600 Pennsylvania Avenue NW
Washington, DC 20580

Standard reference number	Title	Referenced in code section number
CFR Title 16 (May 31, 2005)	R-value Rule	R303.1.4

WDMA

Window and Door Manufacturers Association
2025 M Street, NW Suite 800
Washington, DC 20036-3309

Standard reference number	Title	Referenced in code section number
AAMA/WDMA/CSA 101/I.S.2/A440—11	North American Fenestration Standard/Specification for Windows, Doors and Unit Skylights	R402.4.3

Appendix RA: Recommended Procedure for Worst-case Testing of Atmospheric Venting Systems Under R402.4 or R405 Conditions ≤ $5ACH_{50}$

(This appendix is informative and is not part of the code.)

General Comments

Energy efficiency improvements often have a direct impact on the building pressure boundary affecting the safe operation of combustion equipment. Under certain conditions, reduced natural-air leakage coupled with the installation of atmospheric combustion appliances will reduce air exchange to the outside of a building and potentially contribute to poor indoor air quality and health problems due to spillage, inadequate draft, or carbon monoxide concerns.

Purpose

This appendix is intended to provide guidance to builders, code officials and home performance contractors for worst-case testing of atmospheric venting systems to identify problems that weaken draft and restrict combustion air. Worst-case vent testing uses the home's exhaust fans, air-handling appliances and chimneys to create worst-case depressurization in the combustion appliance zone (CAZ). This appendix is basically a distilled version of predominant combustion safety test procedures for atmospherically vented appliances found in readily available home performance programs across the country, such as EPA's *Healthy Indoor Environments Protocols*, EPA's *Home Performance with Energy Star*, DOE's *Workforce Guidelines for Home Energy Upgrades*, HUD's *Community Development Block Grants and Weatherization Assistance Programs*, BPI's *Technical Standards for the Building Analyst Professional*, and RESNET's *Interim Guidelines for Combustion Appliance Testing and Writing Work Scopes*. This is intended to take the combustion safety test procedures that are used most commonly by these home performance, weatherization, and beyond-code programs, and reduce them to their simplest and most straightforward form for the purpose of combustion safety in field assessment through the use of building diagnostic tools.

SECTION RA101 SCOPE

RA101.1 General. This appendix is intended to provide guidelines for worst-case testing of atmospheric venting systems. Worst-case testing is recommended to identify problems that weaken draft and restrict combustion air.

SECTION RA201 GENERAL DEFINITIONS

COMBUSTION APPLIANCE ZONE (CAZ). A contiguous air volume within a building that contains a Category I or II atmospherically vented appliance or a Category III or IV direct-vent or integral vent appliance drawing combustion air from inside the building or dwelling unit. The CAZ includes, but is not limited to, a mechanical closet, a mechanical room, or the main body of a house or dwelling unit.

DRAFT. The pressure difference existing between the *appliance* or any component part and the atmosphere that causes a continuous flow of air and products of *combustion* through the gas passages of the *appliance* to the atmosphere.

Mechanical or induced draft. The pressure difference created by the action of a fan, blower or ejector that is located between the *appliance* and the *chimney* or vent termination.

Natural draft. The pressure difference created by a vent or *chimney* because of its height and the temperature difference between the *flue* gases and the atmosphere.

SPILLAGE. Combustion gases emerging from an appliance or venting system into the combustion appliance zone during burner operation.

SECTION RA301 TESTING PROCEDURE

RA301.1 Worst-case testing of atmospheric venting systems. Buildings or dwelling units containing a Category I or II atmospherically vented appliance; or a Category III or IV direct-vent or integral vent appliance drawing combustion air from inside of the building or dwelling unit, shall have the Combustion Appliance Zone (CAZ) tested for spillage, acceptable draft and carbon monoxide (CO) in accordance with this section. Where required by the *code official*, testing shall be conducted by an *approved* third party. A written report of the results of the test shall be signed by the party conducting the test and provided to the *code official*. Testing shall be performed at any time after creation of all penetrations of the *building thermal* envelope and prior to final inspection.

Exception: Buildings or dwelling units containing only Category III or IV direct-vent or integral vent appliances that do not draw combustion air from inside of the building or dwelling unit.

The enumerated test procedure as follows shall be complied with during testing:

1. Set combustion appliances to the pilot setting or turn off the service disconnects for combustion appliances. Close exterior doors and windows and the fireplace damper. With the building or dwelling unit in this configuration, measure and record the baseline ambient pressure inside the building or dwelling unit CAZ. Compare the baseline ambient pressure of the CAZ to that of the outside ambient pressure and record the difference (Pa).

2. Establish worst case by turning on the *clothes dryer* and all exhaust fans. Close all interior doors that make the CAZ pressure more negative. Turn on the air handler, where present, and leave on if, as a result, the pressure in the CAZ becomes more negative. Check interior door positions again, closing only the interior doors that make the CAZ pressure more negative. Measure net change in pressure from the CAZ to outdoor ambient pressure, correcting for the base ambient pressure inside the home. Record "worst case depressurization" pressure and compare to Table RA301.1(1).

   Where CAZ depressurization limits are exceeded under worst-case conditions in accordance with Table RA301.1(1), additional combustion air shall be provided or other modifications to building air-leakage performance or exhaust appliances such that depressurization is brought within the limits prescribed in Table RA301.1(1).

3. Measure worst-case spillage, acceptable draft and carbon monoxide (CO) by firing the fuel-fired appliance with the smallest Btu capacity first.

   a. Test for spillage at the draft diverter with a mirror or smoke puffer. An appliance that continues to spill flue gases for more than 60 seconds fails the spillage test.

   b. Test for CO measuring undiluted flue gases in the throat or flue of the appliance using a digital gauge in parts per million (ppm) at the 10-minute mark. Record CO ppm readings to be compared with Table RA301.1(3) upon completion of Step 4. Where the spillage test fails under worst case, go to Step 4.

   c. Where spillage ends within 60 seconds, test for acceptable draft in the connector not less than 1 foot (305 mm), but not more than 2 feet (610 mm) downstream of the draft diverter. Record draft pressure and compare to Table RA301.1(2).

   d. Fire all other connected appliances simultaneously and test again at the draft diverter of each appliance

TABLE RA301.1(1)
CAZ DEPRESSURIZATION LIMITS

VENTING CONDITION	LIMIT (Pa)
Category I, atmospherically vented water heater	–2.0
Category I or II atmospherically vented boiler or furnace common-vented with a Category I atmospherically vented water heater	–3.0
Category I or II atmospherically vented boiler or furnace, equipped with a flue damper, and common vented with a Category I atmospherically vented water heater	–5.0
Category I or II atmospherically vented boiler or furnace alone	
Category I or II atmospherically vented, fan-assisted boiler or furnace common vented with a Category I atmospherically vented water heater	
Decorative vented, gas appliance	
Power-vented or induced-draft boiler or furnace alone, or fan-assisted water heater alone	–15.0
Category IV direct-vented appliances and sealed combustion appliances	–50.0

For SI: 6894.76 Pa = 1.0 psi.

TABLE RA301.1(2)
ACCEPTABLE DRAFT TEST CORRECTION

OUTSIDE TEMPERATURE (°F)	MINIMUM DRAFT PRESSURE REQUIRED (Pa)
< 10	–2.5
10 – 90	(Outside Temperature ÷ 40) – 2.75
> 90	–0.5

For SI: 6894.76 Pa = 1.0 psi.

for spillage, CO and acceptable draft using procedures 3a through 3c.

4. Measure spillage, acceptable draft, and carbon monoxide (CO) under natural conditions—without *clothes dryer* and exhaust fans on—in accordance with the procedure outlined in Step 3, measuring the net change in pressure from worst case condition in Step 3 to natural in the CAZ to confirm the worst case depressurization taken in Step 2. Repeat the process for each appliance, allowing each vent system to cool between tests.
5. Monitor indoor ambient CO in the breathing zone continuously during testing, and abort the test where indoor ambient CO exceeds 35 ppm by turning off the appliance, ventilating the space, and evacuating the building. The CO problem shall be corrected prior to completing combustion safety diagnostics.
6. Make recommendations based on test results and the retrofit action prescribed in Table RA301.1(3).

TABLE RA301.1(3)
ACCEPTABLE DRAFT TEST CORRECTION

CARBON MONOXIDE LEVEL (ppm)	AND OR	SPILLAGE AND ACCEPTABLE DRAFT TEST RESULTS	RETROFIT ACTION
0 – 25	and	Passes	Proceed with work
25 < X ≤ 100	and	Passes	Recommend that CO problem be resolved
25 < X ≤ 100	and	Fails in worst case only	Recommend an appliance service call and repairs to resolve the problem
100 < X ≤ 400	or	Fails under natural conditions	**Stop!** Work shall not proceed until appliance is serviced and problem resolved
> 400	and	Passes	**Stop!** Work shall not proceed until appliance is serviced and problem resolved
> 400	and	Fails under any condition	**Emergency!** Shut off fuel to appliance and call for service immediately

Appendix RB: Solar-ready Provisions—Detached One- and Two-family Dwellings, Multiple Single-family Dwellings (Townhouses)

(The provisions contained in this appendix are not mandatory unless specifically referenced in the adopting ordinance.)

General Comments

This appendix is intended to support future potential improvements for detached one- and two-family dwellings, and multiple single-family dwellings for solar electric and solar thermal systems. This appendix does not require the installation of conduit, prewiring, or preplumbing. It does not require any specific physical orientation of the residential building. It does not require any increased load capacities for residential roofing systems. It does not require the redesign of plans.

Many building departments have been mandated by local regulations to accelerate permits and inspections for solar installation. Having important information and documentation available to the building department, solar contractor and homeowner will assist in supporting the accelerated working environment many municipalities have mandated.

The U.S. Department of Energy's (DOE) SunShot Initiative has set a goal to make solar energy cost competitive with other forms of energy by the end of the decade, which would reduce installed costs of solar energy systems by about 75 percent. This initiative, combined with increased pressures on our energy supply and demand, is expected to drive greater adoption of renewable energy systems on residential buildings.

Purpose

This appendix is intended to identify the areas of a residential building roof, called the solar-ready zone, for potential future installation of renewable energy systems. The ability to plan ahead for possible future solar equipment starts with documenting necessary solar-ready zone information on the plans, some of which may already be required in permit construction requirements. This appendix also requires the builder to post specific information about the home for use by the homeowner(s).

The documentation of solar-ready zones and roof load calculations (already performed during the design phase) will assist building departments as well as any future solar contractors seeking to install renewable energy systems on the roof. The builder/designer is knowledgeable on the intricacies of each model and plan and can easily identify unobstructed roof areas as well as spaces where conduit, wiring and plumbing can be routed from the roof to the respective utility areas. This will save building departments and solar designers time and effort when installing future solar systems. If a homeowner wishes to install a solar energy system later, this documentation can save thousands of dollars in labor, installation, design and integration of the solar system into the house.

SECTION RB101 SCOPE

RB101.1 General. These provisions shall be applicable for new construction where solar-ready provisions are required.

SECTION RB102 GENERAL DEFINITION

SOLAR-READY ZONE. A section or sections of the roof or building overhang designated and reserved for the future installation of a solar photovoltaic or solar thermal system.

SECTION RB103 SOLAR-READY ZONE

RB103.1 General. New detached one- and two-family dwellings, and multiple single-family dwellings (townhouses) with not less than 600 square feet (55.74 m^2) of roof area oriented between 110 degrees and 270 degrees of true north shall comply with Sections RB103.2 through RB103.8.

Exceptions:

1. New residential buildings with a permanently installed on-site renewable energy system.

2. A building with a solar-ready zone that is shaded for more than 70 percent of daylight hours annually.

RB103.2 Construction document requirements for solar-ready zone. Construction documents shall indicate the solar-ready zone.

RB103.3 Solar-ready zone area. The total solar-ready zone area shall be not less than 300 square feet (27.87 m^2) exclusive of mandatory access or set back areas as required by the *International Fire Code*. New multiple single-family dwellings (townhouses) three stories or less in height above grade plane and with a total floor area less than or equal to 2,000 square feet (185.8 m^2) per dwelling shall have a solar-ready zone area of not less than 150 square feet (13.94 m^2). The solar-ready zone shall be composed of areas not less than 5 feet (1524 mm) in width and not less than 80 square feet (7.44 m^2) exclusive of access or set back areas as required by the *International Fire Code*.

RB103.4 Obstructions. Solar-ready zones shall be free from obstructions, including but not limited to vents, chimneys, and roof-mounted equipment.

RB103.5 Roof load documentation. The structural design loads for roof dead load and roof live load shall be clearly indicated on the construction documents.

RB103.6 Interconnection pathway. Construction documents shall indicate pathways for routing of conduit or plumbing from the solar-ready zone to the electrical service panel or service hot water system.

RB103.7 Electrical service reserved space. The main electrical service panel shall have a reserved space to allow installation of a dual pole circuit breaker for future solar electric installation and shall be labeled "For Future Solar Electric." The reserved space shall be positioned at the opposite (load) end from the input feeder location or main circuit location.

RB103.8 Construction documentation certificate. A permanent certificate, indicating the solar-ready zone and other requirements of this section, shall be posted near the electrical distribution panel, water heater or other conspicuous location by the builder or registered design professional.

INDEX

D

E

F

G

H

I

L

M

O

P